# ACTIVE OXYGEN IN BIOCHEMISTRY

*Dedicated to the memory of
Teddy G. Traylor (1925–1993)
and
Bertram Selverstone (1917–1993)*

# ACTIVE OXYGEN IN BIOCHEMISTRY

EDITORS:
**JOAN SELVERSTONE VALENTINE**
University of California, Los Angeles

**CHRISTOPHER S. FOOTE**
University of California, Los Angeles

**ARTHUR GREENBERG**
University of North Carolina at Charlotte

**JOEL F. LIEBMAN**
University of Maryland Baltimore County

## BLACKIE ACADEMIC & PROFESSIONAL
An Imprint of Chapman & Hall

London · Glasgow · Weinheim · New York · Tokyo · Melbourne · Madras

Published by
**Blackie Academic & Professional, an imprint of Chapman & Hall,
Wester Cleddens Road, Bishopbriggs, Glasgow G64 2NZ**

Chapman & Hall, 2–6 Boundary Row, London SE1 8HN, UK

Blackie Academic & Professional, Wester Cleddens Road, Bishopbriggs, Glasgow G64 2NZ, UK

Chapman & Hall GmbH, Pappelallee 3, 69469 Weinheim, Germany

Chapman & Hall USA, 115 Fifth Avenue, New York, NY 10003, USA

Chapman & Hall Japan, ITP-Japan, Kyowa Building, 3F, 2-2-1 Hirakawacho, Chiyoda-ku, Tokyo 102, Japan

DA Book (Aust.) Pty Ltd, 648 Whitehorse Road, Mitcham 3132, Victoria, Australia

Chapman & Hall India, R. Seshadri, 32 Second Main Road, CIT East, Madras 600 035, India

First edition 1995

© 1995 Chapman & Hall

Typeset in 10/12 pt Palatino by Best-set Typesetter Ltd., Hong Kong
Printed in Great Britain by St Edmundsbury Press, Bury St Edmunds, Suffolk

ISBN 0 7514 0293 1

Apart from any fair dealing for the purposes of research or private study, or criticism or review, as permitted under the UK Copyright Designs and Patents Act, 1988, this publication may not be reproduced, stored, or transmitted, in any form or by any means, without the prior permission in writing of the publishers, or in the case of reprographic reproduction only in accordance with the terms of the licences issued by the Copyright Licensing Agency in the UK, or in accordance with the terms of licences issued by the appropriate Reproduction Rights Organization outside the UK. Enquiries concerning reproduction outside the terms stated here should be sent to the publishers at the Glasgow address printed on this page.
 The publisher makes no representation, express or implied, with regard to the accuracy of the information contained in this book and cannot accept any legal responsibility or liability for any errors or omissions that may be made.

A catalogue record for this book is available from the British Library
Library of Congress Catalog Card Number : **95-080020**

∞ Printed on acid-free text paper, manufactured in accordance with ANSI/NISO Z39.48-1992 (Permanence of Paper)

# Contents

    Preface      *vii*

    Series Preface      *ix*

    Structure Energetics and Reactivity in Chemistry Series (SEARCH series)      *xi*

    Editorial Advisory Board      *xii*

    Contributors      *xv*

1. Biological Reactions of Dioxygen: An Introduction      *1*
   Raymond Y. N. Ho, Joel F. Liebman, and
   Joan Selverstone Valentine

2. Oxygen Activation by Flavins and Pterins      *37*
   Bruce A. Palfey, David P. Ballou, and Vincent Massey

3. Reactions of Dioxygen and Its Reduced Forms with Heme Proteins and Model Porphyrin Complexes      *84*
   Teddy G. Traylor and Patricia S. Traylor

4. Dioxygen Reactivity in Copper Proteins and Complexes      *188*
   Stephen Fox and Kenneth D. Karlin

| 5 | Oxygen Activation at Nonheme Iron Centers<br>*Lawrence Que, Jr.* | 232 |
| 6 | The Mechanism of Lipoxygenases<br>*Mark J. Nelson and Steven P. Seitz* | 276 |
| 7 | The Biological Significance of Oxygen-Derived Species<br>*Barry Halliwell* | 313 |
| 8 | Metal-Complex-Catalyzed Cleavage of Biopolymers<br>*Rosemary A. Marusak and Claude F. Meares* | 336 |
| 9 | Exploration of Selected Pathways for Metabolic Oxidative Ring Opening of Benzene Based on Estimates of Molecular Energetics<br>*Arthur Greenberg* | 401 |
| 10 | The Role of Oxidized Lipids in Cardiovascular Disease<br>*Judith A. Berliner and Andrew D. Watson* | 433 |
|  | Index | 450 |

# Preface

The field of "Oxygen Activation" has attracted considerable interest recently, not only because it presents challenges in those fields of basic research that aim to understand the fundamental aspects of chemical and biological reactions that involve dioxygen, but also because of its wide range of practical implications in such diverse fields as medicine, synthesis of pharmaceuticals and other organic compounds, materials science, and atmospheric science. This is the second of two volumes that focus on the subject of oxygen activation, the first slanted toward chemistry and the second toward biological chemistry. We planned these volumes to be more general than many monographs of this sort, not as detailed summaries of the authors' own research but rather as general overviews of the field. Our choice of topics was strongly influenced by our syllabus for a course entitled "Oxygen Chemistry," which two of us have twice taught jointly at UCLA. Definition of important issues, horizons, and future prospects was an important goal, and, although totally comprehensive coverage was not possible, we believe that we have chosen a representative selection of research topics current to the field. We have targeted this work to a diverse audience ranging from professionals in fields from physics to medicine to beginning graduate students who are interested in rapidly acquiring the basics of this field. Intending that these volumes will find use as textbooks, we have edited the chapters to be fairly uniform, at a level accessible to chemically literate readers with a bachelor's degree in science or engineering. We hope that readers will find these volumes useful in providing both an introduction to and an overview of a field that we have found to be both challenging and rewarding.

We also wish to acknowledge the expert help of Margaret Williams in assembling and editing these two volumes.

Joan Selverstone Valentine<br>
Christopher S. Foote<br>
Arthur Greenberg<br>
Joel F. Liebman

# Series Preface

The purpose of this series is the presentation of the most significant research areas in organic chemistry from the perspective of the interplay and inseparability of structure, energetics, and reactivity. Each volume will be modeled as a text for a one-semester graduate course and will thus provide groundwork, coherence, and reasonable completeness. In this context, we have made the editorial decision to defer to the authors the choice of the desired blend of theory and experiment, rigor and intuition, practice and perception. However, we asked them to engage in the spirit of this venture and to explain to the reader the basis of their understanding and not just the highlights of their findings. For each volume, and each chapter therein, we have aimed for both a review and a tutorial of a major research area. Each volume will have a single theme, unified by the common threads of structure, energy, and reactivity for the understanding of chemical phenomena.

*Structure, energetics,* and *reactivity* are three of the most fundamental, ubiquitous, and therefore seminal concepts in organic chemistry. The concept of structure arises as soon as even two atoms are said to be bonded, since it is there that the concept of bond length and interatomic separation begins. Three-atom molecules already introduce bond angle into our functioning vocabulary, while four atoms are needed for the introduction of the terms *planarity, nonplanarity,* and *dihedral angles.* Of course, most organic chemists are interested in molecules of more than four atoms, so new shapes (tetrahedra, cubes, dodecahedra, prisms, and numerous exotic polyhedra) and new degrees of complexity arise. These new molecular shapes, in turn, function as templates for the next molecular generation. Still, the basic assumption remains: molecular structure determines energy and reactivity, and even though van't Hoff, LeBel, and Sachse explained chemical reality with palpable molecular models over a century ago, we still do much the same thing on the screens of personal computers.

The concept of energetics arises in chemistry as soon as there is a proton and an electron and remains with us throughout our discipline. There are the fundamental, experimentally measurable quantities of bond energies, proton affinities, ionization potentials, $pK_a$ values, and heats of formation. There are the derived quantities such as strain and resonance energies, acidity, and basicity. There are also the widely used, generally understood, and rather amorphous concepts such as *delocalization, conjugation,* and *aromaticity.* Indeed, the shape, conformation, and therefore function of a protein are determined by a balance of energetics contributions—resonance in the peptide linkage, hydrogen bonding, hindered rotation of a disulfide, van der Waals forces, steric repulsion, Coulombic interactions and salt bridges, and solvent interactions.

The concept of reactivity inseparably combines structure and energetics and introduces more concepts and words: *stereospecificity, intramolecularity, nucleophilicity, catalysis, entrophy of activation, steric hindrance, polarizability, hard and soft acids and bases, Hammett/Taft parameters.* Reactivity is a more difficult concept than structure and energetics. One must specify reaction conditions and usually accompanying reagents, since reactivity generally refers to two species or at least two seemingly disjoint parts of the same molecule.

Structure, energetics and reactivity in chemistry have been probed by a plethora of experimental and theoretical methods. These tools have different degrees of accuracy and applicability, and consensus is rare as to when our understanding is deemed adequate. Indeed, diverse approaches—heats of hydrogenation and Hartree–Fock calculations, line intensities and $LD_{50}$ values, ease of substitution and of sublimation, coupling constants and color—all contribute to the special blend of rigor and intuition that characterizes modern organic chemistry.

As people, and not just as scientists and editors, we wish to acknowledge the unity of the intellect and the emotions. We are grateful for the love and support of our families and for the inspiration, agitation, and stimulation from our mentors, colleagues, and students, and so we dedicate these volumes *"To Research and to Reason, To Family and to Friendship."*

JOEL F. LIEBMAN  ARTHUR GREENBERG
*Baltimore, Maryland*  *Charlotte, North Carolina*

# Structure Energetics and Reactivity in Chemistry Series (SEARCH series)

**Series Editors**

JOEL F. LIEBMAN
Department of Chemistry and Biochemistry
University of Maryland Baltimore County
Baltimore, MD 21228

ARTHUR GREENBERG
Department of Chemistry
University of North Carolina at Charlotte
Charlotte, NC 28223

The volumes in this series are comprised of state-of-the-art reviews, explicitly pedagogical in nature, in which specific topics are treated in depth. The series acronym SEARCH reflects the interplay between Structure, Energy And Reactivity in CHemistry and how these are also manifested in physical properties and biological activities.

*Other titles in the Series*

1. **Mesomolecules**: From Molecules to Materials
   Edited by G. David Mendenhall, Arthur Greenberg and
   Joel F. Liebman

2. **Active Oxygen in Chemistry**
   Edited by Christopher S. Foote, Joan Selverstone Valentine,
   Arthur Greenberg and Joel F. Liebman

# Editorial Advisory Board

WESTON T. BORDEN
Department of Chemistry
University of Washington
Seattle, WA 98195

JULIAN A. DAVIES
Department of Chemistry
University of Toledo
Toledo, OH 43606

GAUTAM R. DESIRAJU
School of Chemistry
University of Hyderabad
Hyderabad, India 500134

FRANÇOIS N. DIEDERICH
Laboratorium für Organische Chemie
ETH-Zentrum
Universitätstrasse 16
CH-8092 Zürich, Switzerland

DENNIS A. DOUGHERTY
Department of Chemistry
California Institute of Technology
Pasadena, CA 91125

RICHARD D. GANDOUR
Department of Chemistry
Louisiana State University
Baton Rouge, LA 70803

SHARON G. LIAS
Chemical Kinetics and Thermodynamics and Division
National Institute of Standards and Technology
Gaithersburg, MD 20899

ALAN P. MARCHAND
Department of Chemistry
University of North Texas
Denton, TX 76203

JOSÉ ARTUR MARTINHO SIMÕES
Departamento de Química e Bioquímica
Faculdade de Ciencias
Universidade de Lisboa
1700 Lisboa, Portugal

JOAN MASON
Department of Chemistry
Open University
Milton Keynes, UK MK7 6AA

ROBERT A. MOSS
Department of Chemistry
Rutgers University
Piscataway, NJ 08855

BRUCE E. SMART
E.I. Du Pont de Nemours & Co., Inc.
Central Research and Development Experimental Station
Wilmington, DE 19880-0328

JOAN SELVERSTONE VALENTINE
Department of Chemistry and Biochemistry
University of California, Los Angeles
Los Angeles, CA 90095

DEBORAH VAN VECHTEN
Space Sciences Division
Naval Research Laboratory
Washington, DC 20375

# Contributors

Judith A. Berliner
Departments of Pathology and Medicine/Cardiology
UCLA Medical Center
Los Angeles, CA 90095

David P. Ballou
Department of Biological Chemistry
University of Michigan Medical School
Ann Arbor, MI 48109

Stephen Fox
Department of Chemistry
The Johns Hopkins University
Baltimore, MD 21218

Arthur Greenberg
Department of Chemistry
University of North Carolina at Charlotte
Charlotte, NC 28223

Barry Halliwell
Pulmonary Medicine
University of California, Davis
Medical Center
Sacramento, CA 95817
and
Pharmacology Group
University of London, Kings College
Chelsea Campus
London SW3 5LX
United Kingdom

Raymond Y. N. Ho
Department of Chemistry and Biochemistry
University of California, Los Angeles
Los Angeles, CA 90095-1569

Kenneth D. Karlin
Department of Chemistry
The Johns Hopkins University
Baltimore, MD 21218

Joel F. Liebman
Department of Chemistry and Biochemistry
University of Maryland Baltimore County
Baltimore, MD 21228-5398

Rosemary Marusak
Department of Chemistry
Philip Mather Hall
Kenyon College
Gambier, OH 43022-9623

Vincent Massey
Department of Biological Chemistry
University of Michigan Medical School
Ann Arbor, MI 48109

Claude E. Meares
Department of Chemistry
University of California, Davis
Davis, CA 95616

Mark J. Nelson
Experimental Station
DuPont Central Research and Development
Wilmington, DE 19880

Bruce A. Palfey
Department of Biological Chemistry
University of Michigan Medical School
Ann Arbor, MI 48109

Lawrence Que, Jr.
Department of Chemistry
University of Minnesota
Minneapolis, MN 55455-0431

Stephen P. Seitz
Experimental Station
DuPont Central Research and Development
Wilmington, DE 19880

Patricia S. Traylor
Department of Chemistry
University of San Diego
San Diego, CA 92110-2492

Teddy G. Traylor (deceased)
Department of Chemistry
University of California, San Diego
La Jolla, CA 92093-0506

Joan Selverstone Valentine
Department of Chemistry and Biochemistry
University of California, Los Angeles
Los Angeles, CA 90095-1569

Andrew D. Watson
Department of Pathology
UCLA Medical Center
Los Angeles, CA 90095

# 1
# Biological Reactions of Dioxygen: An Introduction

RAYMOND Y. N. HO, JOEL F. LIEBMAN,
AND JOAN SELVERSTONE VALENTINE

## INTRODUCTION

Life on earth originated during a time when the atmosphere contained little or no gaseous oxygen. Primitive cells obtained the energy for their metabolism from glycolysis rather than respiration, and the fact that they were rich in thiols and other reducing agents did not present a problem because they were not normally confronted with appreciable levels of dioxygen ($O_2$) or other strong oxidants. The advent of photosynthesis changed this situation dramatically by introducing gaseous dioxygen into the atmosphere, initiating what has been referred to as the most dramatic example of environmental pollution that has ever occurred on earth (Levine, 1988). Ultimately, the level of dioxygen reached its modern level of 21%, an environment that is toxic to strict anaerobic bacteria, the modern-day descendants of those first primitive organisms. By contrast, modern aerobic organisms, which, like anaerobic organisms, also consist of cells rich in reducing agents, evolved to use the powerful oxidizing potential of dioxygen to their benefit by developing respiration and, at the same time, elaborate systems to protect, repair, or replace their components that might be damaged by the oxidation reactions that are the inevitable by-product of dioxygen metabolism (Bilinski, 1991). The fact that aerobic organisms need dioxygen to survive and yet must constantly guard against its toxicity is frequently referred to as the *oxygen paradox* (Koppenol, 1988).

It is estimated that respiration by aerobic organisms consumes about

90% of all of the $O_2$ utilized in the biosphere. In these organisms, roughly 80% of the consumed $O_2$ is used during the conversion of "food", e.g., glucose, to a viable energy source, e.g., ATP. This conversion proceeds via a series of enzyme-catalyzed reactions which are coupled to the synthesis of ATP. The terminal step of these reactions is the four-electron reduction of $O_2$ to give two molecules of water (eq. 1) (Sawyer, 1991).

$$O_2 + 4H^+ + 4e^- \rightarrow 2H_2O \qquad E° = +0.815\,V \qquad (1)$$

In many organisms, this final step is catalyzed by the enzyme cytochrome c oxidase (Malmström, 1982). Much of the remainder of the consumed $O_2$ is used in the biosynthesis of various biological molecules or in the conversion of water-insoluble molecules into water-soluble molecules for the purpose of excretion. These reactions are also enzyme catalyzed, and one or both oxygen atoms from $O_2$ are incorporated into the final products. The enzymes involved are referred to as *monooxygenases* or *dioxygenases*, respectively. This chapter discusses some of the general characteristics of the reactions of dioxygen and of species derived from it that may be relevant to its reactions in biological systems, particularly reactions of dioxygen with organic substrates, in order to provide a background for the other chapters in this volume.

## CHEMICAL REACTIVITY OF DIOXYGEN

Before addressing the biochemical reactivity of $O_2$, it is useful to review briefly some of the factors that determine its chemistry. (This topic is discussed in more detail in Ho et al., 1995). As mentioned above, one of the most important reactions of $O_2$ is its four-electron reduction to give two equivalents of water (reaction 1). The overall reduction potential for this reaction is $+0.815\,V$ vs. the normal hydrogen electrode (NHE), indicating that this reduction is strongly favored thermodynamically. Although the overall reduction is favored, concerted four-electron reductions of dioxygen are very rare, and reduction typically proceeds via a series of one- or two-electron steps. The reduction potentials for these steps are given in Table 1-1.

Examination of the reduction potentials shows that these dioxygen reduction reactions are all strongly favored except for the one-electron reduction of dioxygen to superoxide, which has a potential of $-0.33\,V$ vs. NHE. The relatively low one-electron reduction potential of dioxygen is one of the factors limiting its kinetic reactivity. In effect,

**TABLE 1-1** Standard Reduction Potential for One- and Two-Electron Reduction of Dioxygen Species in Water

| | | | | | | | E° vs. NHE, pH 7.25 |
|---|---|---|---|---|---|---|---|
| | $O_2$ | + | $e^-$ | $\longrightarrow$ | $O_2^-$ | | $-0.33$ V |
| $O_2^-$ | + | $e^-$ | + $2H^+$ | $\longrightarrow$ | $H_2O_2$ | | $+0.89$ V |
| $H_2O_2$ | + | $e^-$ | + $H^+$ | $\longrightarrow$ | $H_2O$ | + OH | $+0.38$ V |
| OH | + | $e^-$ | + $H^+$ | $\longrightarrow$ | $H_2O$ | | $+2.31$ V |
| $O_2$ | + | $2e^-$ | + $2H^+$ | $\longrightarrow$ | $H_2O_2$ | | $+0.281$ V |
| $H_2O_2$ | + | $2e^-$ | + $2H^+$ | $\longrightarrow$ | $2\ H_2O$ | | $+1.349$ V |

*Source:* Sawyer (1988).

there is a high oxidizing ability stored up in the dioxygen molecule that cannot be used until after the first one-electron reduction. Thus dioxygen can coexist with many reducing agents, such as those that occur within the interior of a living cell, without reacting rapidly with them (George, 1965). This property is crucial in biology since it allows the cell to control the consumption of $O_2$ and to help prevent random $O_2$ reactions that can destroy essential components of the living cell. The enzymes that enable dioxygen to circumvent the unfavorable one-electron step generally do so either by stabilizing superoxide-containing intermediates or by providing two-electron reaction pathways.

Although the first one-electron reduction of dioxygen to give superoxide is relatively unfavorable thermodynamically, it is by no means forbidden. In order for this reaction to occur, all that is needed is a powerful enough reducing agent. Within the cell, reducing agents such as reduced flavins and hydroquinones are strong enough to cause this reduction (Halliwell and Gutteridge, 1989). In fact, the activation of $O_2$ by flavin monooxygenases appears to involve the reduction of $O_2$ to $O_2^-$, followed by rapid reaction between superoxide and the oxidized flavin to form a flavin hydroperoxide intermediate (see Chapter 2).

An additional kinetic barrier causes most direct reactions of dioxygen with organic substrates to be slow. This barrier is generally termed *spin restriction*. The heats of formation for the direct oxygenation of some simple organic molecules with $O_2$ given in Table 1-2 demonstrate that such reactions are generally quite favorable thermodynamically, yet they generally occur very slowly at room temperature in the absence of initiators or catalysts.

The problem arises from the fact that dioxygen has a triplet ground state, i.e., that the most stable form of $O_2$ has two unpaired electrons with parallel spins. Since the majority of ground state organic

**TABLE 1-2** Heats of Formation of the Oxygenation of Simple Organic Compounds with $O_2$

|  |  |  | ΔH, kcal/mol |
|---|---|---|---|
| $CH_4(g)$ + $1/2 O_2(g)$ | ⟶ | $CH_3OH(g)$ | −30 |
| $C_6H_6(g)$ + $1/2 O_2(g)$ | ⟶ | $C_6H_5OH(g)$ | −43 |
| $C_6H_5OH(g)$ + $1/2 O_2(g)$ | ⟶ | $C_6H_4(OH)_2(g)$ | −42 |
| $C_2H_4(g)$ + $1/2 O_2(g)$ | ⟶ | $C_2H_4O(g)$ | −25 |

*Source:* Holm (1987).

molecules have single configurations, i.e., no unpaired electrons, and since the final products of the reactions between $O_2$ and these molecules also have singlet configurations, the direct reaction (eq. 2, where the arrows represent electron spins) requires the number of unpaired electrons to change during the reaction. It is impossible for such a reaction to occur in one fast, concerted step since such a pathway would violate the spin conservation law, which states that the overall spin state must be the same before and after each elementary step of a chemical reaction (Taube, 1965; Hamilton, 1974). Catalysts can break down this kinetic spin restriction, as discussed below.

$$O_2 (\uparrow\uparrow) + 2X(\uparrow\downarrow + \uparrow\downarrow) \longrightarrow \longrightarrow 2XO(\uparrow\downarrow + \uparrow\downarrow) \qquad (2)$$

## METAL–DIOXYGEN CHEMISTRY

From the above discussion of the chemistry of $O_2$, it is apparent that direct, uncatalyzed oxygenation of organic substrates by $O_2$ is not a major reaction pathway. In biology, the majority of nontoxic oxygenation reactions require the action of enzymes to occur at appreciable rates. This requirement allows the living cell to control when and where $O_2$ is used. Many of these enzymes that catalyze reactions of $O_2$ are metalloenzymes in which dioxygen interacts directly with a metal center, usually iron or copper, in the active site. The advantage of using redox-active transition metal ions as the active centers in oxygenating enzymes is that many of the transition metal ions can bind $O_2$ as a ligand and the bound dioxygen ligand can be reduced *in situ* (Sheldon and Kochi, 1981). Intermediates resulting from reduction of the dioxygen ligand, i.e., superoxide ($O_2^-$), peroxide ($O_2^{2-}$), or oxide ($O^{2-}$), can be stabilized by coordination to the metal center, which can direct their subsequent reactions with substrates. Such reaction pathways can overcome the kinetic and thermodynamic barriers that normally inhibit uncatalyzed $O_2$ reactions. Aside from activating $O_2$ for oxygenation reactions, transition metals are found in proteins that transport and store $O_2$ and in enzymes that protect the cell from oxidative damage by $O_2^-$ and $H_2O_2$. Because of the importance of transition metal–dioxygen interactions in biology, the possible binding motifs between transition metals and dioxygen or species derived from it are discussed below, along with a discussion of which motifs are currently believed to occur in the biological systems (Bytheway and Hall, 1994).

## Metal–Dioxygen Complexes

Dioxygen does not react with or complex strongly to metal centers that are highly oxidized. Thus complexes of $Fe^{III}$ or $Cu^{II}$, for example, even when coordinatively unsaturated, i.e., having an empty coordination position available for binding of a ligand, do not bind to or otherwise react with $O_2$. The fact that the $O_2$ molecule is not a good ligand for such metal ions is not surprising because good ligands for high oxidation state metal ions are usually relatively strong bases, and $O_2$ is not. Dioxygen does react readily with a variety of low oxidation state metal centers, usually giving irreversible oxidation of the metal center (Collman, 1977). But sometimes the reaction stops after dioxygen has reacted with the metal center but before complete oxidation has occurred, and a stable dioxygen complex can then be isolated. Such complexes are best understood as the products of oxidative addition, i.e., the metal center has been oxidized and the dioxygen ligand has been reduced. In some cases, the reduced dioxygen ligand is superoxide-like in its properties and in some cases it is peroxide-like; a variety of geometries have been observed (Figure 1-1).

The best indicators of the degree of reduction of the bound dioxygen ligand are the O—O bond lengths and the O—O stretching frequencies which tend to cluster in two groups, corresponding to either superoxo- or peroxo-like complexes (Figure 1-2) (Vaska, 1976; Summerville et al., 1979; Drago and Cordon, 1980; Jameson and Ibers, 1994).

In addition to the O—O bond lengths, the geometries of bonding of the mononuclear dioxygen complexes tend to be different for the superoxo- versus the peroxo-type complexes. As can be seen in Figure 1-1, there are two types of geometries for such complexes: either "end on," with binding of only one oxygen atom to the metal center

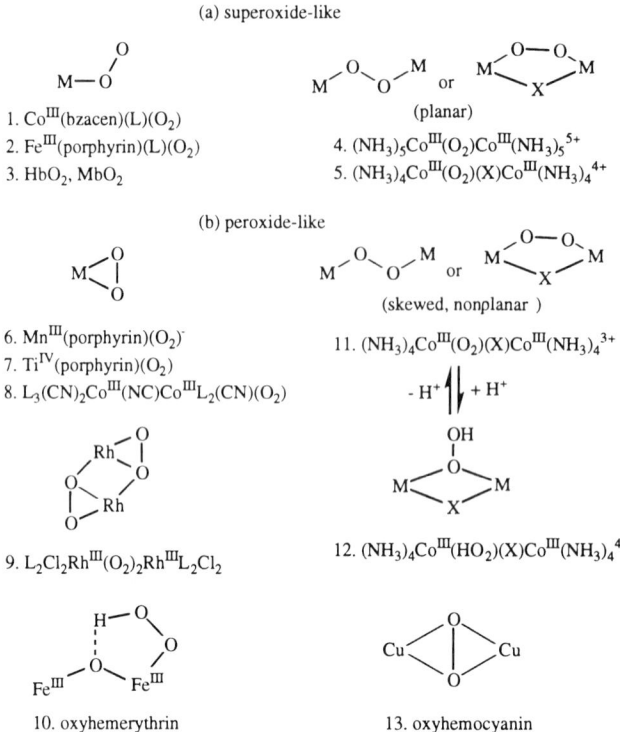

**FIGURE 1-1** Examples of the geometries found for the dioxygen moiety in metal–dioxygen complexes and in metalloproteins (bzacen = benzoylacetoneethylenediimine, L = axial ligand, X = $NH_2$, OH). References: compounds 1, 4, 5, 8, 9, 10 (Basolo et al., 1975; McLendon and Martell, 1976); compound 2 (Collman et al., 1976); compound 3 (Barlow et al., 1973); compound 6 (Hanson and Hoffman, 1980; Urban et al., 1982); compound 7 (Guilard et al., 1976); compound 9 (Raynor et al., 1982); compound 10 (Dunn et al., 1973); compound 13 (Freedman et al., 1976).

(Pauling, 1964; Torrog et al., 1976; Summerville et al., 1979), or "side on," with binding of both oxygen atoms (Griffith, 1956).

The end-on configuration is most commonly seen in superoxo-like complexes, the side-on one in peroxo-like complexes. The bonding between the metal center and the dioxygen ligand is commonly

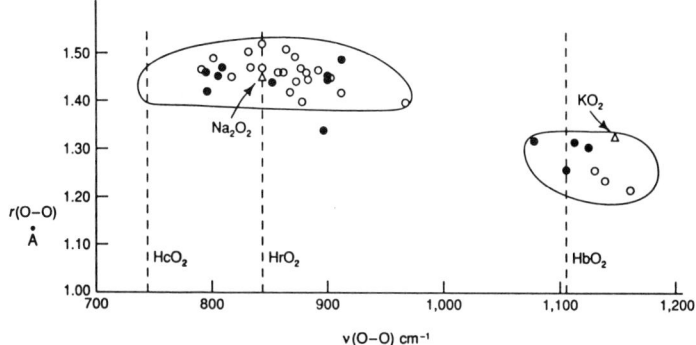

**FIGURE 1-2** Scatter diagram showing the distribution of O—O stretching frequencies and separations in ionic superoxides and peroxides (Δ) and in coordination compounds. An open circle denotes $O_2$ coordinated to one metal; a filled circle denotes $O_2$ bridging two metals. The O—O stretching frequencies of oxyhemoglobin, oxyhemocyanin, and oxyhemerythrin are marked by dashed lines. (Jameson and Ibers, 1994)

described using as a starting point $^1\Delta_g$ $O_2$, i.e., the lowest excited state of dioxygen, in which all the electrons are paired. This species is isoelectronic with ethylene, $H_2C=CH_2$, after removing the protons and placing them in the carbon nuclei, thus converting each carbon atom to an oxygen atom.

ethylene          $^1\Delta_g$ dioxygen

In the end-on case, several different bonding interactions between the metal ion and the dioxygen ligand are important (Summerville et al., 1979). First, a lone pair of electrons from an $sp^2$ hybrid orbital on $O_2$ is donated to an empty σ-symmetry orbital on the metal, thereby forming a ligand-to-metal σ bond (Figure 1-3A). The geometry of the $sp^2$ orbitals is presumably responsible for the bent M—O—O configuration. Such bonding can be thought of as analogous to an organic vinyl group bonding to a metal in an organometallic complex.

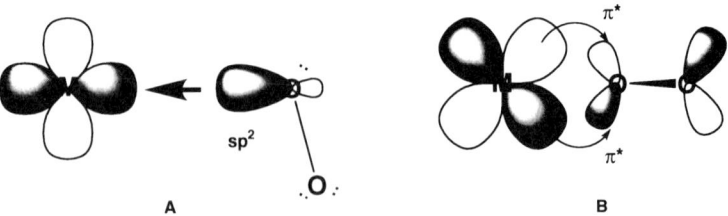

**FIGURE 1-3** Some bonding interactions in "end-on" dioxygen bonding. **A** represents a ligand-to-metal σ-bonding interaction from a lone pair of electrons in an sp²-hybridized oxygen orbital. **B** represents a metal-to-ligand π-backbonding interaction into the π* antibonding O—O molecular orbital. (Note that the M—O—O configuration in B is bent with the terminal oxygen atom coming forward.)

Secondly, the π* molecular orbitals of O₂ have the proper symmetry for metal-to-ligand donation to the t₂g-symmetry orbitals of the metal (Figure 1-3B). This metal-to-ligand π backbonding, which is similar to that in metal carbonyl complexes, provides the M—O bond with some double bond character, with a concomitant weakening of the O—O bond from the transfer of electron density into antibonding dioxygen orbitals.

Thirdly, because O₂ is bound at an angle to the metal, the π* orbitals of oxygen can interact with the metal $d_{z^2}$ orbital (Torrog et al., 1976), providing an additional ligand-to-metal σ bonding interaction. The end result of this bonding scheme is a fairly strong M—O bond and a weakening of the O—O double bond relative to free O₂.

The description for the bonding of dioxygen in its side-on geometry is different and resembles that of a triangularly bonded organometallic ethylene complex.

The dioxygen moiety is rotated about its cylindrical axis so that its π-bonding orbital rather than an sp²-hybridized lone pair orbital is forming a ligand-to-metal σ-bond with the metal center (Griffith, 1956). The bonding in this case results from (1) ligand-to-metal bonding of the π-bonding molecular orbital of O₂ with an empty d orbital of σ-symmetry on the metal (Figure 1-4A) and (2) metal-to-ligand π-backbonding of filled metal orbitals with π* orbitals of dioxygen (Figure 1-4B).

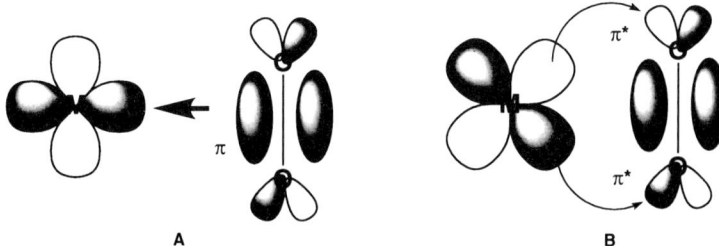

**FIGURE 1-4** Some bonding interactions in side-on dioxygen bonding. **A** represents a ligand-to-metal σ-bonding interaction from the π-bonding O—O molecular orbital and the metal. **B** represents a metal-to-ligand π-backbonding interaction into the π* antibonding O—O molecular orbital. (Note that the sp²-hybrid orbitals containing the electron lone pairs used in end-on bonding are in a plane perpendicular to the plane of the page and are not shown here.)

Like end-on bonding, side-on binding results in a weakening of the O—O bond because of transfer of electron density out of the O—O π-bonding molecular orbital in forming the ligand-to-metal σ-bond and the π-backbonding of election density from the metal into the O—O π*-antibonding molecular orbital.

It should be noted here that the starting point for this bonding description places the metal center in the same oxidation state that was present prior to the reaction of the metal center with dioxygen and starts with the neutral, i.e., uncharged dioxygen ligand, $O_2^0$. The prediction of substantial transfer of electron density to the dioxygen ligand via orbitals of π-symmetry corresponds to the observation that the dioxygen ligand in such complexes has been reduced practically to the level of superoxide or peroxide. This convention of describing the bonding of these complexes starting with complexes of $O_2^0$ can be confusing since it differs from the convention used to describe the bonding of metal–oxo complexes, which starts with the $O^{2-}$ ligand rather than with a neutral ¹D oxygen atom (see below). To avoid confusion, it is important to realize that the starting point in one case is $M^{n+}$ and $L^0$ and in the other case is $M^{(n+x)+}$ and $L^{x-}$. Properly applied, however, either convention can lead to accurate descriptions of the bonding.

## Dioxygen Reduction by Metal Ions or Complexes

The formation of stable dioxygen complexes from the reaction of a reduced metal center with dioxygen is the exception rather than the rule. It is much more common for a reduced metal center to react irreversibly with dioxygen, either to give hydrogen peroxide or water and the oxidized metal center or to give metal–oxo complexes. Such reactions undoubtedly proceed via metal–dioxygen complexes, but the reduction of the bound dioxygen ligand proceeds further to the point where the O—O bond is broken. At that point, the further oxidation of the metal center is highly favored and usually proceeds rapidly.

The mechanistic routes generally proposed for dioxygen reduction by metal ions and complexes are given in reactions 4–10 (Hammond and Wu, 1968).

$$M^{(n-1)+} + O_2 \rightleftharpoons M^{n+}-O_2^{1-} \quad (4)$$

$$M^{n+}-O_2^{1-} + M^{(n-1)+} \longrightarrow M^{n+}-O_2^{2-}-M^{n+} \quad (5)$$

$$M^{n+}-O_2^{2-}-M^{n+} \xrightarrow{2H^+} 2M^{n+} + H_2O_2 \quad (6)$$

$$\xrightarrow{2M^{(n-1)+}} 2M^{n+}-O^{2-}-M^{n+} \quad (7)$$

$$\searrow 2M^{(n+1)+}(O^{2-}) \xrightarrow{2M^{(n-1)+}} \quad (8)$$

$$2M^{(n-1)+} + H_2O_2 \xrightarrow{2H^+} 2M^{n+} + 2H_2O \quad (9)$$

$$M^{n+}-O^{2-}-M^{n+} \xrightarrow{H^+} 2M^{n+} + H_2O \quad (10)$$

(Note that formal charges on superoxo, peroxo, and oxo ligands have been indicated in reactions 4–10 to aid in keeping track of electrons and overall oxidation states.)

The first step of the reaction is probably reversible formation of a metal–superoxo complex (reaction 4). The superoxo ligand, $O_2^-$, is stabilized in such a complex by bonding to the metal ion and in most cases will not dissociate. Protonation of a superoxo complex could lead to dissociation of $HO_2$, which would then disproportionate (reactions 11 and 12), but such a pathway is not common.

$$M^{n+}(O_2^-) + H^+ \rightleftharpoons M^{n+} + HO_2 \quad (11)$$

$$2HO_2 \rightarrow O_2 + H_2O_2 \quad (12)$$

The predominant reaction after formation of the superoxo complex is reaction with another reduced metal ion or complex to form a bridging μ–peroxo complex (reaction 5). This type of μ–peroxo complex can then react by several different pathways. It may release hydrogen peroxide (reaction 6), which will then proceed to react with more reduced metal centers (reaction 9), or it may react directly with reduced metal centers to form a μ–oxo binuclear complex (reaction 7). Such oxo-bridged complexes are particularly stable when the metal ion is ferric, $Fe^{3+}$, and are frequently referred to as the *thermodynamic sink* of iron chemistry. The stability of oxo-bridged iron species contributes to the stability of solid iron oxides, e.g., rust, and provides much of the thermodynamic driving force for irreversible oxidation of ferrous, $Fe^{2+}$, complexes. Another pathway to μ-oxo complexes involves O—O bond cleavage to form a high-valent metal oxo complex followed by reaction with the reduced metal ions to form the final product (reaction 8). This reaction is particularly well characterized in the case of the reaction of ferrous porphyrin complexes with dioxygen, and the $Fe^{IV}$ oxo complex that is produced from the O—O bond cleavage has been observed and characterized (Traylor and Ciccone, 1989; Murata et al., 1990; Ostovic and Bruice, 1992). Finally, μ-oxo complexes may release water and the oxidized metal ion or complex on exposure to a source of protons (reaction 10). Thus the overall reaction is the four-electron reduction of dioxygen by the reduced metal ion or complex to form either the oxidized metal ion plus water (reaction 13) or, if it is particularly stable or no source of protons is present, the μ-oxo complex (reaction 14).

$$4\,M^{(n-1)+} + 4\,H^+ + O_2 \rightarrow 4\,M^{n+} + 2\,H_2O \qquad (13)$$

$$4\,M^{(n-1)+} + O_2 \rightarrow 2\,M^{n+}\!-\!O\!-\!M^{n+} \qquad (14)$$

The intermediacy of high-valent metal oxo complexes in reactions of reduced metal ions with dioxygen is an important aspect of metal–dioxygen chemistry. Such species have been postulated to be key intermediates in many metalloenzyme-catalyzed reactions of dioxygen. Before discussing such reactions, it is useful to describe some of the aspects of their bonding that may be useful in predicting the relative stabilities and reactivities of such species.

### Metal–Oxo Complexes

Metal oxides, i.e., compounds containing the oxide anion, $O^{2-}$, are found quite widely throughout the periodic table. Many of these metal oxides are ionic salts whose stabilities are due to the very high lattice energies that result from the high charge and relatively small size of the $O^{2-}$ anion. Metal oxide salts may be quite stable, but they are frequently powerful bases, e.g., $K_2O$ or MgO, and may deprotonate water or other protic materials when exposed to them. The high electronegativity of oxygen makes the oxide anion very difficult to oxidize. Therefore, it is possible to make oxide salts and complexes with metal ions in extremely high oxidation states, e.g., $CrO_4^{2-}$ or $MnO_4^-$, many of which are powerful oxidants. The only other simple anion or ligand with a similar resistance to oxidation is fluoride, $F^-$.

Metal–oxo complexes are coordination complexes in which the oxide anion is a ligand bound directly to the metal center of a metal ion or complex. Metal–oxo compounds may be described formally as resulting from the deprotonation of water ligands in an aquo complex (reaction 15).

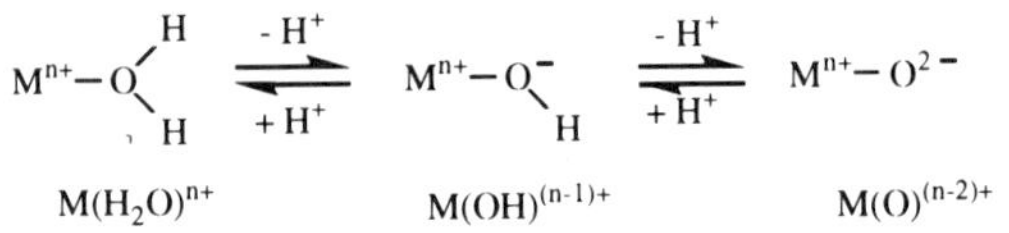

$$M^{n+}-O\begin{smallmatrix}H\\H\end{smallmatrix} \underset{+H^+}{\overset{-H^+}{\rightleftharpoons}} M^{n+}-O\begin{smallmatrix}\\H\end{smallmatrix}^- \underset{+H^+}{\overset{-H^+}{\rightleftharpoons}} M^{n+}-O^{2-} \qquad (15)$$

$$M(H_2O)^{n+} \qquad\qquad M(OH)^{(n-1)+} \qquad\qquad M(O)^{(n-2)+}$$

Reaction 15 illustrates an important point concerning some chemical conventions that are often used differently by inorganic and organic chemists. The bond between the metal and the oxygen atom in the aquo complex shown on the left of reaction 15 is drawn as though it were a normal covalent bond. In the convention used by inorganic chemists, it is understood that this bond is a coordinate covalent bond in which the electron density of the donor electron pair on the aquo ligand remains predominantly on the oxygen atom, and no formal positive charge is written on the oxygen atom, despite the fact that there are three bonds to the oxygen atom. Likewise, the second species in the equation is understood to be a coordination complex of hydroxide, $OH^-$, and the third species a coordination complex of oxide, $O^{2-}$, with the negative charges remaining on the ligand. The source of the confusion can be better appreciated on comparing the formal charges in reactions 15 and 16, the analogous protonation and deprotonation reactions of an organic alcohol. (It is important to remember that these formal charges are just that, i.e., a formality. In fact, there is every reason to believe that there is much less charge separation in all of these molecules than is suggested by the formal charges because such high degrees of charge separation would be very costly energetically.)

$$\text{R}-\overset{\text{H}}{\underset{\text{H}}{\text{O}^+}} \underset{+\text{H}^+}{\overset{-\text{H}^+}{\rightleftharpoons}} \text{R}-\underset{\text{H}}{\text{O}} \underset{+\text{H}^+}{\overset{-\text{H}^+}{\rightleftharpoons}} \text{R}-\text{O}^- \quad (16)$$

$$(\text{ROH}_2)^+ \qquad \text{ROH} \qquad (\text{RO})^-$$

Reaction 15 tends to proceed further to the right, forming hydroxo and oxo complexes when the charge on the metal center, $M^{n+}$, is high, since the higher positive charge on the metal center stabilizes the negative charge that develops when $OH^-$ and $O^{2-}$ are formed. Thus $Al^{3+}$ and $Fe^{3+}$ react readily with water and form solid $Al_2O_3$ and $Fe_2O_3 \cdot H_2O$.

In addition to stability due to predominantly electrostatic interactions between positively charged metal centers and oxo ligands, transition metal centers provide an important additional stabilizing interaction by means of ligand-to-metal $\pi$-bonding into empty metal d orbitals. In fact, the requirement for empty $\pi$-symmetry d orbitals for this type of bonding virtually restricts highly stable transition metal–oxo complexes to the left-hand side of the transition series. As pointed out by Taube (1965), the importance of this type of $\pi$-bonding in stabilizing metal–oxo complexes is evident from the fact that stable mixed ligand oxo–aquo complexes rather than hydroxo complexes exist for many of the transition metals on the left (reaction 17).

$$\underset{\text{OH}}{\overset{\text{OH}}{\text{L}_n\text{M}-\text{OH}}} \rightleftharpoons \underset{}{\overset{\text{O}}{\underset{\parallel}{\text{L}_n\text{M}}}-\text{OH}_2} \rightleftharpoons \underset{}{\overset{\text{O}}{\underset{\parallel}{\text{L}_n\text{M}}}} + \text{H}_2\text{O} \quad (17)$$

Examples of such complexes are aquo complexes containing the titanyl or vanadyl ions, $(TiO)^{2+}$ (Dwyer et al., 1975) and $(VO)^{2+}$ (Rice et al., 1976; Riechel et al., 1976).

Here it is interesting to compare the organic analogy, reaction 18, where the doubly bonded oxygen is also favored energetically over the diol.

$$\text{R}-\underset{\text{OH}}{\overset{\text{OH}}{\underset{|}{\overset{|}{\text{C}}}}}-\text{R}' \rightleftharpoons \text{R}-\overset{\text{O}}{\underset{\parallel}{\text{C}}}-\text{R}' + \text{H}_2\text{O} \quad (18)$$

As mentioned in the discussion of the bonding of dioxygen complexes above, the model typically used to describe the bonding of the oxo ligand to the metal center of such complexes uses as a starting point an oxide anion, $O^{2-}$, rather than a neutral oxygen atom, $O^0$. The

**14** *Biological Reactions of Dioxygen: An Introduction*

ligand-to-metal σ-bonding is attributed to the overlap of σ-symmetry 2s and 2p orbitals on oxygen with empty σ-symmetry metal d orbitals. In an octahedral complex, such orbitals would be the higher-energy $e_g$ orbitals, i.e., $d_{z^2}$ and $d_{x^2-y^2}$. The π-bonding is also ligand-to-metal since the species bonded to the metal ion is assigned as oxide, $O^{2-}$. The orbitals with the correct symmetry for this π-bonding are the lower-energy $t_{2g}$ orbitals, i.e., the $d_{xy}$, $d_{xz}$, and $d_{yz}$ orbitals. These lower-energy $t_{2g}$ orbitals are the first d orbitals to fill and therefore only have vacancies for metal centers in relatively high oxidation states or to the left of the periodic table.

examples of σ-bonding interactions

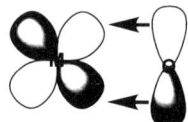

example of a π-bonding interaction

Vacancies in the $t_{2g}$ orbitals are necessary in order to have ligand-to-metal π-bonding interactions with the filled orbitals of π-symmetry on the oxide ligand. Because of this requirement, the formation of doubly bonded oxides of the first row transition metals is generally found in the earlier transition metals and at higher oxidation states (Wilkinson et al., 1987; Cotton and Wilkinson, 1988). With Sc, Ti, and V, which have one to three 3d electrons, the lower +2 and +3 oxidation state cations have empty 3d orbitals available for π bonding and are known to form doubly bonded oxides at these states. Of course, along with these lower oxidation states, the higher +4 and +5 states of Ti and V also form doubly bonded oxides. Moving to the right in the periodic table, the elements with more d electrons will form a double bond with the oxide ligand only in their higher oxidation states. For example, with Cr and Mn, doubly bonded oxo complexes are known for only the +4 to +7 oxidation states. The next transition metal, Fe, has six 3d electrons and should form a double bond only in the higher oxidation states, but because the stability of the higher oxidation states decreases in moving to the right of the periodic table, Fe generally does not form a double bond with the oxide ligand. Exceptions are found in the case of Fe–heme and Fe–porphyrin complexes, where high-valent (+4 or +5) iron–oxo complexes are believed to be the active intermediates in the oxygenation reactions. With the remaining first row transition metals, Co, Ni, Cu, and Zn, the combination of the high number of d electrons and the instability of higher oxidation states prevents the formation of doubly bonded oxo complexes.

In addition to the mononuclear, monodentate metal–oxo complexes described above, oxide ligands are found as bridging ligands between two or more transition metal ions in various different types of complexes (Wilkinson, 1987). Π-bonding may also play a role in μ–oxo complexes.

## METAL–DIOXYGEN BIOCHEMISTRY

### Dioxygen Transport and Storage Proteins

Several common features are observed for dioxygen transport and storage proteins, each of which can be explained in terms of the chemical reactivity of dioxygen. A reversible binder of dioxygen is needed for dioxygen storage and transport, but relatively few chemical reactions of dioxygen are in fact readily reversible. The most common type of reversible reaction of dioxygen is reversible binding of dioxygen with transition metal coordination complexes that have open coordination positions for dioxygen binding. It is therefore not surprising that proteins that evolved to carry out this function contain the low-valent metal ions $Fe^{2+}$ or $Cu^+$, with open coordination positions available to provide a site of interaction for the dioxygen molecule. The protein also plays an essential role in these dioxygen-binding proteins in ensuring that the reaction of the deoxy form with dioxygen to give the oxy form is reversible. Thus they typically provide the metal ion with a bulky, hydrophobic surrounding that protects the metal ion–dioxygen complexes from further reaction with either water, protons, or other metal ions, thus blocking mechanistic pathways that would lead to irreversible oxidation of the metal ion (Collman, 1977; Busch and Alcock, 1994; Momenteau and Reed, 1994).

One of the fascinating aspects of studying the biological reactions of $O_2$ with metalloproteins is realizing how greatly the functions of these proteins vary despite the fact that the initial $O_2$ binding is very similar. For example, heme-containing proteins are known to perform many functions in biology, including the transport and storage of $O_2$ (hemoglobin, myoglobin, and hemerythrin), the oxidation of biological compounds using $O_2$ (oxygenases), and the four-electron reduction of $O_2$ coupled with ATP syntheses (cytochrome c oxidase). Yet in each of these biological reactions, it is believed that $O_2$ initially binds end on in a very similar fashion to the binding of iron in a protein-bound heme moiety (Ingraham and Meyer, 1985). Subsequent reactions then lead to the execution of each of their particular and different functions. In part, the differences in reactivity can be attributed to the environment surrounding the heme, including the accessibility of biological substrates to the $Fe^{II}$ center, the availability of reducing agents, and the presence of other transition metal centers in the same metalloprotein.

For most multicellular organisms, simple unfacilitated diffusion of dioxygen into the organism is not fast enough to meet their metabolic needs. Therefore systems have evolved to bind, transport, and store dioxygen throughout the organism; these systems generally consist of metalloproteins that contain either iron or copper at the $O_2$-binding site. In humans and many other higher organisms, the protein responsible for $O_2$ transport is hemoglobin, a four-subunit protein, and for $O_2$ storage it is myoglobin, its one-subunit analog. For both of these proteins, the $O_2$-binding site is the iron center contained in a heme group (ferrous protoporphyrin IX).

protoporphyrin IX

Deoxy hemoglobin and myoglobin contain an $Fe^{II}$ center bound to the four-coordinate heme macrocyclic ligand and one imidazole ligand from a histidyl residue on the protein (Buchler, 1978; Brunori et al.,

1982). The Fe$^{II}$ center is thus five-coordinate, and consequently coordinatively unsaturated, and binds dioxygen at the sixth coordination position. The heme is buried within a nonpolar cavity in the protein such that the binding site is isolated from the surrounding environment. This isolation is essential for the function of these proteins in order to prevent μ–peroxo complex formation, as in reactions 4 and 5 above, and thus to ensure that the binding of O$_2$ is reversible. The interaction of dioxygen with hemoglobin, myoglobin, and several other heme-containing proteins (Momenteau and Reed, 1994; Springer et al., 1994) is discussed further in Chapter 3.

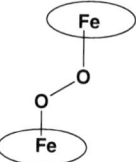

μ-peroxo diheme complex

In certain invertebrate marine organisms, the O$_2$ transport protein is hemerythrin, a multisubunit protein that contains a nonheme binuclear diiron unit at the O$_2$-binding site of each subunit (Klotz and Kurtz, 1984; Stenkamp, 1994). The two iron centers are bound to the protein through the side chains of five histidyl residues—three bound to one iron and two to the other—and the side chains of an aspartate and a glutamate residue, which with an oxo or hydroxo ion act as bridging ligands to the two iron centers. It is believed that O$_2$ binding occurs at the five-coordinate iron center, forming a complex with the hydroperoxo ligand, HO$_2^-$, hydrogen bonded to the bridging oxo ligand. The oxygen-binding properties of hemerythrin are discussed further in Chapter 5.

proposed structure for oxyhemerythrin

One final dioxygen transport protein that will be briefly mentioned here is hemocyanin, a blue dicopper protein found in the hemolymph ("blood") of crabs, snails, spiders, and related species (Brunori et al.,

1982; Magnus et al., 1994). Hemocyanin is also a multisubunit protein, each subunit of which contains a dioxygen-binding site consisting of a binuclear dicopper center bound to histidyl imidazole side chains. As discussed in detail in Chapter 4, it is now believed that $O_2$ binds side on to both copper atoms.

$$(His)_2N-Cu(\mu-O)_2Cu-N(His)_2$$

proposed structure for oxyhemocyanin

## Metal-Containing Oxygenase Enzymes

Dioxygen-activating enzymes, such as cytochrome P-450, methane monooxygenase, or tyrosinase, have metal-binding sites that are similar to those of their dioxygen-binding counterparts, e.g., myoglobin, hemerythrin, or hemocyanin. In those cases, the metalloenzyme contains a substrate-binding site in close proximity to its dioxygen-binding site, and the bound dioxygen ligand is activated by reduction for reaction with the substrate. The oxygenase enzymes catalyze the reaction of dioxygen with organic or inorganic substrates. Such enzymes lower the kinetic barriers to such reactions by binding and activating either dioxygen or the substrate. There are two types of oxygenase enzymes: monooxygenase enzymes, which incorporate one of the oxygen atoms in $O_2$ into the oxygenated products and reduce the other to water, and dioxygenase enzymes, which incorporate both oxygens into the oxygenated products. Aside from the flavin-containing oxygenases (see below and Chapter 2), most of the oxygenase enzymes contain metal ions as required cofactors. The structures of the protein and of the metal-binding sites are known for many of these oxygenase enzymes but, with a few exceptions, detailed information concerning the mechanisms of substrate or $O_2$ activation in these enzymes is quite limited. The mechanistic information that is available for many of these systems is presented in Chapters 3–6.

The similarity of the active sites of some of the monooxygenase enzymes to the $O_2$-binding sites of the dioxygen transport and storage proteins suggests that $O_2$ bonding to the reduced metal center is an essential first step in the mechanism. One of the most thoroughly studied classes of monooxygenase enzymes is cytochrome P-450, which contains heme at its active site (Ortiz de Montellano, 1985). In comparing cytochrome P-450 enzymes with hemoglobin and myoglobin, the first thing to note is that the iron is bound to the same heme (protoporphyrin IX) group in both systems but that the axial ligand in the case of the former is cysteine, whereas it is histidine in the latter. The substitution of a thiolate for an imidazole as an axial ligand to the iron center is likely to change its reactivity substantially, but the significance of these changes is not yet known. The environments around the hemes are also quite different. In the cytochrome P-450 enzymes, the heme group is held in place by being sandwiched between two helices in the protein and a substrate-binding site can be localized near the heme iron center. Mechanisms proposed for cytochrome P-450 are discussed further in the section on "Mechanisms of Dioxygen Activation" and in Chapter 3.

Nonheme metal-containing oxygenase enzymes exist that resemble cytochrome P-450 in that they also catalyze oxygenation of organic substrates, but little is known about the chemical nature of the active intermediates in these nonheme systems (see Chapter 5). While investigators have speculated that the mechanisms may be similar to those of the heme-containing systems, there is no direct evidence that high-valent metal oxo intermediates either are or are not involved in the enzymatic systems in the absence of the porphyrin prosthetic group.

### Cytochrome *c* Oxidase

The final enzyme that will be discussed here is perhaps one of the most important, cytochrome *c* oxidase (Palmer, 1987; Vanngard, 1988). As mentioned at the beginning of this chapter, cytochrome c oxidase is responsible for the terminal step in the synthesis of ATP that catalyzes the reduction of $O_2$ by four electrons, giving $H_2O$. This enzyme contains two or three copper ions and two hemes. The $Cu_A$ site, which contains two copper ions, and the heme A site appear to function only as electron-transferring agents; they do not interact directly with $O_2$. The two remaining metal sites, $Cu_B$ and cytochrome $a_3$, are found in close proximity to each other and appear to act as a binuclear center for binding and reduction of $O_2$ to the peroxide, $O_2^{2-}$, level (Figure 1-5). This peroxo ligand may be bound to one of the metal centers or it may bind to both in a bridging configuration. Addition of the third electron and two protons causes O–O bond cleavage, giving a high-valent iron oxo intermediate, $Fe^{IV}P(O^{2-})$ and a cupric–water complex. The iron oxo intermediate then accepts the fourth electron and a proton, giving a ferric hydroxo center, $Fe^{III}P(OH)$, which then is protonated again and releases water.

The evidence for this mechanism comes from the comparison of spectroscopic data obtain during the reaction with data obtained from

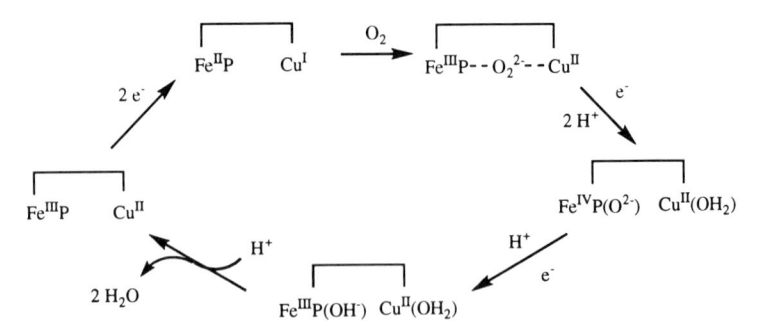

**FIGURE 1-5** Proposed mechanism of the four-electron reduction of dioxygen by cytochrome $c$ oxidase.

reactions of model complexes (Han et al., 1990; Wikström and Babcock, 1990). The binuclear nature of the dioxygen-binding site presumably lowers the kinetic barrier to dioxygen reduction by bypassing the unfavorable one-electron reduction step. It is interesting to note that no leakage of reduced dioxygen intermediates such as hydrogen peroxide or hydroxyl radical from the active site of this enzyme has been detected. Thus the rapid, clean processing of dioxygen by cytochrome $c$ oxidase does not appear to lead to toxic reactions of dioxygen. In fact, the rapid processing of dioxygen tends to keep its concentration low and thus prevents other reactions, some of which may be deleterious.

## MECHANISMS OF DIOXYGEN ACTIVATION

If thermodynamics rather than kinetics governed the reactivity of dioxygen, most of the biosphere would be converted rapidly to $CO_2$ and $H_2O$, and aerobic life as we know it would be impossible. The enzymes that catalyze reactions of dioxygen with substrates not only increase the rates of reaction of substrates with dioxygen but also guide the reactions of dioxygen and substrates along specific reaction pathways that give very specific products. There is every indication that these enzymes generate extremely reactive intermediates that are selective only because the exact site of reaction of oxygen with each substrate is controlled by the mode of substrate binding at the active site. Efforts to mimic biological oxygen activation using model compounds have suggested a number of different reactive intermediates, but, as described below, it is not yet possible to conclude with certainty which one or ones are operating in the case of each individual enzyme.

By *dioxygen activation*, we mean a mechanistic pathway that causes the properties of the dioxygen molecule to be modified in such a way that the kinetic barriers to its reduction or its reaction with a substrate are lowered. The observation that the rate of reaction of a substrate with dioxygen is enhanced in the presence of a reagent does not indicate necessarily that the reagent is an oxygenation catalyst since that reagent could instead be acting as an initiator of free radical autoxidation of the substrate (Pryor, 1976; Halliwell and Gutteridge, 1989). Even when it is ascertained that the reagent is a true oxygenation catalyst, one should not assume that the mechanism involves dioxygen activation since it may be the substrate rather than the dioxygen molecule that is being activated (see below). Most oxygenase enzymes that have been characterized contain transition metal ions as essential cofactors, and it is believed that the interaction of the metal ion with the dioxygen molecule or the substrate is in each case a crucial aspect of its mechanism of reaction. The one major class of oxygenase enzymes that does not contain at least one metal center is the external flavin monooxygenase enzymes. Reaction mechanisms for this class of enzymes appear to involve direct reaction of $O_2$ with the flavin cofactor to give a flavin hydroperoxide intermediate, which then reacts with the substrate. This mechanism is discussed in detail in Chapter 2.

flavin hydroperoxide

## Heme-Containing Enzymes

In the case of dioxygen activation by a metalloenzyme or another metal-containing catalyst, it is clear that the first step is dioxygen binding, but there is not general agreement about the details of the subsequent steps that generate intermediates with sufficient reactivity to oxygenate whatever substrate is present. Our understanding of dioxygen activation by metalloenzymes is most advanced in the case of cytochrome P-450 (Ortiz de Montellano, 1985), and many of the proposals concerning intermediates occurring in other systems are based on this enzyme as a model. But there is no reason to assume that the intermediates in all dioxygen-activating metalloenzymes must be similar, in particular since such widely differing types of metal centers are found in the different enzymes, e.g., mononuclear or binuclear, heme-containing, non-heme iron-containing, copper-containing, etc.

In the case of cytochrome P-450, the first step of the enzymatic reaction is binding of dioxygen to a deoxy heme center to give a heme–dioxygen complex similar to that found in oxyhemoglobin and oxymyoglobin and probably best described as a superoxo-like complex, i.e., $Fe^{III}P(O_2^-)$. Based on studies of model complexes, such heme–dioxygen complexes do not have sufficient reactivity to react directly with most substrates at this point. The heme–dioxygen complex is then reduced by one electron, presumably on protonation, giving a ferric–peroxo or hydroperoxo species (reaction 19). There are three possible mechanistic pathways by which the oxygen-containing species that attacks the substrate may be generated from the ferric–peroxo or hydroperoxo species. Firstly, the ferric–peroxo or hydroperoxo species themselves might react directly with a substrate. Secondly, an iron(IV) oxo complex formed by homolytic O—O bond cleavage of a ferric–hydroperoxo complex (reaction 20) might be the reactive species. Thirdly, a complex at the oxidation level of an iron(V) oxo complex, formed by heterolytic O—O bond cleavage of a ferric–hydroperoxo complex (reaction 21), might be the reactive species. The free hydroxyl radical, HO·, generated in the homolytic cleavage (reaction 20), although highly reactive and capable of attacking P-450 substrates, is considered to be an unlikely candidate to be an intermediate in cytochrome P-450 reactions because of the indiscriminate character of its reactivity (Anbar et al., 1966; Heicklen, 1981; Jeong and Kaufman, 1982).

$$Fe^{II}P + O_2 \rightarrow \underset{\text{ferric superoxo}}{Fe^{III}P(O_2^-)} \overset{e^-}{\rightarrow} \underset{\text{ferric peroxo}}{[Fe^{III}P(O_2^{2-})]^-} \overset{H^+}{\rightarrow} \underset{\text{ferric hydroperoxo}}{Fe^{III}P(OOH^-)} \quad (19)$$

homolytic O—O bond cleavage: $Fe^{III}P(OOH^-) \rightarrow Fe^{IV}P(O^{2-}) + HO·$ (20)

heterolytic O—O bond cleavage: $Fe^{III}P(OOH^-) \rightarrow [Fe^{IV}(P·)(O)^{2-})]^+ + HO^-$ (21)

(P = porphyrin ligand; it has a $2^-$ charge. P· = the radical produced by one-electron oxidation of the porphyrin; it has a $1^-$ charge.)

The high-valent iron–oxo complex, formed via heterolytic cleavage (reaction 21), is the favored intermediate for direct reaction with substrate in cytochrome P-450. This conclusion was initially drawn from studies of reactions of the enzyme with organic peroxides and other single-oxygen atom donors (McMurry and Groves, 1985). Single-oxygen atom (*oxene*) donors are reagents capable of donating a neutral oxygen atom to an acceptor, forming a stable product in the process. It was discovered that ferric cytochrome P-450 could catalyze oxygenation reactions using single-oxygen atom donors in place of dioxygen and reducing agents. This reaction pathway is termed the *peroxide shunt*. This result implies that the same intermediate was generated in each reaction, and the fact that single-oxygen atom donors could drive this reaction implied that this species contained only one oxygen atom, i.e., was generated subsequent to O—O bond cleavage. The mechanism suggested for the peroxide shunt reaction is given in reactions 22 and 23 (see Chapter 3).

$$[Fe^{III}P]^+ + OX \rightarrow [Fe^{IV}(P\cdot)(O^{2-})]^+ + X \qquad (22)$$

$$[Fe^{IV}(P\cdot)(O^{2-})]^+ + \text{substrate} \rightarrow [Fe^{III}P]^+ + \text{substrate(O)} \qquad (23)$$

The observation of the peroxide shunt reaction and modeling experiments that demonstrated similar reactivity for synthetic porphyrin systems (see Chapter 4) led to the proposal that the species responsible for oxygenation of substrate in most, if not all, cytochrome P-450-catalyzed reactions is this high-valent heme–oxo complex, $[Fe^{IV}(P\cdot)(O^{2-})]^+$. The successful generation of members of this class of compounds, using synthetic iron–porphyrin complexes at low temperatures in organic solvents, and the observation that their reactivity paralleled that observed for the enzyme, i.e., alkane hydroxylation, olefin epoxidation, heteroatom oxygenation, etc., added considerable support to this hypothesis. The mechanism shown in Figure 1-6 was therefore proposed for cytochrome P-450 (Ortiz de Montellano, 1985):

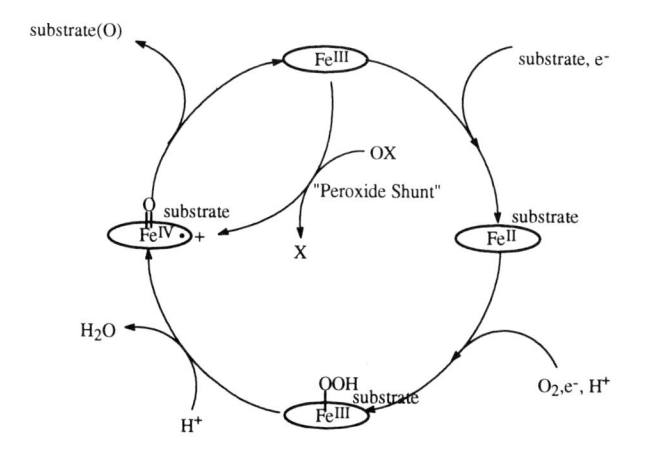

**FIGURE 1-6** Proposed mechanism of dioxygen activation by cytochrome P-450.

The most kinetically challenging of the reactions of cytochrome P-450 and other monooxygenases is the hydroxylation of unactivated alkanes since a C—H bond must be broken in order for the reaction to proceed. The mechanism proposed for this reaction by Groves and coworkers (Groves, 1985) is the *oxygen rebound mechanism*, reactions 24 and 25.

$$[Fe^{IV}(P^\cdot)(O^{2-})]^+ + H-C\diagup \longrightarrow [Fe^{IV}(P)(OH^-)]^+ + \cdot C\diagup \quad (24)$$

$$[Fe^{IV}(P)(OH^-)]^+ + \cdot C\diagup \longrightarrow [Fe^{III}(P)]^+ + HO-C\diagup \quad (25)$$

The mechanisms shown in Figure 1-6 and reactions 24 and 25 are widely accepted. However, there remain several significant uncertainties concerning this mechanism. For example, the structures of those cytochrome P-450s whose X-ray crystal structures are available seem to lack the elaborate machinery to facilitate O—O bond cleavage that is contained in catalase and peroxidase enzymes (Poulos, 1985). The structures of these latter enzymes have been shown to contain conserved residues near the active sites that are oriented in such a way that they stabilize the charges that develop during heterolytic cleavage of the O—O bond (Morrison and Schönbaum, 1976; Quinn et al., 1982, 1984). Cytochrome P-450 contains no such residues, and it has therefore been proposed that the cysteine axial ligand in cytochrome P-450 facilitates O—O bond cleavage and thus makes such machinery for the generation of a high-valent oxo intermediate unnecessary. However, this proposal is difficult to verify in a model system due to the highly reactive character of thiolate ligands.

An alternative to invoking high-valent metal oxo intermediates for monooxygenase enzyme reactions is to consider the possibility that some of the reactions proceed by direct reactions of peroxidic intermediates with substrates, i.e., that O—O bond breaking may be concerted with the attack on the substrate. Such a mechanism should probably be considered even in the case of cytochrome P-450 (reactions 26 and 27).

$$Fe^{III}(P)(OOH^-) + H-C\diagup \longrightarrow Fe^{IV}(P)(O^{2-}) + \cdot C\diagup + H_2O \quad (26)$$

$$Fe^{IV}(P)(O^{2-}) + \cdot C\diagup \longrightarrow (P)Fe^{III}-O-C\diagup \xrightarrow{H^+} [Fe^{III}(P)]^+ + HO-C\diagup \quad (27)$$

Certain observations for cytochrome P-450 are actually better explained by this mechanism. For example, cytochrome P-450$_{cam}$ gives hydroxylation of its substrate d-camphor only in the 5-exo position, but deuterium labeling studies show that either the 5-exo or the 5-endo hydrogen is lost (reaction 28) (Bruice, 1988).

$$ (28) $$

One possible explanation that has been proposed to account for the observation that hydrogen can be removed from either position, but that the oxygen can be delivered only to the exo position, is that the substrate changes position after the initial hydrogen atom abstraction. An alternative explanation is that the initial hydrogen atom abstraction is concerted with a homolytic O—O bond cleavage of a ferric–heme hydroperoxide intermediate, and that this abstraction can occur at either position because the terminal oxygen atom reaches farther than the heme Fe$^{IV}$–oxo intermediate, which can only reach the nearer position.

For other reactions catalyzed by cytochrome P-450 enzymes, direct reaction of substrate with ferric–hydroperoxide intermediates is also a possibility. Many peroxidic species are known to epoxidize olefins, for example, and require no prior O—O bond bond cleavage (Sheldon and Kochi, 1981). It seems quite reasonable that an olefinic substrate bound to the enzyme might react directly with such an intermediate (reaction 29).

$$Fe^{III}(P)(OOH^-) + \underset{/}{\overset{\backslash}{C}}=\underset{\backslash}{\overset{/}{C}} \longrightarrow Fe^{III}(P)(OH^-) + \underset{/}{\overset{\backslash}{C}}\overset{O}{\underset{\backslash}{-}}\underset{\backslash}{\overset{/}{C}} \quad (29)$$

In addition, direct reactions of a peroxidic intermediate with substrate have been proposed for the enzyme aromatase, which is a cytochrome P-450, as well as for another related enzyme, heme oxygenase (Cole and Robinson, 1991).

## Nonheme Metalloenzymes

Nonheme metal-containing oxygenase enzymes exist that resemble cytochrome P-450 in that they also catalyze hydroxylation of aliphatic or aromatic hydrocarbons (see Chapter 7; Feig and Lippard, 1994). But much less is known about the chemical nature of the active intermediates that react with the organic substrates in these nonheme systems. While investigators have speculated that the mechanisms may be similar to those of the heme-containing systems, there is no direct evidence that high-valent metal–oxo intermediates either are or are not involved in the enzymatic systems in the absence of the iron porphyrin prosthetic group. (Possible mechanisms for such nonheme enzymes are discussed in Chapters 4–6, along with a consideration of relevant model systems using low molecular weight coordination complexes of iron or copper.)

In our opinion, the possibility that a metal–peroxo or hydroperoxo intermediate is reacting directly with the substrate without prior O—O bond breaking should be considered for these systems as well. In support of this hypothesis is the observation that high-valent metal–oxo complexes of copper and of nonheme iron are highly unstable and, when it has been possible to generate them, they have appeared to have a high degree of radical character which is not consistent with the nature of the products of many of the reactions catalyzed by these enzymes (Leising et al., 1991). For example, nonporphyrin, high-valent

(30)

metal–oxo complexes typically carry out allylic oxidation in preference to epoxidation, whereas epoxidation is the favored pathway for most of these enzymes.

One major problem that arises in evaluating mechanisms in which direct reactions of ferric and cupric–peroxo, hydroperoxo, or high-valent metal–oxo complexes with substrates are invoked is that few synthetic analogs of such intermediates have been prepared and characterized with respect to their potential reactivities with substrates (Feig and Lippard, 1994; Kitajima and Moro-oka, 1994). In addition, much of the modeling that has been done to date fails to mimic the tight binding of the substrate at the active site that often occurs before dioxygen binding but instead relies on trapping of reactive intermediates by substrates in bimolecular reactions. In the case of the iron porphyrin system, iron–peroxo (McCandlish et al., 1980; Burstyn et al., 1988) and high-valent iron–oxo (McMurry and Groves, 1985) species have been characterized, but the iron–hydroperoxo species have not. In the case of the nonporphyrin iron systems, high-valent iron–oxo species have been partially characterized (Stassinopoulos and Caradonna, 1990; Leising et al., 1991), but the iron–peroxo and –hydroperoxo species have not. In the case of copper, only peroxo complexes have been characterized (see Chapter 6). An additional complicating factor in extending the approaches used in iron porphyrin chemistry to the nonporphyrin systems is that the single-oxygen atom donors such as iodosylbenzene and peracids that react with iron porphyrins to produce high-valent metal–oxo species (see Chapter 3) have much more complex reactions with nonporphyrin metal complexes. For example, iodosylbenzene complexes apparently are formed in the reactions of OIPh with nonporphyrin metal complexes, and these complexes are capable of direct reaction with substrates without the intermediacy of high-valent metal–oxo complexes (Yang et al., 1990, 1991). For this reason, it may be useful to consider the reactivity of other peroxidic species, both organic and inorganic, in evaluating the feasibility of mechanisms in which metal peroxo or hydroperoxo complexes react directly with substrate.

Peroxo complexes of the $d^0$ metals such as $Ti^{IV}$, $V^V$, or $Mo^{VI}$ react smoothly with unactivated olefins to give epoxides in high yields. By contrast, peroxo complexes of low-spin $d^6$ metal such as $Ir^{III}$, $Rh^{III}$ and $Co^{III}$ do not react with such olefins (Sheldon and Kochi, 1981; Butler et al., 1994; Dickman and Pope, 1994). A likely explanation for the ability of $d^0$ metals to activate the bound peroxo ligand is that ligand-to-metal $\pi$-bonding into the empty orbitals of $\pi$ symmetry on the metal lowers the activation energy for oxygen atom transfer to the substrate and stabilizes the resulting oxo ligand on the resulting metal complex as it is formed (reaction 31; Conte et al., 1992).

$$\text{Mo}^{VI}(\text{O}_2)^{2-} + \text{C}=\text{C} \longrightarrow \text{Mo}^{VI}=\text{O} + \text{C}-\text{C}-\text{O} \quad (31)$$

Such a mechanism is not available to the low-spin $d^6$ metals since the π-symmetry orbitals on the metal ($t_{2g}$ in octahedral symmetry) are entirely filled. Similar activation of alkyl hydroperoxides is found for $d^0$ metals but not for low-spin $d^6$ metals. In this case, ligand-to-metal π-bonding in the starting ROO⁻ complex may also be invoked to explain the apparent enhancement in electrophilicity of the bound ROO⁻ ligand relative to ROOH, but the formation of a metal–oxo product is not involved since the product is a metal alkoxide (reaction 32).

$$\text{V}^V(\text{O}-\text{O}\text{R})^{1-} + \text{C}=\text{C} \longrightarrow \text{V}^V-\text{O}\text{R}^{1-} + \text{C}-\text{C}-\text{O} \quad (32)$$

Note that these reactions represent direct reactions of olefins with coordinated peroxide or alkyl hydroperoxides. It is not necessary to invoke high-valent metal–oxo intermediates to explain the electrophilic reactions of peroxidic ligands coordinated to $d^0$ metals.

The peroxo complexes of the low-spin $d^6$ metals show entirely different reactivity patterns than those of the $d^0$ metals. In the case of Co$^{III}$, peroxo complexes are remarkably stable and show no evidence of electrophilic reactivity toward nucleophilic substrates such as olefins (Mori et al., 1968; Roberts and Symes, 1968; Crump et al., 1977; Sheldon and Kochi, 1981). The low-spin $d^6$ metal–peroxo complexes of the heavier group VIII metal ions are found to be more reactive with substrates than the Co$^{III}$ peroxo complexes described above, but the reactivity patterns are quite different from those of the $d^0$ peroxo complexes, i.e., the peroxo ligand is nucleophilic rather than electrophilic (e.g., reaction 33; Sheldon and Kochi, 1981; Conte et al., 1992).

$$\text{Rh}^{III}(\text{O}_2) + \text{R}\text{HC}=\text{CH}_2 \longrightarrow \text{Rh}^{III}\text{-peroxo intermediate} \longrightarrow \text{R}(\text{H}_3\text{C})\text{C}=\text{O} \quad (33)$$

A particularly important issue in the case of the nonheme iron monooxygenase enzymes is what reactivity to expect from either high-spin or low-spin $d^5$ ferric hydroperoxide complexes, both of which have vacancies in π-symmetry orbitals. In other words, is there sufficient ligand-to-metal π-bonding in a Fe$^{III}$—O—O—H complex, either high-spin d$^5$ or low-spin d$^5$, to make one of the oxygen atoms in that complex electrophilic enough to react directly with an olefin, in a fashion similar to that observed for a V$^V$–O–O–R complex?

Unlike metal–peroxo and hydroperoxo complexes, where our knowledge of the mechanism of reaction with substrates is limited, organic peracids are well known to be capable of direct reaction with olefins without prior O—O bond cleavage (Bartlett, 1950; Sawaki, 1992). The higher degree of electrophilic reactivity of peracids relative to alkyl hydroperoxides has been explained by the enhanced stability of the carboxylate-leaving group relative to the alkoxides, as can be demonstrated by the dramatically different p$K_a$ values of carboxylic acids (p$K_a$ ~4–5) relative to those of alcohols (p$K_a$ ~ 15). This is illustrated by the formal reactions 34 and 35.

$$\text{(34)}$$

$$\text{(35)}$$

Similar types of formal reactions may be written to illustrate why HOO$^-$ bound to Fe$^{III}$ might be electrophilic (reactions 36 and 37).

$$\text{(36)}$$

$$\text{(37)}$$

Recent *ab initio* calculations of the structure and reactivity of diamidoiron(III) hydroperoxide predict that a high-spin ferric peroxide would have a cyclic FeOOH structure and would be electrophilic enough to react with some substrates prior to O—O bond cleavage (Bach et al., 1993). This prediction remains to be tested experimentally.

Another fact that should be noted in assessing the possible electrophilic reactivity of ferric–hydroperoxide complexes is that activated organic peracids such as *p*-nitroperbenzoic acid are capable not only of epoxidizing olefins in a concerted fashion but also of inserting oxygen atoms into C—H bonds in a regio- and stereoselective fashion (Müller and Schneider, 1979; Schneider and Müller, 1985).

$$\text{cyclohexane with CH}_3, \text{H, CH}_3, \text{H substituents} \xrightarrow[\text{CH}_3\text{Cl, reflux 6-12 h}]{p\text{-nitroperbenzoic acid}} \text{cyclohexane with CH}_3, \text{H, CH}_3, \text{OH substituents}$$

52% conversion
97% retention

Quoting from the more recent paper (Schneider and Müller, 1985), "The involvement of peracid-related structures in monooxygenases seemed so far to be unlikely in view of the assumed low reactivity of peracids toward carbon-hydrogen bonds. The present work, however, shows that with moderately activated peracids the alkane oxidation is to a large degree limited by entropic factors, which could be easily overcome in the active site of a protein." If organic peracids are electrophilic enough to react directly with the C—H bond of alkanes, it certainly seems possible that a ferric hydroperoxide is sufficiently electrophilic to epoxidize olefins or possibly even to hydroxylate aliphatic or aromatic hydrocarbons without prior O—O bond cleavage.

### Substrate Activation

Enzymes that catalyze oxygenation of substrates by dioxygen may do so by activating dioxygen, as described above or, alternatively, they may activate substrate. Two prominent examples of substrate activation by oxygenase enzymes are the catechol dioxygenase enzymes and lipoxygenase. In both, binding of substrate to or reaction with the metal center at the active site of the enzyme produces modifications in the substrate that make it more reactive to dioxygen. The mechanisms proposed for these enzymes are described in detail in Chapters 5 and 6, respectively.

## DIOXYGEN TOXICITY

As stated earlier in this chapter there is a delicate balance between constructive and destructive oxidation by $O_2$ in aerobic organisms

(Valentine, 1994). The relatively slow rate of chemical reactions of dioxygen in the absence of catalysts helps to maintain this balance, and enzymes are used to direct the use of $O_2$ for oxidation of substrates. Unfortunately, this factor limits the rates or deleterious reactions but does not prevent them; biological systems have therefore been forced to evolve to scavenge and detoxify toxic species that originate from dioxygen as well as to repair and replace damaged biological compounds. Thus, the classification of an organism as aerobic only indicates that it uses dioxygen for its metabolism and that it is capable of surviving under normal atmospheric concentrations of $O_2$. In fact, there is considerable evidence that $O_2$ is toxic even to aerobic organisms. This topic is discussed in detail in Chapter 7.

## REFERENCES

ANBAR, M., MEYERSTEIN, D., and NETA, P. (1966) Reactivity of Aliphatic Compounds Towards Hydroxyl Radicals. *J. Chem. Soc., B, Phys. Org.*, 88, 742–747.

BACH, R. D., SU, M. D., ANDRÉ, J. L., and SCHLEGEL, H. B. (1993) Structure and Reactivity of Diamidoiron(III) Hydroperoxide. The Mechanism of Oxygen Atom Transfer to Ammonia. *J. Am. Chem. Soc.*, 115, 8763–8769.

BARLOW, C. H., MAXWELL, J. C., WALLACE, W. J., and CAUGHEY, W. S. (1973) Elucidation of the Binding Mode of Oxygen to Iron in Oxyhemoglobin by Infrared Spectroscopy. *Biochem. Biophys. Res. Commun.*, 55, 91–95.

BARTLETT, P. D. (1950) Recent Work on the Mechanisms of Peroxide Reactions. *Rec. Chem. Prog.*, 11, 47–51.

BASOLO, F., HOFFMAN, B. M., and IBERS, J. A. (1975) Synthetic Oxygen Carriers of Biological Interest. *Acc. Chem. Res.*, 8, 384–392.

BILINSKI, T. (1991) Oxygen Toxicity and Microbial Evolution. *Biosystems*, 24, 305–312.

BRUICE, T. C. (1988) Chemical Studies Related to Iron Protoporphyrin-IX Mixed-Function Oxidases, in *Mechanistic Principles of Enzyme Activity* (J. F. Liebman and A. Greenberg, Eds.), VCH Publishers, New York, 227–277.

BRUNORI, M., GIARDINA, B., and KUIPER, H. A. (1982) Oxygen-transport Proteins. *Inorg. Biochem.*, 3, 126–182.

BUCHLER, J. W. (1978) Hemoglobin—An Inspiration for Research in Coordination Chemistry. *Ang. Chem. Int. Ed. Engl.*, 17, 407–423.

BURSTYN, J. N., ROE, J. A., MIKSZTAL, A. R., SHAEVITZ, B. A., LANG, G., and VALENTINE, J. S. (1988) Magnetic and Spectroscopic Characterization of an Iron Porphyrin Peroxide Complex. Peroxoferrioctaethylporphyrin($1^-$). *J. Am. Chem. Soc.*, 110, 1382–1388.

BUSCH, D. H., and ALCOCK, N. W. (1994) Iron and Cobalt "Lacunar" Complexes as Dioxygen Carriers. *Chem. Rev.*, 94, 585–623.

BUTLER, A., CLAGUE, M. J., and MEISTER, G. E. (1994) Vanadium Peroxide Complexes. *Chem. Rev.*, 94, 625–638.

BYTHEWAY, I., and HALL, M. B. (1994) Theoretical Calculations of Metal–Dioxygen Complexes. *Chem. Rev.*, 94, 639–658.

COLE, P. A., and ROBINSON, C. H. (1991) Mechanistic Studies on a Placental Aromatase Model Reaction. *J. Am. Chem. Soc.*, 113, 8130–8137.

COLLMAN, J. P. (1977) Synthetic Models for the Oxygen-Binding Hemoproteins. *Acc. Chem. Res.*, 10, 265–272.

COLLMAN, J. P., BRAUMAN, J. I., HALBERT, T. R., and SUSLICK, K. S. (1976) Nature of $O_2$ and CO Binding to Metalloporphyrins and Heme Proteins. *Proc. Natl. Acad. Sci. U.S.A.*, 73, 3333–3337.

CONTE, V., DI FURIA, F., and MODENA, G. (1992) Transition Metal Catalyzed Oxidation. The Role of Peroxometal Complexes, in *Organic Peroxides* (W. Ando, Ed.), John Wiley & Sons, New York, 559–598.

COTTON, A. F., and WILKINSON, G. (1988) *Advanced Inorganic Chemistry*, 5th edition, John Wiley and Sons, New York.

CRUMP, D. B., STEPANIAK, R. F., and PAYNE, N. C. (1977) Charge Distribution in Dioxygen Complexes of Cobalt(III). The Crystal Structure and Absolute Configuration of $(+)_{546}$-Δ-*cis*-β-[{2,13-Dimethyl-6,9-diphenyl-2,6,9,13-tetraarsatetradecane}(Dioxygen)Cobalt(III)] Perchlorate. *Can. J. Chem.*, 55, 438–446.

DICKMAN, M. H., and POPE, M. T. (1994) Peroxo and Superoxo Complexes of Chromium, Molybdenum, and Tungsten. *Chem. Rev.*, 94, 569–584.

DRAGO, R. S., and CORDEN, B. B. (1980) Spin-Pairing Model of Dioxygen Binding and Its Application to Various Transition-Metal Systems as well as Hemoglobin Cooperativity. *Acc. Chem. Res.*, 13, 353–360.

DUNN, J. B., SHRIVER, D. F., and KLOTZ, I. M. (1973) Resonance Raman Studies of the Electronic State of Oxygen in Hemerythrin. *Proc. Natl. Acad. Sci. U.S.A.*, 70, 2582–2584.

DWYER, P. N., PUPPE, L., BUCHLER, J. W., and SCHEIDT, W. R. (1975) Molecular Stereochemistry of (α,γ-Dimethyl-α,γ-dihydrooctaethylporphinato)-oxotitanium(IV). *Inorg. Chem.*, 14, 1782–1785.

FEIG, A. L., and LIPPARD, S. J. (1994) Reactions of Non-Heme Iron(II) Centers with Dioxygen in Biology and Chemistry. *Chem. Rev.*, 94, 759–805.

FREEDMAN, T. B., LOEHR, J. S., and LOEHR, T. M. (1976) A Resonance Raman Study of the Copper Protein Hemocyanin: New Evidence for the Structure of the Oxygen-Binding Site. *J. Am. Chem. Soc.*, 98, 2809–2815.

GEORGE, P. (1965) The Fitness of Oxygen, in *Oxidases and Related Redox Systems* (T. E. King, H. S. Mason, and M. Morrison, Eds.), John Wiley and Sons, New York, pp. 3–36.

GRIFFITH, J. S. (1956) On the Magnetic Properties of Some Hemoglobin Complexes. *Proc. R. Soc. London Ser. A*, 235, 23–36.

GROVES, J. T. (1985) Key Elements of the Chemistry of Cytochrome P-450. The Oxygen Rebound Mechanism. *J. Chem. Ed.*, 62, 928–931.

GUILARD, R., FONTESSE, M., and FOURNARI, P. (1976) The First Peroxometalloporphyrin with Dioxygen Symmetrically Bonded by Both Atoms; Synthesis and X-Ray Crystal Structure of Peroxotitaniumoctaethylporphyrin. *J. Chem. Soc., Chem. Commun.*, 161–162.

HALLIWELL, B., and GUTTERIDGE, J. M. C. (1985) *Free Radicals in Biology and Medicine*, Clarendon Press, Oxford.

HAMILTON, G. A. (1974) Chemical Models and Mechanisms for Oxygenases, in *Molecular Mechanisms of Oxygen Activation* (O. Hayaishi, Ed.), Academic Press, New York, pp. 405–451.

HAMMOND, G. S., and WU, C.-H. S. (1968) Oxidation of Iron(II) Chloride in Nonaqueous Solvents, in *Oxidation of Organic Compounds* (R. F. Gould, Ed.), Advances in Chemistry Series, vol. 3, American Chemical Society, Washington, DC, pp. 186–207.

HAN, S., CHING, Y.-C., and ROUSSEAU, D. L. (1990) Ferryl and Hydroxy Intermediates in the Reaction of Oxygen with Reduced Cytochrome C Oxidase. *Nature*, 348, 89–90.

HANSON, L. K., and HOFFMAN, B. M. (1980) Griffith Model Bonding in Dioxygen Complexes of Manganese Porphyrins. *J. Am. Chem. Soc.*, 102, 4602–4609.

HEICKLEN, J. (1981) The Correlation of Rate Coefficients for H-Atom Abstraction by HO Radicals with C—H Bond Dissociation Enthalpies. *Int. J. Chem. Kinet.*, 13, 651–665.

HO, R. Y. N., LIEBMAN, J. F., and VALENTINE, J. S. (1995) Overview of the Energetics and Reactivity of Oxygen, in *Active Oxygen in Chemistry* (C. S. Foote, A. Greenberg, J. F. Liebman, and J. S. Valentine, Eds.), Chapman & Hall, New York, pp. 1–23.

HOLM, H. (1987) Metal-Centered Oxygen Atom Transfer Reactions. *Chem. Rev.*, 87, 1401–1449.

INGRAHAM, L. L., and MEYER, D. L. (1985) *Biochemistry of Dioxygen*, Plenum, New York, pp. 55–64.

JAMESON, G. B., and IBERS, J. A. (1994) Biological and Synthetic Dioxygen Carriers, in *Bioinorganic Chemistry* (I. Bertini, H. B. Gray, S. J. Lippard, and J. S. Valentine, Eds.), University Science Books, Mill Valley, CA, pp. 167–252.

JEONG, K.-M., and KAUFMAN, F. (1982) Kinetics of the Reaction of Hydroxyl Radical with Methane and with Nine Cl- and F-substituted Methanes. 2. Calculation of Rate Parameters as a Test of Transiton State Theory. *J. Phys. Chem.*, 86, 1816–1821.

KITAJIMA, N., and MORO-OKA, Y. (1994) Copper–Dioxygen Complexes. Inorganic and Bioinorganic Perspectives. *Chem. Rev.*, 94, 737–757.

KLOTZ, I. M., and KURTZ, D. M., JR. (1984) Binuclear Oxygen Carriers: Hemerythrin. *Acc. Chem. Res.*, 17, 16–22.

KOPPENOL, W. H. (1988) The Paradox of Oxygen: Thermodynamics *versus* Toxicity, in *Oxidases and Related Redox Systems*, 4th ed. (T. E. King, H. S. Mason, and M. Morrison, Eds.), Alan R. Liss, New York, pp. 93–109.

LEISING, R. A., BRENNAN, B. A., QUE, L., JR., FOX, B. G., and MÜNCK, E. (1991) Models for Non-heme Iron Oxygenases: A High-valent Iron-oxo Intermediate. *J. Am. Chem. Soc.*, 113, 3988–3990.

LEVINE, J. S. (1988) The Origin and Evolution of Atmospheric Oxygen, in *Oxidases and Related Redox Systems*, 4th ed. (T. E. King, H. S. Mason, and M. Morrison, Eds.), Alan R. Liss, New York, pp. 111–126.

MAGNUS, K. A., TON-THAT, H., and CARPENTER, J. E. (1994) Recent Structural Work on the Oxygen Transport Protein Hemocyanin. *Chem. Rev.*, 94, 727–735.

MALMSTRÖM, B. G. (1982) Enzymology of Oxygen. *Annu. Rev. Biochem.*, 51, 21–59.

MCCANDLISH, E., MIKSZTAL, A. R., NAPPA, M., SPRENGER, A. Q., VALENTINE, J. S., STONG, J. D., and SPIRO, T. G. (1980) Reactions of Superoxide with Iron Porphyrins in Aprotic Solvents. A High Spin Ferric Porphyrin Peroxo Complex. *J. Am. Chem. Soc.*, 102, 4268–4271.

MCLENDON, G., and MARTELL, A. E. (1976) Inorganic Oxygen Carriers as Models for Biological Systems. *Coord. Chem. Rev.*, 19, 1–39.

MCMURRY, T. J., and GROVES, J. T. (1986) Metalloporphyrin Models for Cytochrome P-450, in *Cytochrome P-450: Structure, Mechanism, and Biochemistry* (P. R. Ortiz de Montellano, Ed.), Plenum, New York, pp. 1–28.

MOMENTEAU, M., and REED, C. A. (1994) Synthetic Heme Dioxygen Complexes. *Chem. Rev.*, 94, 659–698.

MORI, M., WEIL, J. A., and ISHIGURO, M. (1968) The Formation of and Interrelation Between Some μ-Peroxo Binuclear Cobalt Complexes. II. *J. Am. Chem. Soc.*, 90, 615–621.

MORRISON, M., and SCHÖNBAUM, G. R. (1976) Peroxidase-Catalyzed Halogenation. *Ann. Rev. Biochem.*, 45, 861–888.

MÜLLER, W., and SCHNEIDER, H.-J. (1979) Regio- and Stereospecific Hydroxylation of Alicyclic Hydrocarbons with Substituted Perbenzoic Acids. *Angew. Chem. Int. Ed. Engl.*, 18, 407–408.

MURATA, K., PANICUCCI R., GOPINATH, E., and BRUICE, T. C. (1990) 5,10,15,20-Tetrakis (2,6-dichloro-3-sulfonatophenyl) Porphinato-iron(III) Hydrate with Alkyl and Acyl Hydroperoxides. The Dynamics of Reaction of Water-soluble and Non-μ-oxo Dimer Forming Iron(III) Porphyrins in Aqueous Solution. 7. *J. Am. Chem. Soc.*, 112, 6072–6083.

ORTIZ DE MONTELLANO, P. R. (1986) *Cytochrome P-450. Structure, Mechanism, and Biochemistry*, Plenum, New York.

OSTOVIC, D., and BRUICE, T. C. (1992) Mechanism of Alkene Epoxidation by

Iron, Chromium, and Manganese Higher Valent Oxo-metalloporphyrins. *Acc. Chem. Res.*, 25, 314–320.

PALMER, G. (1987) Cytochrome Oxidase: A Perspective. *Pure Appl. Chem*, 59, 749–758.

PAULING, L. (1964) Nature of the Iron–Oxygen Bond in Oxyhaemoglobin. *Nature*, 203, 182–183.

POULOS, T. L. (1986) The Crystal Structure of Cytochrome P-450$_{cam}$, in *Cytochrome P-450: Structure, Mechanism, and Biochemistry* (P. R. Ortiz de Montellano, Ed.), Plenum, New York, pp. 505–523.

PRYOR, W. A. (1976) *Free Radicals in Biology*, New York, Academic Press.

QUINN, R., MERCER-SMITH, J., BURSTYN, J. N., and VALENTINE, J. S. (1984) The Influence of Hydrogen Bonding on the Properties of Iron Porphyrin Imidazole Complexes. An Internally Hydrogen Bonded Imidazole Ligand. *J. Am. Chem. Soc*, 106, 4136–4144.

QUINN, R., NAPPA, M., and VALENTINE, J. S. (1982) New Five- and Six-coordinate Imidazole and Imidazolate Complexes of Ferric Tetraphenylporphyrin. *J. Am. Chem. Soc.*, 104, 2588–2595.

RAYNOR, J. B., GILLARD, R. D., and PEDROSA DE JESUS, J. D. (1982) Paramagnetic Dioxygen Complexes of Rhodium. *J. Chem. Soc., Dalton Trans.*, 1165–1166.

RICE, C. E., ROBINSON, W. R., and TOFIELD, B. C. (1976) Crystal Structure of a Condensed Phosphatosilicate, Oxovanadium(IV) Diphosphatomonosilicate, VO(SiP$_2$O$_8$). *Inorg. Chem.*, 15, 345–348.

RIECHEL, T. L., DE HAYES, L. J., and SAWYER, D. T. (1976) Electrochemical Studies of Vanadium(III), -(IV), and -(V) Complexes of Diethyldithiocarbamate in Acetonitrile. *Inorg. Chem.*, 15, 1900–1904.

ROBERTS, H. L., and SYMES, W. R. (1968) The Reaction of Dipotassium Hydridotetracyanoaquorhodate(III) with Oxygen Dipotassium Hydroperoxotetracyanoaquorhodate(III). *J. Chem. Soc.*, 1450–1453.

SAWAKI, Y. (1992) Peroxy Acids and Peroxy Esters, in *Organic Peroxides* (W. Ando, Ed.), John Wiley & Sons, New York, pp. 425–477.

SAWYER, D. T. (1988) The Chemistry and Activation of Dioxygen Species (O$_2$, O$_2^-$·, and HOOH) in Biology, in *Oxygen Complexes and Oxygen Activation by Transition Metals* (A. E. Martell and D. T. Sawyer, Eds.), Plenum, New York, pp. 131–148.

SAWYER, D. T. (1991) *Oxygen Chemistry*, Oxford University Press, New York.

SCHNEIDER, H.-J. and MÜLLER, W. (1985) Mechanistic and Preparative Studies on the Regio- and Stereoselective Paraffin Hydroxylation with Peracids. *J. Org. Chem.*, 50, 4609–4615.

SHELDON, R. A., and KOCHI, J. K. (1981) *Metal-catalyzed Oxidations of Organic Compounds*, Academic Press, New York.

SPRINGER, B. A., SLIGAR, S. G., OLSON, J. S., and PHILLIPS, G. N., JR. (1994)

Mechanisms of Ligand Recognition in Myoglobin. *Chem. Rev.*, 94, 699–714.

STASSINOPOULOS, A., and CARADONNA, J. P. (1990) Binuclear Non-heme Iron Oxo Transfer Analogue Reaction System: Observations and Biological Implications. *J. Am. Chem. Soc.*, 112, 7071–7073.

STENKAMP, R. E. (1994) Dioxygen and Hemerythrin. *Chem. Rev.*, 94, 715–726.

SUMMERVILLE, D. A., JONES, R. D., HOFFMAN, B. M., and BASOLO, F. (1979) Assigning Oxidation States to Some Metal Dioxygen Complexes of Biological Interest. *J. Chem. Ed.*, 56, 157–162.

TAUBE, H. (1965) Mechanisms of Oxidations with Oxygen. *J. Gen. Physiol*, 49, 29–52.

TOVROG, B. S., KITKO, D. J., and DRAGO, R. S. (1976) Nature of Bound $O_2$ in a Series of Cobalt Dioxygen Adducts. *J. Am. Chem. Soc.*, 98, 5144–5153.

TRAYLOR, T. G., and CICCONE, J. P. (1989) Mechanism of Reactions of Hydrogen Peroxide and Hydroperoxides with Iron(III) Porphyrins. Effects of Hydroperoxide Structure on Kinetics. *J. Am. Chem. Soc.*, 111, 8413–8420.

URBAN, M. W., NAKAMOTO, K., and BASOLO, F. (1982) Infrared Spectra of Molecular Oxygen Adducts of (Tetraphenylporphyrinato) Manganese(II) in Argon Matrices. *Inorg. Chem.* 21, 3406–3408.

VALENTINE, J. S. (1994) Dioxygen Reactions, in *Bioinorganic Chemistry* (I. Bertini, H. B. Gray, S. J. Lippard, and J. S. Valentine, Eds.), University Science Books, Mill Valley, CA, pp. 253–313.

VANNGARD, T. (1988) *Biophysical Chemistry of Dioxygen Reactions in Respiration and Photosynthesis*, Cambridge University Press, Cambridge.

VASKA L. (1976) Dioxygen–Metal Complexes: Toward a Unified View. *Acc. Chem. Res.*, 9, 175–183.

WIKSTRÖM, M., and BABCOCK, G. T. (1990) Catalytic Intermediates. *Nature*, 348, 16–17.

WILKINSON, G., GILLARD, R. D., and MCCLEVERTY, J. A. (1987) *Comprehensive Coordination Chemistry*, Pergamon, New York.

YANG, Y., DIEDERICH, F., and VALENTINE, J. S. (1990) Reaction of Cyclohexene with Iodoxylbenzene Catalyzed by Non-porphyrin Complexes of Iron(III) and Aluminum(III). Newly Discovered Products and a New Mechanistic Proposal. *J. Am. Chem. Soc.*, 112, 7826–7828.

YANG, Y., DIEDERICH, F., and VALENTINE, J. S. (1991) Lewis Acidic Catalysts for Olefin Epoxidation by Iodosylbenzene. *J. Am. Chem. Soc.*, 113, 7195–7205.

# 2
# Oxygen Activation by Flavins and Pterins

BRUCE A. PALFEY, DAVID P. BALLOU,
AND VINCENT MASSEY

## INTRODUCTION

As discussed in other chapters in this volume, the high reduction potential of $O_2$ makes it thermodynamically an excellent oxidizing agent for most biochemicals. However, the activation energy for these reactions is rather high, and $O_2$ is kinetically unreactive without the intervention of a catalyst. Furthermore, the products generated by the uncontrolled reduction of $O_2$ are often toxic. Such reductions usually produce highly reactive radicals, which react nonspecifically with the many reducing cellular compounds. Thus, the reactions of $O_2$ must be strictly controlled. The task of oxygen-utilizing enzymes, then, is to accelerate the reaction of $O_2$ at the enzyme active site and control the fate of the activated oxygen intermediate so that only a specific organic compound is oxidized. Enzymes that activate $O_2$ generally require the participation of either metals or organic cofactors. Metals are discussed in other chapters. The most widely occurring and best-studied organic cofactors that activate $O_2$ are the flavins, derivatives of vitamin $B_2$ (riboflavin; FMN without the phosphate). The structures of the two most common flavins, flavin adenine dinucleotide (FAD) and flavin mononucleotide (FMN), are shown overleaf.

The isoalloxazine ring system is the actual redox-active portion of the molecule and has two prochiral faces, the *re* and *si* faces. In the chiral environment of enzyme active sites, reactions with the flavin will occur exclusively on only one of these faces. In some enzymes, the isoalloxazine ring is covalently attached to the protein, but usually flavins are tightly bound to enzymes by noncovalent interactions (hydrogen bonds, salt bridges, and hydrophobic forces) and remain on the enzyme throughout catalysis. Most of the prosthetic group interactions with the protein are from the ribityl phosphate chain, which is attached to N(10) of the isoalloxazine system of both FMN and FAD. In the case of FAD, an adenosine monophosphate moiety is attached to the ribityl chain by a pyrophosphate linkage and also serves to anchor the flavin to the protein.

In this chapter, we will discuss how flavins and flavoproteins react with $O_2$. General chemical properties of model flavins will be considered. The reactions of these model compounds in solution will be compared with the reactions of various flavoproteins, highlighting what is known about the means that flavoproteins use to control the outcome of reactions with $O_2$. By providing specific interactions with transition states and reaction intermediates, oxygen-utilizing flavoenzymes ensure that only a particular reaction sequence occurs and damaging nonspecific reactions are avoided. These effects are illustrated quantitatively by the rate constants of the reactions. We will also examine the fates of the enzyme-bound reactive oxygen species and again will compare these with model reactions when available.

Flavins are versatile catalysts and are involved in a wide variety of redox reactions central to metabolism. Their ability to take part in either one- or two-electron transformations is one of their most important properties. Thus, flavin-utilizing enzymes are found in electron transport chains, where electrons from two-electron donors such as NADH are donated to one-electron acceptors such as cytochromes. The $O_2$ reactivity of a few of these proteins will be considered, with attention focused on how some enzymes retard or prevent reactions with $O_2$.

Three classes of $O_2$-utilizing flavoproteins will be discussed. First, we will focus on the oxidases, which oxidize a substrate by two electrons and reduce $O_2$ to $H_2O_2$ (Table 2-1). The second class of $O_2$-utilizing flavoproteins to be considered is the internal monooxygenases, which utilize the $H_2O_2$ generated in an oxidase reaction to oxidize further the organic substrate. Neither group of enzymes utilizes an activated flavin–oxygen intermediate. The last group of flavoenzymes, considered in the most detail, is the external monooxygenases, which do utilize a reactive flavin–oxygen intermediate.

The other oxygen-activating cofactor considered in this chapter is tetrahydrobiopterin. The pterin heterocycle resembles the isoalloxazine ring system and is involved in several hydroxylation reactions of physiological importance, most notably the hydroxylations of phenylalanine and tyrosine. The pterin hydroxylases are not as well understood as the flavin hydroxylases. These will be discussed at the end of the chapter.

### Redox States and Reactivity of Flavins

As stated above, an important property of flavins is their ability to exist in a one- or two-electron reduced state. The fully reduced flavin is pale yellow and is ionizable at N(1), which has a $pK_a$ near 7 (see the scheme below). The two-electron reduction potential of this species at pH 7.0 is $-210\,mV$, indicating that its oxidation to oxidized flavin by $O_2$ is quite favorable. Protein-binding interactions can significantly increase or decrease this value. It is thought that when bound to enzymes, most reduced flavins exist as the anion (FlH$^-$) rather than as the neutral molecule (FlH$_2$) (Müller, 1991b). The one-electron reduced flavin, the flavin semiquinone, can also exist as an anion or as a neutral molecule. The stability of the semiquinone depends on the relative reduction potential for each one-electron step. In solution, semiquinones of free flavins are thermodynamically unstable and react bimolecularly very rapidly (i.e., disproportionate) to form an equilibrium mixture consisting primarily of fully oxidized and fully reduced flavin. Interactions with proteins can significantly alter the position of the

**TABLE 2-1** Classes of Flavoenzymes Utilizing $O_2$

| | |
|---|---|
| Oxidases | RH-XH + $O_2$ → R = X + $H_2O_2$; X = NH, O |
| Internal monooxygenases | HX-RH-CO$_2$H + $O_2$ → R'C(=O)-XH + $CO_2$ + $H_2O$; X = NH, O |
| External monooxygenases | Sub-H + $O_2$ + NAD(P)H + H$^+$ → Sub-OH + $H_2O$ + NAD(P)$^+$ |

equilibrium, as well as the rate at which it is attained. Thus, one general characteristic of the flavoprotein oxidases is stabilization of the anionic semiquinone, which appears red, while the electron transferases stabilize the blue neutral semiquinone (Massey et al., 1979). The oxidized isoalloxazine ring is bright yellow and can be quite fluorescent, depending on the environment. For example, some flavoproteins exhibit strong flavin fluorescence, while others are essentially non-fluorescent. The distinctive visible spectral properties of the various forms of flavin are a major reason why so much information has been collected on the various flavoenzymes.

## Reactions with $O_2$

The electronic ground state of $O_2$ is a triplet with an unpaired electron in each of two degenerate orbitals. Most stable organic molecules, including reduced flavins, are singlets in the ground state; all the electrons are paired. Reactions of free reduced flavins with $O_2$ usually occur in less than 1 second, forming $H_2O_2$ and oxidized flavin. This happens in spite of the fact that the direct reaction between singlet reduced flavin and triplet $O_2$ to form a flavin-oxygen adduct is forbidden on the basis of spin conservation. Therefore, the first step in the reaction is presumed to be the transfer of a single electron from the reduced flavin to $O_2$, forming a semiquinone–superoxide pair (Figure 2-1, path a). A number of fates for this radical pair may be imagined. The radicals can couple, yielding a singlet flavin–oxygen adduct (path

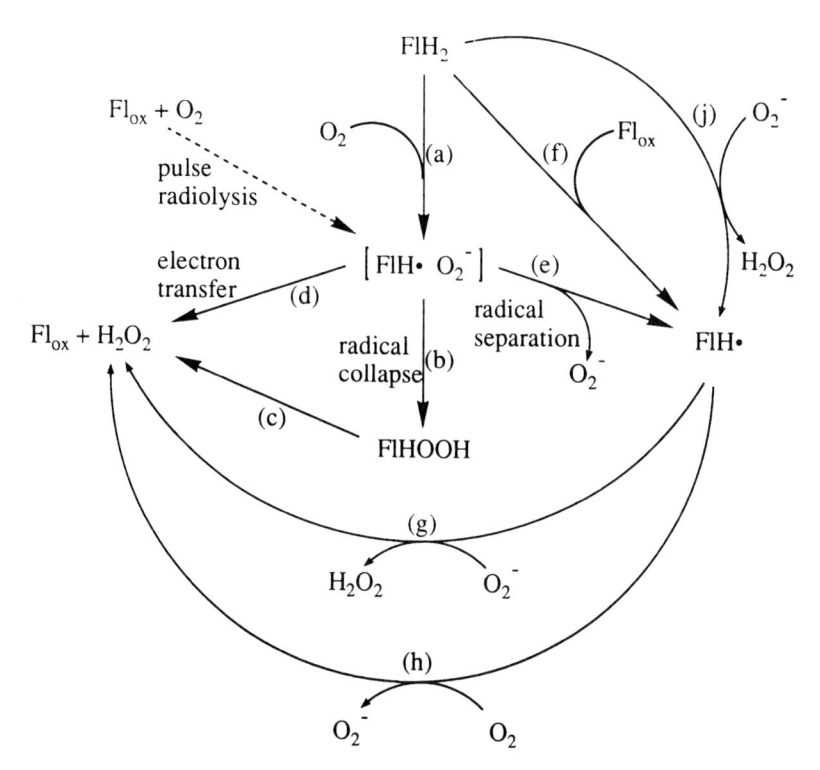

**FIGURE 2-1** Possible reactions involved in the oxidation of reduced flavin. See text for details.

b). Electron paramagnetic resonance (EPR) spectroscopy on semiquinones indicates that there is a high spin density on C(4a), making a likely product the C(4a) hydroperoxide (Müller, 1991a). Flavin hydroperoxides are known compounds and are discussed below. This adduct would then break down to form oxidized flavin and $H_2O_2$ (path c). Another possible fate for the semiquinone–superoxide pair is a second reaction by electron transfer, without the intermediacy of a covalent adduct, to form oxidized flavin and $H_2O_2$ directly (path d). An alternative to these two paths is the separation of the initial radical pair by diffusion (path e). These radicals would then form products by bimolecular reactions such as paths g, h, and j.

The reaction of reduced flavins with $O_2$ has been studied by techniques suitable for rapid reactions: stopped-flow spectrophotometry and rapid quenching. In both methods, a reaction is initiated by driving two reactant solutions through a mixing chamber. In a stopped-flow

Flavin C(4a)-Hydroperoxide

spectrophotometer, the reaction flows from the mixing chamber through an observation cell and into the stopping syringe. When a predetermined volume of solution has been mixed, the plunger of the stopping syringe is physically stopped and a micro-switch triggers the collection of absorbance or fluorescence data. In a rapid-quench device, the reacting solution is allowed to age for a certain length of time after the initial mixing and is then quenched by mixing with a quenching agent or by rapidly freezing with isopentane in a liquid nitrogen bath. The quenched samples may then be analyzed by the appropriate spectroscopic or chemical methods. Flow techniques such as these are rapid enough to allow observation of reactions after about the first millisecond of reaction.

The reaction of a photochemically reduced flavin, tetraacetyl riboflavin (TARF), with $O_2$ has been studied by rapid reaction methods (Massey et al., 1971, 1973; Kemal et al., 1977a). In a stopped-flow spectrophotometer, the formation of oxidized flavin may be monitored by observing the increase in absorbance at 450 nm, where oxidized flavin has an absorbance maximum. There is a clear lag at the start of the reaction, followed by an exponential formation of oxidized flavin. This lag indicates the formation of at least one reaction intermediate. When superoxide dismutase (SOD), an enzyme that catalyzes the reaction of superoxide to form $O_2$ and $H_2O_2$, is included in flavin–$O_2$ reaction mixtures, the rate of oxidized flavin formation slows by a factor of 4, indicating that superoxide is a reactive intermediate during flavin oxidation and that there are also superoxide-independent pathways leading to oxidized flavin. The formation and decay of superoxide were directly observed by freeze-quenching reactions of $TARFH_2$ and $O_2$ and measuring the superoxide EPR signal. A catalytic role for oxidized flavin was demonstrated by including oxidized TARF in the $O_2$-containing buffer. TARF and $TARFH_2$ react rapidly to form a small amount of TARFH· (path f), which is highly reactive with $O_2$ (path h). The inclusion of TARF shortens but does not eliminate the lag phase. Apparently, then, the lag represents the formation of both oxidized flavin and superoxide, and these species react individually to consume reduced flavin. Thus, much of the reaction being observed is not the reaction of reduced flavin with $O_2$ but instead a complicated chain involving the products of the initial reaction (Figure 2-1, paths d–j).

Model studies of the reaction of $O_2$ with various model alkylated flavins have been especially helpful in understanding the events of chain initiation in flavin oxidation (Eberlein and Bruice, 1983; Bruice, 1984). Electrochemical and kinetic measurements and calculations over a range of pH values were used to construct free energy profiles for possible reaction schemes. The initial reaction of $O_2$ with $FlH_2$ is the reversible formation of a semiquinone–superoxide pair. Kinetic considerations led to the conclusion that this pair can either revert to the reactants or collapse to form the C(4a)-flavin peroxide anion. The possibility that a second electron transfer reaction occurs (Figure 2-1, path d) was rejected because the superoxide ion must be protonated during this reaction; otherwise, the immediate product of the second electron transfer would be the energetically inaccessible $O_2^{2-}$ ion. It was reasoned that the rate of proton transfer would be too slow to account for the observed reaction kinetics. The flavin peroxide formed by the radical coupling is then protonated, and the reactive C(4a)-hydroperoxide (FlHOOH) eliminates $H_2O_2$, generating oxidized flavin.

When reduced flavins are bound to enzymes, the protein exerts an important influence on the reactions available to $O_2$. Many electron transferases will reduce $O_2$ by single electron transfers, much as they will their natural electron acceptors, without forming oxygenated flavin adducts (Massey et al., 1969). The electron transferring protein flavodoxin, for example, rapidly reduces $O_2$ to superoxide, forming a neutral semiquinone. The three-dimensional structures of several flavodoxins are known from X-ray crystallography and nuclear magnetic resonance (NMR) spectroscopy, showing that only the methyl edge of the flavin is exposed to solvent; the C(4a) position is not accessible (Ludwig and Luschinsky, 1992). Thus, the protein allows electron transfer, but unlike model compounds, a covalent adduct with oxygen is sterically prohibited.

Another interesting example of the influence that the protein structure has on $O_2$ reactivity is provided by the acyl CoA dehydrogenases, which oxidize the coenzyme A esters of fatty acids to form *trans*-α,β unsaturated compounds (Engel, 1992). The physiological electron acceptor is another flavoprotein, electron transfer flavoprotein. Artificially reduced medium-chain acyl CoA dehydrogenase from pig kidney, in the absence of ligands, will also reduce a variety of nonphysiological one-electron acceptors, including $O_2$, at comparably fast rates. When the complex of enoyl-CoA and the reduced enzyme is exposed to $O_2$, however, the oxidation by $O_2$ is extremely slow, taking hours to complete. As an alternative to steric control of $O_2$ reactivity, it has been proposed that ligand binding to the active site of the medium-chain acyl CoA dehydrogenase prevents the reaction of $O_2$ by suppressing the formation of the superoxide product, which could be solvated in an open active site but would be formed in a less stable, desolvated form when ligand is bound (Wang and Thorpe, 1991).

## OXIDASES AND INTERNAL MONOOXYGENASES

Flavoprotein oxidases are enzymes that oxidize an organic substrate by two electrons and then reduce $O_2$, giving the following overall reaction: RH-XH + $O_2$ → R = X + $H_2O_2$, X = NH, O. The oxidized substrate is frequently unstable and may be hydrolyzed, e.g., the imino acid products of D-amino acid oxidase are hydrolyzed to α-ketoacids. Several characteristics are shared by the members of this class of enzymes. The N(5) of the flavin of these enzymes is activated for attack by nucleophiles by a positively charged amino acid residue near N(1). Thus the formation of N(5) sulfite adducts, easily detected spectroscopically, is diagnostic for this class of enzymes. During catalysis, it is thought that the flavin is reduced by attack at N(5) by a carbanion formed from the deprotonation of the α-carbon of the substrate to produce an adduct, followed by the elimination of reduced flavin to yield the oxidized substrate. This is illustrated by the first three steps of Figure 2-2.

The reduced oxidase reacts with $O_2$ to form hydrogen peroxide. This reaction has been studied by stopped-flow methods for several oxidases. When reduced glucose oxidase is mixed with buffer containing $O_2$, the rapid monophasic formation of oxidized enzyme is observed, with no indication of intermediates or the production of superoxide. In 2 mM phosphate buffer, pH 6.0, 20 °C, the observed pseudo–first-order rate constant is linear in $O_2$ concentration up to the maximum attainable $O_2$ concentration, with a bimolecular rate constant of $2.2 \times 10^6$ $M^{-1}s^{-1}$ (Massey et al., 1988), much faster than the reaction of free reduced flavin with $O_2$, about 250 $M^{-1}s^{-1}$ at 30 °C, pH 6.4 (Kemal et al., 1977). Therefore, the protein activates the reduced flavin for the reaction with $O_2$. Since no intermediates are detected, it is not known whether a C(4a)-FlHOOH is on the reaction pathway to $H_2O_2$ formation. It is possible that this intermediate forms by the collapse of a semiquinone–superoxide pair but decays with a rate constant much larger than the rate constant for its formation, making it impossible to observe. The

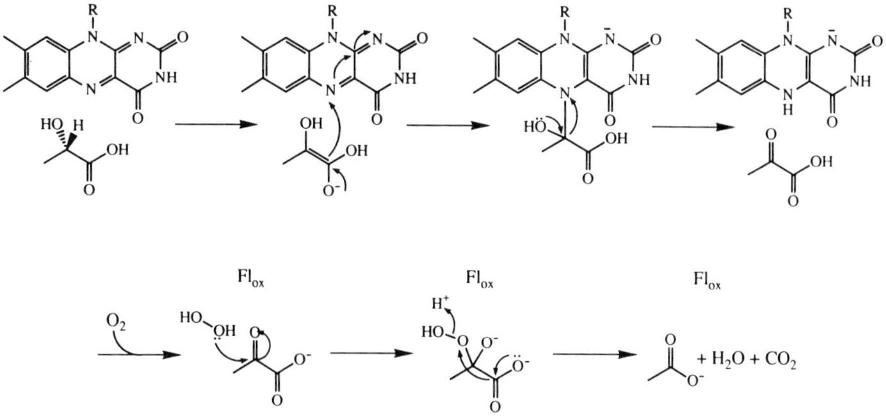

**FIGURE 2-2** The reaction mechanism of L-lactate oxidase. The α-proton of L-lactate is removed by an active site base. The resulting anion forms an adduct at N(5) of the flavin, which breaks down to pyruvate and reduced flavin. The reduced flavin reacts with $O_2$, forming $H_2O_2$ at the active site, which reacts with enzyme-bound pyruvate, forming acetate and $CO_2$.

alternative possibility is that there are two successive one-electron transfers to form $H_2O_2$ without any covalent intermediates (Figure 2-1, path d). This possible reaction for free flavin was rejected because the necessary proton transfers were expected to be slower than the radical recombination (Figure 2-1, path b). However, by having a properly positioned acidic group, the enzyme might protonate superoxide faster than could solvent.

Stopped-flow methods are limited by the time needed to mix the reactants, so that reactions faster than about $1000\,s^{-1}$ are too fast to observe. Pulse radiolysis is a technique that generates the reactants *in situ* in a period of microseconds, allowing faster reactions to be studied. This technique has been used to try to detect FlHOOH formation with glucose oxidase (Massey et al., 1988). The two postulated intermediates in the formation of the enzyme-bound flavin hydroperoxide (EFlHOOH), blue-semiquinone glucose oxidase and superoxide ion, were generated by pulse radiolysis. About 80–90% of the enzyme was observed to form EFlHOOH, while the other 10–20% formed oxidized enzyme by a second electron transfer. The observed rate constant for the formation of EFlHOOH was linear in superoxide concentration, with a second-order rate constant of $10^9\,M^{-1}\,s^{-1}$, near the diffusion limit. The decay of EFlHOOH was a first-order reaction with $k = 350\,s^{-1}$. If reduced glucose oxidase normally reacts with $O_2$ to form a radical pair and then

EFlHOOH, at high concentrations of $O_2$ the flavin hydroperoxide should be detected by stopped-flow methods, since a maximum decay rate constant of $350\,s^{-1}$ is predicted by the pulse radiolysis experiments. However, EFlHOOH was not detected by stopped-flow experiments. Therefore, the pulse radiolysis-generated EFlHOOH was not a kinetically competent intermediate for normal turnover. It is possible that pulse radiolysis generates EFlHOOH in an improper conformation. At present, the question of how $H_2O_2$ is formed by the oxidases remains unsettled.

The precise control that the protein exerts over the reactions of reduced flavin is illustrated by three closely related enzymes: glycolate oxidase, lactate oxidase, and flavocytochrome $b_2$. These enzymes have highly homologous sequences, especially at their active sites. Glycolate oxidase appears to be a typical oxidase, catalyzing the conversion of glycolate to glyoxalate: $HOCH_2CO_2^- + O_2 \rightarrow OCHCO_2^- + H_2O_2$. The crystal structure of glycolate oxidase has been determined (Lindqvist, 1992). There is a pocket above the *re* side of the flavin near the C(4a) position occupied by a water molecule. This pocket is large enough to accommodate the oxygens of a hydroperoxide and may be present for this purpose.

Flavocytochrome $b_2$ catalyzes the oxidation of another α-hydroxy acid, lactate, to pyruvate, but the flavin reduces a heme on another protein domain rather than reducing $O_2$ (Lederer, 1991). The three-dimensional structure of this enzyme is also known. The portion of the active site involved in α-hydroxy acid oxidation is similar to that of glycolate oxidase. However, unlike glycolate oxidase, there is no space over the *re* face of the flavin for oxygen atoms. Apparently this enzyme prevents the oxidation of the flavin by $O_2$ by barring its access to the critical reaction site.

Lactate oxidase catalyzes the oxidative decarboxylation of lactate to acetate (Ghisla and Massey, 1991): $CH_3CH(OH)CO_2^- + O_2 \rightarrow CH_3CO_2^- + CO_2 + H_2O$, a reaction that is superficially different from an oxidase reaction. The enzyme does exhibit properties typical of oxidases, such as sulfite adduct formation. It has been shown that this enzyme first oxidizes lactate to pyruvate in a manner typical of an oxidase, as suggested in Figure 2-2. While the pyruvate intermediate is still bound to the enzyme, the flavin reacts with $O_2$, presumably forming $H_2O_2$ bound to the active site. The $H_2O_2$ then reacts with pyruvate to form water and acetate. The oxidative decarboxylation of α-ketoacids has precedent from solution chemistry. That the reactive peroxide species is $H_2O_2$ rather than FlHOOH is not known, but there is no evidence for the intermediacy of a flavin hydroperoxide. The three-dimensional structure of lactate oxidase has not yet been determined. Thus the similarities in the protein structure to glycolate oxidase which allow $O_2$ to react, and the differences which catalyze the further reaction of lactate oxidase, remain to be seen.

Lactate oxidase is the best-studied of the internal monooxygenases, enzymes that reduce $O_2$ by four electrons derived from the "internal" source of the organic substrate. Other examples of these enzymes include lysine monooxygenase and tryptophan monooxygenase. Both of these enzymes catalyze the oxidative decarboxylation of α-amino acids, presumably by mechanisms analogous to those of lactate oxidase.

Thus far, we have seen that most flavoproteins that do not use $O_2$ as a natural oxidant react with $O_2$ to form superoxide. Apparently, specific interactions with enzymes are required to prevent the initial one-electron product from diffusing from the active site. The classic oxidases reduce $O_2$ to form hydrogen peroxide, and the active sites of these enzymes appear to be optimized for reaction with $O_2$. A further level of sophistication is exhibited by the internal monooxygenases, which catalyze the additional reaction of the products of an oxidase reaction, thus preventing the release of $H_2O_2$. As noted above, there is no evidence for any activated flavin-oxygen intermediates; it appears that the reactive oxygen agent is $H_2O_2$.

## EXTERNAL MONOOXYGENASES

Next, we will consider the external monooxygenases, which oxygenate their substrates with the reactive flavin C(4a) hydroperoxide. They are called *external* oxygenases because they reduce $O_2$ by four electrons, two from an organic substrate and two from a pyridine nucleotide-reducing agent (NADH or NADPH). Most of these enzymes have a noncovalently bound FAD at the active site. Also in this class is bacterial luciferase, an unusual enzyme that uses reduced FMN as a substrate and emits light as a result of the chemical reaction.

### Flavin Hydroperoxide Models

In solution, flavin hydroperoxides are quite unstable and undergo hydrogen peroxide elimination rather quickly. The hydroperoxide generated from riboflavin by pulse radiolysis at 25 °C, pH 6.5, decays at a rate of $260 \, s^{-1}$ (Anderson, 1982). More recent pulse radiolysis studies have generated flavin C(4a)-alkylperoxides in a variety of solvents (Merény and Lind, 1991a,b). It was found that the elimination rates

48   Oxygen Activation by Flavins and Pterins

in protic solvents were much faster than in aprotic solvents, demonstrating a key role for proton transfers in these reactions. Solvents of higher polarity also tended to increase the rates. These results suggest that an enzyme can stabilize a FlHOOH by controlling the solvent accessibility and polarity of the bound flavin.

Flavins blocked at N(5) by alkylation can form stable hydroperoxides and have been synthesized by the reaction of 1,5-dihydro-5-ethyl-3-methyl lumiflavin with $O_2$ in anhydrous dimethylformamide or anhydrous t-butanol (Figure 2-3). Alternatively, the cationic oxidized N(5) alkylated flavin reacts with $H_2O_2$ to form the hydroperoxide (Kemal and Bruice, 1976). The model hydroperoxy flavins react as hydroxylating agents (Bruice et al., 1983; Bruice, 1984), transferring the terminal oxygen of the hydroperoxide to form a flavin C(4a)-hydroxide, converting sulfides to sulfoxides and amines to the corresponding N-oxides (3° amines) or hydroxylamines (2° amines). Tertiary amines are more reactive than secondary amines, and no reaction is detectable with

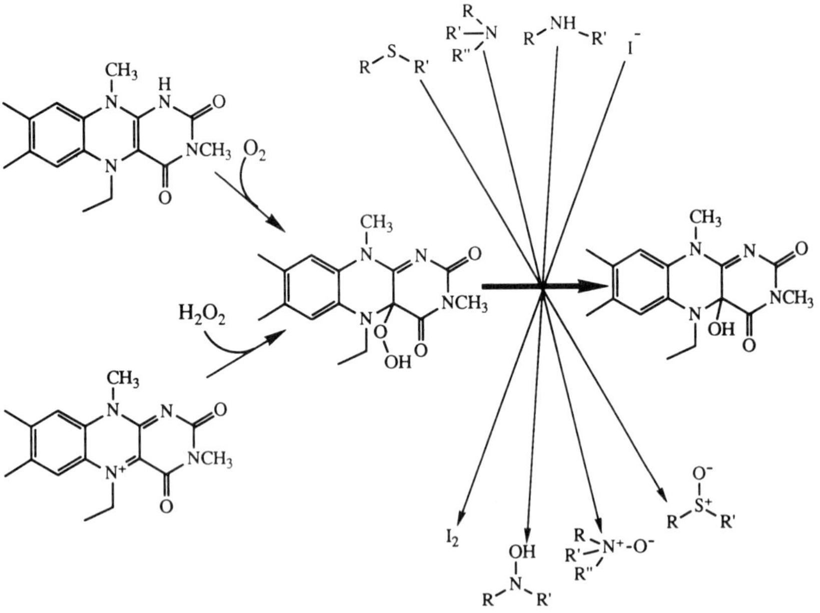

**FIGURE 2-3** Synthesis and reactions of FlEtOOH. The N(5)-blocked flavin hydroperoxide of lumiflavin (N(10)-methyl isoalloxazine) may be synthesized by oxidizing the reduced molecule with $O_2$ or by adding $H_2O_2$ across the C(4a)-N(5) double bond of the cationic flavin. The model flavin hydroperoxide is capable of oxidizing a large number of molecules, including sulfides, amines, and iodide.

primary amines. The reaction rates within a class of amines depend on the nucleophilicity of the amine, as evidenced by the linear relation between the logarithm of the oxidation rate constant and the p$K_a$ of the amine. The reactivity of the flavin hydroperoxide in oxygen transfer reactions is less than that of *m*-chloroperbenzoic acid, a common reagent used in organic syntheses for these reactions, but is greater than that of hydrogen peroxide.

A study of the rates of sulfoxidation of a model sulfide by a number of hydroperoxides revealed that an important factor in determining the reactivity of the hydroperoxide is the leaving group ability of its corresponding alcohol or acid (Bruice et al., 1983). A plot of the logarithm of the second-order rate constants as a function of the p$K_a$ of the resultant alcohols or acids decreased linearly. Although the p$K_a$ of the C(4a)-hydroxyflavin model could not be measured directly due to its instability at high pH, it was estimated by empirical substituent calculations to be about 9.2. Using this value, the flavin hydroperoxide oxygen transfer rate also fell on the correlation line. It can be concluded, then, that flavin hydroperoxides are effective electrophilic oxygen donors. This ability is the result of the electron-withdrawing nature of the isoalloxazine ring, which produces a low p$K_a$ for the product hydroxyflavin.

## Microsomal FAD-containing Monooxygenase (FMO)

Xenobiotics are processed in mammals by oxidative enzymes such as the cytochromes P-450. The FAD-containing monooxygenase (FMO) from microsomes is another such enzyme (Poulson, 1991). It is a membrane-bound enzyme containing about 5% phospholipid and is an oligomeric protein composed of subunits of 64,000 daltons. Each subunit has an active site that contains a noncovalently bound FAD. FMO oxidizes a large number of compounds: thiols to disulfides (presumably via a sulfenic acid intermediate), sulfides to sulfoxides, tertiary amines to N-oxides, secondary and primary amines to hydroxylamines, and primary and secondary hydroxylamines to oximes and nitrones. This type of oxidation represents an important pathway for the metabolism of many drugs.

The catalytic cycle has been studied in some detail by steady-state and rapid-reaction methods. The free oxidized enzyme first binds NADPH, and the flavin is reduced. NADPH is the natural reducing agent for the flavin, although NADH at high concentrations will achieve the same rates (Beaty and Ballou, 1981a). While NADP is still bound, $O_2$ reacts to form a C(4a)-flavin hydroperoxide (Poulsen and Ziegler,

1979). The substrate then binds to the enzyme and is hydroxylated, and products are released. An absorbance peak at 370 nm characterizes the spectrum of FlHOOH of FMO and is similiar to the spectra of model flavin hydroperoxides (Beaty and Ballou, 1981b). In the absence of NADP, the reduced enzyme is oxidized without any detectable hydroperoxide intermediate, but at the same rate as the formation of EFlHOOH in the presence of NADP. The presence of NADP at the active site causes a remarkable stabilization of the flavin hydroperoxide, which has a half-life of about 2 hours at 4 °C. Alterations in the pyridine ring decrease or eliminate this stabilization, suggesting that the pyridine ring is directly blocking access to the N(5)–C(4a) area of the flavin or causes a protein conformational change having the same effect. The decay of the flavin hydroperoxide increases with pH, which could be consistent with either facilitating the deprotonation of N(5) to cause hydrogen peroxide elimination or promoting a faster rate of NADP release. When a substrate is mixed with the enzyme in its hydroperoxide form, an oxygen transfer reaction forms an intermediate with a blue-shifted spectrum characteristic of a C(4a)-hydroxyflavin (Jones and Ballou, 1984, 1986).

The enzyme-bound flavin hydroperoxide is quite similar to the alkyl flavin hydroperoxide model in its reactivity toward some substrates (Jones, 1985). The FMO-bound hydroperoxide shows a linearly increasing correlation between log $k_{ox}$ and the p$K_a$ of the aromatic amine substrate (Jones, 1985), as does the model (Bruice et al., 1983). However, for amines of a particular p$K_a$, the enzymatic reaction is frequently $10^3$ to $10^6$ times faster than the model reaction. FMO and the model oxidize thiols at similar rates. Other factors seem to be important for determining substrate specificity. Nonpolar compounds tend to be better substrates, and the presence of negative charges on a compound near the oxygenation site prevents the reaction.

After product release, the enzyme is in the C(4a)-hydroxyflavin form. The catalytic cycle is completed when water is eliminated and NADP is released in the rate-determining step in catalysis. It has been shown that NADP is still bound to this intermediate and may be stabilizing it somewhat. The slow decay of the C(4a)-hydroxyflavin accounts for the low $K_m$ values observed with most substrates, i.e., the preceding reactions are comparatively fast so that rate saturation occurs at low substrate concentrations.

Interestingly, the hydroperoxide of FMO is also capable of slowly oxidizing boronates. With these compounds, the hydroperoxide reacts as a nucleophile, as hydroperoxides are known to do, thus demonstrating the versatility of the FlHOOH species.

## Cyclohexanone Monooxygenase

Hydroperoxides used in organic synthesis are capable of reacting as nucleophiles as well as electrophiles. FMO utilizes an electrophilic flavin hydroperoxide in the metabolism of xenobiotic substrates. Several enzymatic flavin hydroperoxides are thought to react as nucleophiles. The ketosteroid monooxygenase from *Cylindrocarpon radicicola* allows this fungus to cleave the ketone side chain from steroids (Katagiri and Itagaki, 1991). Cyclohexanone monooxygenase is a similar enzyme isolated from *Acinetobacter* NCIB 9871 and is part of the pathway for hydrocarbon utilization (Donoghue et al., 1976). These enzymes catalyze what appears to be a Baeyer–Villiger cleavage of substrate ketones.

The flavin hydroperoxide of cyclohexanone monooxygenase has been identified by stopped-flow spectroscopy (Ryerson et al., 1982). When the reduced enzyme–substrate complex is mixed with $O_2$-containing buffer in a stopped-flow instrument, the flavin hydroperoxide is formed too quickly for the reaction to be observed, even at low $O_2$ concentrations, setting a lower limit of $3 \times 10^7 M^{-1} s^{-1}$ for the bimolecular rate constant. This is an unusually fast reaction; apparently, cyclohexanone monooxygenase is optimally constructed to stabilize the putative semiquinone–superoxide pair and accelerates this reaction by around $10^6$-fold over the model reaction.

Baeyer–Villiger Oxidation of Cyclohexanone to ε-Caprolactone

The available evidence is consistent with a Baeyer–Villiger mechanism for the oxidation of ketones but is not conclusive. Chiral 2-$^2$H-cyclohexanone yields ε-caprolactone that has retained the configuration at the chiral carbon, as do authentic Baeyer–Villiger reactions (Schwab et al., 1983). The enzyme-catalyzed reaction exhibits the expected migratory group preference; the secondary alkyl substituent of 2-methyl cyclohexanone migrates exclusively, yielding 6-methyl-ε-caprolactone. Additionally, it has been shown in related enzymes that the carbonyl oxygen of the product is derived from the substrate, and the other comes from $O_2$ (Prairie and Talalay, 1963; Britton et al., 1974).

Interestingly, Baeyer–Villiger reactions of model flavin hydroperoxides have not been observed. An alternative mechanism, in which the hydroperoxide reacts as an electrophile with the enol(ate) of cyclohexanone, has also been considered (Ryerson et al., 1982). No label exchange was observed, in either the product or the remaining reactant, when 2,2,6,6-$^2$H$_4$-cyclohexanone was used as a substrate. This result is predicted by a Baeyer–Villiger mechanism, but can also be explained by the epoxidation mechanism (see the discussion below) if the base that abstracts the 2-proton of cyclohexanone in forming the enolate is inaccessible to the solvent and thus unable to exchange. It should be noted that epoxidations by flavin hydroperoxides are also unknown, so that neither mechanism has a model precedent (Bruice et al., 1983).

Epoxidation Mechanism for Cyclohexanone Monooxygenase

Cyclohexanone monooxygenase will catalyze the oxidation of a number of ketones and aldehydes, as well as the very electrophilic boronic acids, forming the corresponding alcohol and boric acid. The enzyme is also capable of oxygenating nucleophiles. Sulfides, sulfoxides, selenides, iodide, and triethyl phosphite are all oxygenated (Latham and Walsh, 1986; Walsh and Chen, 1988).

A C(4a)-hydroxyflavin is the intermediate observed after the oxygen transfer reaction (Ryerson et al., 1982). This eliminates water to regenerate the oxidized enzyme at the determining rate for the catalytic cycle. There is no observable kinetic phase that can be assigned to the breakdown of the Baeyer-Villiger tetrahedral intermediate. Presumably, then, this reaction is fast. The hydroxyflavin intermediate has a peak near 360 nm, characteristic of a C(4a)-flavin adduct. However, a small peak is also seen at 530 nm. The nature of this absorbance is not yet known.

## Bacterial Luciferases

The luciferases from luminescent bacteria are additional examples of flavoenzymes that oxidize electrophiles. In this case the natural substrates oxidized are long chain aldehydes, which are converted to carboxylic acids. Luciferases are unusual in many respects, most notably for the production of light, which is their biological function. This reaction is also unique among the external monooxygenases because reduced FMN is a substrate and is not tightly bound to the enzyme. Luciferase is a heterodimer (Baldwin and Ziegler, 1992) with an α subunit containing the active site, and a β subunit that is required for efficient light generation. Some of the luciferase systems have also been found to employ accessory proteins containing chromophores that alter the color of the emitted light.

Kinetic studies of the enzyme show that reduced FMN is the first substrate to bind to the enzyme (Abu-Soud et al., 1992). This is followed by a quick reaction with $O_2$ (about $10^6 M^{-1} s^{-1}$ at 25 °C), forming a flavin hydroperoxide. Aldehyde binding takes place next, followed by the oxygenating and light-producing reactions. Although the aldehyde substrate is actually able to bind to the free enzyme or to the enzyme-$FMNH_2$ complex before $O_2$, these are dead-end complexes, and the aldehyde must then dissociate for the oxygen reaction to proceed further.

Luciferase has proved particularly valuable in demonstrating that the initial intermediate observed after the reaction with $O_2$ really is a C(4a)-flavin hydroperoxide. This intermediate is extraordinarily stable at 2 °C. Without ligands present, it has a half-life of about 1 hour (Hastings et al., 1973). In the presence of long chain alcohols, which compete for the aldehyde-binding site, the intermediate has a half-life of about 12 hours. Long chain amines, amides, and nitriles also stabilize the hydroperoxide (Makemson et al., 1992). This stability has allowed the characterization of the intermediate by $^{13}$C-NMR (Vervoort et al., 1986a). $^{13}$C-labeled FMN was used to generate the intermediate in the presence of dodecanol. Unbound FMN was removed by gel filtration, and the spectrum of the intermediate and its decay were observed. The chemical shift of the C(4a) of the intermediate was 82.5 ppm, nearly the same as for a model flavin C(4a) adduct, and has been interpreted to indicate a flattened sp$^3$ geometry. The absorption spectrum of this intermediate is similar to those of other presumed enzymatic C(4a)-hydroperoxyflavins. Together with the information gained from model compounds, this provides a strong basis for assigning the C(4a)-hydroperoxyflavin structure to the first oxygenated intermediate observed for the external monooxygenases. A likely explanation for the stability of the hydroperoxide on luciferase is that the flavin is buried. $^{15}$N-NMR spectra of the reduced FMN–enzyme complex are consistent with this (Vervoort et al., 1986b), as is the observation that the enzyme is unable to bind N(5)-methyl-C(4a)-hydroperoxyflavin (Kemal et al., 1977b).

Most of the current research effort on luciferase is directed toward elucidating the events causing chemiluminescence. Light emission occurs after the aldehyde substrate binds to the C(4a)-hydroperoxy flavin enzyme and is simultaneous with the disappearance of the hydroperoxide. A highly fluorescent C(4a)-hydroxide is the next observable intermediate. The fluorescence emission spectrum of this species is essentially the same as the chemiluminescence spectrum, leading to the assignment of the hydroxyflavin as the emitting species (Kurfürst et al., 1984a). This is corroborated by the observation that the luminescence spectrum shifts when FMN analogs with different spectra are used as substrate. Superficially, then, the reaction resembles another Baeyer-Villiger oxygenation mechanism similar to that of cyclohexanone monooxygenase, producing a carboxylic acid and hydroxyflavin. However, a simple Baeyer-Villiger oxygenation mechanism has been rejected because it does not explain how the hydroxyflavin is produced in an excited state, which is required for subsequent light emission. Furthermore, alkyl boronates, which can be oxidized by nucleophilic peroxides, are not oxidized by luciferase (Ahrens et al., 1991).

Three proposed mechanisms are shown in Figure 2-4. The initial step of each is the nucleophilic attack of the flavin hydroperoxide on the aldehyde to produce the flavin peroxyhemiacetal. In mechanism A, the hydroxyflavin is displaced from the flavin peroxyhemiacetal by the attack of the oxygen from the substrate, producing a highly strained dioxirane (Raushel and Baldwin, 1989). The oxygen–oxygen bond of this intermediate homolytically cleaves, forming a diradical. Migration of a hydrogen atom from carbon to oxygen creates the carboxylic acid in an excited state. The primary excited molecule then transfers its

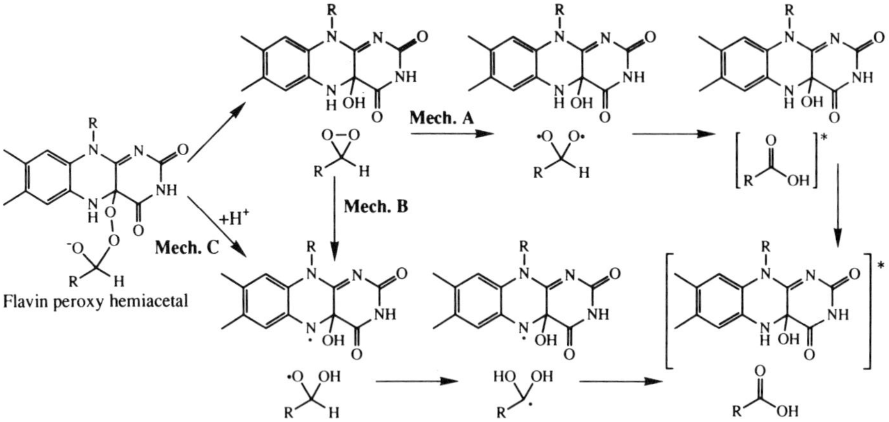

FIGURE 2-4  Possible mechanisms for producing an excited state in bacterial luciferase. Three mechanisms for the production of an electronically excited state are shown. The excited state is marked by *. Each presumes that a flavin peroxyhemiacetal is the initial product of the reaction of the flavin hydroperoxide and the aldehyde. In mechanisms A and B, the peroxyhemiacetal forms a dioxirane and hydroxyflavin. In mechanism A, the dioxirane rearranges to form an excited carboxylic acid, which can transfer energy to the hydroxyflavin via Förster dipole interactions. Mechanisms B and C propose that the excited state is produced by a CIEEL mechanism. In mechanism B, the hydroxyflavin serves as an electron donor in the absence of accessory proteins; in their presence, the electron donor is the accessory chromophore. The initial excited state is presumed to always be the hydroxyflavin in mechanism C; accessory chromophores are subsequently excited by Förster-type interactions.

energy to the hydroxyflavin by a Förster dipole–dipole interaction mechanism, as in fluorescence energy transfer. Mechanism B also postulates a dioxirane intermediate (Raushel and Baldwin, 1989). However, it decomposes by an electron exchange mechanism that has precedent in model chemiluminescent systems. In this chemically induced electron exchange luminescence (CIEEL) reaction, the dioxirane is reduced by a hydrogen atom transfer from a donor, such as the hydroxyflavin. After hydrogen atom migration, the substrate radical transfers an electron back to an excited-state orbital of the original flavin donor molecule (the chemically induced electron exchange), and emission occurs when the donor returns to the ground state. A variation on this proposal is shown in mechanism C, also a CIEEL mechanism. In this case, the peroxyhemiacetal is reduced by the flavin, forming a hydroxyflavin radical and substrate radical pair. Hydrogen atom migration followed by hydrogen atom transfer similarly results in the product carboxylate and an excited hydroxyflavin.

Supporting evidence for mechanism C has been provided by experiments using a number of 8-substituted flavins and the corresponding 4a,5-propano analogs, which serve as C(4a)-hydroxyflavin models (Eckstein et al., 1993). Linear correlations between the logarithms of the observed rate constants for light emission and the 1-electron oxidation potentials for the corresponding propano analogs were observed. The reaction rate increases as the energy required to remove an electron from the C(4a)-flavin adduct decreases. This linear correlation holds for luciferases isolated from three bacterial species and is strong evidence against mechanisms A and B. As flavins with more electron-withdrawing substituents (which would raise the reduction potential) are used, one would expect the rate of formation of the dioxirane required in mechanisms A and B, and thus light production, to increase. The opposite was found.

Lumazine

An objection to mechanism C comes from the study of the luminescence in the presence of the accessory lumazine protein. The chromophores of this protein, derivatives of lumazine, are highly fluorescent-truncated flavins. If hydroxy FMN of luciferase is the primary excited species, then it must transfer its energy to any secondary emitter such as lumazine by a Förster mechanism, which normally requires significant overlap between the emission spectrum of the energy donor and the excitation spectrum of the acceptor. Hydroxy FMN emits at 490 nm. Lumazine has a fluorescence excitation maximum near 417 nm and emits at 470 nm, so that there is poor overlap between the emission spectrum of the hydroxyflavin donor and the excitation spectrum of the lumazine acceptor. Efficient energy transfer may still be possible, however, if the donor and acceptor are optimally oriented, as has been suggested by a recent study of fluorescence lifetimes (Lee et al., 1991). Mechanism B explains the involvement of accessory emitters by considering these as alternate electron donors for the CIEEL. However, an oxidized flavin (or analog) would have to be even further oxidized for the CIEEL mechanism to be operative with the accessory emitters. This possibility seems unlikely, but it has not been tested by model reactions.

It should be noted that only weak chemiluminescence has been generated by the decomposition of model flavin peroxides (Bruice, 1984). The relevance of the model systems to the enzymatic reaction is not firmly established, since the quantum yield from the model is so low compared to that from luciferase (<0.001 vs. 0.1–0.2). Efficient light production, therefore, is a consequence of the strict control exerted by the enzyme in the oxidation of aldehydes. Like FMO and cyclohexanone monooxygenase, luciferase is capable of oxidizing methyl sulfides to sulfoxides, but without the production of light (McCapra and Hart, 1976). Presumably with sulfides, the reaction is a simple nucleophilic attack by the sulfide on the C(4a)-flavin hydroperoxide.

The hydroxyflavin-enzyme produced from aldehyde oxidation decays slowly, with a half-life of 7 minutes at 9 °C (Kurfürst et al., 1984a). An activation energy of about 20 kcal mol$^{-1}$ has been estimated for this elimination reaction. The substrate may bind to this enzyme form as a dead-end complex, causing a more than 10-fold inhibition of the dehydration of the hydroxyflavin.

## Aromatic Hydroxylases

The largest and most extensively studied group of external monooxygenases, the flavoprotein hydroxylases, transfer oxygen to activated aromatic rings either *ortho* or *para* to an activating substituent. These microbial enzymes are involved in the catabolism of a wide variety of aromatic compounds, such as those produced from lignin degradation. All known examples have a tightly bound FAD at the active site. It appears that all enzymes in this class hydroxylate their substrates by a similar mechanism, which we will consider in detail. Initially, a nonaromatic product is formed, and this frequently rearomatizes to yield the final product. With several hydroxylases, however, the nonaromatic intermediate undergoes some other reaction, causing ring cleavage or substituent loss. A few of these reactions will also be discussed.

The hydroxylation reaction of *p*-hydroxybenzoate hydroxylase (PHBH) has been extensively studied and is typical of the hydroxylases. This enzyme hydroxylates its natural substrate, *p*-hydroxybenzoate (pOHB), to form 3,4-dihydroxybenzoate and also hydroxylates several other aromatic acids, though not as efficiently. The catalytic cycle is shown in Figure 2-5 and can be divided into two half reactions, which can be studied independently. The free oxidized enzyme binds the substrate and NADPH, and NADPH reduces the flavin in this complex. NADP is then released, completing the reductive half reaction.

The oxidative half reaction may be observed directly by mixing in a

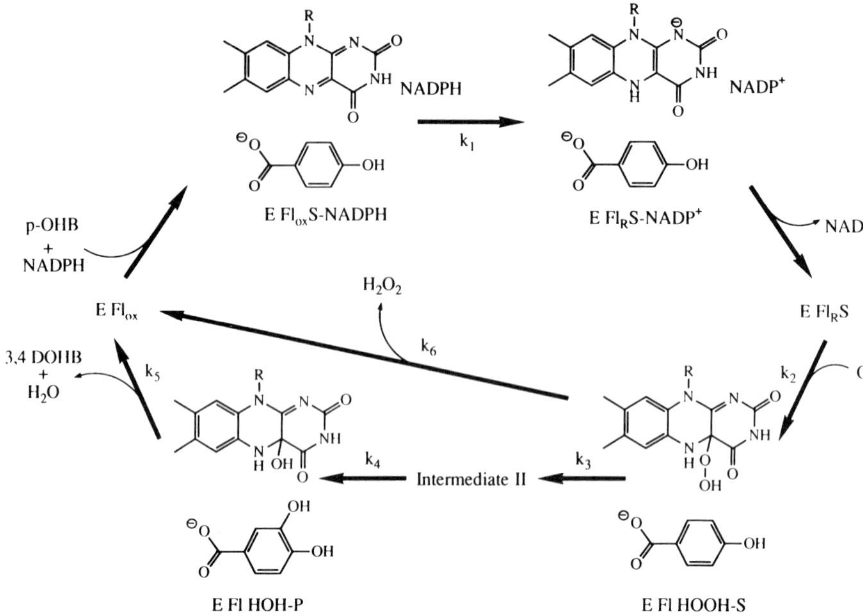

**FIGURE 2-5** The catalytic cycle of p-hydroxybenzoate hydroxylase. The reductive half-reaction is shown in the top half of the figure, where the enzyme binds NADPH and p-hydroxybenzoate, and the flavin is reduced. After the release of NADP, the reduced flavin reacts with $O_2$, forming the hydroperoxyflavin. Normally, this intermediate hydroxylates the substrate ($k_3$), leading to the formation of hydroxyflavin. With poor substrates, the flavin hydroperoxide can eliminate $H_2O_2$ ($k_6$).

stopped-flow instrument the anaerobic reduced enzyme–substrate complex with buffer containing $O_2$ and observing the absorbance or fluorescence changes (Figure 2-6). When this is done at pH 6.5 and 4 °C, intermediates in the reaction are readily resolved (Entsch et al., 1976; Entsch and Ballou, 1989). The first observable intermediate is a flavin hydroperoxide (EFlHOOH), which has a typical C(4a)-hydroperoxide spectrum with a maximum near 390 nm. Presumably the initial reaction steps are the quick formation and collapse of a semiquinone–superoxide pair. A plot of the observed EFlHOOH formation rate constant as a function of $O_2$ concentration is a straight line with a nonzero intercept, giving a second-order rate constant of $1-6 \times 10^5 \, M^{-1} s^{-1}$, depending on the substrate, and a reverse rate of about $10-50 \, s^{-1}$ (the y-intercept). In the absence of bound ligands, oxidized flavin is generated in a single reaction without the observation of any intermediates. Interestingly, this second-order reaction is irreversible

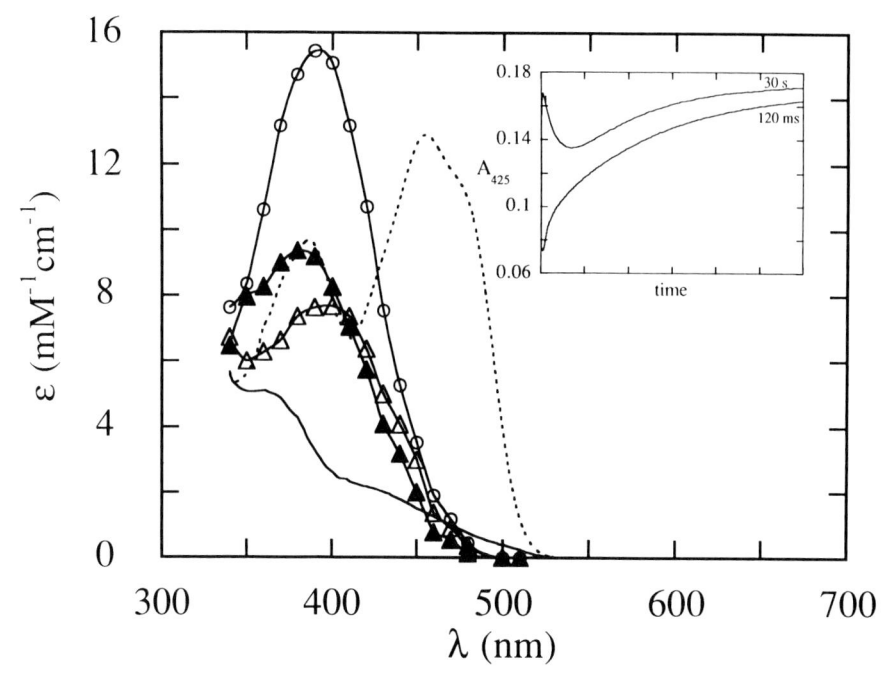

**FIGURE 2-6** Intermediates in the hydroxylation of 2,4-dihydroxybenzoate. When the reduced PHBH–2,4-dihydroxybenzoate complex (solid line) is mixed with $O_2$-containing buffer in a stopped-flow spectrophotometer, the flavin hydroperoxide (intermediate I) rapidly forms (open triangles). Intermediate II forms as the substrate is hydroxylated (circles) and then decays to the enzyme–hydroxyflavin–product complex (intermediate III, solid triangles). Finally, water is eliminated, forming oxidized enzyme (dashed line). Inset: A kinetic trace produced by following the absorbance changes at 425 nm in a stopped-flow spectrophotometer. The same reaction is shown on two time scales. Within the first 120 milliseconds, the hydroperoxide is formed and the substrate is hydroxylated. On the longer time scale, intermediate II decays to intermediate III, which then decays to oxidized enzyme. Reactions were at 4 °C, pH 6.55.

and an order of magnitude slower than when an aromatic ligand is present. Therefore, the presence of a substrate facilitates the reaction of reduced flavin with $O_2$ and stabilizes the hydroperoxide. These effects are presumably mediated through a protein conformational change.

The crystal structures of a few PHBH–substrate complexes have been elucidated (Schreuder et al., 1988, 1989, 1992), giving direct structural information about the active site (Figure 2-7). A pocket above the *re* face of the flavin could accommodate the oxygen atoms. In modeling studies, it was found that two orientations of the oxygen atoms were not overly crowded (Schreuder et al., 1990). In one orientation, the

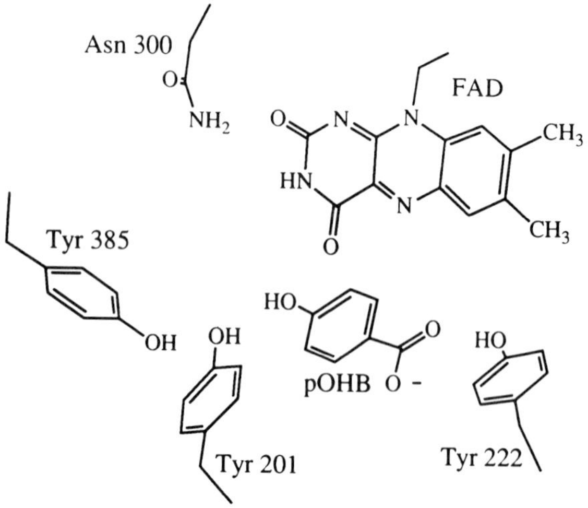

**FIGURE 2-7** Active site of PHBH. Several residues involved with substrate and flavin activation are shown. The view is from the *re* side of the flavin.

terminal oxygen atom is near the C(10a) of the isoalloxazine ring. It was suggested that $O_2$ reacts with the reduced flavin in this orientation to produce the initial flavin peroxide ion. A water molecule near N(10) of the flavin is appropriately positioned to protonate this anion. The hydroperoxide could then rotate about the C(4a)—oxygen bond so that the terminal oxygen atom moves next to the substrate. Several hydrogen-bonding partners are available to stabilize this orientation. In this position, the terminal oxygen atom would be at the proper location for hydroxylation to occur and nearly coincidental with the 3-hydroxyl of 3,4-dihydroxy benzoate seen in the enzyme–product complex. Furthermore, it was reasoned that the enzyme assists in the formation of flavin hydroperoxide by forming new, strong hydrogen bonds between the peptide backbone and the carbonyl group at the 4-position of the flavin. When the flavin is in its nearly planar, oxidized form, these interactions are weak; the loss of planarity by forming the C(4a) adduct would move the carbonyl oxygen toward the amide hydrogens of glycine 46 and valine 47.

The flavin hydroperoxide that is formed is quite unstable and reacts either to form products or to eliminate hydrogen peroxide without hydroxylating the substrate. The partitioning of this intermediate through steps involving $k_3$ and $k_6$ of Figure 2-5 is determined by the reactivity of the aromatic ligand. Some compounds, such as 6-

hydroxynicotinate, are not hydroxylated at all and quantitatively form $H_2O_2$. At the other extreme, no $H_2O_2$ is detected when pOHB is the substrate. Substrates such as 2,4-dihydroxybenzoate and *p*-aminobenzoate are hydroxylated less efficiently, with about 30% of the enzyme forming $H_2O_2$.

When either 2,4-dihydroxybenzoate or *p*-aminobenzoate is hydroxylated, the next observed intermediate, intermediate II, has a comparatively high extinction coefficient and an absorbance maximum near 400 nm. This decays to intermediate III, which has the spectrum of a C(4a)-hydroxyflavin, with an absorbance maximum at 380 nm. pOHB and several fluorinated pOHB derivatives do not form an observable intermediate II, presumably for kinetic reasons, and intermediate III is the immediately observed product of hydroxylation. This intermediate then eliminates water to form the oxidized enzyme.

There has been extensive speculation on the mechanism of hydroxylation and the identity of intermediate II, and a consensus is only now beginning to emerge. Chemical quenching showed that the oxygen atom is transferred at the same time that intermediate II is formed, leading to the early proposal (Entsch et al., 1974) that the reaction is a simple electrophilic aromatic substitution of the aromatic ring, yielding a nonaromatic, unstable cyclohexadienone and the flavin as the C(4a)-hydroxide. The characteristic spectrum of the intermediate, then, would be the result of the sum of the spectra of the cyclohexadienone and the hydroxyflavin.

Electrophilic Aromatic Substitution Mechanism

At that time, it was reasoned that a flavin hydroperoxide would not be a sufficiently reactive hydroxylating agent (Hamilton, 1971, 1974), although subsequent model studies showed that flavin hydroperoxides can be effective oxygen transfer reagents, as described above. Therefore, a carbonyl oxide, which would be extremely reactive, was

Carbonyl oxide, ring opened mechanism

proposed to form by opening the ring between N(5) and C(4a) of the EFlHOOH. The resultant carbonyl oxide would transfer oxygen very quickly to the substrate, leaving the flavin in a ring-opened form. The ring-opened flavin was thought to give intermediate II its unique spectrum. The amine of the ring-opened flavin would then attack the carbonyl at C(4a), generating the hydroxyflavin of intermediate III (Entsch et al., 1976). The putative ring-opened intermediate, alkylated at the amino nitrogen to prevent cyclization, was synthesized (Wessiak and Bruice, 1983). Its spectrum, in a number of different solvents or bound to riboflavin-binding protein, did not match the spectrum of intermediate II, casting doubt on this mechanism (Wessiak et al., 1984a, 1984b).

More recently, a diradical intermediate II has been proposed. According to this mechanism, the terminal OH of the hydroperoxide is homolytically transferred to the aromatic ring, forming a C(4a)-flavinoxy radical and a hydroxycyclohexadienyl radical. The formation of intermediate III would occur when an electron is transferred from the hydroxycyclohexadienyl radical to the flavinoxy radical, forming the aromatic product and the hydroxyflavin. This proposal was based on pulse radiolysis model studies (Anderson et al., 1987, 1991), in which

Diradical mechanism

the cyclohexadienyl radicals of several substrates of *p*-hydroxybenzoate hydroxylase and phenol hydroxylase were generated, and the spectra of these radicals were determined. Assuming that the flavinoxy radical spectrum is essentially the same as the spectrum of a hydroxyflavin, the spectrum of intermediate III should give a good approximation of this contribution. Adding the measured product radical spectra from pulse radiolysis with the assumed flavinoxy radical spectrum gave spectra for intermediate II that agree remarkably well with those obtained by stopped-flow experiments for both enzymes.

There are a number of problems with this proposal. Attempts to detect radicals directly in PHBH were unsuccessful (L. M. Schopfer, E. P. Day, J. Peterson, R. H. Sands, W. R. Dunham, R. F. Anderson, D. P. Ballou, and V. Massey, unpublished data). The rate-determining step in catalysis becomes the decay of intermediate II at high pH when 2,4-dihydroxybenzoate is the substrate. Thus, in steady-state turnover at high pH, nearly all of the enzyme is in the intermediate II form. Intermediate II was generated under turnover conditions using an NADPH regenerating system and then frozen. EPR spectroscopy of such a sample showed no radicals, either because there were none or because the radical pair interacted and broadened the signals so that they were undetectable. However, magnetic susceptibility measurements, which are independent of radical pair interactions, did not reveal radicals either. Furthermore, recent fluorescence stopped-flow experiments on phenol hydroxylase and *p*-hydroxybenzoate hydroxylase show that intermediate II is quite fluorescent, which is unprecedented for organic free radicals (Maeda-Yorita and Massey, 1993; B. Palfey, B., Entsch, D. Ballou, and V. Massey, unpublished results).

A comparison of isotope effects obtained by absorbance and fluorescence stopped-flow techniques for phenol hydroxylase in complex with deuterated resorcinol yielded important information (Maeda-Yorita and Massey, 1993). A very fluorescent intermediate II is observed when resorcinol is hydroxylated, and the decay of intermediate II fluorescence has no measurable deuterium isotope effect. However, when the same reaction is observed by absorbance, an isotope effect is observed at short wavelengths, although none is observed at long wavelengths. The fluorescence from intermediate II comes from the hydroxyflavin only. In absorbance, the behavior of all of the chromophores is observed—both the flavin and the substrate/product. Deuterium substitution is only expected to affect the rearomatization of the product (see the scheme for electrophilic aromatic substitution above). If the flavin reacts independently of the rearomatization reaction, then no isotope effect is expected when the signal originates from the flavin, which is the case for fluorescence and for longer wavelength absorbance. At the shorter wavelengths, the nonaromatic form of the substrate is expected to contribute to the absorbance, and therefore an isotope effect is observable. The spectrum of this intermediate product form has an absorbance maximum at 340 nm, with an extinction coefficient of 4800 M$^{-1}$ cm$^{-1}$.

There is direct evidence that the spectrum of PHBH intermediate II is the sum of a hydroxyflavin contribution and a contribution from some form of the product (Schopfer et al., 1991). The flavin of PHBH can be removed and replaced with analogs containing different substituents at the C(8)-position. A series of 8-substituted-FAD enzymes was prepared, and the spectra of intermediates II and III were determined by stopped-flow spectrophotometry using 2,4-dihydroxybenzoate as substrate at several pH values. The spectra of the intermediates formed for the different C(8)-substituted flavins were different from each other. However, subtracting the spectrum of intermediate III from the spectrum of intermediate II for a particular flavin gave the same spectrum in each case. These results argue strongly that the spectrum of intermediate II is the sum of contributions from the product (with a p$K_a$ of 7.8) and a C(4a)-flavin derivative. This is consistent with either the simple electrophilic mechanism or the diradical mechanism. Recent pulse radiolysis studies have generated the nonradical, nonaromatic form of the hydroxylation product of 2,4-dihydroxybenzoate (Merényi et al., 1991). It was seen that the spectrum of this species is essentially the same as the corresponding radical spectrum. Thus, there is no need to invoke the involvement of radicals, and after 20 years of speculation and debate, it seems that one of the earliest proposals is correct—the reaction is a simple electrophilic aromatic substitution.

One reason for the slow acceptance of the most obvious mechanism is the lack of precedents from model chemistry. Although hydroperoxyflavins were shown to be effective oxygenating agents for some substrates, the reactions of model hydroperoxyflavins with phenols were quite unlike the enzymatic reaction. The difficulty in modeling this reaction stems from the fact that a phenol will be more reactive for hydroxylation when it is in the phenolate form. At basicities high enough to deprotonate a phenol, however, the electrophilic flavin hydroperoxide will be deprotonated to become a nucleophilic flavin peroxide. Thus, conditions which activate the aromatic substrate deactivate the hydroxylating agent. An unusual hydroxylation of phenols has been observed under basic conditions (Muto and Bruice, 1982). While interesting, these reaction intermediates have the wrong spectral properties to be considered relevant to the enzymatic reaction.

The inability to model enzymatic aromatic hydroxylations highlights the importance of the protein in this process. The aromatic hydroxylases regulate the reaction outcome by controlling the reactivities of both the hydroperoxide and the substrate. It appears that PHBH activates pOHB by lowering the phenolic $pK_a$ (Shoun et al., 1979; Entsch et al., 1991). In the oxidized enzyme, the ionization of the phenol to the phenolate can be observed near 290 nm. Of course, it is the $pK_a$ of the substrate on the enzyme, which is in the hydroperoxide form, that is important and has been inaccessible to direct determination so far. However, there is evidence from site-directed mutagenesis studies that the behavior of the hydroperoxide enzyme–substrate complex mimics the behavior of the substrate bound to the oxidized enzyme (Entsch et al., 1991). Replacing tyrosine 201, which hydrogen bonds to pOHB, with phenylalanine removes any detectable lowering of the phenolic $pK_a$ of pOHB bound to the oxidized mutant enzyme. This mutant enzyme forms the flavin hydroperoxide normally but is nearly incapable of hydroxylating the substrate, with about 95% of the flavin hydroperoxide eliminating $H_2O_2$. The importance of substrate activation has also been emphasized by a recent theoretical study (Vervoort et al., 1992). More activated compounds will have higher energies for the highest occupied molecular orbital (HOMO), the relevant orbital for the interaction of the compound as a nucleophile. In this study, there was a linear correlation between the energy of the HOMO of the phenolate of several fluorine-substituted substrates and the logarithm of $k_{cat}$ measured by steady-state kinetics. A possible difficulty with such an approach, however, is the use of $k_{cat}$ rather than the actual hydroxylation rates, thus relying on the assumption that other reactions in the catalytic cycle are not partially rate-determining and conrtibuting to $k_{cat}$.

The identity of intermediate III was firmly established a number of years ago. The spectrum of the intermediate implicated a C(4a)-hydroxyflavin. N(5)-ethyl-C(4a)-hydroxy FAD was synthesized and introduced onto the apoprotein of PHBH (Ghisla et al., 1977). The spectrum of the resulting enzyme was nearly identical to that of the observed intermediate, providing very strong evidence for the structural assignment. Although the identity of the intermediate is well established, the events leading to its decay appear complicated. To continue turnover, water must be eliminated from the hydroxyflavin, the hydroxylation product must dissociate, and another substrate must bind. There appears to be some complexity in the ordering of these events, depending on the particular hydroxylase and the particular substrate. The decay of the hydroxyflavin has been studied most thoroughly with phenol hydroxylase (Taylor and Massey, 1990; Maeda-

Yorita and Massey, 1993). In the presence of very high concentrations of substrate, it is possible to displace the hydroxylation product from intermediate III and replace it with the substrate. These enzyme–hydroxyflavin–substrate complexes are rather stable and can have half-lives of more than 100 minutes at 4 °C. The trapping of the hydroxyflavin at substrate concentrations much higher than physiological ones is probably the basis of the substrate inhibition observed for a number of hydroxylases.

While most flavoprotein hydroxylases are composed of identical, nondissociating subunits, an interesting exception is $p$-hydroxyphenyl acetate 3-hydroxylase isolated from *Pseudomonas putida* (Arunachalam et al., 1992). The chemical mechanism appears to be essentially the same as that already discussed. However, the enzyme is composed of two protein components, each isolated independently. One component is a flavoprotein containing FAD. The other component has no prosthetic group and has been termed the *coupling protein*. The flavoprotein component binds the substrate, $p$-hydroxyphenyl acetate, is reducible by NADH and is reoxidized by $O_2$. However, no hydroxylation products are detectable unless the coupling protein is included in the reaction mixture. By measuring the extent of hydroxylation as the ratio of coupling protein to flavoprotein is increased, it has been concluded that the coupling protein forms a 1:1 complex with the flavoprotein. Therefore, it appears that the coupling protein stabilizes the flavin hydroperoxide and could possibly have a role in substrate activation. Dramatic evidence for its role in the stabilization of the hydroperoxide comes from stopped-flow experiments in which the reduced flavoprotein–coupling protein complex was mixed with $O_2$-containing buffer in the absence of substrate. The flavin hydroperoxide is quickly formed and is extremely stable, with a half-life of about 6 minutes at 4 °C. Such a stable hydroperoxide is unprecedented among the other aromatic hydroxylases.

The chemical events of the hydroxylation reaction are similar for all the hydroxylases studies so far. Several hydroxylases perform additional transformations on the substrate after it has been hydroxylated. Salicylate hydroxylase converts salicylate to catechol and $CO_2$. Apparently, the enzyme hydroxylates the aromatic ring *ipso* to the carboxylate and *ortho* to the activating hydroxy substituent, as is typical of hydroxylases.

Rearomatizing the ring after hydroxylation requires the loss of an electrophile, in this case $CO_2$. Thus this enzyme has probably retained the same reaction mechanism as PHBH but changed the identity of the departing electrophile.

Salicylate Hydroxylase Reaction

Anthranilate hydroxylase appears to take advantage of the reactivity of the nonaromatic tautomer generated by hydroxylation and catalyzes a reaction of this compound before allowing rearomatization. It hydroxylates anthranilate (2-aminobenzoate) at the 3-position. The immediate hydroxylation product is the conjugated iminium diene.

Authranilate Hydroxylase Reaction

Conceivably, this intermediate could rearomatize by deprotonation at this point, as is the case with other hydroxylases that use aromatic amines as substrates. However, the iminium intermediate is hydrolyzed to give ammonia and a ketone, which tautomerizes to form the product, 2,3-dihydroxybenzoate. This mechanism is supported by $^{18}O$ labeling studies, which show that the 3-hydroxyl oxygen of the product is derived from $O_2$ and the 2-hydroxyl oxygen is derived from water (Powlowski et al., 1987). Furthermore, the products from several substrate analogs support this mechanism (Powlowski et al., 1990). 4-Fluoro and 5-fluoro anthranilates are hydroxylated and give mixtures of the 2-hydroxy- and 2-amino-aromatic products. The proportion of the amino product increases with pH, suggesting that the iminium ion is required for hydrolysis. A nonaromatic product has also been trapped from the hydroxylation of 3-methyl anthranilate.

In the case of 2-methyl-3-hydroxy-5-carboxypyridine (MHPC) oxygenase, the aromatic ring is hydroxylated at the 2-position, *ipso* to the methyl group. Since the loss of a methyl cation would be enormously unfavorable energetically, there is no route to the formation of an aromatic product. Instead, the intermediate ketone is attacked by

MHPC Oxygenase Reaction

water and the ring is cleaved, giving an acyclic product. Labeling patterns with $^{18}$O in water or in $O_2$ are consistent with this mechanism. Interestingly, when $^{18}O_2$ is used, about 5% of the product contains two atoms of $^{18}$O, suggesting that the water molecule generated by the decay of the hydroxyflavin does not exchange quickly with bulk solvent and that the attack of water on the cyclic ketone occurs at the same time as or after the elimination of water from the hydroxyflavin (Brissette, 1990).

## PTERIN HYDROXYLASES

In mammalian systems, aromatic amino acids are hydroxylated by enzymes that use tetrahydrobiopterin and a metal ion to hydroxylate the substrate. The flavoprotein aromatic hydroxylases not only catalyze substrate hydroxylation but also carry out the reactions needed to regenerate the hydroxylating species: the elimination of water to reform the oxidized flavin and the reduction of the flavin by NAD(P)H. In the pterin-dependent hydroxylating systems a similar set of reactions occurs, but each of these reactions is catalyzed by separate enzymes. In this section, we will focus on the reactions performed by the hydroxylase components of these systems. Unlike most flavin-dependent hydroxylases, the pterin is not tightly bound by the enzyme. It is in fact a substrate rather than a prosthetic group in hydroxylation reactions, and the C(4a)-hydroxypterin is released as a product (Dix and Benkovic, 1988). The naturally occurring pterin substrate is tetrahydrobiopterin, but a variety of analogs are also effective.

Phenylalanine hydroxylase (PAH), which hydroxylates phenylalanine at the 4-position to form tyrosine, is the most thoroughly studied of these enzymes. The tetrameric enzyme is isolated with one ferric ion per subunit, which must be reduced to the ferrous state for activity. Tyrosine hydroxylase, also an iron-containing enzyme, catalyzes the formation of 3,4-dihydroxyphenylalanine (L-DOPA), a neurotransmitter. This enzyme has received less attention than PAH because it has been difficult to isolate. These difficulties have recently been overcome by the cloning of the gene for the enzyme and its overexpression in tissue culture (Fitzpatrick et al., 1990). Tetrahydrobiopterin might also be involved in the hydroxylation of nonaromatic compounds. Nitric oxide synthase oxidizes arginine to citrulline and nitric oxide and contains tetrahydrobiopterin, which apparently does not dissociate during turnover (Hevel and Marletta, 1992). However, it is not yet clear how the pterin is involved in catalysis, since the enzyme also contains FMN, FAD, and a P-450-like heme component (McMillan et al., 1992; White and Marletta, 1992).

The two-ringed heterocyclic pterin system in its fully oxidized, two-electron reduced, and four-electron reduced states is shown below. The resemblance to the flavin ring system is apparent, the notable differences being the replacement of the oxygen at the 2-position with an amino group and the lack of a fused benzene ring. Pterin-dependent hydroxylases use the four-electron reduced form (tetrahydropterin) in reactions with oxygen, finally forming the C(4a)-hydroxypterin. Free in solution, this species dehydrates to produce the two-electron reduced quinonoid species, and, within seconds or minutes (depending on the pH), the quinonoid form tautomerizes, since the equilibrium at the two-electron level of reduction lies toward the biologically inactive 7,8-dihydropterin. *In vivo*, however, the quinonoid form is quickly reduced enzymatically by dihydropterin reductase, thus avoiding the formation of the 7,8-dihydropterin (Craine et al., 1972).

Free tetrahydrobiopterin is spontaneously oxidized by $O_2$, although more slowly than is dihydroflavin. Electrochemical, kinetic, and spectral studies of model pterins are consistent with a reaction of tetrahydropterin with $O_2$ occurring by the formation of a semiquinone–

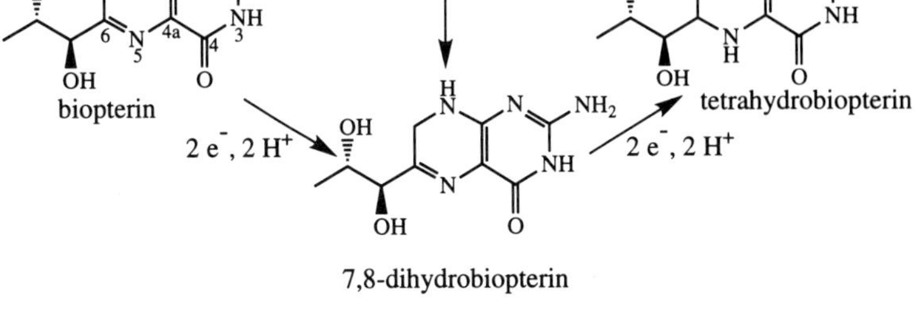

superoxide pair, followed by radical collapse to the C(4a)-hydroperoxide and elimination of $H_2O_2$ (Eberlein et al., 1984). The initial electron transfer reaction to $O_2$ is slower in pterin oxidation than in flavin oxidation because the pterin semiquinone is much less stable. As with the oxidation of flavin, there is a lag before a steady oxidation rate is seen. Autocatalytic reactions analogous to those observed in the flavin case also appear to occur in pterin oxidation. These reactions have been suppressed by including SOD to scavenge superoxide and dihydropterin reductase and NADPH to prevent the reaction of oxidized and reduced pterin (Bailey et al., 1991). When such chain-suppressing agents are present, a reaction rate that is first order both in tetrahydropterin and in $O_2$ is observed. A rate of $0.2\,M^{-1}s^{-1}$ was estimated for the reaction of tetrahydropterin with $O_2$, two to four orders of magnitude slower than the reactions of free reduced flavins with $O_2$.

The low reactivity of tetrahydropterins with $O_2$ has been a barrier to the study of the mechanisms of the pterin hydroxylases. Since the subsequent steps are much faster, the direct observation of reaction intermediates has not been possible for pterin hydroxylases. In stopped-flow studies of the hydroxylation of phenylalanine by PAH, the production of tyrosine was observed, with the simultaneous formation of C(4a)-hydroxypterin; no direct evidence for earlier intermediates was reported, although the time resolution of the instrument employed was insufficient to detect intermediates occurring before 300 milliseconds (Haavik et al., 1986). Only a small deuterium isotope effect has been observed with PAH when deuterated phenylalanine is used as substrate, indicating that reactions other than substrate proton loss are significantly rate determining (Abita et al., 1984), consistent with a slow $O_2$ reaction. The reaction of $O_2$ might also be the rate-determining step in catalysis in tyrosine hydroxylase (Fitzpatrick, 1991).

Research into the reaction mechanism has attempted to delineate the roles played by the metal and the reduced pterin in activating $O_2$ and catalyzing the oxygenation reaction. One proposal is that the hydroxylation occurs analogously to the flavoprotein hydroxylases (Figure 2-8). The reduced pterin is thought to react with $O_2$, forming a C(4a)-hydroperoxide, which hydroxylates the substrate. In this case, the metal might serve to bind and activate the pterin and/or $O_2$, helping to overcome the spin inversion barrier. The iron atom appears to be vital for the function of mouse liver PAH, as demonstrated in a study of site-directed mutants of PAH which could no longer bind iron and were also inactive catalysts (Gibbs et al., 1993). Since, in contrast to flavin hydroxylases, the pterin hydroxylases can hydroxylate un-

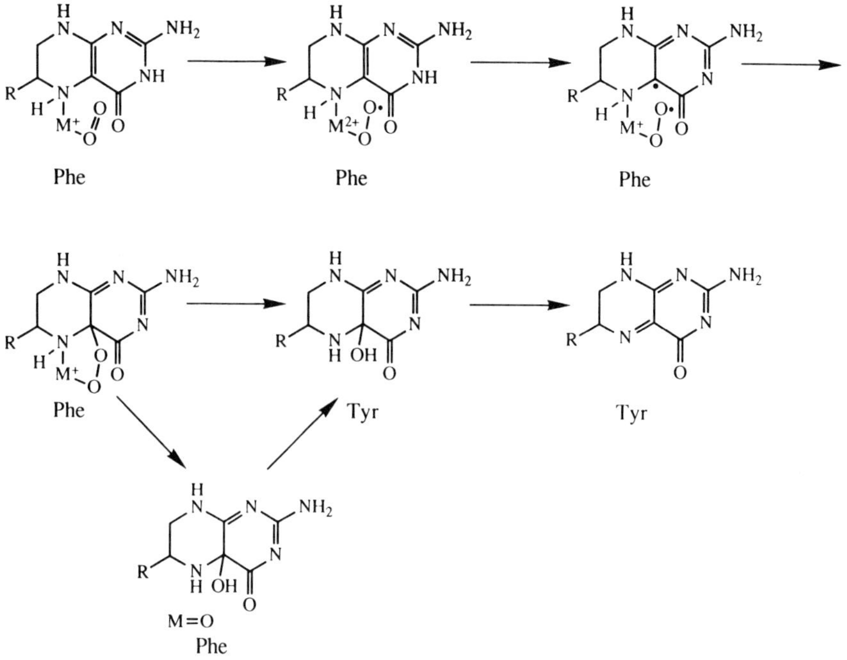

**FIGURE 2-8** Possible tyrosine-forming pathways of PAH. The formation of a pterin hydroperoxide may be assisted by the active site metal, either by forming enzyme-bound superoxide or by forming a metal-stabilized pterin semiquinone, or by both means, as shown. The peroxypterin might hydroxylate phenylalanine directly, or a metal–oxo intermediate may form first. In either case, the C(4a)-hydroxypterin is formed, and dehydrates to the quinonoid dihydropterin.

activated aromatic rings, the postulated pterin hydroperoxide would probably have to be activated. It has been speculated that this could be accomplished by protonating N(5) of the pterin hydroperoxide, forming a cation (Ayling and Bailey, 1990). Alternatively, the same effect could be provided by coordination of a metal acting as a Lewis acid. The other distinct possibility is that the metal ion acts as the oxygenating species through a metal–oxo intermediate, analogous to cytochrome P-450. This metal–oxo species could be formed from a pterin hydroperoxide donor. At present, no conclusive evidence has been reported to exclude either mechanism.

It has been established that the pterin is not simply a two-electron reductant but is also involved with $O_2$ activation in phenylalanine hydroxylase. The probability that this involves the formation and reaction of a pterin C(4a)-hydroperoxide is supported by the following experiments. The pterin product released from the enzyme after the hydroxylation of phenylalanine is a C(4a)-hydroxypterin, analogous to the hydroxyflavin intermediate of the flavoprotein hydroxylases (Kaufman, 1975; Lazarus et al., 1981). The formation of this compound is certainly enzyme catalyzed since the tetrahedral C(4a)-adduct is formed exclusively as the S-isomer (Dix et al., 1985). This adduct is significantly more stable ($t_{1/2}$ = 30 minutes) than hydroxyflavin, and its structure has been determined by $^{13}$C-NMR (Lazarus et al., 1982a). $^{18}$O-labeling studies have shown that the hydroxyl oxygen comes from $O_2$ (Dix et al., 1985). The hydroxypterin product is released from the enzyme and is the substrate for another enzyme, a pterin C(4a)-carbinolamine dehydratase, which catalyzes the elimination of water to form the quinonoid dihydrobiopterin (Lazarus et al., 1982b). Tyrosine hydroxylase also produces C(4a)-hydroxypterin as a reaction product, suggesting a similarity in mechanisms (Dix et al., 1987).

The possibility that the metal ion is involved in the initial oxygen reaction has received support from experiments using PAH isolated from *Chromobacterium violaceum* (Pember et al., 1989). This enzyme has a copper at the active site rather than iron. Both the Fe and Cu enzymes have been reported to require the metal to be in the reduced form for activity. PAH from liver is activated in a time-dependent manner by phenylalanine (Tourian, 1971); the bacterial enzyme is not, making it a better subject for steady-state kinetic analysis. Steady-state kinetic studies indicate that $O_2$ binds to the enzyme before the pterin and amino acid substrates bind. Evidence for a PAH–$O_2$ complex was also obtained by quenched-flow experiments by observing a stoichiometric burst of tyrosine production when a solution of PAH and $O_2$ was mixed with a solution of the other substrates. Preincubation with a lower $O_2$ concentration decreased the size of the burst. Magnetic susceptibility measurements demonstrated independently that $O_2$ binds to the diamagnetic $Cu^I$ of PAH.

The metal might have a role in activating the pterin in addition to a role in $O_2$ binding. EPR studies on the bacterial PAH in its $Cu^{II}$ form have shown that the N(5) of the pterin is coordinated to the copper ion. If this coordination is retained when the metal is reduced to its active state and during turnover, it might assist both in the formation of the pterin hydroperoxide and in the subsequent transfer of oxygen to the substrate (Figure 2-8). The metal may facilitate the formation of a pterin hydroperoxide by both reducing $O_2$ to $O_2^-$ and oxidizing the pterin to the semiquinone. It is known that flavin semiquinones are stabilized by chelation to metals (Müller, 1991a). The metal-bound radicals could then couple, yielding the pterin hydroperoxide. Note that in this process no net change in the metal oxidation state occurs. Coordination of the N(5) of the pterin hydroperoxide to the metal might also increase the reactivity of the hydroperoxide by inductive effects, since the positive charge of the metal would be expected to improve the leaving-group ability of the pterin hydroxide by lowering its p$K_a$ (Ayling and Bailey, 1990).

It seems likely that a pterin hydroperoxide is generated during catalysis. Still unresolved, however, is the fate of this species. It may be the actual hydroxylating species. Objections similar to those raised about the hydroxylating ability of flavin hydroperoxides have also been raised for the analogous pterin species. As a result, a ring-opened pterin carbonyl oxide has been considered. Experiments utilizing pterin analogs have cast doubt on a ring-opened species, but the evidence is inconclusive (Bailey and Ayling, 1980; Lazarus et al., 1981; Bailey et al., 1982).

Instead of the direct transfer of oxygen from an oxygenated pterin to the substrate, the possibility that the active-site metal is the oxygenating species has also been considered. In this model, the terminal oxygen of the C(4a)-hydroperoxypterin would be heterolytically transferred to the metal ion, forming the observed hydroxypterin and a metal–oxo species analogous to the iron–oxo intermediate thought to be involved in hydroxylations by cytochromes P-450. Such a metal–oxo species would be very reactive and could transfer the oxygen to the relatively unactivated substrate quite quickly. Evidence supporting this mechanism has been derived both from studies of hydroxylation of unnatural substrates and of the unproductive tetrahydropterin oxidation. PAH has been reported to hydroxylate the alkyl side chain of norleucine and the methyl group of 4-methylphenylalanine (Kaufman and Mason, 1982; Siegmund and Kaufman, 1991). Such reactions would almost certainly require the intermediacy of radicals, which is possible in a P-450-like mechanism. PAH also catalyzes the uncoupled oxidation of tetrahydrobiopterin when tyrosine is included in the reaction mixtures. It was reported that when liver PAH was activated by lysolecithin, about one-third of the tetrahydrobiopterin was converted to the C(4a)-hydroxy pterin in the presence of tyrosine, although tyrosine was not hydroxylated (Davis and Kaufman, 1989). The rest of the pterin was converted directly to the quinonoid dihydrobiopterin. These results were explained in terms of an intermediate hydroperoxypterin, which can decay either by eliminating $H_2O_2$ to form the quinonoid dihydropterin or by homolytically transferring an oxygen atom to the iron atom to form an iron–oxo intermediate. This intermediate would then be reduced by tetrahydrobiopterin in the uncoupled reaction.

This interpretation contradicts an earlier study of the uncoupled reactions of PAH, in which no hydroxypterin was observed to form from tetrahydropterin consumed nonproductively (Dix and Benkovic, 1985). Furthermore, evidence that PAH cleaves peroxide bonds homolytically rather than heterolytically has been provided by the study of its reactions with several peroxy compounds (Benkovic et al., 1986). The behavior of tyrosine hydroxylase is also consistent with the formation of an iron–oxo intermediate. Tyrosine hydroxylase can hydroxylate tyrosine when $H_2O_2$ is substituted for tetrahydropterin (Dix et al., 1987). The mechanism for this reaction is unclear; the iron of the enzyme might activate $H_2O_2$ by serving as a Lewis acid, or it might cleave the oxygen bond, forming a highly reactive iron–oxo intermediate, which hydroxylates tyrosine.

The oxygenation reactions catalyzed by flavin and pterin hydroxylases appear to have many similarities, but clearly there are distinct differences. Whereas the oxygenating agent with all of the flavoprotein external monooxygenases has been shown to be the C(4a)-flavin hydroperoxide by characterization of the intermediates, the nature of the hydroxylating species with pterins remains unknown and is one of the major questions to be resolved. Although a C(4a)-hydroxypterin has been demonstrated, the formation of the putative C(4a)-pterin hydroperoxide, either as the active oxygen agent or as a precursor to the formation of activated oxygen, is not yet established. Finally, the role of the iron, which is apparently important to catalysis, is also unknown.

In this chapter, we have seen that singlet reduced flavin initially reacts with $^3O_2$, probably by a single electron transfer, followed by the rapid formation and decay of a covalent intermediate. A common characteristic of the oxygen-utilizing flavoenzymes is the dramatic rate increase for the reduction of $^3O_2$. The external monooxygenases are able to stabilize the flavin hydroperoxide and use this reactive molecule as either a nucleophile or an electrophile. The particular chemical task for which a flavin hydroperoxide is used is controlled by the protein structure in ways that are beginning to be understood. While electron-rich aromatic rings are readily hydroxylated by flavin hydroperoxides, unactivated rings appear to require a more powerful hydroxylating species, generated by the reaction of tetrahydrobiopterin, a metal, and $O_2$ in an unknown manner. In the case of all the $O_2$-utilizing enzymes considered in this chapter, however, the reaction of $O_2$ is controlled, and the release of potentially destructive oxygen species, such as superoxide, is avoided.

# REFERENCES

ABITA, J. -P., PARNIAK, M., and KAUFMAN, S. (1984) The Activation of Rat Liver Phenylalanine Hydroxylase by Limited Proteolysis, Lysolecithin, and Tocopherol Phosphate. Changes in Conformation and Catalytic Properties. *J. Biol. Chem.*, 259, 14560–14566.

ABU-SOUD, H., MULLINS, L. S., BALDWIN, T. O., and RAUSHEL, F. M. (1992) Stopped-Flow Kinetic Analysis of the Bacterial Luciferase Reaction. *Biochemistry*, 31, 3807–3813.

AHRENS, M., MACHEROUX, P., EBERHARD, A., GHISLA, S., BRANCHAUD, B., and HASTINGS, J. W. (1991) Boronic Acids as Mechanistic Probes for the Bacterial Luciferase Reaction. *Photochem. Photobiol.*, 54, 295–299.

ANDERSON, R. F. (1982) Flavin–Oxygen Complex Formed on the Reaction of Superoxide Ions with Flavosemiquinone Radicals, In *Flavins and Flavoproteins* (V. Massey and C. H. Williams, Eds.), Elsevier North-Holland, New York, pp. 278–283.

ANDERSON, R. F., PATEL, K. B., and STRATFORD, M. R. L. (1987) Absorption Spectra of Radicals of Substrates for p-Hydroxybenzoate Hydroxylase Following Electrophilic Attack of the ·OH Radical in the 3 Position. *J. Biol. Chem.*, 262, 17475–17479.

ANDERSON, R. F., PATEL, K. B., and VOJNOVIC, B. (1991) Absorption Spectra of Radical Forms of 2,4-Dihydrobenzoic Acid, a Substrate for p-Hydroxybenzoate Hydroxylase. *J. Biol. Chem.*, 266, 13086–13090.

ARUNACHALAM, U., MASSEY, V., and VAIDYANATHAN, C. S. (1992) p-Hydroxyphenylacetate-3-hydroxylase: A Two-protein Component Enzyme. *J. Biol. Chem.*, 267, 25848–25855.

AYLING, J. E., and BAILEY, S. W. (1990) Why Is the Cofactor for Tetrahydrobiopterin Dependent Enzymes not a Dihydroflavin?, in *Biological Oxidation Systems* (C. C. Reddy, G. A. Hamilton, and K. M. Madyastha, Eds.), Academic Press, San Diego, 1, pp. 221–236.

BAILEY, S. W., and AYLING, J. E. (1980) Cleavage of the 5-Amino Substituent of Pyrimidine Cofactors by Phenylalanine Hydroxylase. *J. Biol. Chem.*, 255, 7774–7781.

BAILEY, S. W., CROW, J. P., and AYLING, J. E. (1991) The Cofactor Dependent Interaction of Molecular Oxygen with Phenylalanine Hydroxylase, in *Flavins and Flavoproteins 1990* (B. Curti, S. Ronchi, and G. Zanetti, Eds.), Walter de Gruyter, New York, pp. 247–250.

BAILEY, S. W., WEINTRAUB, S. T., HAMILTON, S. M., and Ayling, J. E. (1982) Incorporation of Molecular Oxygen into Pyrimidine Cofactors by Phenylalanine Hydroxylase. *J. Biol. Chem.*, 257, 8253–8260.

BALDWIN, T. O., and ZIEGLER, M. M. (1992) The Biochemistry and Molecular Biology of Bacterial Bioluminescence, in *Chemistry and Biochemistry of Flavoenzymes* (F. Müller, Ed.), CRC Press, Boca Raton, FL, III, pp. 467–530.

BEATY, N. B., and BALLOU, D. P. (1981a) The Reductive Half-reaction of Liver Microsomal FAD-containing Monooxygenase. *J. Biol. Chem.*, 256, 4611–4618.

BEATY, N. B., and BALLOU, D. P. (1981b) The Oxidative Half-reaction of Liver Microsomal FAD-containing Monooxygenase. *J. Biol. Chem.*, 256, 4619–4625.

BENKOVIC, S. J., BLOOM, L. M., BOLLAG, G., DIX, T. A., GAFFNEY, B. J., and PEMBER, S. (1986) The Mechanism of Action of Phenylalanine Hydroxylase. *Ann. N.Y. Acad, Sci.*, 471, 226–232.

BLOOM, L. M., BENKOVIC, S. J., and GAFFNEY, B. J. (1986) Characterization of Phenylalanine Hydroxylase. *Biochemistry*, 25, 4204–4210.

BRISSETTE, P. (1990) Mechanisms of Two Flavoenzymes Involved in the Degradation of Vitamin B-6 in Bacteria. Ph.D. Thesis, University of Michigan.

BRITTON, L. N. BRAND, J. M., and MARKOVETZ, A. J. (1974) Sources of Oxygen in the Conversion of 2-Tridecanone to Undecyl Acetate by *Pseudomonas cepacia* and *Nocardia* Sp. *Biochim. Biophys. Acta*, 369, 45–49.

BRUICE, T. C. (1984) Oxygen-Flavin Chemistry. *Isr. J. Chem.*, 24, 54–61.

BRUICE, T. C., NOAR, J. B., BALL, S. S., and VEKATARAM, U. V. (1983) Monoxygen Donation Potential of 4a-Hydroperoxyflavins as Compared with Those of a Percarboxylic Acid and Other Hydroperoxides. Monooxygen Donation to Olefin, Tertiary Amine, Alkyl Sulfide, and Iodide Ion. *J. Am. Chem. Soc.*, 105, 2452–2463.

CRAINE, J. E., HALL, E. S., and KAUFMAN, S. (1972) The Isolation and Characterization of Dihydropteridine Reductase from Sheep Liver. *J. Biol. Chem.*, 247, 6082–6091.

DAVIS, M. D., and KAUFMAN, S. (1989) Evidence for the Formation of the 4a-Carbinolamine During the Tyrosine-dependent Oxidation of Tetrahydrobiopterin by Rat Liver Phenylalanine Hydroxylase. *J. Biol. Chem.*, 264, 8585–8596.

DIX, T. A., and BENKOVIC, S. J. (1982) Mechanism of "Uncoupled" Tetrahydropterin Oxidation by Phenylalanine Hydroxylase. *Biochemistry*, 24, 5839–5846.

DIX, T. A., and BENKOVIC, S. J. (1988) Mechanism of Oxygen Activation by Pteridine-dependent Monooxygenases. *Acc. Chem. Res.*, 21, 101–107.

DIX, T. A., BOLLAG, G., DOMANICO, P. L., and BENKOVIC, S. J. (1985) Phenylalanine Hydroxylase: Absolute Configuration and Source of Oxygen of the 4a-Hydroxypterin Species. *Biochemistry*, 24, 2955–2958.

DIX, T. A., KUHN, D. M., and BENKOVIC, S. J. (1987) Mechanism of Oxygen Activation by Tyrosine Hydroxylase. *Biochemistry*, 26, 3354–3361.

DONOGHUE, N. A., NORRIS, D. B., and TRUDGILL, P. W. (1976) The Purification of Cyclohexanone Oxygenase from *Nocardia globerula* CL1 and *Acinetobacter* NCIB 9871. *Eur. J. Biochem.*, 63, 175–192.

EBERLEIN, G., and BRUICE, T. C. (1983) The Chemistry of a 1,5-Diblocked Flavin. 2. Proton and Electron Transfer Steps in the Reaction of Dihydroflavins with Oxygen. *J. Am. Chem. Soc.*, 105, 6685–6697.

EBERLEIN, G., BRUICE, T. C., LAZARUS, R. A., HENRIE, R., and BENKOVIC, S. J. (1984) The Interconversion of the 5,6,7,8-Tetrahydro-, 7,8-Dihydro-, and Radical Forms of 6,6,7,7-Tetramethyldihydropterin. A Model for the Biopterin Center of Aromatic Amino Acid Mixed Function Oxidases. *J. Am. Chem. Soc.*, 106, 7916–7924.

ECKSTEIN, J. W., HASTINGS, J. W., and GHISLA, S. (1993) Mechanism of Bacterial Bioluminescence: 4a,5-Dihydroflavin Analogs as Models for Luciferase Hydroperoxide Intermediates and the Effects of Substituents at the 8-Position of Flavin on Luciferase Kinetics. *Biochemistry*, 32, 404–411.

ENGEL, P. C. (1992) Acyl-coenzyme A Dehydrogenases, in *Chemistry and Biochemistry of Flavoenzymes* (F. Müller, Ed.), CRC Press, Boca Raton, FL, III, pp. 597–655.

ENTSCH, B., and BALLOU, D. P. (1989) Purification, Properties, and Oxygen Reactivity of p-Hydroxybenzoate Hydroxylase from *Pseudomonas aeruginosa*. *Biochim. Biophys. Acta*, 999, 313–322.

ENTSCH, B., BALLOU, D. P., and MASSEY, V. (1976) Flavin-oxygen Derivatives in Hydroxylation by p-Hydroxybenzoate Hydroxylase. *J. Biol. Chem.*, 251, 2550–2563.

ENTSCH, B., MASSEY, V., and BALLOU, D. P. (1974) Intermediates in Flavoprotein Catalyzed Hydroxylations. *Biochem. Biophys. Res. Commun.*, 57, 1018–1025.

ENTSCH, B., PALFEY, B. A., BALLOU, D. P., and MASSEY, V. (1991) Catalytic Function of Tyrosine Residues in *para*-Hydroxybenzoate Hydroxylase as Determined by the Study of Site-directed Mutants. *J. Biol. Chem.*, 266, 17341–17349.

FITZPATRICK, F. P. (1991) Studies of the Rate-limiting Step in the Tyrosine Hydroxylase Reaction: Alternate Substrates, Solvent Isotope Effects, and Transition-state Analogues. *Biochemistry*, 30, 6386–6391.

FITZPATRICK, P. F., CHLUMSKY, L. J., DAUBNER, S. C., and O'MALLEY, K. (1990) Expression of Rat Tyrosine Hydroxylase in Insect Tissue Culture Cells and Purificiation and Characterization of the Cloned Enzyme. *J. Biol. Chem.*, 265, 2042–2047.

GHISLA, S., ENTSCH, B., MASSEY, V., and HUSAIN, M. (1977) On the Structure of Flavin–Oxygen Intermediates Involved in Enzymatic Reactions. *Eur. J. Biochem.*, 76, 139–148.

GHISLA, S., and MASSEY, V. (1991) L-Lactate Oxidase, in *Chemistry and Biochemistry of Flavoenzymes* (F. Müller, Ed.), CRC Press, Boca Raton, FL, II, pp. 243–289.

GIBBS, B. S., WOJCHOWSKI, D., and BENKOVIC, S. J. (1993) Expression of Rat Liver Phenylalanine Hydroxylase in Insect Cells and Site-directed

Mutagenesis of Putative Non-Heme Iron-binding Sites. *J. Biol. Chem.*, 268, 8046–8052.

HAAVIK, J., DØSKELAND, A. P., and FLATMARK, T. (1986) Stereoselective Effects in the Interactions of Pterin Cofactors with Rat-liver Phenylalanine 4-Monooxygenase. *Eur. J. Bioch.*, 160, 1–8.

HAMILTON, G. A. (1971) The Proton in Biological Redox Reactions. *Progr. Bioorg. Chem.*, 1, 83–157.

HAMILTON, G. A. (1974) Chemical Models and Mechanisms for Oxygenases, in *Molecular Mechanisms of Oxygen Activation* (O. Hayaishi, Ed.), Academic Press, New York, pp. 405–451.

HASTINGS, J. W., BALNY, C., LePEUCH, C., and DOUZOU, P. (1973) Spectral Properties of an Oxygenated Luciferase–Flavin Intermediate Isolated by Low-temperature Chromatography *Proc. Natl. Acad. Sci. U.S.A.*, 70, 3468–3472.

HEVEL, J. M., and MARLETTA, M. A. (1992) Macrophage Nitric Oxide Synthase: Relationship Between Enzyme-Bound Tetrahydrobiopterin and Synthase Activity. *Biochemistry*, 31, 7160–7165.

JONES, K. (1985) Nature of the 4a-Flavinhydroperoxide of Microsomal Flavin-containing Monooxygenase. Ph.D. Thesis, University of Michigan.

JONES, K., and BALLOU, D. P. (1984) The Nature of the 4a-Hydroxyflavin in the Mammalian Flavin Containing Monooxygenase, in *Flavins and Flavoproteins* (R. C. Bray, P. C. Engel, and S. G. Mayhew, Eds.), Walter de Gruyter, New York, pp. 619–622.

JONES, K. C., and BALLOU, D. P. (1986) Reactions of the 4a-Hydroperoxide of Liver Microsomal Flavin-containing Monooxygenase with Nucleophilic and Electrophilic Substrates. *J. Biol. Chem.*, 261, 2553–2559.

KATAGIRI, M., and ITAGAKI, E. (1991) A Steroid Ketone Monooxygenase from *Cylindrocarpon radicicola*, in *Chemistry and Biochemistry of Flavoenzymes* (F. Müller, Ed.), CRC Press, Boca Raton, FL, II, pp. 101–108.

KAUFMAN, S. (1975) Studies on the Mechanism of Phenylalanine Hydroxylase: Detection of an Intermediate in *Chemistry and Biology of Pteridines* (Pfleiderer W., Ed.), Walter de Gruyter, Berlin, pp. 291–304.

KAUFMAN, S., and MASON, K. (1982) Spectificity of Amino Acids as Activators and Substrates for Phenylalanine Hydroxylase. *J. Biol. Chem.*, 257, 14667–14678.

KEMAL, C., and BRUICE, T. C. (1976) Simple Synthesis of a 4a-Hydroperoxy Adduct of a 1,5-Dihydroflavine: Preliminary Studies of a Model for Bacterial Luciferase. *Proc. Natl. Acad. Sci. U.S.A.*, 73, 995–999.

KEMAL, C., CHAN, T. W., and BRUICE, T. C. (1977a) Reactions of $^3O_2$ with Dihydroflavins. 1. $N^{3,5}$-Dimethyl-1,5-dihydrolumiflavin and 1,5-Dihydroisoalloxazines. *J. Am. Chem. Soc.*, 99, 7272–7286.

KEMAL, C., CHAN, T. W., and BRUICE, T. C. (1977b) Chemiluminescent Reactions and Electrophilic Oxygen Donating Ability of 4a-Hydroperoxyflavins: General Synthetic Method for the Preparation of $N^5$-Alkyl-1,5-dihydroflavins. *Proc. Natl. Acad. Sci. U.S.A.*, 74, 405–409.

KÜRFURST, M., GHISLA, S., and HASTINGS, J. W. (1984a) Characterization and Postulated Structure of the Primary Emitter in the Bacterial Luciferase Reaction *Proc. Natl. Acad. Sci. U.S.A.*, 81, 2990–2994.

KÜRFÜRST, M., HASTINGS, J. W., GHISLA, S., and MACHEROUX, P. (1984b) Identification of the Luciferase-bound Flavin-4a-hydroxide as the Primary Emitter in the Bacterial Bioluminescence Reaction, in *Flavins and Flavoproteins* (R. C. Bray, P. C. Engel, and S. G. Mayhew, Eds.), Walter de Gruyter, New York, pp. 657–667.

LATHAM, J., and WALSH, C. T. (1986) Bacterial Cyclohexanone Oxygenase. A Versatile Flavoprotein Oxygen Transfer Catalyst. *Ann. N.Y. Acad. Sci.*, 471, 208–216.

LAZARUS, R. A., BENKOVIC, S. J., and KAUFMAN, S. (1983) Phenylalanine Hydroxylase Stimulator Protein is a 4a-Carbinolamine Dehydratase. *J. Biol. Chem.*, 258, 10960–10962.

LAZARUS, R. A., DEBROSSE, C. W., and BENKOVIC, S. J. (1982a) Phenylalanine Hydroxylase: Structural Determination of the Tetrahydropterin Intermediates by $^{13}$C NMR Spectroscopy. *J. Am. Chem. Soc.*, 104, 6869–6871.

LAZARUS, R. A., DEBROSSE, C. W., and BENKOVIC, S. J. (1982b) Structural Determination of Quinonoid Dihydropterins. *J. Am. Chem. Soc.*, 104, 6871–6872.

LAZARUS, R. A., DIETRICH, R. F., WALLICK, D. E., and BENKOVIC, S. J. (1981) On the Mechanism of Action of Phenylalanine Hydroxylase. *Biochemistry*, 20, 6834–6841.

LEDERER, F. (1991) Flavocytochrome $b_2$, in *Chemistry and Biochemistry of Flavoenzymes* (F. Müller, Ed.), CRC Press, Boca Raton, FL, II, pp. 153–242.

LEE, J., WANG, Y., and GIBSON, B. G. (1991) Electronic Excitation Transfer in the Complex of Lumazine Protein with Bacterial Bioluminescence Intermediates. *Biochemistry*, 30, 6825–6835.

LINDQVIST, Y. (1992) The Structure and Mechanism of Spinach Glycolate Oxidase, in *Chemistry and Biochemistry of Flavoenzymes* (F. Müller, Ed.), CRC Press, Boca Raton, FL, III, pp. 367–387.

LUDWIG, M. L., and LUSCHINSKY, C. L. (1992) Structure and Redox Properties of Clostridial Flavodoxins, in *Chemistry and Biochemistry of Flavoenzymes* (F. Müller, Ed.), CRC Press, Boca Raton, FL, III, pp. 427–466.

MAEDA-YORITA, K., and MASSEY, V. (1993) On the Reaction Mechanism of Phenol Hydroxylase. New Information Obtained by Correlation of Fluorescence and Absorbance Stopped Flow Studies. *J. Biol. Chem.*, 266, 4134–4144.

MAKEMSOM, J. C., HASTINGS, J. W., and QUIRKE, M. E. (1992) Stabilization of

Luciferase Intermediates by Fatty Amines, Amides, and Nitriles. *Arch. Biochem. Biophys.*, 249, 361–366.

MAROTA, J. J., and SHIMAN, R. (1984) Stoichiometric Reduction of Phenylalanine Hydroxylase by Its Cofactor: A Requirement for Enzymatic Activity. *Biochemistry*, 23, 1303–1311.

MASSEY, V., GHISLA, S., and MOORE, E. G. (1979) 8-Mercaptoflavins as Active Site Probes of Flavoenzymes. *J. Biol. Chem.*, 254, 9640–9650.

MASSEY, V., PALMER, G., and BALLOU, D. (1971) On the Reaction of Reduced Flavins and Flavoproteins with Molecular Oxygen, in *Flavins and Flavoproteins* (H. Kamin, Ed.), University Park Press, Baltimore, pp. 349–361.

MASSEY, V., PALMER, G., and BALLOU, D. (1973) On the Reaction of Reduced Flavins with Molecular Oxygen, in *Oxidases and Related Redox Systems* (T. E. King, H. S. Mason, and M. Morrison, Eds.), University Park Press, Baltimore, pp. 25–49.

MASSEY, V., SCHOPFER, L. M., and ANDERSON, R. F. (1988) Structural Determinants of the Oxygen Reactivity of Different Classes of Flavoproteins, in *Oxidases and Related Redox Systems* (T. E. King, H. S. Mason, and M. Morrison, Eds.), Alan R. Liss, New York, pp. 147–166.

MASSEY, V., STRICKLAND, S., MAYHEW, S. G., HOWELL, L. G., ENGEL, P. C., MATTHEWS, R. G., SCHUMAN, M., and SULLIVAN, P. A. (1969) The Production of Superoxide Anion Radicals in the Reaction of Reduced Flavins and Flavoproteins with Molecular Oxygen. *Biochem. Biophys. Res. Commun.*, 36, 891–897.

MCCAPRA, F., and HART, R. (1976) Oxidation of Dialkyl Sulphides by Bacterial Luciferase. *J. Chem. Soc., Chem. Commun.*, 273–274.

MCMILLAN, K., BREDT, D. S., HIRSCH, D. J., SNYDER, S. H., CLARK, J. E., and MASTERS, B. S. S. (1992) Cloned, Expressed Rat Cerebellar Nitric Oxide Synthase Contains Stoichiometric Amounts of Heme, Which Binds Carbon Monoxide. *Proc. Natl. Acad. Sci. U.S.A.*, 89, 11141–11145.

MERÉNYI, G., and LIND, J. (1991a) Chemistry of Peroxidic Tetrahedral Intermediates of Flavin. *J. Am. Chem. Soc.*, 113, 3146–3153.

MERÉNYI, G., and LIND, J. (1991b) The Breakdown of C(4a)-flavin-peroxides into Flavin and a Hydroperoxide in Different Environments, in *Flavins and Flavoproteins 1990* (B. CURTI, S. RONCHI, and G. ZANETTI, Eds.), Walter de Gruyter, New York, pp. 37–40.

MERÉNYI, G., LIND, J., and ANDERSON, R. F. (1991) Spectral Characterization of 4-Carboxy-5,6-dihydro-2,4-cyclohexadienone, a Likely Component of Intermediate II in *p*-Hydroxy Benzoate Hydroxylase. *J. Am. Chem. Soc.*, 113, 9371–9372.

MÜLLER, F. (1991a) Free Flavins: Synthesis, Chemical and Physical Properties, in *Chemistry and Biochemistry of Flavoenzymes* (F. MÜLLER, Ed.), CRC Press, Boca Raton, FL, I, pp. 1–71.

MÜLLER, F. (1991b) Nuclear Magnetic Resonance Studies on Flavoproteins, in *Chemistry and Biochemistry of Flavoenzymes* (F. MÜLLER, Ed.), CRC Press, Boca Raton, FL, III, pp. 557–595.

MUTO, S., and BRUICE, T. C. (1982) Dioxygen Transfer from 4a-Hydroperoxyflavin Anion. 4. Dioxygen Transfer to Phenolate Anion as a Means of Aromatic Hydroxylation. *J. Am. Chem. Soc.*, 104, 2284–2290.

PEMBER, S. O., JOHNSON, K. A., VILLAFRANCA, J. J., and BENKOVIC, S. J., (1989) Mechanistic Studies on Phenylalanine Hydroxylase from *Chromobacterium violaceum*. Evidence for the Formation of an Enzyme–Oxygen Complex. *Biochemistry*, 28, 2124–2130.

POULSEN, L. L. (1991) The Multisubstrate FAD-containing Monooxygenase, in *Chemistry and Biochemistry of Flavoenzymes* (F. MÜLLER, Ed.), CRC Press, Boca Raton, FL, II, pp. 87–100.

POULSEN, L. L., and ZIEGLER, D. M. (1979) The Liver Microsomal FAD-containing Monooxygenase. Spectral Characterization and Kinetic Studies. *J. Biol. Chem.*, 254, 6449–6455.

POWLOWSKI, J. B., BALLOU, D. P., and MASSEY, V. (1990) Studies of the Oxidative Half-reaction of Anthranilate Hydroxylase (Deaminating) with Native and Modified Substrates. *J. Biol. Chem.*, 265, 4969–4975.

POWLOWSIKI, J. B., DAGLEY, S., MASSEY, V., and BALLOU, D. P. (1987) Properties of Anthranilate Hydroxylase (Deaminating), a Flavoprotein from *Trichosporon Cutaneum*. *J. Biol. Chem.*, 262, 69–74.

PRAIRIE, R. L., and TALALAY, P. (1963) Enzymatic Formation of Testolactone. *Biochemistry*, 2, 203–208.

RAUSHEL, F. M., and BALDWIN, T. O. (1989) Proposed Mechanism for the Bacterial Bioluminescence Reaction Involving a Dioxirane Intermediate. *Biochem. Biophys. Res. Commun.*, 164, 1137–1142.

RYERSON, C. C., BALLOU, D. P., and WALSH, C. (1982) Mechanistic Studies on Cyclohexanone Oxygenase. *Biochemistry*, 21, 2644–2655.

SCHOPFER, L. M., WESSIAK, A., and MASSEY, V. (1991) Interpretation of the Spectra Observed During Oxidation of *p*-Hydroxybenzoate Hydroxylase Reconstituted with Modified Flavins. *J. Biol. Chem.*, 266, 13080–13085.

SCHREUDER, H. A., HOL, W. G. J., and DRENTH, J. (1990) Analysis of the Active Site of the Flavoprotein *p*-Hydroxybenzoate Hydroxylase and Some Ideas with Respect to Its Reaction Mechanism. *Biochemistry*, 29, 3101–3108.

SCHREUDER, H. A., PRICK, P. A. J., WIERENGA, R. K., VRIEND, G., WILSON, K. S., HOL, W. G. J., and DRENTH, J. (1989) Crystal Structure of the *p*-Hydroxybenzoate Hydroxylase–Substrate Complex Refined at 1.9 Å Resolution: Analysis of the Enzyme–Substrate and Enzyme–Product Complexes. *J. Mol. Biol.*, 208, 679–696.

SCHREUDER, H. A., van der LANN, J. M., HOL, W. G. J., and DRENTH, J. (1988) Crystal Structure of *p*-hydroxybenzoate Hydroxylase Complexed with Its Reaction Product, 3,4-Dihydroxybenzoate. *J. Mol. Biol.*, 199, 637–648.

SCHREUDER, H. A., VAN DER LANN, J. M., SWARTE, M. B. A., KALK, K. H., HOL, W. G. J., and DRENTH, J. (1992) Crystal Structure of the Reduced Form of p-Hydroxybenzoate Hydroxylase Refined at 2.3 Å Resolution. *Proteins: Structure, Function, and Genetics*, 14, 178–190.

SCHWAB, J. M., LI, W. B., and THOMAS, L. P. (1983) Cyclohexanone Oxygenase: Stereochemistry, Enantioselectivity, and Regioselectivity of an Enzyme-catalyzed Baeyer–Villiger Reaction *J. Am. Chem. Soc.*, 105, 4800–4808.

SHOUN, H., BEPPU, T., and ARIMA, K. (1979) On the Stable Enzyme·Substrate Complex of p-Hydroxybenzoate Hydroxylase. Evidences for the Proton Uptake from the Substrate. *J. Biol. Chem.*, 254, 899–904.

SIEGMUND, H., and KAUFMAN, S. (1991) Hydroxylation of 4-Methylphenylalanine by Rat Liver Phenylalanine Hydroxylase. *J. Biol. Chem.*, 266, 2903–2910.

TAYLOR, M. G., and MASSEY, V. (1990) Decay of the 4a-Hydroxy-FAD Intermediate of Phenol Hydroxylase. *J. Biol. Chem.*, 265, 13687–13694.

TOURIAN, A. (1971) Activation of Phenylalanine Hydroxylase by Phenylalanine. *Biochim. Biophys. Acta*, 242, 345–354.

VERVOORT, J., MÜLLER, F., LEE, J., VAN DEN BERG, W. A. M., and MOONEN, C. T. (1986a) Identifications of the True Carbon-13 Nuclear Magnetic Resonance Spectrum of the Stable Intermediate II in Bacterial Luciferase. *Biochemistry*, 25, 8062–8067.

VERVOORT, J., MÜLLER, F., O'KANE, D. J., LEE, J., and BACHER, A. (1986b) Bacterial Luciferase: A Carbon-13, Nitrogen-15, and Phosphorus-31 Nuclear Magnetic Resonance Investigation. *Biochemistry*, 25, 8067–8075.

VERVOORT, J., RIETJENS, I. M. C. M., VAN BERKEL, W. J. H., and VEEGER, C. (1992) Frontier Orbital Study on the 4-Hydroxybenzoate-3-hydroxylase-dependent Activity with Benzoate Derivatives. *Eur. J. Biochem.*, 206, 479–484.

WALLICK, D. E., BLOOM, L. M., GAFFNEY, B. J., and BENKOVIC, S. J. (1984) Reductive Activation of Phenylalanine Hydroxylase and Its Effect on the Redox State of the Non-Heme Iron. *Biochemistry*, 23, 1295–1302.

WALSH, C. T., and CHEN, Y.-C. J. (1988) Enzymic Baeyer–Villiger Oxidations by Flavin-dependent Monooxygenases. *Angew. Chem. Int. Ed. Engl.*, 27, 333–343.

WANG, R., and THORPE, C. (1991) Reactivity of Medium-chain Acyl-CoA Dehydrogenase Toward Molecular Oxygen. *Biochemistry*, 30, 1895–1901.

WESSIAK, A., and BRUICE, T. C. (1983) Synthesis and Study of a 6-Amino-5-oxo-3H,5H-uracil and Derivatives. The Structure of an Intermediate Proposed in Mechanisms of Flavin and Pterin Oxygenases. *J. Am. Chem. Soc.*, 105, 4809–4825.

WESSIAK, A., NOAR, J. B., and BRUICE, T. C. (1984a) The Possibility That the Spectrum of Intermediate Two, Seen in the Course of Reaction of

Flavoenzyme Phenol Hydroxylases, May Be Attributable to Iminol Isomers of a Flavin-dervived 6-Arylamino-5-oxo(3H,5H) Uracil. *Proc. Natl. Acad. Sci. U.S.A.*, 81, 332–336.

WESSIAK, A., SCHOPFER, L. M., YUAN, L. C., BRUICE, T. C., and MASSEY, V. (1984b) Use of Riboflavin-binding Protein to Investigate Steric and Electronic Relationships in Flavin Analogs and Models. *Proc. Natl. Acad. Sci. U.S.A.*, 81, 4246–4249.

WHITE, K. A., and MARLETTA, M. A. (1992) Nitric Oxide Synthase is a Cytochrome P450 Type Hemoprotein. *Biochemistry*, 31, 6627–6631.

# 3
# Reactions of Dioxygen and Its Reduced Forms with Heme Proteins and Model Porphyrin Complexes

TEDDY G. TRAYLOR AND PATRICIA S. TRAYLOR

### INTRODUCTION

Among the many kinds of biological complexes of transition metals that interact with dioxygen or its variously reduced forms, heme proteins are the most common and the most studied. With only a few exceptions, heme proteins are globular proteins which sequester and spatially isolate the active porphyrin–iron complex, protoheme (the iron(II) form) or protohemin (the iron(III) form; Figure 3-1). The iron porphyrin is tightly bound to the protein through numerous hydrophobic interactions and by a single coordinate bond between a base from a so-called proximal amino acid residue, such as an imidazole nitrogen from histidine or phenolate from tyrosine, and the iron atom. This is shown in Figure 3-2, where B represents the proximal base. Dioxygen or other small molecules bind to the opposite (distal) side of the porphyrin.

These heme proteins carry out a wide variety of processes, including dioxygen binding and transport (Wang, 1962; Antonini and Brunori, 1971; Rifkind, 1988), dioxygen reduction to water (Wikström et al., 1981; Brunori and Chance, 1988; Cusanovich et al., 1988; Capaldi, 1990a 1990b; Cooper, 1990; Malmström, 1990a, 1990b; Babcock et al., 1992; Babcock and Wikström, 1992), catalysis of oxidation by dioxygen or hydrogen peroxide (Chance, 1951; Wittenberg et al., 1967; Dunford

**FIGURE 3-1** Protohemin.

and Stillman, 1976; Ortiz de Montellano, 1992), electron transfer (Quagliariello et al., 1975; Wikström, 1989; Onuchic et al., 1992; Winkler and Gray, 1992), and nitric oxide binding, which activates the conversion of guanosine triphosphate to cyclic guanosine monophosphate, leading to wide-ranging physiological effects (Murad, 1986; Waldman and Murad, 1988; Nakatsu and Diamond, 1989; Walter, 1989; Ignarro, 1990; Goy, 1991; Lincoln and Cornwell, 1991; Hofmann et al., 1992).

A fascinating challenge to biochemical and chemical researchers is to understand the mechanisms by which different proteins induce this particular iron porphyrin moiety to carry out specific functions from among a variety of chemical processes. This review describes chemical approaches in the study of one area in this field, the binding

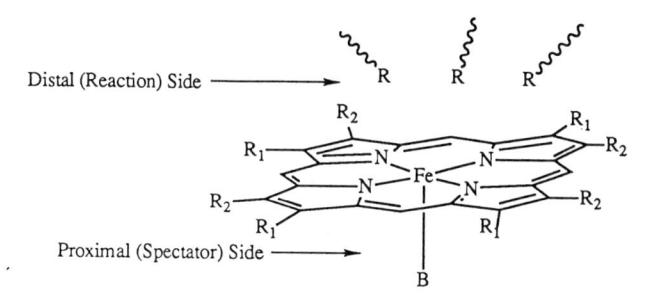

**FIGURE 3-2** Heme protein active sites. B = spectator base: imidazole, phenoxide anion, or mercaptide anion.

and activation of dioxygen. Emphasis is placed on the mechanistic description of two aspects in particular: (1) how the affinities and dynamics of dioxygen binding are controlled and (2) the mechanisms by which substrates are catalytically oxidized by the O—O bond in dioxygen or hydrogen peroxide.

This chapter is not intended to be a comprehensive review of biological dioxygen utilization. Many thorough reviews are referenced (Collman, 1977; Traylor, 1981; Guengerich and MacDonald, 1984; David et al., 1986; Momenteau, 1986; Ortiz de Montellano, 1986a; Mansuy, 1987, 1990; Bruice, 1991; Meunier, 1992). Rather, we concentrate on those chemical studies that have been directed at the mechanistic aspects of dioxygen transport and activation.

## DIOXYGEN CARRIERS

### Myoglobin

The simplest and one of the most thoroughly studied dioxygen-binding heme proteins is myoglobin (Antonini and Brunori, 1971; Stryer, 1988), a globular protein of 18,000 daltons having one protoheme molecule in a "pocket" in which the two carboxylic acid groups reach just to the surface of the protein. The heme is held in the pocket by a single bond between the iron and the nitrogen of a histidine residue, along with several van der Waals contacts and hydrogen bonds. This molecule carries out the simple process shown in eq. (1).

$$\text{Im·Fe}^{II} + O_2 \rightleftharpoons \text{Im·Fe·O–O} \qquad (1)$$

The myoglobins in different species vary somewhat in protein sequence and binding properties. However, the general behavior of, e.g., sperm whale, human, and horse myoglobins are very similar (see Table 3-1, later, for the binding characteristics of heme proteins).

### Hemoglobins

Generally, hemoglobins (Perutz, 1979) are aggregates of single heme proteins. For example, human hemoglobin is a tetramer of myoglobin-like monomers held together by salt bridges so that the heme molecules are about 300 nm apart (Shaanan, 1988). Other hemoglobins exist as dimers or as large aggregates. The aggregation of heme protein monomers allows cooperative interaction between the hemes, which results in efficient loading and unloading of dioxygen (eq. 2) by having the affinity increase with increasing oxygenation of hemoglobin ($K_4 > K_1$) (Baldwin, 1975; Parkhurst, 1979).

$$Hb + O_2 \overset{K_1}{\rightleftharpoons} HbO_2 \overset{K_2}{\rightleftharpoons} Hb(O_2)_2 \overset{K_3}{\rightleftharpoons} Hb(O_2)_3 \overset{K_4}{\rightleftharpoons} Hb(O_2)_4 \qquad (2)$$

The lower-affinity deoxyhemoglobin with "domed" (nonplanar) heme groups is called the *T-state* and the higher-affinity oxyhemoglobin with planar heme groups is called the *R-state*.

## Peroxidases

The most thoroughly studied menber of this class, horseradish peroxidase (Saunders et al., 1964; Dunford, 1991; Everse et al., 1991; Ortiz de Montellano, 1992), comprises isozymes which, like myoglobin, are single globular proteins having one protohemin per molecule. Like cytochrome *c* peroxidase (Bosshard et al., 1991), this peroxidase converts hydrogen peroxide to water by oxidizing a substrate. With horseradish peroxidase, the substrate is typically an organic compound such as phenol and, with cytochrome *c* peroxidase, as the name suggests, the substrate is ferrous cytochrome *c*. Typical reactions are shown in eqs. (3) and (4).

$$H_2O_2 + HO-C_6H_4-OH \xrightarrow{HRP} O=C_6H_4=O + 2H_2O \qquad (3)$$

$$2H^+ + H_2O_2 + 2\,Cyt\,c \xrightarrow{CCP} 2\,Cyt\,c^+ + 2H_2O \qquad (4)$$

Although carrying out processes very different from that of myoglobin, these peroxidases have structures (Adediran and Dunford, 1983; Finzel et al., 1984; Penner-Hahn et al., 1986; Edwards et al., 1987) around the heme rather similar to that of myoglobin.

## Catalase

Catalases (Schonbaum and Chance, 1976), like peroxidases, react with hydrogen peroxide (eq. 5) (Frew and Jones, 1984; Fita and Rossmann, 1985). There are two significant differences. Unlike peroxidases, catalases exclusively disproportionate hydrogen peroxide, and the proximal ligand on the iron is a phenolate from a tyrosine.

$$2\ H_2O_2 \xrightarrow{\text{Catalases}} 2\ H_2O + O_2 \tag{5}$$

### Cytochromes

The heme in these proteins (Mathews, 1985), which function in electron transport (Golly et al., 1988; Davies and Lawther, 1989; Whitford et al., 1991), is usually six-coordinated. The most common examples—cytochrome $c$ (Dickerson, 1972, 1980; Lemberg, 1973; Salemme, 1977; Moore and Pettigrew, 1990) and cytochromes $b_5$ (Hildebrandt and Estabrook 1971; Noshiro et al., 1981; Pompon and Coon, 1984; Altman et al., 1989)—have molecular weights of about 13,000 daltons, with one heme per molecule. The two ligands to iron in cytochrome $c$ are the imidazole ring of histidine and the thioether of methionine, whereas the iron in cytochrome $b_5$ binds two histidine residues. These heme proteins usually act as oxidizing or reducing agents for other metalloproteins (eq. 6) and do not enter into covalent iron chemistry (eq. 1).

$$\text{Cyt } c^+ + S \rightarrow S^+ + \text{Cyt } c \tag{6}$$

### Cytochrome Oxidase

This, the most complex heme protein, catalyzes the reduction of dioxygen to water. To achieve this reaction and supply the energy for other biological processes, this large membranous protein uses two heme molecules and two copper atoms. The considerable progress made toward understanding this complex process is reviewed in detail elsewhere (Wikström et al., 1981; Scott, 1989; Capaldi, 1990a; Chan and Li, 1990; Malmström, 1990b; Babcock et al., 1992; Babcock and Wikström, 1992; Stern and Groves, 1992) and will not be discussed here.

### Cytochromes P-450

These enzymes (Ortiz de Montellano, 1986a; Mansuy et al., 1989), named *cytochrome P-450* for the absorption maximum of the carbon monoxide complex at 450 nm, are single-protein molecules of ~45,000 daltons and contain one protohemin molecule with a cysteine thiolate group bound to the iron. In common with peroxidases, these enzymes catalyze oxidations (White and Coon, 1980; Guengerich and MacDonald, 1984, 1990; Ortiz de Montellano, 1989; Guengerich, 1991; Porter and Coon, 1991). In contrast to peroxidases, these enzymes have the $RS^-$ proximal base, use dioxygen as oxidant, and transfer oxygen atoms to almost all organic compounds (eqs. 7 and 8).

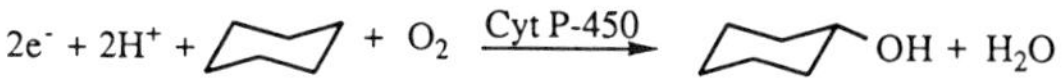

$$2e^- + 2H^+ + \text{[cyclohexene]} + O_2 \xrightarrow{\text{Cyt P-450}} \text{[cyclohexanol]} \cdot OH + H_2O \quad (7)$$

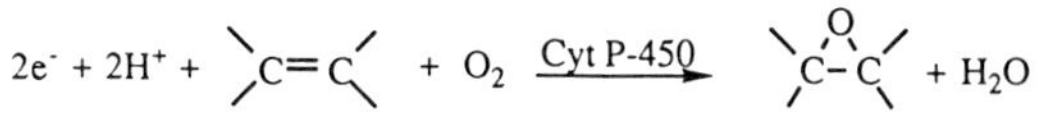

$$2e^- + 2H^+ + \overset{\textstyle\diagdown}{\underset{\textstyle\diagup}{C}}=\overset{\textstyle\diagup}{\underset{\textstyle\diagdown}{C}} + O_2 \xrightarrow{\text{Cyt P-450}} \overset{\textstyle\diagdown}{\underset{\textstyle\diagup}{C}}\overset{O}{\underset{}{\triangle}}\overset{\textstyle\diagup}{\underset{\textstyle\diagdown}{C}} + H_2O \quad (8)$$

## COMMON PROPERTIES OF HEME PROTEINS

All of the heme proteins which engage in covalent iron ligation have some common chemical properties. These are summarized in Scheme 3-1. While four of the heme proteins bind dioxygen, two of these

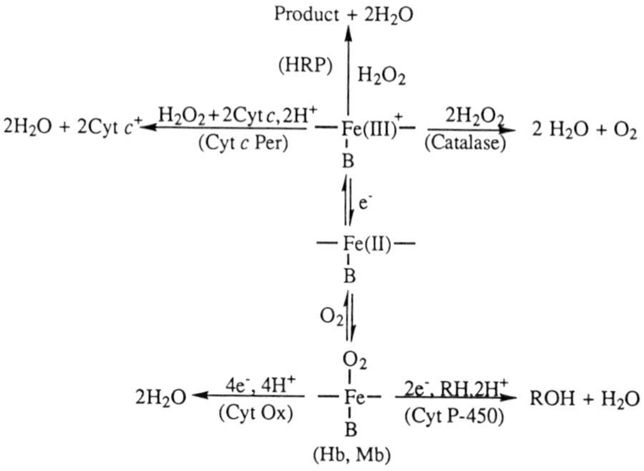

$$\begin{array}{c}
\text{Product} + 2H_2O \\
\uparrow (HRP) \bigg| H_2O_2 \\
2H_2O + 2\text{Cyt } c^+ \xleftarrow{H_2O_2 + 2\text{Cyt }c, 2H^+}_{(\text{Cyt } c \text{ Per})} -\underset{B}{Fe(III)}- \xrightarrow{\quad 2H_2O_2 \quad}_{(\text{Catalase})} 2H_2O + O_2 \\
\updownarrow e^- \\
-\underset{B}{Fe(II)}- \\
O_2 \updownarrow \\
O_2 \\
2H_2O \xleftarrow{4e^-, 4H^+}_{(\text{Cyt Ox})} -\underset{B}{Fe}- \xrightarrow{2e^-, RH, 2H^+}_{(\text{Cyt P-450})} ROH + H_2O \\
(Hb, Mb)
\end{array}$$

**Scheme 3-1**

reduce the bound dioxygen, suggesting that a two-electron reduced intermediate might be involved (eq. 9).

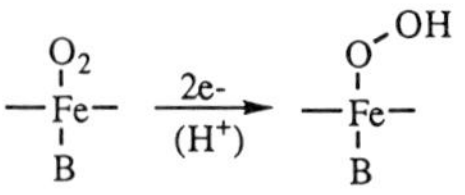

$$-\underset{B}{\overset{O_2}{Fe}}- \xrightarrow{\quad 2e^- \quad}_{(H^+)} -\underset{B}{\overset{O^{-OH}}{Fe}}- \quad (9)$$

This possibility has prompted the suggestion (Dunford and Stillman, 1976; Nordblum et al., 1976; Poulos and Kraut, 1980; Fita and Rossman, 1985; Babcock et al., 1992) that all of these catalytic heme proteins have a further commonality: the iron(III) form could react with hydrogen peroxide to reach the same intermediate (eq. 10).

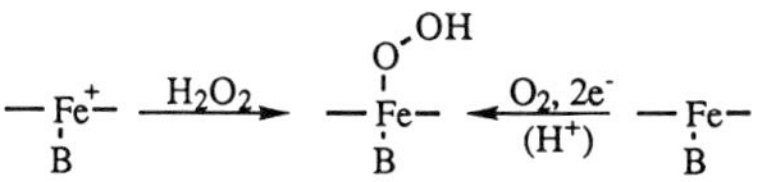

$$-\underset{B}{Fe^+}- \xrightarrow{\quad H_2O_2 \quad} -\underset{B}{\overset{O^{-OH}}{Fe}}- \xleftarrow{\quad O_2, 2e^- \quad}_{(H^+)} -\underset{B}{Fe}- \quad (10)$$

It has been attractive to carry this analogy among these five classes of enzymes to the next, and perhaps crucial, step—that of breaking the O—O bond (eq. 11) to give a strongly oxidizing species, referred to as an *oxene*. Recent results (Scott, 1989; Capaldi, 1990a; Chan and Li, 1990; Babcock and Wikström, 1992) suggest that cytochrome oxidase might be an exception, but all the other oxygen-activating heme proteins seem to fit this scheme.

$$\underset{B}{\overset{O^{OH}}{-\underset{|}{Fe}-}} \xrightarrow{HA^+} \underset{B}{\overset{O}{\underset{|}{-\overset{\|}{Fe^+}-}}} + H_2O + A \tag{11}$$

The chemistry of these latter processes will be reviewed in detail before turning to the final step, the delivery of the oxygen atom to (or removal of electrons from) the substrate (eq. 12).

$$\underset{B}{\overset{O}{-\underset{|}{\overset{\|}{Fe}}-}} \xleftarrow{\overset{S^+}{\frown}\overset{S}{\frown}} \underset{B}{\overset{O}{-\underset{|}{\overset{\|}{Fe^+}}-}} \xrightarrow{\overset{O}{\frown}\overset{S}{\frown}} \underset{B}{\overset{SO}{-\underset{|}{Fe^+}-}} \tag{12}$$

But our first objective is to examine the possible differences that might account for the incredibly different chemical properties of heme proteins. Active-site sections, taken from X-ray crystal data, are shown in Figure 3-3.

## STRUCTURE AND FUNCTION IN HEME PROTEINS

The first question that arises is why some heme proteins bind dioxygen reversibly and others do not. As an example, compare myoglobin (Figure 3-3a) with cytochrome *c* peroxidase (Figure 3-3c). The proximal bases are identical, and the active sites differ only in the presence of an arginine on the distal (reactive) side and polar groups on the proximal (spectator) side. However, the former carries dioxygen and reacts very inefficiently with hydrogen peroxide, while the peroxidase does not generally bind dioxygen reversibly.

The subtle differences in active-site structure account for very large differences in catalytic properties, as well as in the stability of the dioxygen complexes. To understand these heme proteins, a number of model reaction studies have been undertaken. As we shall see, the

*Structure and Function in Heme Proteins* 91

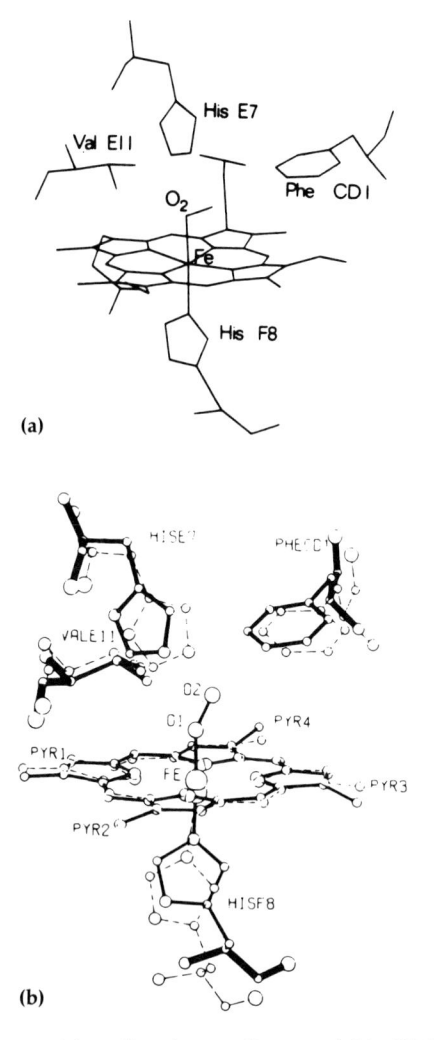

**FIGURE 3-3** Heme protein active sites. **a.** Oxymyoglobin (Phillips, 1980). **b.** α subunits of hemoglobin; HbO$_2$ (thick lines) and Hb (thin lines) (Shaanan, 1983).

heme or hemin removed from the protein still carries out all protein functions in limited ways: e.g., reversible dioxygen binding; hydrogen peroxide disproportionation; catalysis of phenol, arene, alkene, and alkane oxidations; and electron-transfer processes.

We first address the chemical studies of reversible dioxygen binding that have been directed at duplicating and understanding those

92    Reactions of Dioxygen

(c)

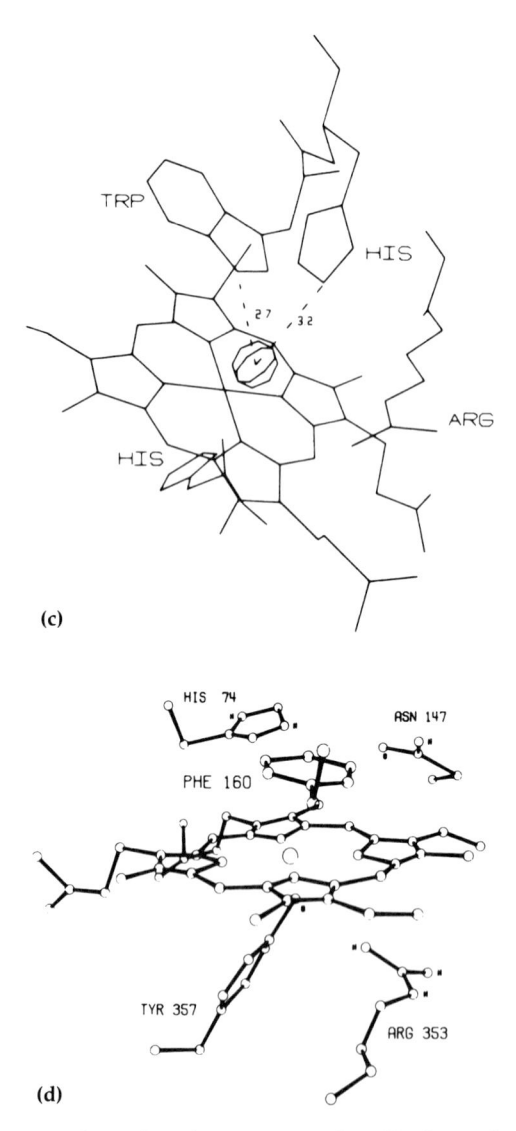

(d)

**FIGURE 3-3** (*continued*)   **c.** Cytochrome *c* peroxidase (Poulos et al., 1980). **d.** Catalase (Murthy et al., 1981).

factors controlling dioxygen and other ligand dynamics with heme proteins. Since model studies (Collman, 1977; Traylor, 1981; Traylor and Traylor, 1982) and site-directed mutagenesis (Olson et al., 1988; Morikis et al., 1989; Egeberg et al., 1990; Rohlfs et al., 1990; Stayton

*Structure and Function in Heme Proteins* 93

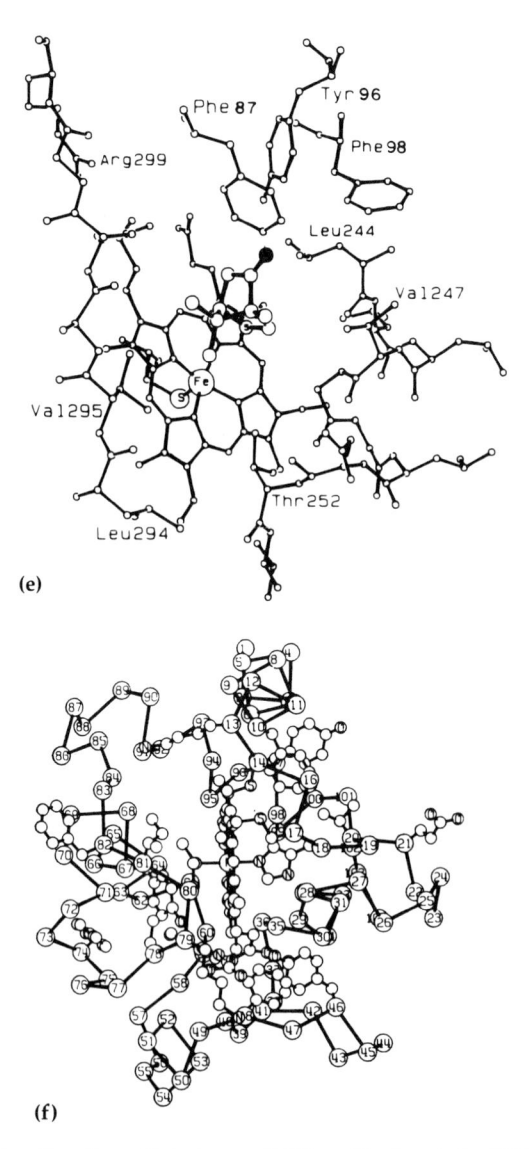

(e)

(f)

**FIGURE 3-3** *(continued)* **e.** Cytochrome P-450$_{cam}$ (Poulos et al., 1987). **f.** Cytochrome *c* (Takano et al., 1973). Figures used with permission of the authors.

and Sligar, 1990) week to determine those factors that control the dynamics and equilibria of oxygen binding in heme proteins from various organisms, we present Table 3-1, which lists dioxygen and carbon monoxide binding kinetics and equilibria observed with various

**TABLE 3-1** CO and $O_2$ Kinetic and Equilibrium Parameters for Several Heme Proteins in Water

| Heme Protein | $k^{CO}$ ($M^{-1}s^{-1}$) | $k^{-CO}$ ($s^{-1}$) | $K^{CO}$ ($M^{-1}$) | $K^{O_2}$ ($M^{-1}s^{-1}$) | $k^{-O_2}$ ($s^{-1}$) | $K_2^{O_2}$ ($M^{-1}$) |
|---|---|---|---|---|---|---|
| Hb(R-state)[a] | $6.0 \times 10^6$ | 0.009 | $6.7 \times 10^8$ | $5.9 \times 10^7$ | 12–21 | $2.8–4.9 \times 10^6$ |
| Hb(T-state) α chain[b] | $2.7 \times 10^5$ | 0.09 | $3.0 \times 10^6$ | $2.9 \times 10^6$ | 180 | $1.4 \times 10^4$ |
| Hb(T-state) β chain[b] | $2.7 \times 10^5$ | 0.09 | $3.0 \times 10^6$ | $1.2 \times 10^7$ | 2,500 | $3.9 \times 10^3$ |
| Hb, Zürich[c] | $2.2 \times 10^7$ | | | $6.5 \times 10^7$ | 34 | $1.9 \times 10^6$ |
| E7 His → Arg | | | | | | |
| Glycera Hb[d] | $2.2 \times 10^7$ | 0.055 | $2.5 \times 10^9$ | $1.9 \times 10^8$ | 2,800 | $6.8 \times 10^4$ |
| E7 His → Leu | | | | | | |
| Chironomus Hb[e] | $2.7 \times 10^7$ | 0.095 | $2.7 \times 10^8$ | $3.0 \times 10^8$ | 218 | $1.4 \times 10^5$ |
| E7 His out | | | | | | |
| LegHb[f] | $1.35 \times 10^7$ | 0.012 | $1.13 \times 10^9$ | $1.5 \times 10^8$ | 11 | $1.36 \times 10^7$ |
| Mobile E7 His | | | | | | |
| Ascaris Hb[g] | $2.1 \times 10^5$ | 0.018 | $1.2 \times 10^7$ | $1.5 \times 10^6$ | $4.1 \times 10^{-3}$ | $3.6 \times 10^8$ |
| Carp Hb[h] | $1.5 \times 10^5$ | 0.06 | $2.5 \times 10^6$ | $7.6 \times 10^6$ | 300 | $2.5 \times 10^4$ |
| Sperm whale Mb[i] | $5.0 \times 10^5$ | 0.02 | $3.1 \times 10^7$ | $1.5 \times 10^7$ | 11 | $1.3 \times 10^6$ |
| Elephant Mb[j] | $5.3 \times 10^5$ | 0.0068 | $7.8 \times 10^6$ | $1.8 \times 10^7$ | 18.4 | $9.8 \times 10^5$ |

[a] Wittenberg et al. (1972).
[b] Steinmeier and Parkhurst (1975); Sharma et al. (1976); Sawicki and Gibson (1977).
[c] Giacometti et al. (1980); Tucker et al. (1978).
[d] Parkhurst et al. (1980).
[e] Amiconi et al. (1972).
[f] Imamura et al. (1972).
[g] Gibson and Smith (1965).
[h] Noble et al. (1970).
[i] Antonini and Brunori (1971, pp. 225, 263).
[j] Romero-Herrera et al. (1981).

heme proteins. There are clearly very large differences even in these simple ligation properties.

## Designs of Model Systems

The concept in the first step of model design was as follows: excise a proposed viable active-site section from the X-ray crystal structures and, using synthetic chemistry, arrange a scaffolding to hold that section in its natural geometry. It was thought (St. George and Pauling, 1951) that such an approach—termed *active-site synthesis*—would allow a small molecule to duplicate the binding properties of the heme protein. In designing these systems, it was important to consider both the spectator and reactive sides of the heme (Figure 3-2).

The chelated heme model system concentrated on duplicating the porphyrin and the spectator ligand (Chang and Traylor, 1973b). The focus of the protected heme approach—i.e., the picket fence (Collman et al., 1973, 1975) and cyclophane (Diekmann et al., 1971; Almog et al., 1975a, 1975b; Traylor et al., 1979b, 1980b, 1981a; Battersby and Hamilton, 1980) systems—was to model and protect the heme's reactive side to prevent oxidation to $Fe^{III}$ (see eq. 15). In the chelated model approach, it was considered sufficient to have proper electronic properties in order to observe reversible dioxygen binding. Both approaches were theoretically reasonable, and both methods were used in the structure-reactivity studies which followed.

## Model Systems for Myoglobin and Hemoglobin

The first report of reversible dioxygen binding to an iron(II) porphyrin was by Wang (1958). An immobile film of polystyrene containing ferrous iron protoheme and $N$-($\beta$ phenethyl)imidazole was prepared in the hope of inhibiting bimolecular oxidation by protecting the heme (eq. 13).

$$-\underset{\underset{\underset{CH_2CH_2Ph}{N}}{\overset{|}{\underset{N}{\bigcirc}}}}{Fe}- + O_2 \xrightleftharpoons{film} -\underset{\underset{\underset{CH_2CH_2Ph}{N}}{\overset{|}{\underset{N}{\bigcirc}}}}{\overset{\overset{O_2}{|}}{Fe}}- \qquad (13)$$

The reversible spectroscopic change observed on addition of dioxygen resembled that of myoglobin oxygenation. This observation has been confirmed (Chang and Traylor, unpublished).

The first example of reversible oxygenation of a model heme compound in solution was reported in 1973 (Chang and Traylor, 1973b), using a chelated heme system similar to those described earlier by Warme and Hager (1970).

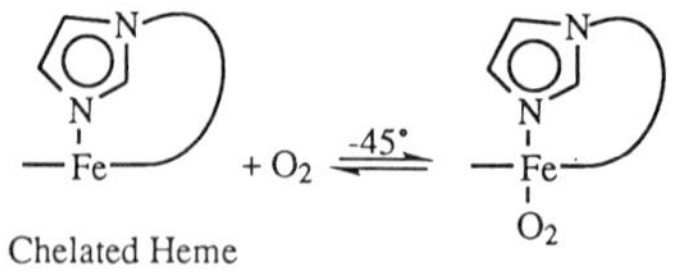

$$\text{Chelated Heme} + O_2 \overset{-45°}{\rightleftharpoons} \text{—Fe—} \quad O_2 \tag{14}$$

This finding was confirmed very shortly (Almog et al., 1974).

A crystalline iron(II) porphyrin–dioxygen complex, using meso-tetra (α,α,α,α)-*o*–pivalamidophenyl)porphyrin iron(II)-1-methylimidazole, was reported by Collman et al. (1973). This structure, sometimes called *picket-fence* porphyrin, is shown in Figure 3-4.

With these developments and the application of kinetic methods—developed earlier by Gibson (1959)—to model heme systems, it became possible to examine both the structural and dynamic processes of model systems. Structure-reactivity studies (Collman et al., 1973, 1977, 1978; Wagner and Kassner, 1974; Almog et al., 1975b; Basolo et al., 1975; Chang and Traylor, 1975a, 1975b; Weschler et al., 1975) rapidly unfolded, along with the development of site-directed mutagenesis of proteins (Morikis et al., 1989; Egeberg et al., 1990; Stayton and Sligar, 1990), making possible cross-comparison between the dynamics of model and heme protein systems.

### Mimics of Heme Proteins

Both structural and dynamic properties of heme proteins have been successfully modeled. Table 3-2 compares the binding dynamics of

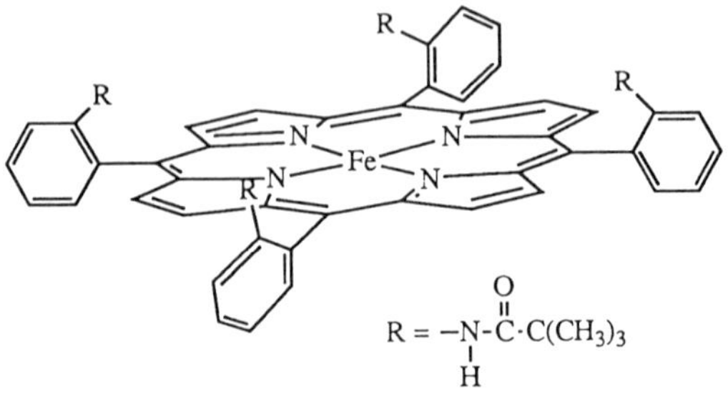

**FIGURE 3-4** *Meso*-tetra(α,α,α,α-*o*-pivalamidophenyl)porphyrin iron(II) (Collman et al., 1975).

**TABLE 3-2** Comparison of the Binding Dynamics of R-State Hemoglobin and Chelated Hemes at 20°C

| Heme | $k^{O_2} \times 10^{-7}$ $M^{-1}s^{-1}$ | $k^{-O_2}$ $s^{-1}$ | $K^{O_2} \times 10^{-6}$ $M^{-1}$ | $k^{CO} \times 10^{-6}$ $M^{-1}s^{-1}$ | $k^{-CO}$ $s^{-1}$ | $K^{CO} \times 10^{-8}$ $M^{-1}$ | $\Delta H_{CO}$ | $\Delta S_{CO}$ | $\Delta H_{O_2}$ | $\Delta S_{O_2}$ |
|---|---|---|---|---|---|---|---|---|---|---|
| Chelated protoheme[a] | 2.6 | 47 | 5.5 | 3.6 | 0.0089 | 4.0 | −17.5 | −34 | −14.0 | −35 |
| Hemoglobin (R-state)[a,b] | 5.9 | 12<br>21 | 4.9<br>2.8 | 6 | 0.006 to 0.009 | 7 to 10 | −17.4 | — | −13.5 to −14.5 | — |
| Chelated[c] picket fence heme | 43 | 2,900 | 0.15 | 36 | 0.0078 | 46 | — | — | −16.3 ±0.8 | −40 ±3 |

[a] Protoheme monomethyl ester mono-3-imidazolylpropylamide, pH 7.3, 2% MTAB (myristyltrimethylammonium bromide) suspension at 20°C (Traylor and Berzinis, 1980).
[b] pH 7, 0.05–0.1 M phosphate at 20°C (Traylor et al., 1980a).
[c] FePiv$_3$(5ClImP)Por, toluene (Collman et al., 1983b).

R-state hemoglobin with those of the model employing chelated protoheme in an aqueous suspension in cetyltrimethyl ammonium bromide and the corresponding picket fence heme in toluene. The kinetics, equilibria, and thermodynamics are very similar when the same heme is involved, leading to the conclusion that in R-state hemoglobin the protein is simply a suspending and immobilizing agent for the heme (Chang and Traylor, 1975b). Other heme proteins are very different (Table 3-1) and are not modeled by chelated protoheme under the conditions studied.

The geometries of R- and T-state hemoglobin have been modeled rather accurately with complexes of picket fence heme (Collman et al., 1978; Jameson et al., 1978, 1980) using 1-methylimidazole (R-state) and 1,2-dimethylimidazole (T-state) (Figure 3-5).

## Reversible Dioxygen Binding

When the heme is removed from myoglobin and covalently attached to an alkylimidazole to ensure a five-coordinated species, it mimics the spectroscopic properties of myoglobin (Traylor et al., 1979c). When exposed to dioxygen in solution, the heme is oxidized to the iron(III) form in less than 1 minute. This reaction is second-order in heme and inverse first-order in dioxygen, suggesting the events in eq. (15).

$$\text{—Fe—} + O_2 \rightleftharpoons \text{—Fe—OO} \rightarrow \text{—Fe—} \rightarrow \text{B—Fe—OO—Fe—B} \rightarrow \text{B—Fe=O} \rightarrow \text{B—FeOFe—B} \quad (15)$$

This process has been thoroughly documented by Balch and coworkers (Chin et al., 1977, 1980a, 1980b, 1980c; Latos-Grazynski et al., 1982).

To avoid this process and observe a reversible oxygenation sufficiently stable for titrations, spectroscopic and crystallographic measurements, etc., three approaches have been successfully implemented: (1) low-temperature (Chang and Traylor, 1973a, 1977b; Anderson et al., 1974; Weschler et al., 1975; Ansari et al., 1986); (2) steric protection of iron (Diekmann et al., 1971; Collman et al., 1973, 1975, 1980, 1983a, 1983b; Almog et al., 1975a, 1975b); and (3) fast observations (Chang and Traylor, 1975a, 1975b). These approaches were subsequently used for other systems as well.

*Structure and Function in Heme Proteins* 99

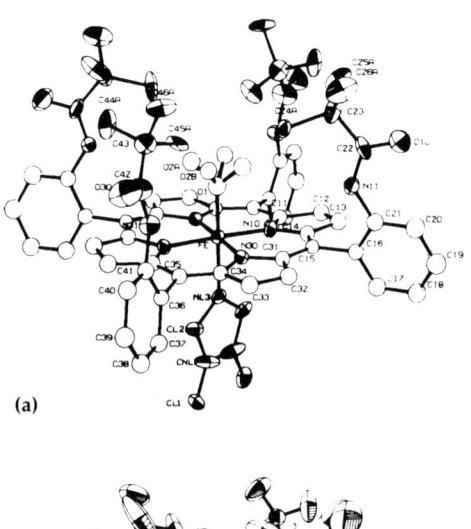

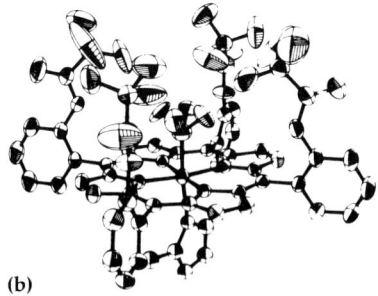

**FIGURE 3-5** Crystal structures: (a) Fe(TpivPP) (1-MeIm)(O$_2$) and (b) Fe(TpivPP) (2-MeIm)(O$_2$). R- and T-state hemoglobin models, respectively (Jameson et al., 1978, 1980). Used with permission of the author.

1. The low-temperature method relies on the fact that the oxidation process is slower than the binding of dioxygen. The oxygenation of a simple heme–base complex can be observed at room temperature, but the complex is oxidized in a few seconds. At −45 °C the oxygen complex is stable for hours in solution (Chang et al., 1977b).

2. Steric blocking of the iron in heme model compound was introduced in 1971 (Diekmann et al.), first with the cyclophane heme, then in 1973 (Collman et al.), with the more stable and versatile picket fence heme, and in 1975 (Almog et al., 1975a,b) with the capped class of cyclophane hemes. Both peripherally protected and cyclophane-protected systems were widely and effectively elaborated for many purposes in subsequent years (Baldwin et al., 1976, 1982, 1984; Chang et al., 1977a; Collman et al., 1977, 1983a, 1983b; Battersby et al., 1978; Budge et al., 1979; Traylor et al., 1979b; Ellis et al., 1980; Linard et al., 1980; Traylor,

1981; Hiom et al., 1983; Momenteau et al., 1983a, 1983b; Chang and Abdalmuhdi, 1984; Momenteau, 1986; Maillard et al., 1988, 1989; Wijesekera et al., 1988; Osuka et al., 1990; Kim and Ibers, 1991; Karaman et al., 1992; Kreysel and Voegtle, 1992; Lee and Lee, 1992; Nagata, 1992; Reddy and Chandrashekar, 1992). Oxy–picket fence heme remains the most stable oxygen complex and is generally preferred for structural studies.

3. The fast observation method (Chang and Traylor, 1975a, 1975b), sometimes referred to as the *kinetic method*, is based, like the low-temperature studies, on the observation that oxygenation typically occurs with on and off rate constants (at 1 atm dioxygen) in excess of 100 sec$^{-1}$, whereas oxidation is much slower. Therefore, both equilibrium and kinetic measurements can be made. The rates of oxygen and carbon monoxide binding to adamantane-6,6-cyclophane heme-1,5-DCI in toluene (Traylor et al., 1984a) are as follows: $k_B^{O_2} = 1.5 \times 10^5 \, M^{-1} s^{-1}$; $k_B^{-O_2} = 690 \, s^{-1}$; $k_B^{CO} = 9.2 \times 10^3 \, M^{-1} s^{-1}$; and $k_B^{-CO} = 0.05 \, s^{-1}$. Thus, a stable complex of the BHmCO (i.e. B-heme-CO) complex can be made in the presence of an equimolar mixture of CO and $O_2$ and flashed to remove the CO. Subsequently, the equilibrium is rapidly established (eq. 16).

$$O_2 + BHmCO \underset{k_B^{-CO}}{\overset{h\nu}{\rightleftharpoons}} O_2 + BHm + CO \underset{k_B^{-O_2}}{\overset{k_B^{O_2}}{\rightleftharpoons}} BHmO_2 + CO \qquad (16)$$

Figure 3-6 shows a series of spectra taken after such a flash. Observing the ratio of the absorbances at $\lambda_{max}$ for BHmO$_2$ (i.e. B-Heme-O$_2$) and BHm (i.e. B-Heme) allows the equilibrium constant to be calculated by the usual (not kinetic) method. Observation of the changes in absorbances versus time in the same experiment gives the individual rate constants. Using this procedure, the dioxygen equilibrium can be simultaneously measured by two independent methods for any heme, and steric protection becomes unnecessary. This allows a great deal of freedom in designing structure-reactivity studies. Most of the structure-reactivity studies have employed the steric protection (Collman et al., 1973, 1975, 1976, 1981; Jameson et al., 1978, 1980) and/or fast observation methods (Chang et al., 1975a; Traylor and Berzinis, 1980; Traylor et al., 1981b). It should be noted that steric protection necessarily introduces some ambiguity into structure-reactivity studies since steric protection itself is a structural change.

In summary, although reversible oxygenation of heme complexes in solution at first seemed impossible, it soon became clear that the commercially available protohemin chloride could be mixed with imidazole and reversibly oxygenated at room temperature using method (1) or (3) described above (Chang and Traylor, 1975a).

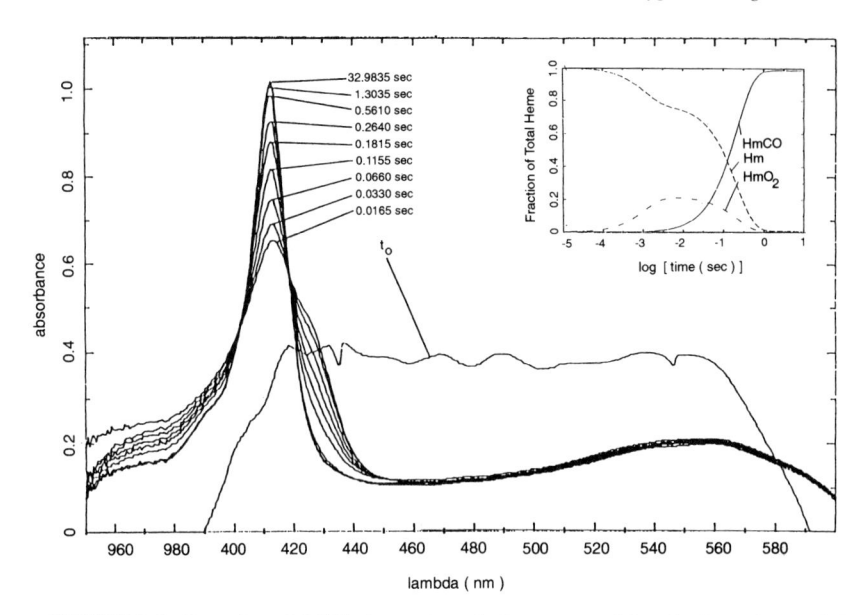

**FIGURE 3-6** Experimental UV-vis spectra of adamantane-6,6-cyclophane heme-(1,5-DCI = 1,5-dicyclohexylimidazole) recorded following photolysis of the CO complex. Spectra labeled with times relative to $t_0$. Inset: concentrations vs. log time. Conditions: pCO = 64.7 torr, pO$_2$ = 109 torr, 20 °C, toluene, [1,5-DCI] = 0.5 M (Lopez, 1987).

## STRUCTURAL AND ENVIRONMENTAL EFFECTS ON DIOXYGEN BINDING

### Electronic Effects

Electron-donating groups on the porphyrin ring (Traylor et al., 1981b) or on the spectator base (Lavalette et al., 1979; Lavalette and Momenteau, 1982; Mispelter et al., 1983; Momenteau and Lavalette, 1983a) increase oxygen affinity, principally by slowing the dissociation rates (Traylor et al., 1981b). The kinetics and equilibria of dioxygen and carbon monoxide binding to some chelated iron porphyrins are shown in Table 3-3. Included are both pyridine and imidazole bases. These data show that electronic effects are rather large for dioxygen binding but small for carbon monoxide binding. This electronic effect constitutes one explanation for the differing behavior of different heme proteins toward these ligands.

**TABLE 3-3** Reaction of Chelated Hemes with Carbon Monoxide and Dioxygen[a,b]

| Compound (Substituent) | pK₃[c] | $10^{-6}k^{CO}$ $M^{-1}s^{-1}$ | $k^{-CO}$ $s^{-1}$ | $10^{-8}K^{CO}$ $M^{-1}$ | $10^{-7}k^{O_2}$ $M^{-1}s^{-1}$ | $k^{-O_2}$ $s^{-1}$ | $10^{-5}K^{O_2}$ $M^{-1}$ |
|---|---|---|---|---|---|---|---|
| Chelated mesoheme (ethyl) | 5.8 | 8.2 | 0.014 | 5.8 | 3.5 | 22 | 16 |
| Chelated protoheme (vinyl) | 4.8 | 3.6 | 0.005 | 7.2 | 2.6 | 47 | 5.5 |
| Chelated diacetyldeuterohemе (acetyl) | 3.3 | 5.6 | 0.0085 | 7.0 | 3.4 | 400 | 0.85 |
| FeTPP-pyridine[d] | — | 6.2 | — | — | 5.9 | 110,000 | 0.0065 |
| Pyridine chelated mesoheme[e] | — | 12 | — | — | 1.7 | 380 | 0.0045 |

[a] pH 7.3, 2% MTAB, 0.05–0.10 potassium buffer, 20 °C.
[b] Traylor et al. (1981b).
[c] pK for the addition of a proton to a neutral porphyrin ring (Hambright, 1975).
[d] In toluene. FeTPP = tetraphenylporphyrin (Lavalette and Momenteau, 1982).
[e] pH 7.3, 2% CTAB (cetyltrimethylammonium bromide), 22 °C (Traylor et al., 1979c).

## Solvent and Internal Polar Effects

Early studies (Brinigar et al., 1974; Traylor et al., 1979a) of dioxygen binding to chelated protoheme in polar and nonpolar solvents such as dimethylformamide and toluene, as well as studies in aqueous suspension, revealed a very large stabilization of the dioxygen complex under polar conditions. This effect has also been demonstrated with the highly protected 5,10,15,20-tetrakis(2,4,6-triphenylphenyl)porphinato iron(II) complex (Suslick and Fox, 1983). Such stabilization of dioxygen complexes and destabilization of carbon monoxide complexes by polar solvents is illustrated in Table 3-4. These observations led to the suggestion (Brinigar et al., 1974) that the dipolar nature of the $Fe^{\delta+}-O_2^{\delta-}$ complex causes interaction with polar solvents, whereas the Fe–CO complex, having a much lower charge separation, is not as stabilized by a polar environment.

Stabilization of the dioxygen complex can also be enhanced by polar groups covalently bound near the reactive iron center. For example, compare the dioxygen and carbon monoxide affinities for the picket-fence heme (FeTpivPP)-1,2-dimethylimidazole and the 5,10,15,20-tetrakis(2,4,6-triphenylphenyl porphinato)(FeTTPPP)-1,2-dimethylimidazole complexes (Table 3-4). These two complexes are rather similar with regard to steric effects but vastly different in local polarity due to the presence of four amide groups in the picket-fence heme. Again, dioxygen affinity is increased by the polar environment but CO affinity is not.

This local polar effect, as well as a direct hydrogen-bonding effect, has been quantitatively demonstrated in a study of the compounds shown in Figure 3-7 (Chang et al., 1988). Hydrogen bonding and local polar effects have also been demonstrated in studies of the cyclophanes (Momenteau et al., 1983a, 1983b; Momenteau, 1986) shown in Figure 3-8, where the binding constants for dioxygen are given in parentheses as $P_{1/2}^{O_2}$ in toluene solution. All of these studies indicate that dioxygen affinities are greatly increased by polar environments and by hydrogen bonding to neighboring XH groups.

## STERIC EFFECTS

### Proximal (Spectator) Base Steric Effects— The Hemoglobin T-State

The two superimposed views of hemoglobin shown in Figure 3-9 (also see Figure 3-3) illustrate the differences between its T-state

**TABLE 3-4** Binding of Dioxygen and Carbon Monoxide by Hemes and Heme Proteins in Several Solvents

| System | Solvent | $P^{CO}_{1/2}$ (torr) | $P^{O_2}_{1/2}$ (torr) | $\dfrac{P^{O_2}_{1/2}}{P^{CO}_{1/2}}$ | Reference |
|---|---|---|---|---|---|
| Chelated protoheme | H$_2$O$^a$ | 0.0018 | 1.0 | 560 | b |
|  | Benzene | 0.00023 | 5.6 | 24,000 | c |
| Chelated mesoheme | Tol/10% CH$_2$Cl$_2$ | 0.0005 | 2.8 | 5,600 | b |
|  | Toluene |  | 4.9 |  | d |
| Hb (R-state) | H$_2$O (pH 7) | 0.001 | 0.16 | 200 | e |
| Mb | H$_2$O (pH 7–7.5) | 0.024 | 0.42 | 17 | f |
| FeTTPPP (1,2-Me$_2$Im) | Mesitylene | 0.0080 | 640 | 80,000$^g$ | h |
|  | Toluene | 0.0091 | 508 | 55,800$^g$ | h |
|  | Benzene | 0.0092 | 473 | 51,400$^g$ | h |
|  | Chlorobenzene | 0.012 | 299 | 24,900$^g$ | h |
|  | o-Dichlorobenzene | 0.016 | 227 | 14,200$^g$ | h |
| FeTpivPP (1,2-Me$_2$Im) | Toluene | 0.0089 | 38 | 4,280 | i |
| FeTPP (1,2-Me$_2$Im) | Mesitylene | 0.23 | — | — | h |
|  | Toluene | 0.27 | — | — | h |
|  | Chlorobenzene | 0.38 | — | — | h |
|  | o-Dichlorobenzene | 0.68 | — | — | h |

$^a$ 2% aqueous CTAB or MTAB.
$^b$ Traylor et al. (1979c).
$^c$ Traylor et al. (1981a).
$^d$ Traylor et al. (1984a).
$^e$ Wittenberg et al. (1972).
$^f$ Antonini and Brunori (1971, pp. 225, 263).
$^g$ Suslick and Fox (1983).
$^h$ Suslick et al. (1984).
$^i$ Collman et al. (1976, 1979).

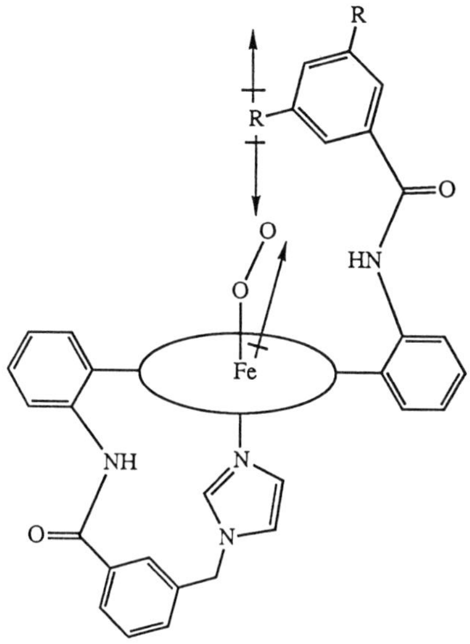

**FIGURE 3-7** Relative orientation of FeO–O dipole and the dipole of a *meta*-substituted benzamide (Chang et al., 1988).

(completely unligated) and its R-state (partially saturated) forms, showing that the protein moves toward an enforced doming of the heme in the T-state, whereas in the R-state the heme is not restricted. This kind of steric deformation has been introduced into model systems in two ways. The use of 2-methylimidazole or 1,2-dimethylimidazole instead of imidazole or 1-methylimidazole destabilizes the planar heme geometry as a result of the steric interaction of the 2-methyl group with the porphyrin nitrogens (Rougée and Brault, 1975; Collman et al., 1978; White et al., 1979). The same effect results from shortening of the chain in a chelated heme (Geibel et al., 1978). This tilts the imidazole and introduces strain in the iron–imidazole bond. The effects of these kinds of strain on the kinetics and equilibria of binding are shown in Table 3-5, along with data for R- and T-state hemoglobin.

These data confirm the postulate that such strain decreases association rates and increases dissociation rates, with a consequent decrease in the affinities of both CO and $O_2$, suggesting that this effect operates in the bond-making or -breaking step. This is discussed below.

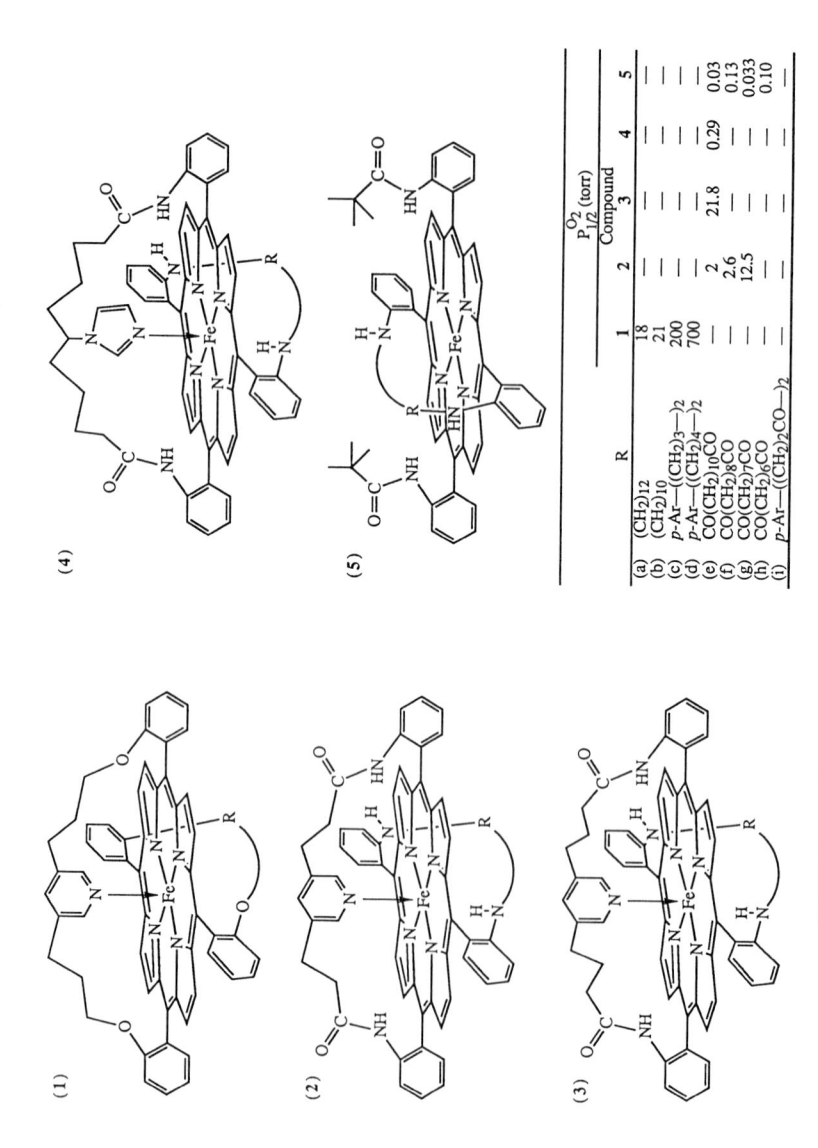

**FIGURE 3-8** Series of amide basket-handle hemes (Momenteau, 1986).

| | R | $P_{1/2}^{O_2}$ (torr) Compound | | | | |
|---|---|---|---|---|---|---|
| | | 1 | 2 | 3 | 4 | 5 |
| (a) | $(CH_2)_{12}$ | 18 | — | — | — | — |
| (b) | $(CH_2)_{10}$ | 21 | — | — | — | — |
| (c) | $p$-Ar—$((CH_2)_3$—$)_2$ | 200 | — | — | — | — |
| (d) | $p$-Ar—$((CH_2)_4$—$)_2$ | 700 | — | — | — | — |
| (e) | $CO(CH_2)_{10}CO$ | — | 2 | — | — | 0.03 |
| (f) | $CO(CH_2)_8CO$ | — | 2.6 | 21.8 | 0.29 | 0.13 |
| (g) | $CO(CH_2)_7CO$ | — | 12.5 | — | — | 0.033 |
| (h) | $CO(CH_2)_6CO$ | — | — | — | — | 0.10 |
| (i) | $p$-Ar—$((CH_2)_2CO$—$)_2$ | — | — | — | — | — |

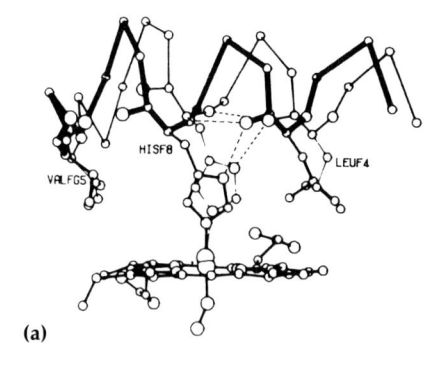

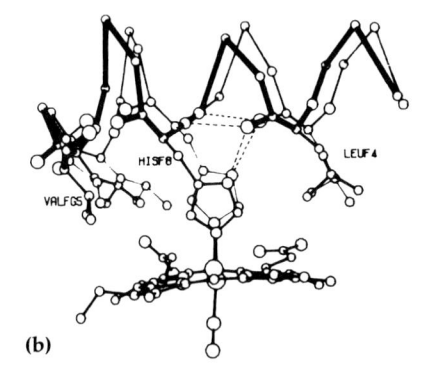

**FIGURE 3-9** Superimposed view of hemoglobin R- and T-states. HbO$_2$ is represented by thick lines, Hb by thin lines in the α (**a**) and β (**b**) subunits (Shaanan, 1983). Used with permission of the author.

## Ortho Effect

Although not directly related to heme protein behavior, this ortho effect is of some importance since many model compounds are affected (Collman et al., 1976, 1983b; Jones et al., 1977; Traylor and Traylor, 1982; Lee et al., 1984; Momenteau, 1986; Mansuy, 1987; Portela et al., 1993). When tetraphenylhemes have large substituents in the 2- and 6-positions of the phenyl group, they do not bind ligands less strongly—as expected from steric effects—but more strongly. The effect, which appears to be general for both iron(II) and iron(III) porphyrins (Nakamura, 1989), is illustrated by comparing the binding of the second 1,2-dimethylimidazole molecule to different hemes (eq. 17).

## Reactions of Dioxygen

**TABLE 3-5** Steric Strain Effects on Kinetic and Equilibrium Data for Several Heme Compounds[a]

| Heme Compound | $k^{CO}$ ($M^{-1}s^{-1}$) | $k^{-CO}$ ($s^{-1}$) | $P^{CO}_{1/2}$ (torr) | $k^{O_2}$ ($M^{-1}s^{-1}$) | $k^{-O_2}$ ($s^{-1}$) | $P^{O_2}_{1/2}$ (torr) | $\dfrac{P^{O_2}_{1/2}}{P^{CO}_{1/2}}$ |
|---|---|---|---|---|---|---|---|
| HbA, R-state[b,c,d] [0.05–0.1 M KP$_i$] $\alpha$ | $4.6 \times 10^6$ | 0.009 | 0.0014[e] | $3.3 \times 10^7$ | 13.1 | 0.22 | 150 |
| $\beta$ | — | — | — | — | 12.1 | 0.36 | 190 |
| Hb, T-state[b] [0.1 M KP$_i$] $\alpha^d$ | $2.2 \times 10^5$ | 0.09 | 0.30 | $2.9 \times 10^6$ | 180 | 40 | 135 |
| $\beta^f$ | $2.2 \times 10^5$ | 0.09 | 0.30 | $1.18 \times 10^7$ | 2,500 | 140 | 460 |
| Chelated mesoheme[g,h,i] | $1.04 \times 10^7$ | 0.05 | 0.00049[e] | $5.3 \times 10^{7i}$ | 1,700[i] | 2.8[e] | 5,700[e] |
| —NH(CH$_2$)$_3$-Im[j] | $1.1 \times 10^7$ | — | — | $2.2 \times 10^7$ | 23 | — | — |
| —NH(CH$_2$)$_4$-Im[j] | $1.5 \times 10^7$ | — | — | $2.9 \times 10^7$ | 24 | — | — |
| —NH(CH$_2$)$_3$-(2-MeIm)[j] | $12 \times 10^7$ | — | — | $k$ | — | — | — |
| DHD(Im)[l] | $1.2 \times 10^7$ | 0.028 | — | — | — | — | — |
| DHD(2-MeIm) ($3 \times 10^{-4}$ to $1 \times 10^{-3}$)[l] | $9 \times 10^{5e,g}$ | 0.45[e,g] | 0.046[g] | — | — | — | — |
| FePiv$_3$5CIm[m,n] | $3.6 \times 10^7$ | 0.0078[o] | 0.000022[p] | $4.3 \times 10^8$ | 2,900 | 0.58[q] | 26,600[q] |
| FeTPiv P[m,n] | $1.4 \times 10^6$ | 0.14[o] | 0.0089[p] | $1.06 \times 10^8$ | 46,000 | 38[m] | 4,280[m] |
| FePocPiv(1-MeIm)[m,n] [0.1–1 M][m,n] | $5.8 \times 10^5$ | 0.0086[o] | 0.0015[p] | $2.2 \times 10^6$ | 9[e] | 0.36[q] | 270[q] |
| FePocPiv[m,n] (1,2-Me$_2$Im) [0.3–1 M] | $9.8 \times 10^4$ | 0.055[o] | 0.067[q,r] | $1.9 \times 10^6$ | 280[s] | 12.6[q] | 216[q] |

| FeMedPoc$^{m,n}$ (1-MeIm) [0.1–0.5M] | 1.5 × 10$^6$ | 0.0094$^o$ | 0.00065$^p$ | 1.7 × 10$^7$ | 71$^e$ | 0.36$^{q,t}$ | 550$^q$ |
| FeMedPoc$^{m,n}$ (1,2-Me$_2$Im) [0.3–0.5M] | 2.1 × 10$^5$ | 0.053$^o$ | 0.026$^q$ | 5.2 × 10$^6$ | 800 | 12.4$^q$ | 480$^q$ |

[a] For FeTPP4ClIm (R-state model), $P_{1/2}^{CO} = 10^{-3}$ $^r$; for FeTPP(1,2-Me$_2$Im) (T-state model), $P_{1/2}^{CO} = 0.15$. $^r$ (4Cl$_m$ = 4-cyclohexylimidazole).
[b] Aqueous, pH 7–7.4, 20°C.
[c] Gibson (1970), Olson et al. (1971).
[d] Steinmeier and Parkhurst (1975), Sharma et al. (1976).
[e] Derived from kinetic measurements.
[f] Sawicki and Gibson (1977).
[g] 20°C, benzene.
[h] Traylor et al. (1979c).
[i] 9:1 Toluene:CH$_2$Cl$_2$, 22°C.
[j] Geibel et al. (1978).
[k] This compound was very quickly oxidized, even in the presence of CO.
[l] DHD = deuteroheme dimethylester; Rougée and Brault (1975), White et al. (1979).
[m] 25°C, toluene. (5Cl$_m$ = 5-cyclohexylimidazole).
[n] Collman et al. (1983b).
[o] Calculated from $P_{1/2}^{CO}$ and $k_B^{CO}$.
[p] Calculated from $P_{1/2}^{O_2}$ and M value.
[q] Measured by the flow technique.
[r] Measured with the tonometer.
[s] Calculated from $P_{1/2}^{O_2}$ and $k_B^{O_2}$.
[t] ±20% error.

$$\text{---Fe---} + \underset{\underset{Me}{N}}{\overset{\overset{Me}{N}}{\bigvee}}\text{-Me} \underset{}{\overset{K_1}{\rightleftharpoons}} \text{-Fe-} \underset{}{\overset{K_2}{\rightleftharpoons}} \text{-Fe-} \qquad (17)$$

Whereas $K_1 = 3 \times 10^4 \text{M}^{-1}$ for both tetraphenyl heme and tetramesityl heme, $K_2 = 2.5 \text{M}^{-1}$ for tetraphenylheme and $850 \text{M}^{-1}$ for tetramesityl heme (Portella, 1987). In addition, the heme-1,2-DMI complex of tetramesityl heme binds CO 15 times more strongly than does the same complex of tetraphenyl heme. This effect, like the T-state effect (Collman, 1978; Geibel et al., 1978), is reflected in the dissociation rate. Thus, it occurs in the bond-breaking step. The ortho effect is believed to result from destabilization of the domed form of the heme by the ortho substituents.

### Distal (Reactive) Side Steric Effects

X-ray crystal structures of heme proteins reveal that when the ligand ($O_2$, CO, etc.) is absent, a nearby side chain occupies some of the space required by the bound ligand (Alberding et al., 1978b; Tucker et al., 1978). This interferes with ligand rebinding (Perutz, 1979). Thus, the binding rate constants for carbon monoxide or dioxygen (Table 3-1) decrease greatly in the following order: hemoglobin (R-state) > myoglobin > horseradish peroxidase, in accord with the crystal structure description of the sizes of the heme pockets.

### Separation of Distal (Reactive Side), Proximal (Spectator Side) Steric and Polar Effects

In order to separate distal from proximal steric effects in proteins, cyclophanes of various sizes were prepared (Traylor et al., 1979b, 1980b, 1981a, 1982, 1985b; Collman et al., 1983a; Momenteau et al., 1983a,b; Momenteau, 1986). Examples are shown in Figures 3-8, 3-10, and 3-11.

Kinetic and equilibrium data for these cyclophanes are shown in Table 3-6. Several conclusions can be consistently drawn from this table. First, as the cyclophane size gets smaller, the association rates and equilibrium constants for carbon monoxide binding decrease. Secondly, carbon monoxide dissociation rates change vary little. These

*Steric Effects* 111

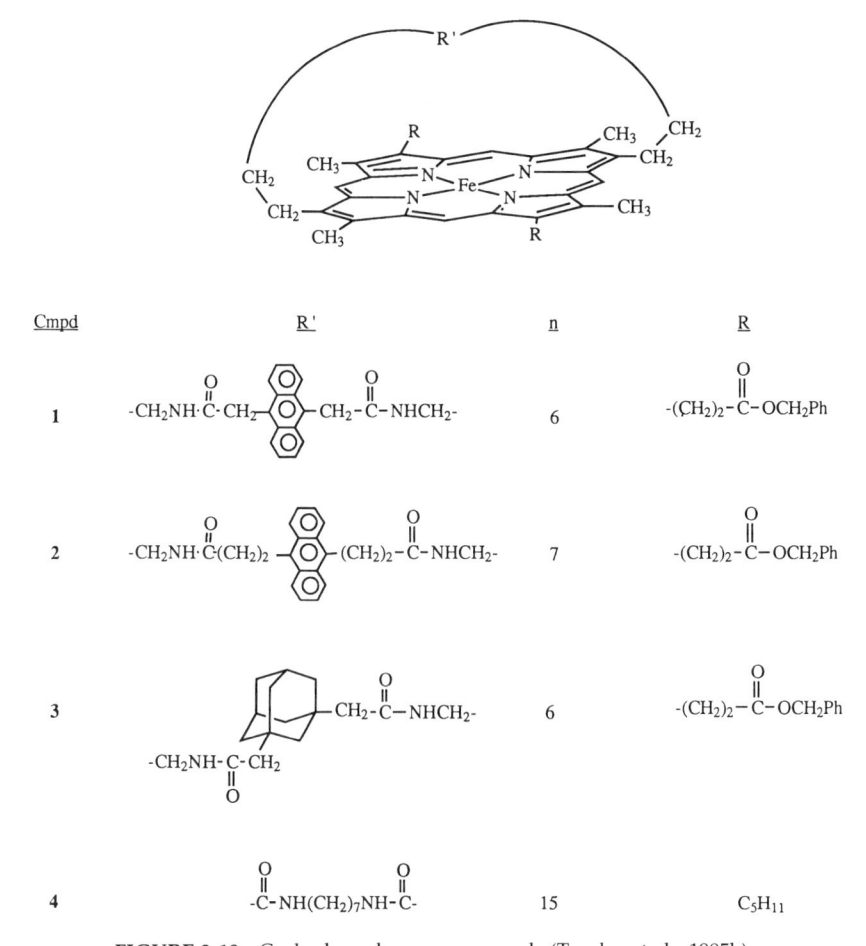

**FIGURE 3-10** Cyclophane heme compounds (Traylor et al., 1985b).

| Cmpd | R' | n | R |
|---|---|---|---|
| 1 | -CH$_2$NH-CO-CH$_2$-(anthracene)-CH$_2$-CO-NHCH$_2$- | 6 | -(CH$_2$)$_2$-CO-OCH$_2$Ph |
| 2 | -CH$_2$NH-CO-(CH$_2$)$_2$-(anthracene)-(CH$_2$)$_2$-CO-NHCH$_2$- | 7 | -(CH$_2$)$_2$-CO-OCH$_2$Ph |
| 3 | -CH$_2$NH-CO-CH$_2$-(adamantyl)-CH$_2$-CO-NHCH$_2$- | 6 | -(CH$_2$)$_2$-CO-OCH$_2$Ph |
| 4 | -CO-NH(CH$_2$)$_7$NH-CO- | 15 | C$_5$H$_{11}$ |

results suggest that the steric effects are realized in the approach to the iron site rather than in the bond-making or -breaking step. This interpretation is confirmed in the following section. Thirdly, changes in the association rates for dioxygen binding accurately parallel those for carbon monoxide, strongly suggesting that carbon monoxide and dioxygen are equally affected by steric restraints. This conclusion would be demanded if the steric effect were on a diffusion process, as we discuss below.

Because of varying dioxygen dissociation rates, several interpretations were presented. In one case, the anthracene cyclophanes (Traylor

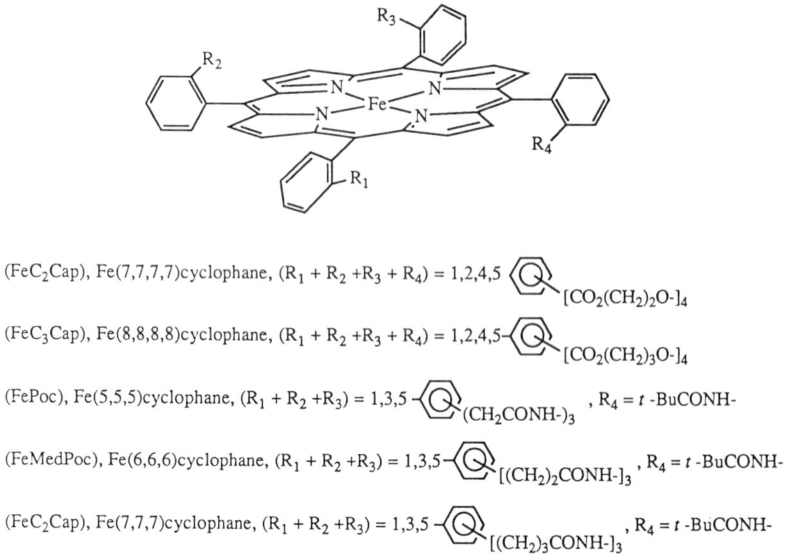

112    Reactions of Dioxygen

(FeC$_2$Cap), Fe(7,7,7,7)cyclophane, (R$_1$ + R$_2$ +R$_3$ + R$_4$) = 1,2,4,5 –⟨Q⟩–[CO$_2$(CH$_2$)$_2$O-]$_4$

(FeC$_3$Cap), Fe(8,8,8,8)cyclophane, (R$_1$ + R$_2$ +R$_3$ + R$_4$) = 1,2,4,5 –⟨Q⟩–[CO$_2$(CH$_2$)$_3$O-]$_4$

(FePoc), Fe(5,5,5)cyclophane, (R$_1$ + R$_2$ +R$_3$) = 1,3,5 –⟨Q⟩–(CH$_2$CONH-)$_3$, R$_4$ = $t$-BuCONH-

(FeMedPoc), Fe(6,6,6)cyclophane, (R$_1$ + R$_2$ +R$_3$) = 1,3,5 –⟨Q⟩–[(CH$_2$)$_2$CONH-]$_3$, R$_4$ = $t$-BuCONH-

(FeC$_2$Cap), Fe(7,7,7)cyclophane, (R$_1$ + R$_2$ +R$_3$) = 1,3,5 –⟨Q⟩–[(CH$_2$)$_3$CONH-]$_3$, R$_4$ = $t$-BuCONH-

**FIGURE 3-11**  Capped and pocket hemes (Almog et al., 1975a,b; Collman et al., 1983a).

et al., 1985b), the dissociation rate is decreased compared to an open model in the same nonpolar solvent, but the on-rate is independent of the steric effect. In the pocket hemes (Collman et al., 1981, 1983a, 1983b; Suslick and Fox, 1983), the dioxygen dissociation rate *decreases* as the steric effect increases, although the dissociation rate for CO is essentially unchanged. In the case of severe steric hindrance, CO dissociation rates increase greatly. The polarity effect discussed above is an alternative explanation for these findings. As the pockets become smaller, the amide groups are brought closer to the polar Fe$^+$—OO bond, thereby stabilizing the bound state by both dipolar and possibly hydrogen bonding effects (Figure 3-12) (Traylor et al., 1981b; Momenteau and Lavalette, 1982; Suslick and Fox, 1983).

One example of deliberately increasing the polarity in a hindered cyclophane, 3,5-pyridine-5,5-heme cyclophane (Figure 3-13), illustrates this effect in the extreme (Traylor et al., 1985a). The dioxygen dissociation rate is decreased to ~35 sec$^{-1}$, compared to 80,000 sec$^{-1}$ for an equally hindered cyclophane having no polar groups. This demonstrates chemical control of dioxygen binding beyond those observed in heme proteins.

An additional mimic of the early study of the binding of alkyl

**TABLE 3-6** Rate Constants for Binding of Dioxygen and Carbon Monoxide with Cyclophanes

| Heme Compound | $k^{CO}$ (M$^{-1}$s$^{-1}$) | $k^{-CO}$ (s$^{-1}$) | $P^{CO}_{1/2}$ (torr) | $k^{O_2}$ (M$^{-1}$s$^{-1}$) | $k^{-O_2}$ (s$^{-1}$) | $P^{O_2}_{1/2}$ (torr) | $\dfrac{P^{O_2}_{1/2}}{P^{CO}_{1/2}}$ |
|---|---|---|---|---|---|---|---|
| Chelated protoheme[a,b] | $3.6 \times 10^6$ | 0.009 | 0.002 | $2.6 \times 10^7$ | 47 | 1.0 | 500 |
| [c,d] | $1.1 \times 10^7$ | 0.025 | 0.00025 | $6.2 \times 10^7$ | 4,000 | 5.6 | 22,000 |
| Chelated TP heme[e,f] | $4.2 \times 10^6$ | ~0.04 | ~0.001 | $2.9 \times 10^7$ | 30,000 | 83 | ~80,000 |
| Chelated TPiv heme[e,g] | $3.6 \times 10^7$ | 0.0078 | $2.2 \times 10^{-5}$ | $4.3 \times 10^8$ | 2,900 | 0.58 | 27,000 |
| 13-Cyclophane heme[c,h] | $6 \times 10^2$ | 0.07 | 12 | — | — | — | — |
| 15-Cyclophane heme[c,h] | $9.1 \times 10^4$ | 0.04 | 0.05 | $1.7 \times 10^6$ | 250 | 15 | 300 |
| 7,7-Anthracene cyclophane heme[c,d] | $6 \times 10^6$ | 0.05 | 0.0009 | $6.5 \times 10^7$ | 1,000 | 1.4 | 1,500 |
| 6,6-Anthracene cyclophane heme[c,d] | $3 \times 10^4$ | 0.05 | 0.17 | $1 \times 10^5$ | 800 | 700 | 4,100 |
| 6,6-Adamantane cyclophane heme[c,i] | $9.2 \times 10^3$ | 0.05 | 0.59 | $1.5 \times 10^5$ | 690 | 300 | 530 |
| 5,5-Pyridine cyclophane heme[e,i] | $6 \times 10^2$ | 0.24 | 37 | $1.1 \times 10^4$ | 68 | 540 | 14 |
| 5,5-Durene(Dclm)[e,j] | $1.1 \times 10^7$ | $0.29^k$ | $1.2 \times 10^{-3}$ | $1 \times 10^{8\ k}$ | $10^{5\ l}$ | $83^m$ | 70,000 |
| 6,6-Cyclophane heme[e,g] | $1.5 \times 10^6$ | 0.0093 | $6.5 \times 10^{-4}$ | $1.7 \times 10^7$ | 71 | 0.36 | 550 |

Steric Effects 113

**TABLE 3-6** Continued

| Heme Compound | $k^{CO}$ $(M^{-1}s^{-1})$ | $k^{-CO}$ $(s^{-1})$ | $P_{1/2}^{CO}$ (torr) | $k^{O_2}$ $(M^{-1}s^{-1})$ | $k^{-O_2}$ $(s^{-1})$ | $P_{1/2}^{O_2}$ (torr) | $\dfrac{P_{1/2}^{O_2}}{P_{1/2}^{CO}}$ |
|---|---|---|---|---|---|---|---|
| 5,5,5-Cyclophane heme[e,g] | $5.8 \times 10^5$ | 0.0086 | $1.5 \times 10^{-3}$ | $2.2 \times 10^6$ | 9 | 0.36 | 270 |
| 7,7,7-Cyclophane heme[e,g] | $9.5 \times 10^5$ | 0.05 | $5.0 \times 10^{-3}$ | — | — | 23 | 4,300 |
| Chelated 18-cyclophane[e,n] | $4.0 \times 10^{7o}$ | 0.0067 | $1.7 \times 10^{-5}$ | $1.8 \times 10^{8o}$ | 620 | 0.29 | 17,000 |
| 5e[e,p] | $6.3 \times 10^7$ | 0.0027 | $4.44 \times 10^{-6}$ | $6.23 \times 10^8$ | 130 | 0.03 | 4,960 |
| 5f[e,p] | $1.8 \times 10^6$ | 0.002 | $1.2 \times 10^{-4}$ | $3.0 \times 10^7$ | 27 | 0.13 | 814 |

[a] H₂O/MTAB.
[b] Traylor and Berzinis (1980).
[c] Benzene.
[d] Traylor et al. (1981a).
[e] Toluene.
[f] Lavalette and Momenteau (1982).
[g] Collman et al. (1983b).
[h] Ward et al. (1981).
[i] Traylor et al. (1985a).
[j] David et al. (1994). DcIm = 1,5-dicyclohexylimidazole.
[k] Measured directly.
[l] $k^{O_2}/k^{O_2}$.
[m] Kinetic determination.
[n] Momenteau et al. (1983a).
[o] These values have been recalculated using the solubilities of $10^{-5}$ torr$^{-1}$ for CO and $1.2 \times 10^{-5}$ torr$^{-1}$ for O₂ (Traylor et al., 1979b).
[p] Momenteau (1986); see Figure 3-8.

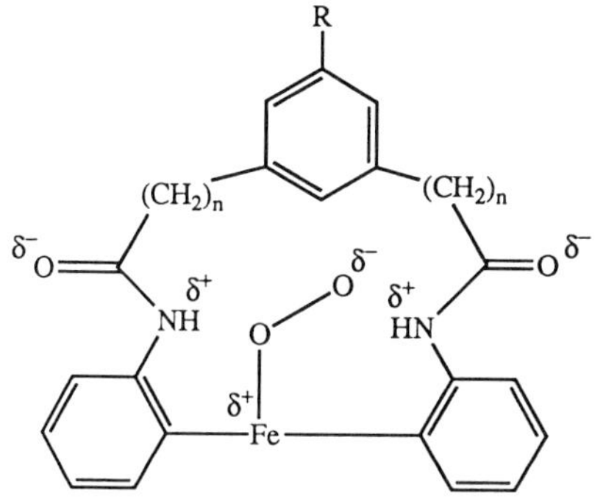

**FIGURE 3-12** Stabilization of the bound state of dioxygen by a pocket heme, where $n = 1$ or 2 (Traylor et al., 1985a).

isocyanides to hemoglobin by St. George and Pauling (1951), which first introduced the idea of a restricted pocket, is shown in the comparison of isocyanide binding to cyclophane hemes and to heme proteins, illustrated in Table 3-7. It is also possible to rotate the plane of the proximal imidazole, thus changing its $\pi$ overlap with the iron d-orbitals, which changes its spectroscopic behavior (LaMar and Walker, 1979). Designs to test this effect on dioxygen binding are still in progress (Safo et al., 1991, 1992; Walker and Simonis, 1991). Figure 3-14 lists all the effects discussed above.

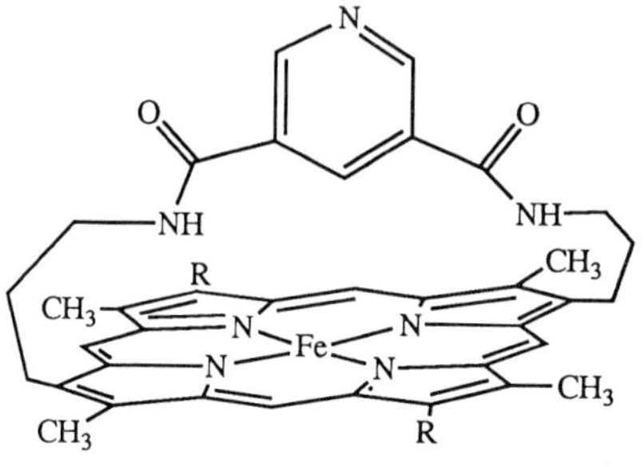

**FIGURE 3-13** Pyridine-5, 5-cyclophane—a highly polar cyclophane (Traylor et al., 1985a).

**TABLE 3-7** Kinetic Data for RNC Binding to Hemes in Benzene at 20 °C

| Compound | RNC | $k^{RNC\,a}$ M$^{-1}$s$^{-1}$ | $k^{-RNC\,b}$ s$^{-1}$ | $K^{RNC\,c}$ M$^{-1}$ | Reference |
|---|---|---|---|---|---|
| Fe(6,6-cyclophane)MeIm | TMIC$^d$ | $1.4 \times 10^4$ | 0.020 | $7 \times 10^5$ | e |
| Fe(7,7-cyclophane)DcIm | TMIC$^d$ | $1.2 \times 10^7$ | 0.015 | $7 \times 10^8$ | e |
| Chelated protoheme | TMIC$^d$ | $1.7 \times 10^8$ | 0.023 | $7 \times 10^9$ | e |
| Chelated protoheme | $n$-BuNC | $2.2 \times 10^8$ | 0.5 (0.7) | $4.4 \times 10^8$ | f |
| Mb (SW) | $n$-BuNC | $3.7 \times 10^{4\,g}$ | $0.7^g$ | — | — |
| Mb (SW)$^h$ | TMIC$^d$ | $2.3 \times 10^{2\,i}$ | 0.010 | $2.5 \times 10^4$ | e |

$^a$ Calculated from $k^{-RNC}$ and $K^{RNC}$.
$^b$ By CO displacement.
$^c$ Calculated from $K^{CO}$ and $K^{RNC,CO}$.
$^d$ TMIC = tosylmethyl isocyanide.
$^e$ Traylor et al. (1985b).
$^f$ Olson et al. (1983).
$^g$ Data obtained by flash photolysis (Reisberg and Olson, 1980; Mims et al., 1983).
$^h$ Experimentally determined by direct reaction of Mb and TMIC.
$^i$ Aqueous phosphate buffer, pH 7.3.

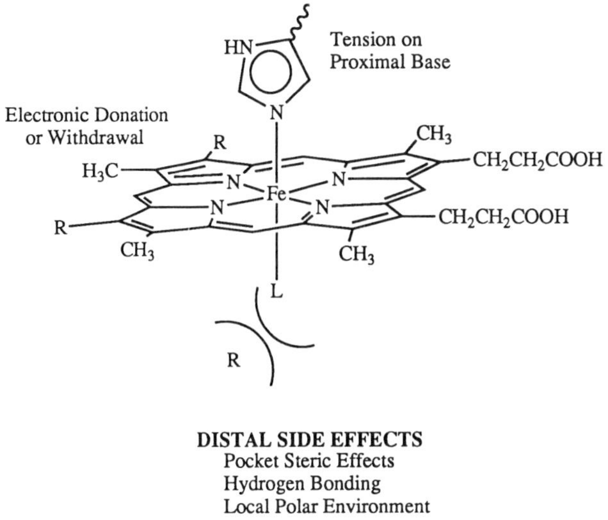

**FIGURE 3-14** List of proposed effects on $O_2$ and CO binding (Traylor, 1981).

## DYNAMICS OF BINDING DIOXYGEN AND OTHER LIGANDS

The use of low-temperature (Alberding et al., 1978a; Marden, 1982; Ansari et al., 1986) kinetic studies and the advent of sub-picosecond laser systems, have revealed very fast processes occurring after photolysis of heme protein–dioxygen complexes (Caldin and Hasinoff, 1975; Greene et al., 1978; Nöe et al., 1978; Duddell et al., 1979; Martin et al., 1982, 1983; Marden et al., 1986; Moore et al., 1987; Traylor et al., 1987a; Jongeward et al., 1988; Petrich et al., 1988; Rohlfs et al., 1988; Anfinrud et al., 1989; Postlewaite et al., 1989; Chance et al., 1990; Miers et al., 1990, 1991). A large number of these studies have now established that the binding of ligands to hemes and heme proteins involves important, and sometimes rate-limiting, diffusional processes which must be considered in order to understand dioxygen binding and transport. The kinetic and equilibrium data presented above, although varying widely with structure, are further complicated by structural effects on the formerly hidden processes.

Equations 18–20 describe the overall binding kinetics and equilibria, where Hm = a heme or heme protein and B is the spectator ligand.

118   Reactions of Dioxygen

$$BHm + O_2 \underset{k_B^{-O_2}}{\overset{k_B^{O_2}}{\rightleftharpoons}} BHmO_2 \tag{18}$$

$$BHm + CO \underset{k_B^{-CO}}{\overset{k_B^{CO}}{\rightleftharpoons}} BHmCO \tag{19}$$

Or in general,

$$BHm + L \underset{k_B^{-L}}{\overset{k_B^{L}}{\rightleftharpoons}} BHmL \tag{20}$$

As one of many examples, we describe isocyanide binding.

Photolysis of the chelated heme-$t$-butyl isocyanide complex releases the isocyanide exclusively, with a quantum yield of about 0.5 (Olson et al., 1983). Picosecond photolysis shows that this quantum yield corresponds to the fraction of photolyzed ligand which fails to return on a picosecond time scale (Jongeward et al., 1988a). The observed picosecond processes, along with slower overall processes, can be described as shown in eq. (21)

$$BHmL \underset{k_{-1}}{\overset{h\nu \atop k_1}{\rightleftharpoons}} \underset{\text{Geminate Pair}}{[BHmL]} \underset{k_{-2}}{\overset{k_2}{\rightleftharpoons}} BHm + L \tag{21}$$

where $k_{psec}^{obs.} = k_{-1} + k_2$ is an accurate first-order process with "infinity" absorbance which remains constant until second-order reactions occur (up to $10^{10}$ psec). The quantum yield for photolysis, $\Phi$, is about one but the observed quantum yield is given by eq. (22)

$$\frac{\Delta Abs_0 - \Delta Abs_\infty}{\Delta Abs_0} = \frac{k_2}{k_{-1} + k_2} \tag{22}$$

from which, along with the expression for $k_{psec}^{obs.}$, $k_{-1}$ and $k_2$ can be calculated. $\Delta Abs_\infty/\Delta Abs_0$ represents the fraction of photolyzed ligand that rebinds with heme in this time domain. The slower overall bimolecular rate constant for recombination is given by eq. (23)

$$k_B^L = k_{-2}\left(\frac{k_{-1}}{k_{-1} + k_{-2}}\right) \tag{23}$$

from which $k_{-2}$ can be calculated. A separate determination of the equilibrium dissociation constant allows $k_1$ to be calculated and completes this picture. To date, all ligands examined show this geminate pair behavior in model systems (Rose and Hoffman, 1983; Jongeward et al., 1986, 1988a; Traylor et al., 1987a, 1990, 1992c, 1992d; Bandyopadhyay et al., 1990; Taube et al., 1990). Table 3-8 shows values of these rate constants determined for several ligands.

The data in this table teach us several things. First, most ligands react at diffusion-controlled rates and have bimolecular rate constants of around $2 \times 10^8$ $M^{-1}sec^{-1}$ (Momenteau and Lavalette, 1978; Rose and Hoffman, 1983; Taube et al., 1990). Carbon monoxide is an exception, but even it can be made to react with diffusion control by increasing both the solvent viscosity and the hydrostatic pressure, thus decreasing $k_2$ and increasing $k_{-1}$ (Traylor et al., 1992b).

Structural effects on the individual rate constants are also instructive. Introduction of spectator ligand strain by changing from 1-methylimidazole to 1,2-dimethylimidazole decreases $k_{-1}$ so much that no geminate return of $t$-butyl isocyanide is observed ($k_2 \gg k_{-1}$; eq. 24) (Traylor et al., 1990).

**TABLE 3-8** Picosecond Kinetic Behavior of Several Ligands with Hemes[a-c]

| | $k_1$ (s$^{-1}$) | $k_{-1}$ (s$^{-1}$) | $k_2$ (s$^{-1}$) | $k_{-2}$ (M$^{-1}$s$^{-1}$) | $k_B^{L\,d}$ (M$^{-1}$s$^{-1}$) |
|---|---|---|---|---|---|
| $t$-BuNC[b,e,f] | 2.3 | $7.6 \times 10^{10}$ | $4.2 \times 10^{10}$ | $2.2 \times 10^8$ | $2.5 \times 10^8$ |
| MeNC[b,e,f] | 2.8 | $7.5 \times 10^{10}$ | $3.8 \times 10^{10}$ | $3.0 \times 10^8$ | $3.9 \times 10^8$ |
| α-ChNC[a,g,h] | 0.84 | $9.2 \times 10^{10}$ | $5.8 \times 10^9$ | $6.6 \times 10^7$ | $0.6 \times 10^8$ |
| NO | $<10^{-4}$ | $7 \times 10^{10\,h}$ | $\sim 3 \times 10^{10}$ | $2.8 \times 10^8$ | $2 \times 10^{8\,i}$ |
| 1-MeIm[b,e,f] | $7.3 \times 10^3$ | $8.4 \times 10^{10}$ | $2.2 \times 10^{10}$ | $2.3 \times 10^8$ | $1.5 \times 10^8$ |
| CO[c,h,k,l] | 0.028 | $1 \times 10^9$ | $3 \times 10^{10}$ | $2.6 \times 10^8$ | $1.1 \times 10^7$ |

[a] Binding to the monostearyl ester chelated protoheme.
[b] Binding to the 1-MeIm complex of protoheme dimethyl ester.
[c] Binding to protoheme-3-(1-imidazolyl) propylamide methyl ester chloride.
[d] Bimolecular rate constant; heme = protoheme dimethylester (Taube et al., 1990).
[e] Solvent: Toluene/1-MeIm, 80:20 by volume.
[f] Traylor et al. (1987).
[g] α-ChNC = 5α-cholestane-3α-isocyanide.
[h] Solvent: Toluene.
[i] K. N. Walda and G.-Z. Wu, unpublished.
[j] Protoheme dimethyl ester in water containing 30% 1-methylimidazole.
[k] Traylor et al. (1992c).
[l] Traylor et al. (1992b).

$$\begin{array}{c}\text{t-Bu}\\|\\ \text{N}\\ \text{C}\\|\\ -\text{Fe}-\\|\\ \text{N}\\ \diagdown\!\diagup\!\text{Me}\\ \text{N}\\ |\\ \text{Me}\end{array} \underset{k_{-1}}{\overset{h\nu\ k_1}{\rightleftharpoons}} \left[\begin{array}{c}\text{t-Bu}\\|\\ \text{N}\\ \text{C}\\|\\ -\text{Fe}-\\|\\ \text{N}\\ \diagdown\!\diagup\!\text{Me}\\ \text{N}\\ |\\ \text{Me}\end{array}\right] \underset{k_{-2}}{\overset{k_2}{\rightleftharpoons}} \begin{array}{c}\text{t-Bu}\\|\\ \text{N}\\ \text{C}\\ +\\ -\text{Fe}-\\|\\ \text{N}\\ \diagdown\!\diagup\!\text{Me}\\ \text{N}\\ |\\ \text{Me}\end{array} \quad (24)$$

Since *t*-BuNC returns about 50% in the absence of the 2-methyl group, there is clearly a decrease in $k_{-1}$, indicating that T-state hemoglobin should, in general, have a higher thermal and photochemical quantum yield than does the R-state. This prediction has been corroborated experimentally (Bandyopadhyay et al., 1990, 1992).

In contrast, the values of $k_{-1}$ and $k_2$ in the adamantane heme cyclophane, which is very hindered on the reaction side, are almost identical to those of chelated protoheme (Traylor et al., 1990). These two observations make the effects of steric constraints quite clear. Spectator (proximal side) steric effects operate on *bond-making or -breaking steps* ($k_1$ and $k_{-1}$), whereas reaction (distal) side steric effects operate exclusively on *diffusional steps* ($k_2$ and $k_{-2}$) and do not appreciably affect $k_1$ and $k_{-1}$. Confirmation of this last conclusion comes from the observation that $k_1$ and $k_2$ for *t*-butyl isocyanide are almost identical for chelated protoheme in toluene and myoglobin in water even though the overall binding rates differ by five orders of magnitude (Taube et al., 1990).

The consequences of this conclusion are far-reaching. If distal steric effects are diffusional, then there is no steric mechanism for differentiating CO and $O_2$ because they are the same size. The theory of distal steric differentiation of CO and $O_2$ is commonly assumed (Caughey, 1970; Antonini and Brunori, 1971; Tucker et al., 1978; Collman et al., 1979), but according to these results, this assumption is without foundation. On the other hand, it is not yet clear whether there are some solvation phenomena in proteins that might differ slightly for $O_2$ and CO.

The effects on eq. (21)—observed during picosecond measurements on model compounds—are summarized in Table 3-9.

## GEMINATE PROCESSES IN HEME PROTEINS

As Frauenfelder (Austin et al., 1975; Alberding et al., 1978a, 1978b) and Hasinoff (Hasinoff, 1974, 1977, 1981; Caldin and Hasinoff, 1975)

**TABLE 3-9** Environmental Effects on Rate Constants Observed During Picosecond Measurements on Model Compounds

| | Effect on: | | | |
|---|---|---|---|---|
| Structural Change | $k_1$ | $k_{-1}$ | $k_2$ | $k_{-2}$ |
| Distal steric effect | None | None | Little | Dec. |
| Proximal steric effect | Inc. | Dec. | None | None |
| Polarity | [a] | Little | | |
| Viscosity | None | None | Dec. | Dec. |
| Hydrostatic pressure | Dec. | Inc. | Dec. | Dec. |
| Ligand size | | Little | Dec. | Dec. |

[a] Large effect for $O_2$; very small effect for CO.

pointed out early on, heme proteins are complicated, and several imtermediate steps are discernible during geminate processes. In particular, there are very important diffusional processes which follow those shown for model systems.

These processes will be exemplified again with isocyanides, which tend to give simple kinetics even though many other ligands—$O_2$ (Jongeward et al., 1988a), NO (Antonini and Brunori, 1971; Jongeward et al., 1986), CO (Greene et al., 1978; Henry et al., 1983; Moore et al., 1987; Anfinrud et al., 1989; Traylor et al., 1992d), etc.—have been studied. Direct comparisons with chemically similar model systems are best known for isocyanides (Traylor et al., 1985b, 1987a). With $t$-butyl isocyanide in myoglobin, the sequence in eq. (25) can be observed (Jongeward et al., 1988a).

$$\text{MbL} \underset{k_{-1}}{\overset{h\nu \; k_1}{\rightleftarrows}} \underset{\text{Contact Geminate Pair}}{[\text{MbL}]} \underset{k_{-2}}{\overset{k_2}{\rightleftarrows}} \underset{\text{Protein Separated Geminate Pair}}{[\text{Mb} \parallel \text{L}]} \underset{k_{-3}}{\overset{k_3}{\rightleftarrows}} \text{Mb} + \text{L} \qquad (25)$$

Instead of the two exponential processes for models, three exponential processes are seen. As above, combining these with the overall association and dissociation rates affords the indicated rate constants. Some examples are seen in Table 3-10.

This table is somewhat misleading in that it presents the kinetic behavior in terms of three discrete steps, whereas in many studies (Alberding et al., 1978b; Marden, 1982; Doster et al., 1987) more than three steps are indicated, and the individual steps are not always

TABLE 3-10  Rate Constants for the Geminate Recombination of Isocyanides to Whale Myoglobin[a-c]

| Sample | $k_{g,ps}$ (s$^{-1}$)[d] × 10$^{-10}$ | $k_{-1}$ (s$^{-1}$) × 10$^{-10}$ | $k_2$ (s$^{-1}$)[d] × 10$^{-10}$ | $k_{g,ns}$ (s$^{-1}$)[e] × 10$^{-7}$ | $k_{-2}$ (s$^{-1}$) × 10$^{-7}$ | $k_3$ (s$^{-1}$) × 10$^{-7}$ |
|---|---|---|---|---|---|---|
| MbCNMe  | 5.3 ± 0.9 | 1.3 ± 0.4   | 4.0 ± 0.8 | 2.45 ± 0.11 | 7.75 ± 2.12 | 0.59 ± 0.20 |
| MbCNEt  | 2.1 ± 0.3 | 0.55 ± 0.13 | 1.6 ± 0.2 | 7.6 ± 1.0   | 24.5 ± 6.2  | 1.22 ± 0.64 |
| MbCN$^t$Bu | 3.4 ± 0.2 | 2.6 ± 0.2 | 8.0 ± 2.8 | 0.75 ± 0.20 | 0.52 ± 0.17 | 0.36 ± 0.12 |

[a] Rate constants over the range $3 \times 10^{11}$ to $5 \times 10^8$ sec$^{-1}$ measured with a picosecond apparatus; rate constants over $10^8$ to $10^5$ sec$^{-1}$ measured with a nanosecond apparatus.
[b] Jongeward et al. (1988a).
[c] $k_{-3}$ is bimolecular and may be ignored if concentrations are adjusted so that the pseudo-first-order rate of the nongeminate rebinding is at least 10 times slower than $k_3$ and $k_{-2}$.
[d] $k_{g,ps}$ = rate constant for disappearance of the contact geminate pair $(k_{-1} + k_2)$.
[e] $k_{g,ns}$ = rate constant for disappearance of the protein separated geminate pair $(k_{-2} + k_3)$.

described for single exponentials. Power law behavior is often seen, and the kinetic behavior is described by terms of diffusion theory. We therefore present a "four-state" model as a minimum description of heme protein-binding phenomena (Jongeward et al., 1988a; Traylor et al., 1992d). In the case of the isocyanides ($t$-butyl isocyanide and methyl isocyanide), simple exponentials are observed, justifying the description in eq. (25) for myoglobin and hemoglobin with these ligands. With these caveats, we can draw several conclusions.

First, and most important, the initial process represented by $k_{-1}$ and $k_2$ in eqs. (21) and (25) have approximately the same values in myoglobin as in model compounds such as the sterically hindered adamantane heme cyclophane (Traylor et al., 1990). This means that in myoglobin, as well as in model systems, the bond-making or -breaking steps are devoid of distal steric effects, which occur instead in the diffusion steps. Additionally, there must be little proximal (spectator ligand) steric strain, for this would greatly decrease the extent of return. But the fraction of return $k_{-1}/(k_{-1} + k_2)$, as well as the individual rate constants, have similar values in myoglobin and in model systems.

We are now in a position to use the comparisons of the dynamics of model systems with those of heme proteins to assign the various structural and environmental effects to the individual steps in the ligand binding pathways. The effects to be considered are shown in Figure 3-14 for general ligand binding and in Figure 3-15 for dioxygen. All of those influences which affect the stability of the Fe—O bond toward stretching, and finally breaking, are seen in the bond-making or -breaking step ($k_1$ and $k_{-1}$). These are the polarity and H-bonding effects, as well as the spectator steric effects. Those influences which

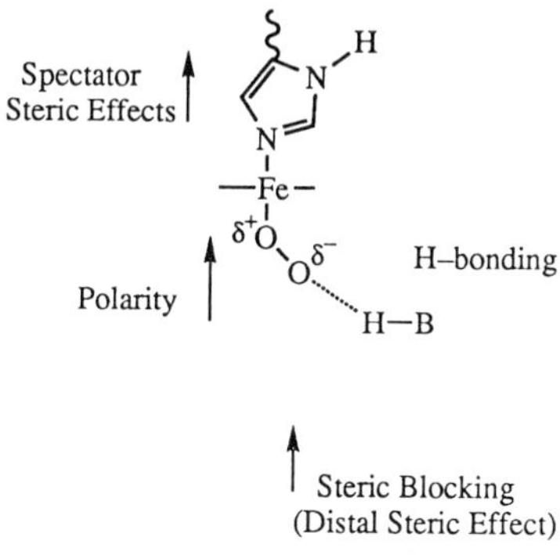

**FIGURE 3-15** Environmental effects on dioxygen binding to a heme model system.

124    Reactions of Dioxygen

affect the assembly of the ligand into a bonding geometry—i.e., the assembly of the contract pair—are seen in the subsequent steps $k_2$, $k_{-2}$, $k_3$, and $k_{-3}$. These are illustrated in Scheme 3-2. Either the $k_1$, $k_{-1}$ step

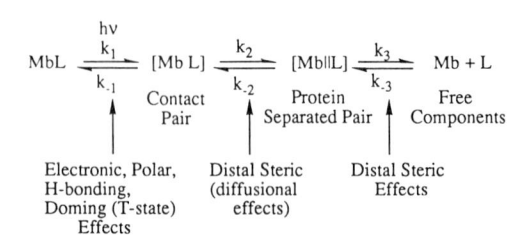

$$\text{MbL} \underset{k_{-1}}{\overset{h\nu \; k_1}{\rightleftharpoons}} [\text{Mb L}] \underset{k_{-2}}{\overset{k_2}{\rightleftharpoons}} [\text{Mb}\|\text{L}] \underset{k_{-3}}{\overset{k_3}{\rightleftharpoons}} \text{Mb} + \text{L}$$

Contact Pair — Electronic, Polar, H-bonding, Doming (T-state) Effects

Protein Separated Pair — Distal Steric (diffusional effects)

Free Components — Distal Steric Effects

**Scheme 3-2**

or the protein entry step ($k_3$, $k_{-3}$) can be dominant. For example, whereas the dissociation rate constant for dioxygen from T-state hemoglobin (Sawicki and Gibson, 1977) is about 2000 sec$^{-1}$, which is partly a result of porphyrin doming, the dissociation rate for the horseradish peroxidase–oxygen complex is less than 0.002 sec$^{-1}$ and is not measurable because oxidation to iron(III) competes. This slow dissociation, as well as the tendency to oxidize, can be attributed to the presence of very polar and strongly hydrogen-bonding (and perhaps proton transfer) groups in the pocket of HRP (eqs. 26 and 27) (Poulos and Kraut, 1980).

$$\begin{array}{c} \text{(imidazole)} \cdots \text{H}_2\text{N-C(NH}_2\text{)-NH-} \\ \text{O} \\ | \\ \text{O} \\ | \\ -\text{Fe}- \\ | \\ \text{B} \end{array} \rightleftharpoons -\overset{|}{\underset{|}{\text{Fe}}}- \quad \quad (26)$$

$$\longrightarrow -\overset{+}{\underset{|}{\text{Fe}}}- + \text{HOO}\cdot \quad \quad (27)$$

These rather large effects are due to the influence of the heme structure and the environment on the first ($k_1$, $k_{-1}$) step, i.e., on the bond-breaking step. But at the same time, there are very large effects on the overall association rate that make the rate constant for carbon monoxide about $6 \times 10^6 \, \text{M}^{-1} \, \text{sec}^{-1}$ for R-state hemoglobin (Antonini and Gibson, 1960; De Young et al., 1976; Sharma et al., 1976; Sawicki and Gibson,

1977) and $10^3 \, M^{-1} \, sec^{-1}$ for horseradish peroxidase (Wittenberg et al., 1967). This difference occurs predominantly in the rate of entry into the protein ($k_3$, $k_{-3}$). Some examination of this difference is in order here since there is still some disagreement concerning the carbon monoxide binding to these proteins.

It is now quite clear that the carbon monoxide-heme bond-making rate constant for model compounds—e.g., chelated protoheme—is around $10^9 \, sec^{-1}$, based on studies in at least three different laboratories (eq. 28) (Traylor et al., 1987a; Petrich et al., 1988; Lingle et al., 1991).

$$BHmCO \underset{10^9 \, sec^{-1}}{\overset{\substack{h\nu \\ k_1 = 0.03}}{\rightleftarrows}} [BHm \; CO] \underset{3 \times 10^8 \, M^{-1} \, sec^{-1}}{\overset{3 \times 10^{10} \, sec^{-1}}{\rightleftarrows}} BHm + CO \qquad (28)$$

Only one concentration-independent rate constant for carbon monoxide binding is observed in myoglobin ($3 \times 10^5 \, sec^{-1}$) or hemoglobin ($\sim 10^7 \, sec^{-1}$). This is usually attributed to the bond-making step. However, with other ligands, such as methyl isocyanide or t-butyl isocyanide, the bond-making rate constant is about the same in models as in proteins. Additionally, a geminate rate constant for the horseradish peroxidase–hydroxamic acid complex with CO of $2 \times 10^9 \, sec^{-1}$ has recently been measured (Berinstain et al., 1990). These observations are most easily accommodated by assuming that the less hindered (faster overall binding) myoglobin also has a contact pair collapse rate constant of $10^9 \, sec^{-1}$, as shown in eq. (29)

$$MbCO \underset{10^9 \, sec^{-1}}{\overset{h\nu}{\rightleftarrows}} [Mb \; CO] \underset{k_{-2}}{\overset{3 \times 10^{10} \, sec^{-1}}{\rightleftarrows}} [Mb \, \| \, CO] \underset{\sim 10^8 \, M^{-1} \, sec^{-1}}{\overset{5 \times 10^6 \, sec^{-1}}{\rightleftarrows}} Mb + CO \qquad (29)$$

The failure to observe the $10^9 \, sec^{-1}$ collapse rate for myoglobin–CO could be traced to the low fraction of return $(10^9)/(3 \times 10^{10} + 10^9) \cong 1/30$, with a resulting small change in absorbance. For this reason, there is still no consensus concerning the exact pathway for CO binding. It is suggested here that Scheme 3-2 represents the pathway and controls for all ligands which bind to iron(II) proteins.

The extensive studies of equilibria and dynamics with both model compounds and heme proteins make possible not only the understanding of the controlling factors, but also the design of models which bind oxygen with a desired affinity and with differentiation between dioxygen and other ligands.

## DIOXYGEN ACTIVATION

The activation of dioxygen could, as mentioned above, take one of two courses: The first is reduction to hydrogen peroxide, followed by reaction of hydrogen peroxide with the catalyst, and finally oxygen transfer (eqs. 30–32).

$$O_2 \xrightarrow[2H^+]{2e^-} H_2O_2 \tag{30}$$

$$H_2O_2 + {-}\underset{B}{Fe^+}{-} \rightarrow {-}\underset{B}{\overset{OOH}{Fe}}{-} + H^+ \tag{31}$$

$${-}\underset{B}{\overset{OOH}{Fe}}{-} \xrightarrow{substrate} Product + {-}\underset{B}{Fe^+}{-} \tag{32}$$

The second course is reduction of the hemin to iron(II), oxygen binding, reduction of the complex with proton transfer (eqs. 33–35), and finally oxygen transfer (eq. 32).

$${-}\underset{B}{Fe^+}{-} \xrightarrow{e^-} {-}\underset{B}{Fe}{-} \tag{33}$$

$${-}\underset{B}{Fe}{-} + O_2 \rightleftharpoons {-}\underset{B}{\overset{O_2}{Fe}}{-} \tag{34}$$

$${-}\underset{B}{\overset{O_2}{Fe}}{-} \xrightarrow[H^+]{e^-} {-}\underset{B}{\overset{OOH}{Fe}}{-} \tag{35}$$

The first pathway is used by peroxidases and catalase (which react with hydrogen peroxide) (Deisseroth and Dounce, 1970; Dunford and Stillman, 1976; Schonbaum and Chance, 1976; Hewson and Hager, 1979). The second pathway is that of cytochrome P-450 (Ortiz de Montellano, 1986b). The important point is that both pathways are postulated to pass through the hydroperoxyiron intermediate. If this is correct, then many of the details concerning the fate of this intermediate, including the transfer of oxygen to substrates, can be studied by preparing this intermediate from the iron(III) systems and hydrogen peroxide or its analogs.

Balch and coworkers (Arasingham et al., 1987, 1988, 1989a, 1989b) have prepared alkyl analogs of this intermediate at low temperature ($-80\,°C$) by oxidizing alkyl iron species (eq. 36).

$$\overset{R}{\underset{|}{-Fe-}} + O_2 \rightarrow \overset{OR}{\underset{|}{\overset{O}{\underset{|}{-Fe-}}}} \tag{36}$$

Decomposition of this complex in a polar medium produces products typically derived from radicals ($RO\cdot$). This is attributed to simple homolysis (eq. 37).

$$\overset{OR}{\underset{|}{\overset{O}{\underset{|}{-Fe-}}}} \rightarrow \overset{O}{\underset{|}{\overset{\|}{-Fe-}}} + RO\cdot \tag{37}$$

The effect of changing the solvent polarity on this decomposition has not been studied. However, it is clear from the kinetics of ligation that bringing this species to neutral pH in methanol, for example, would produce essentially the same mixture as that obtained with a methanol solution of the hydroperoxide (eq. 38).

$$\overset{OOR}{\underset{|}{-Fe-}} + MeOH \overset{BH^+}{\rightleftharpoons} \overset{H \quad Me}{\underset{|}{\overset{O}{\underset{|}{\pm Fe-}}}} + ROOH + B \tag{38}$$

Therefore, decomposition in protic solvents can best be studied by observing the bimolecular reaction of an oxidant with the hemin.

It is interesting that the effective oxidants for reaction with the iron(III) porphyrin are either potentially anionic or dipolar. Examples are peracids, hydroperoxides, iodosylbenzenes (especially in hydroxylic solvents), amine oxides, hypochlorite, persulfate, etc., in which the following equilibria can be set up. Two examples are given in eqs. (39) and (40) for hydroperoxide and in eqs. (41) and (42) for iodosylbenzenes.

$$ROOH \rightleftharpoons ROO^- + H^+ \qquad (39)$$

$$ROO^- + HmFe^+ \rightleftharpoons HmFeOOR \qquad (40)$$

$$RI\begin{smallmatrix}OH\\ \\OR'\end{smallmatrix} \rightleftharpoons RI\begin{smallmatrix}O^-\\ \\OR'\end{smallmatrix} + H^+ \qquad (41)$$

$$RI\begin{smallmatrix}O^-\\ \\OR'\end{smallmatrix} + -Fe^+- \rightleftharpoons \begin{smallmatrix}R\\|\\I-OR'\\|\\O\\|\\-Fe-\end{smallmatrix} \qquad (42)$$

## CONVERTING BOND ENERGY INTO OXIDIZING POWER— MAKING HIGH-VALENT METAL COMPLEXES

The next step in catalytic oxidations involves the breaking of a weak bond (e.g., an O—O or O—I bond). The process has been studied for many years (Groves et al., 1979; Chin et al., 1980a, 1980b, 1980c; Groves et al., 1981; Latos-Grazynski et al., 1982; Traylor et al., 1984b, 1989; Lee and Bruice, 1985; Groves and Gilbert, 1986; Groves and Watanabe, 1986b, 1988; Traylor and Xu, 1987, 1990; Bruice et al., 1988; Balasubramanian et al., 1989; Traylor and Ciccone, 1989; Balch et al., 1990).

Horseradish peroxidase reacts with hydrogen peroxide (Chance, 1952) to produce a two-electron oxidized species called *Compound I* (Theorell, 1941). Many kinds of spectroscopic and chemical studies have established that this complex is an iron(IV) porphyrin cation radical (Dolphin et al., 1971; Felton et al., 1971; Aasa et al., 1975; Chin et al., 1977; LaMar and de Ropp, 1980; Roberts et al., 1981) that can be reduced with one-electron reductants to a one-electron oxidized iron(IV) species, *Compound II* (eq. 43). Recently, the oxoiron form of cytochrome *c* peroxidase has been studied by X-ray crystallography (Poulos et al., 1980; Finzel et al., 1984) and the iron-oxygen structure confirmed.

$$-Fe^+- + H_2O_2 \longrightarrow \underset{\underset{\text{(an oxene)}}{\text{Compound I}}}{\overset{O}{\underset{B}{\overset{\|}{\pm Fe-}}}} \xrightarrow{e^-} \underset{\underset{\text{(an oxoiron)}}{\text{Compound II}}}{\overset{O}{\underset{B}{\overset{\|}{-Fe-}}}} \qquad (43)$$

Elegant model studies in the laboratories of Groves (Groves et al., 1981) and Balch (Chin et al., 1977) have made possible the preparation and characterization of both the oxene and oxo iron species (Figure 3-16). Using iron(III) tetramesitylporphyrin and peracids or iodosylbenzenes, Groves et al. (1979, 1981) prepared the oxene at $-80\,°C$ (eq. 44).

$$-Fe^+- + RIO \xrightarrow{-80\,°C} -\overset{\overset{O}{\|}}{Fe}{}^+- + RI \quad (44)$$

The green species obtained transfers oxygen to alkenes (eq. 45).

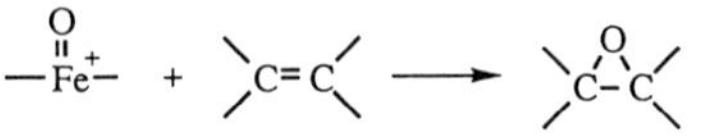

$$-\overset{\overset{O}{\|}}{Fe}{}^+- + \overset{}{\underset{}{}}C=C\overset{}{\underset{}{}} \longrightarrow \overset{}{\underset{}{}}C-C\overset{}{\underset{}{}} \quad (45)$$

The oxoiron species was prepared in a different way by Balch et al. (Chin et al., 1977), based on proposed mechanisms for iron(II) porphyrin oxidations (eqs. 46 and 47)

$$-Fe- + O_2 \underset{}{\overset{-80°C}{\rightleftharpoons}} FeOOFe \quad (46)$$

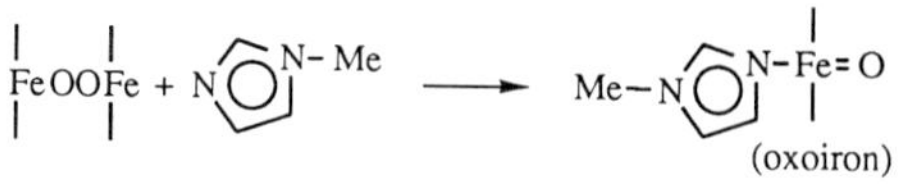

$$FeOOFe + N\!\!\bigcirc\!\!N\text{-}Me \longrightarrow Me-N\!\!\bigcirc\!\!N\text{-}Fe=O \quad (47)$$
(oxoiron)

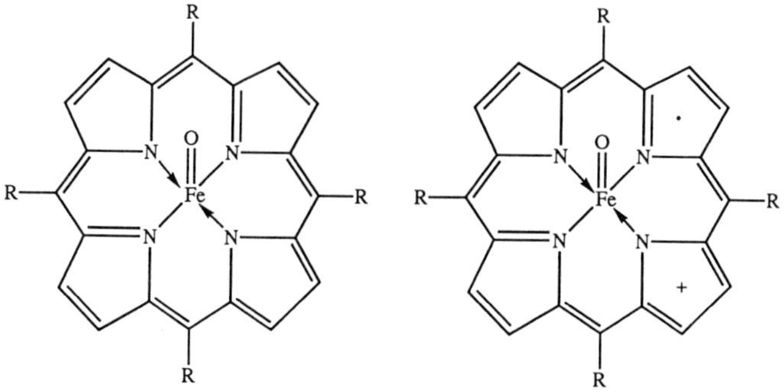

Fe(IV) Species
Oxoiron

Fe(V) Species
Oxene

**FIGURE 3-16** Oxoiron [Fe$^{IV}$] and oxene [Fe$^V$] intermediates.

This oxoiron species transfers oxygen to phosphines but not to alkenes.

Although neither model species has been characterized by X-ray crystallography, practically every other method of characterization has been used (Groves et al., 1981; Boso et al., 1983; Balch et al., 1984, 1985, 1992; Schapper et al., 1985; Gold et al., 1988, 1989; Bill et al., 1990; Fujii and Ichikawa, 1992), and the structures are known with a high degree of confidence. Much is known about their chemical reactivity, providing important tools for overall mechanistic studies of the more biological oxidants, hydrogen peroxide and dioxygen.

## MECHANISMS IN THE REACTION OF OXIDANTS WITH IRON(III) PORPHYRINS

Early work of Jones et al. (1977, 1983) and Kelley et al. (1977) showed that hydrogen peroxide reacts with protohemin in aqueous buffers to produce high-valent intermediates that oxidize the same substrates oxidizable by horseradish peroxidase. Although complicated by aggregation, these studies demonstrated a bimolecular reaction of hydrogen peroxide with the catalyst. More recently, these studies have been continued in several laboratories (Balch et al., 1984, 1985; Traylor et al., 1984b; Bruice et al., 1986, 1988; Koola and Kochi, 1987; Mansuy, 1987; Long and Hecht, 1988; Padburg et al., 1988; Labeque and Marnett, 1989; Traylor and Ciccone, 1989; Traylor and Xu, 1990; Bruice, 1992; Erman et al., 1992).

### Iodosylbenzenes

Kinetic studies (Groves et al., 1979, 1983a; Lindsay Smith and Sleath, 1982; Traylor et al., 1985c, 1989) of the reaction of iodosylbenzenes with iron(III) porphyrins—the reaction which produces the oxene species—are complicated by insolubility of the oxidant. However, by solubilizing the oxidant in a slightly aqueous alcohol solvent, a very potent oxidant with reproducible kinetics is formed (eq. 48) (Schmeisser et al., 1967; Traylor and Xu, 1987).

$$F_5\text{-C}_6\text{H}_4\text{-IO (PFIB)} \xrightarrow{\text{CH}_2\text{Cl}_2/\text{CF}_3\text{CH}_2\text{OH}/\text{H}_2\text{O}, 89/10/1} F_5\text{-C}_6\text{H}_4\text{-I(OH)(OCH}_2\text{CF}_3) \quad (48)$$

In this solvent system, the rate of reaction of tetra-2,6-dichlorophenyl porphyrin iron(III) chloride (TDCPPFeCl) with PFIB (eq. 49) is very fast ($k_2 = 2 \times 10^4 \text{M}^{-1}\text{sec}^{-1}$).

$$\text{PFIB} + -\text{Fe}^+- \xrightarrow{\text{CH}_2\text{Cl}_2/\text{CF}_3\text{CH}_2\text{OH}/\text{H}_2\text{O}} -\overset{\overset{\displaystyle O}{\|}}{\text{Fe}^+}- + \text{F}_5\text{C}_6\text{I} \qquad (49)$$

The rate of this reaction can be measured by following the disappearance of PFIB at 260 nm (Traylor et al., 1985c, 1986b, 1989) or the disappearance of a substrate such as 1,4-diphenylbutadiene at 342 nm (Traylor and Xu, 1987, 1988). Since the reaction of the oxene is orders of magnitude faster than the formation of the oxene, no matter what substrate is used (phenol, alkene, alkane, etc.), the rate of overall oxidation is independent of substrate type or concentration (eq. 50).

$$-\text{Fe}^+- + \text{C}_6\text{H}_4(\text{OR})(\text{I-OH}) \xrightarrow{\text{slow}} -\overset{\overset{\displaystyle O}{\|}}{\text{Fe}^+}- \xrightarrow[\text{fast}]{\text{RH}} \text{ROH} + -\text{Fe}^+- \qquad (50)$$

Although the oxidizing power of the oxene is high, it nevertheless is capable of differentiating among substrates. The mechanism of reaction of iodosylbenzenes with iron(III) porphyrins is now well established to be a heterolytic process (Groves et al., 1981; Traylor et al., 1984b, 1986, 1989; Groves and Watanabe, 1988).

## Peracids

To avoid the aggregation of iron porphyrins in aqueous solution, researchers turned to studies in organic solvents, as described above. Subsequently, highly sulfonated porphyrins (Bruice, 1991) or peptide porphyrins such as microperoxidase (Wang et al., 1991) were used in aqueous solution (Traylor and Xu, 1990). There is a long history of mechanism studies in water, alcohols, and aqueous alcohols; this facilitates interpretation of data in these solvents.

In order to approach as closely as possible the active-site structure of horseradish peroxidase, the model compound, chelated protohemin, was used in methanol solvents in which the pH was controlled with an unreactive buffer, collidine/collidine hydrochoride. The easily oxidized 2,4,6-tri-$t$-butyphenol was used to trap the oxene rapidly (eq. 51) (Traylor et al., 1984b).

$$\text{RCO}_3\text{H} + -\text{Fe}^+- \xrightarrow[\text{limiting}]{k \text{ rate-}} -\overset{\overset{\displaystyle \|}{O}}{\text{Fe}^+}- \xrightarrow{\text{fast}} 2 \text{ (t-Bu)}_3\text{C}_6\text{H}_2\text{-O}\cdot \qquad (51)$$

(Blue)

Under these conditions, phenylperacetic acid gives no $CO_2$. Since the phenylacetoxy radical has essentially a zero lifetime, rapidly producing benzyl radical and $CO_2$ (Bartlett and Hiatt, 1958), this means that peracids cleave by heterolysis (eq. 52).

$$PhCH_2-\overset{O}{\underset{\|}{C}}-OOH + \overset{|}{\underset{|}{Fe^+}} + B \rightarrow PhCH_2-\overset{O}{\underset{\|}{C}}-O^- + \overset{|}{\underset{|}{Fe^+}}=O + BH^+$$

$$\underset{PhCH_2\cdot + CO_2}{\overset{\searrow \;\; X \;\; \rightarrow}{\longleftarrow}} PhCH_2-\overset{O}{\underset{\|}{C}}-O\cdot + \overset{|}{\underset{|}{Fe}}=O + BH^+ \quad (52)$$

The reaction has the stoichiometry shown in eq. (51) and is clearly first order in catalyst and in peracid and zero order in phenol. The reaction is accelerated by increasing the buffer concentration, by making the solvents more aqueous, and by introducing internal (i.e., covalently attached) carboxylic acid groups, and is almost unaffected by salt concentration. The various effects on the rate of reaction are shown in Figure 3-17.

These data, the low-temperature study of Groves et al. (1979, 1981), and other evidence make it clear that the reaction of peracids with iron porphyrins, like that of iodosylbenzenes, is simple heterolysis in polar solvents and resembles the same reaction in horseradish peroxidase. The discovery of internal and external buffer catalysis corroborates the proton transfer mechanisms proposed for cytochrome c peroxidase and catalase (Figure 3–18) (Poulos and Kraut, 1980; Poulos et al., 1980).

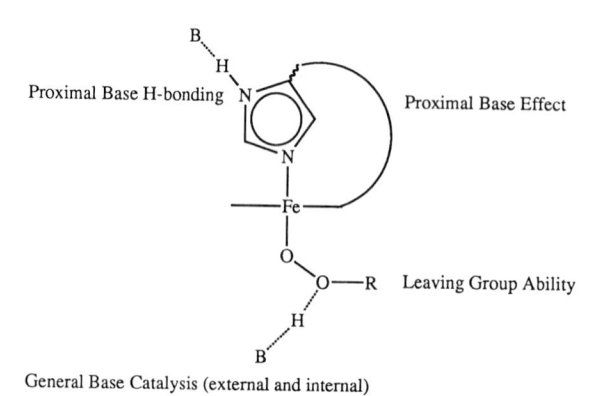

**FIGURE 3-17** Effects on the rate of eq. (51).

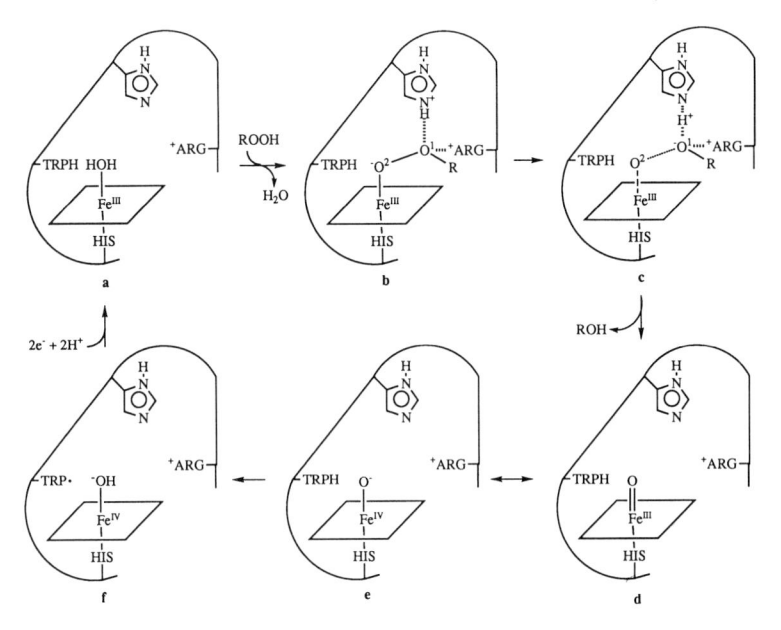

FIGURE 3-18 Proposed cytochrome $c$ peroxidase-catalyzed proton-transfer mechanism (Poulos and Kraut, 1980). **a** Enzyme at rest. **b** Enzyme–substrate complex. **c** Transition state. **d, e** Resonance forms of oxene intermediate. **f** Compound I structure after hydrogen abstraction. Used with permission of the authors.

## Hydrogen Peroxide and Hydroperoxides

Kinetic studies (Portsmouth and Beal, 1971; Jones et al., 1977, 1983; Kelly et al., 1977; Traylor et al., 1984b; Traylor and Ciccone, 1989; Traylor and Xu, 1990) of $t$-butyl hydroperoxide and hydrogen peroxide reactions with chelated protohemin showed the same kinds of dependence on solvent, buffer, and internal catalysis as those outlined above for peracids (Traylor et al., 1984b). Therefore, a similar heterolytic mechanism for these oxidants was proposed (Traylor et al., 1984b), and this introduced a biomimetic model for horseradish peroxidase (eq. 53).

Again, this a simple process, not very sensitive to pH changes over the biological range of 7 to 8.5 but highly dependent on buffer concentration and sensitive to solvent acidity. It also displays a solvent isotope effect ($k_{H_2O}/k_{D_2O} = 2$) consistent with eq. (53). Structural and medium effects are described in Figure 3-17.

Low-temperature studies of a similar biomimetic system (microperoxidase-8, a peptide fragment of cytochrome $c$), recently reported by Wang, Baek, and van Wart (1991), provided definitive evidence for the heterolysis of hydrogen peroxide, as described in eq. (52). The spectroscopic changes were essentially identical to those observed with horseradish peroxidase. Evidently this model system also oxidizes through an oxene (Fe$^+$=O).

These results lead to the rather satisfying conclusion that simple hemins react with oxidizing agents in the same way as do peroxidases, i.e., by a fast pre-equilibrium binding followed by catalytic breakage of the O—O bond (eqs. 54 and 55).

$$-\text{Fe}^+- + \text{XOH} + \text{B} \rightleftharpoons -\overset{\overset{X}{|}}{\underset{|}{\text{Fe}}}- + \text{BH}^+ \qquad (54)$$

$$-\overset{\overset{X}{|}}{\underset{|}{\text{Fe}}}- + \text{HB}^+ \rightarrow \left[-\overset{\overset{X---\text{HB}^+}{|}}{\underset{|}{\text{Fe}}}-\right] -\overset{\overset{O}{\|}}{\text{Fe}^+}- + \text{HX} + \text{B} \qquad (55)$$

This seems to be the reaction mechanism for peracids, hydroperoxides, hydrogen peroxide, hypochlorite, iodosylbenzenes, and other oxidants. The reaction rate depends on the stability of the leaving group, X$^-$, the nature of the buffer or protic catalyst (BH$^+$), and the solvent.

Recently, Bruice et al. (Bruice et al., 1986, 1988; Zipplies et al., 1986; Balasubramanian et al., 1987; Lee et al., 1988; Lindsay-Smith et al., 1988) studied these reactions (using hemins without the proximal base) over wide pH ranges and postulated extraordinarily complex reaction mechanisms involving more than 10 rate constants. They concluded that the reaction of hydroperoxides with hemins predominantly or exclusively results in homolysis of the FeOOR species and therefore does not afford epoxidation of alkenes. This work has been reviewed (Bruice, 1986, 1991; Meunier, 1986, 1992).

## Free-Radical Reactions of Hydroperoxides

Extensive studies of the products of reactions of both heme proteins (Thompson and Wand, 1985; Yumibe and Thompson, 1988) and simple hemins (Dix and Marnett, 1981, 1983, 1985; Labeque and Marnett, 1989; Bruice, 1991) with hydroperoxides have made it clear that free radicals can be produced when hydroperoxides are employed as oxidants. Invariably, alkoxyl and peroxyl radical products seem to be obtained, as is evidenced by nonstereospecific epoxidation and the production of ketones from the alkoxyl radicals (eqs. 56–58).

$$-Fe^+ - + ROOH \rightarrow \rightarrow ROO\cdot + RO\cdot \qquad (56)$$

$$ROO\cdot + \underset{Ph}{\overset{}{\diagup}}\underset{Ph}{\overset{}{\diagdown}} \longrightarrow \underset{Ph}{\overset{H}{\diagup}}\underset{}{\overset{OOR}{\underset{|}{C}}}-\underset{Ph}{\overset{H}{\underset{}{C}}}\underset{}{\diagdown} \longrightarrow \underset{Ph}{\overset{H}{\diagup}}\underset{}{\overset{O}{C}}-\underset{H}{\overset{Ph}{C}} \qquad (57)$$

$$R'-\underset{R''}{\overset{R''}{\underset{|}{C}}}-O\cdot \longrightarrow R'\cdot + R''_2CO \qquad (58)$$

It is clear that the alkoxyl radicals must arise from homolysis of the O—O bond. This homolysis has been commonly assigned to the initial FeOOR species and is usually believed to be in competition with heterolysis, where $BH^+$ can be the solvent or a buffer component (eqs. 59 and 60).

$$\underset{|}{\overset{|}{FeOOR}} \underset{\overset{BH^+}{\underset{k_{het}}{\searrow}}}{\overset{k_{hom}}{\nearrow}} \begin{array}{l} \overset{|}{\underset{|}{Fe=O}} + RO\cdot \qquad (59) \\ \\ \overset{|}{\underset{|}{Fe\overset{+}{=}O}} + ROH + B \qquad (60) \end{array}$$

However, it has been shown that the oxene can be prepared in the presence of hydroperoxides (Traylor et al., 1984b, 1993). The addition of pentafluoroiodosylbenzene to a solution of tetramesityl iron(III) porphyrin chloride and hydroperoxide rapidly produces the oxene before the hydroperoxide can react with the porphyrin. Under these conditions, cis-β-methylstyrene is oxidized with loss of stereospecificity, yielding the same products as those resulting from the reaction of the hydroperoxide alone with porphyrin. Equation (63) shows the production of an alkoxyl radical, through the reaction of hydroperoxide with the oxene, which subsequently reacts with the alkene. Thus, neither the formation of alkoxyl radicals nor the loss of stereospecificity in products represents compelling evidence for homolytic cleavage of the FeOOR species.

$$C_6F_5IO + \overset{|}{\underset{|}{Fe^+}} \xrightarrow{fast} C_6F_5I + \overset{|}{\underset{|}{Fe^+}}=O \qquad (61)$$

$$R_3COOH + \overset{|}{\underset{|}{Fe^+}} \xrightarrow{slower} \overset{|}{\underset{|}{Fe}}=O \text{ or } \overset{|}{\underset{|}{Fe^+}}=O \qquad (62)$$

$$R_3COOH + \overset{|}{\underset{|}{Fe}}=O \longrightarrow R_3COO\cdot + \overset{|}{\underset{|}{Fe}}-OH \qquad (63)$$

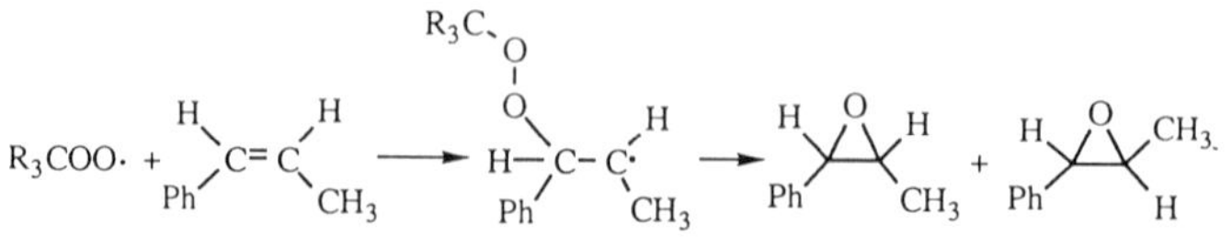

$$R_3COO\cdot + \underset{Ph}{\overset{H}{\diagdown}}C=C\underset{CH_3}{\overset{H}{\diagup}} \longrightarrow \underset{Ph}{\overset{R_3C\diagdown O}{\underset{|}{\overset{|}{O}}}}\underset{CH_3}{\overset{H}{\diagup}} \longrightarrow \underset{Ph}{\overset{H}{\diagdown}}\overset{O}{\bigtriangleup}\underset{CH_3}{\overset{H}{\diagup}} + \underset{Ph}{\overset{H}{\diagdown}}\overset{O}{\bigtriangleup}\underset{H}{\overset{CH_3}{\diagup}} \qquad (64)$$

As a further confirmation of the mechanisms shown above (eqs. 61–64), quantitative yields of alkene epoxidation using *t*-butyl hydroperoxide or hydrogen peroxide and hemin catalyses, an impossibility according to the proposals of homolytic cleavage, have now been reported (Traylor et al., 1993). These were achieved by utilizing conditions favoring reaction of the oxene with alkenes over that of its reaction with the hydroperoxide, a direct competitor. Specifically, the use of electron-deficient iron(III) porphyrins in inert hydroxylic solvents greatly favors epoxidation over hydrogen abstraction from hydroperoxide, and in this way high yields of epoxides were obtained.

Therefore the simple heterolytic eqs. (50) and (60) seem to hold for all hydroxylic oxygen transfer agents, consistent with the mechanisms of reaction of the heme proteins CCP, HRP, and catalase. There are side reactions of hydroperoxides that sometimes lower the yields of epoxide and have effects on stereospecificity and ketone production mentioned above, but the initial "oxygen activation" seems to be the production of the oxene in almost all cases. (For exceptions, see Dix and Marnett, 1981, 1983; Dix et al., 1985; Thompson et al., 1985; Yumibe and Thompson, 1988).

## NONPROTIC VS. PROTIC OXIDANTS

Most of the oxidants generally used can react with iron after loss of a proton (eqs. 65 and 66).

$$B + XOH \rightleftharpoons XO^- + HB^+ \qquad (65)$$

$$XO^- + \overset{|}{\underset{|}{Fe}}{}^+ \rightarrow \overset{|}{\underset{|}{Fe}}OX \qquad (66)$$

Two aprotic oxidants have recently been investigated. Epoxidation by ozone, catalyzed by porphyrins, has been reported, but few details are known (Campestrini et al., 1991). Dialkyldioxiranes are very strong oxidants which react fairly rapidly with alkenes and alkanes to give epoxides and ketones (eqs. 67 and 68) (Adam et al., 1989).

$$\underset{Et}{\underset{|}{Me}}\diagdown\overset{O}{\underset{C-O}{\diagup}} + \underset{R}{\overset{H}{\diagup}}C=C\underset{R}{\overset{H}{\diagdown}} \longrightarrow \underset{R}{\overset{H}{\diagup}}\overset{O}{\underset{C-C}{\diagup\diagdown}}\underset{R}{\overset{H}{\diagdown}} + Me\overset{O}{\underset{||}{C}}Et \qquad (67)$$

$$\underset{Et}{\underset{|}{Me}}\diagdown\overset{O}{\underset{C-O}{\diagup}} + \diagup\!\!\!\diagdown \longrightarrow \diagup\!\!\!\diagdown\!\!^O + Me\overset{O}{\underset{||}{C}}Et \qquad (68)$$

However, a dichloromethane solution of ethyl methyl dioxirane and tetra-(2,6 dichlorophenyl) porphyrin Fe$^{III}$ chloride does not react in 20 minutes at room temperature (Traylor et al., 1995). Addition of norbornene after this time affords only exo-epoxynorbornane, the product of direct reaction with the dioxirane. Evidently, an oxidant which donates an oxygen atom to norbornene cannot donate an oxygen atom to an iron(III) porphyrin. This is interpreted to mean that a pre-equilibrium to form an oxyanion, or at least a dipolar species (such as an amine oxide), is required for reaction with hemin. There is considerable evidence for the formation of such intermediates (Groves and Watanabe, 1988; Yamaguchi et al., 1992).

These obervations explain why hydrogen peroxide, by itself a poorer oxidizer than dioxirane, is nevertheless much more effective in reactions with iron porphyrins, through which a better oxidizer than dioxirane is produced. We can now write a specific mechanism for hydrogen peroxide (eqs. 69–71) and a general mechanism for oxidants.

$$B + H_2O_2 \rightleftharpoons HOO^- + BH^+ \qquad (69)$$

$$HOO^- + \overset{|}{\underset{|}{Fe}}{}^+ \rightleftharpoons HOO\overset{|}{\underset{|}{Fe}} \qquad (70)$$

$$\overset{|}{\underset{|}{Fe}}OOH + BH^+ \rightleftharpoons \overset{|}{\underset{|}{Fe}}O-O^{+}\!\!\diagup^{H}_{\diagdown H} + B \rightarrow \overset{|}{\underset{|}{Fe}}{}^+\!=\!O + H_2O \qquad (71)$$

138  Reactions of Dioxygen

Both the pre-equilibria (eqs. 69–71) and the bond-breaking step will be sensitive to the hydrogen bonding available in HRP, as illustrated for model systems in Figure 3-17.

## SUBSTRATE OXIDATION BY HIGH-VALENT IRON INTERMEDIATES

Since both the oxene and the oxoiron (see Figure 3-16) have been isolated and studied, it is instructive to consider the reactivities of these species. There seem to be very large differences in the reactions which these two species readily carry out, and these differences have an important bearing on mechanistic discussions. It is instructive to describe the limited studies conducted thus far.

The oxene species appears to react rapidly with almost any organic compound, including alkanes, alkenes, and aromatic compounds, whereas the oxoiron species does not react with alkanes or alkenes (Mansuy et al., 1980; Labeque and Marnett, 1989; Hirao et al., 1990; Bruice, 1991; He and Bruice, 1991; Lindsay Smith and Lower, 1991) even though it transfers an oxygen to phosphines (Chin et al., 1980c; Peterson et al., 1985; Traylor et al., 1989). A rough idea of the differences is displayed in Scheme 3-3, where any substrate to the left of the marker reacts with the species at room temperature. The substrates are listed from left to right in decreasing ease of oxidizability.

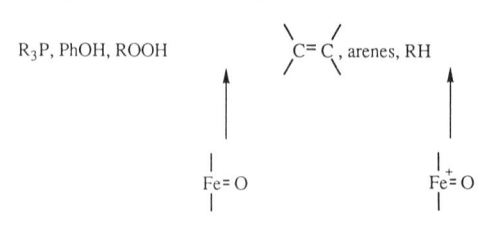

$$R_3P, PhOH, ROOH \qquad\qquad \diagdown C = C \diagup, \text{arenes, RH}$$

$$Fe = O \qquad\qquad\qquad\qquad Fe^+ = O$$

Scheme 3.3

This sequence suggests that, for example, hydroperoxides—and hydrocarbons with C—H bonds of the same strength as the OH bond strength in phenol—should react with both $Fe^+=O$ and $Fe=O$. These estimates are based on studies using tetramesityl hemin. The perhalohemins, having much higher oxidation potentials, might show higher reactivities of the $Fe=O$ species. These questions have not yet been explored.

## OXENE, IRON(IV) CATION RADICAL REACTIONS

The green species derived from iron(III)tetramesitylporphyrin at −80 °C by Groves et al. (1979, 1980) and identified as the oxene species (Figure 3-16) reacts with cyclooctene at low temperature to produce the epoxide. *p*-Substituted styrenes also react with this intermediate at low temperature at rates which correlate with the $\sigma^+$ values of the substituent (Groves and Watanabe, 1986).

Attempts to carry out high-turnover catalytic epoxidation or hydroxylation reactions using tetramesitylporphyrin resulted in catalyst destruction and low turnover. This problem was overcome by preparing a series of electronegatively substituted tetraphenylporphyrins shown in Figure 3-19 (Traylor et al., 1992c).

With these iron(III) porphyrins, high-turnover hydroxylations became

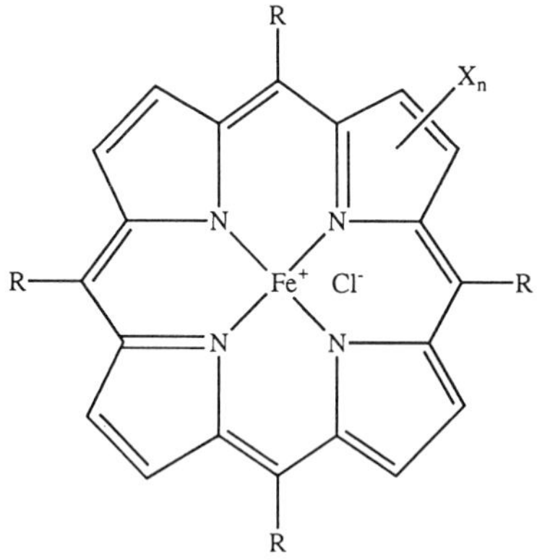

| Catalyst | $X_n$ | R | Reference |
|---|---|---|---|
| $1^+Cl^-$ | $H_8$ | 2,6-diClPh | a |
| $2^+Cl^-$ | $Br_8$ | 2,6-diClPh | b |
| $3^+Cl^-$ | $H_8$ | $C_6F_5$ | -- |
| $4^+Cl^-$ | $F_7H$ | 2,6-diClPh | b, c, d |
| $5^+Cl^-$ | $Cl_8$ | $C_6F_5$ | d |
| $6^+Cl^-$ | $F_8$ | $C_6F_5$ | e |

[a] Traylor, et al., 1984
[b] Traylor and Tsuchiya, 1987
[c] Franzus et al., 1968b
[d] Franzus et al., 1968a
[e] Tsuchiya and Seno, 1989

**FIGURE 3-19** Series of electronegatively substituted tetraphenyl porphyrins.

**TABLE 3-11** Preparative Oxidation of Cyclohexane[a,b]

| Concentration of Cyclohexane (M) | % Yield[c] Cyclohexanol | % Conversion[d] | Alcohol/Ketone |
|---|---|---|---|
| 0.1 | 41 | 25 | 9 |
| 0.5 | 62 | 7 | 13 |
| 1.0[e] | 76 | 4.5 | 23 |
| 2.0[e] | 82 | 2.5 | 31 |
| 5.0[e] | 93 | 1 | 37 |

[a] Conditions: $2^+Cl^-$, $10^{-3}$ M, pentafluoroiodosylbenzene (PFIB), "0.06 M", in dichloromethane at 25 °C.
[b] Traylor et al. (1992a).
[c] Based on $C_6F_5I$ produced.
[d] Based on cyclohexane.
[e] No hemin destruction.

**TABLE 3-12** Yields of Epoxide Obtained from cis-Cyclooctene and Norbornene with Hydrogen Peroxide or t-Butyl Hydroperoxide as Oxidants

| | Oxidant Conc. (M) | Substrate Conc. (M) | Yields (%)[c] Epoxycyclo-octane | Epoxy-norbornane |
|---|---|---|---|---|
| $H_2O_2$ | 0.02 | 1.5 | 86 | 80 |
| t-BuOOH | 0.1 | 0.3 | 89 | |
| | 0.2 | 0.6 (1.5)[a] | 86 | 59[b] |
| | 0.3 | 0.1 | 100[d] | |

[a] First value for cyclooctene; value in parentheses for norbornene.
[b] Includes some rearrangement product (<5%).
[c] Based on oxidant.
[d] Based on alkene.

possible. Table 3-11 shows yields achieved in hydroxylation reactions using the halogenated metalloporphyrin, $2^+Cl^-$.

As the tetraphenylporphyrin is substituted with more electron-withdrawing halogens, it becomes more resistant to oxidative destruction. At $10^{-3}$ M hemin in the presence of 0.03 M PFIB in dichloromethane, $6^+Cl^-$ suffers less than 20% loss of hemin activity in a time frame which completely destroys $1^+Cl^-$, $2^+Cl^-$, and $3^+Cl^-$ (Traylor et al., 1992a).

A comparison of epoxide yields obtained using either hydrogen peroxide or t-butyl hydroperoxide with $3^+Cl^-$ is given in Table 3-12.

Epoxidation turnover rates as high as $10^5$ were reported for catalyst $1^+Cl^-$ (Collman et al., 1986a, 1990a), and at least this amount of turnover is expected for the more robust catalysts $2^+Cl^-$, $4^+Cl^-$, $5^+Cl^-$, and $6^+Cl^-$.

## Relative Reactivities of Substrates with Oxene

Because the formation of the oxene, even at 100 turnovers per second, is still slow compared to the reaction of the oxene with substrates at room temperature, relative rates of oxygen transfer from the oxene have been determined by competition methods. Often gas-liquid chromatographic determination of products is used to establish relative rates (Lindsay Smith et al., 1982, 1983a, 1983b; Meunier et al., 1984; Collman et al., 1985, 1986b). Another simple method employs the effect of a second substrate on the rate of consumption and the total consumption of a substrate (such as 1,4-diphenylbutadiene) which has a convenient ultraviolet peak (Traylor and Xu, 1987, 1988). Table 3-13 shows a series of relative rates of reaction of oxene with a variety of substrates using a combination of these two methods.

In several cases, both methods were used in a single comparison. These data are approximate for two reasons. First, in most cases, only one of the two methods mentioned was used. Second, the data were obtained using catalyst $1^+Cl^-$ (see Figure 3-19) (Traylor et al., 1984). There are indications that the selectivities among substrates are de-

TABLE 3-13  Relative Reactivities of Substrates with TDCPPFe$^+$=O from $1^+Cl^-$

| Substrate | $k_{rel}$ |
|---|---|
| Cyclohexane | $1^a$ |
| Cyclohexanol | 36 |
| 1-Octene | 300 |
| Styrene | $5 \times 10^3$ |
| Norbornene | $10^4$ |
| cis, trans-2,4-Hexadiene | $5 \times 10^4$ |
| trans,trans-1,4-Diphenylbutadiene | $5 \times 10^5$ |
| t-Butyl hydroperoxide | $10^6$ |
| Hydroquinone | $10^{6\ a}$ |
| β-carotene | $10^7$ |

[a] On a per hydrogen basis.

pendent on the structure of the iron(III) porphyrin—in particular, on electron-withdrawing substitutents. As an example of this selectivity, the oxene from iron(III) tetramesitylporphyrin reacts at least 100 times faster with *t*-butyl hydroperoxide than with norbornene, whereas the oxene from iron(III) tetrapentafluorophenylporphyrin reacts at a rather similar rate (Traylor et al., 1993). The consequence of this large difference is that high-yield epoxidations with *t*-butyl hydroperoxide can be obtained with the fluorinated catalyst but not with iron(III) tetramesitylporphyrin. Future studies of relative reactivities of substrates with oxenes—as a function of oxene structure—should greatly improve both the synthetic utility and the mechanistic understanding of such reactions.

## Mechanisms of Oxidation by Oxenes

As in the case of many high-valent, and thus high potential, metal systems, the oxene species has three well-established modes of interaction with substrates (Ortiz de Montellano and Kunze, 1981; Groves and Myers, 1983; Groves and Nemo, 1983; Kunze et al., 1983; Guengerich and MacDonald, 1984; Meunier et al., 1984; Collman et al., 1985; Traylor et al., 1986b): electron transfer, direct oxygen transfer, and hydrogen abstraction (eqs. 72–74).

$$\mathrm{Fe^+{=}O + S \rightarrow [S^{+\cdot}\ Fe{=}O] \rightarrow product} \qquad (72)$$

$$\mathrm{Fe^+{=}O + S \rightarrow [Fe^+{=}O{-}S] \rightarrow Fe^+ + SO} \qquad (73)$$

$$\mathrm{Fe^+{=}O + SH \rightarrow [Fe^+{=}OH \cdot S] \rightarrow product} \qquad (74)$$

A fourth possibility, involving the concerted formation of a metallooxetane (eq. 75), has been considered, although evidence for this route with iron porphyrins is not as clear as it is with small oxoiron species.

$$\mathrm{Fe^+{=}O\ +\ \underset{/}{\overset{\backslash}{C}}{=}\underset{\backslash}{\overset{/}{C}} \longrightarrow \underset{|}{\overset{|}{Fe^+}}\underset{O}{\overset{C}{\diamond}}\underset{\backslash}{\overset{/}{C}}} \qquad (75)$$

We will consider the mechanism of reactions of oxenes with substrates in the order of increasing ionization potential of the substrate.

1. *Phenols and hydroperoxides.* The reaction of an oxene (e.g., horseradish peroxidase compound I) with phenols shows kinetic properties consistent with electron transfer. A simple oxene derived from a chelated hemin reacts with tri-*t*-butyl phenol with a kinetic isotope effect of 1.0 (eq. 76).

$$\text{X-C}_6\text{H}_2\text{X-OH(D)} + \text{Fe}^+{=}\text{O} \rightarrow [\text{X-C}_6\text{H}_2\text{X-OH}]^{+\cdot} + \text{Fe}{=}\text{O} \xrightarrow[-\text{H}^+]{\text{fast}} \text{X-C}_6\text{H}_2\text{X-O}\cdot \quad (76)$$

This observation is consistent with electron transfer followed by proton loss.

In contrast, the oxidation of *t*-butyl hydroperoxide has an isotope effect of about 4.0, and reactions with hydrocarbons show primary isotope effects of up to 11.0. Rate-limiting hydrogen abstraction from the hydroperoxide is therefore indicated (eq. 77).

$$t\text{-BuOOH(D)} + \text{Fe}^+{=}\text{O} \rightarrow t\text{-BuOO}\cdot + \text{Fe}{-}\text{OH} \quad (77)$$

2. *Low-potential hydrocarbons and alkenes.* Electrochemical or other one-electron oxidations of very strained hydrocarbons usually lead to rapid rearrangements, as exemplified by quadricyclene in eq. (78).

$$\text{quadricyclene} \xrightarrow{-e^-} \text{cation radical} \rightarrow \text{rearranged} + \text{etc.} \quad (78)$$

Oxidation of this hydrocarbon does not lead to hydroxylation, in part because all the C—H bonds are high in s-character and therefore strong but mainly because the low ionization potential allows facile oxidation to the cation radical (Stearns and Ortiz de Montellano, 1985). The liver enzyme cytochrome P-450 catalyzes the oxidation of this hydrocarbon to the rearranged aldehyde (eq. 79).

$$\text{quadricyclene} + \text{O}_2 \xrightarrow{\text{P-450}} [\text{cation radical, OFe}] \rightarrow \text{intermediate} \rightarrow \text{H-CHO product} \quad (79)$$

Exactly the same result was obtained using the simple system of hemin (Figure 3-19) and iodosylbenzene (Evans and MacDougall, unpublished). Therefore, electron transfer from hydrocarbons is a viable oxidation process if the ionization potential is low. Other examples of established radical-cation rearrangements catalyzed by the oxene species have been observed, as indicated in eq. (80) (Traylor and Miksztal, 1987; Traylor et al., 1987b).

$$\text{(a one-electron oxidant)} \quad \text{quadricyclene} + \overset{+}{\text{Fe}}=O \longrightarrow \text{rearranged product} + \text{epoxide} \tag{80}$$

However, unlike the case of quadricyclene, this ring closure does not lie along the epoxidation pathway and could therefore be a side reaction unrelated to the mechanism of epoxidation.

3. *Alkenes*. Epoxidation of alkenes (eq. 81) can probably occur by different mechanisms, depending on the ionization potential of the alkene.

$$\overset{\diagdown}{\underset{\diagup}{C}}=\overset{\diagup}{\underset{\diagdown}{C}} + \overset{+}{\text{Fe}}=O \xrightarrow{k_{ep}^{rel}} \overset{O}{\underset{\diagup \diagdown}{C-C}} \tag{81}$$

Relative rate constants, $k_{ep}^{rel}$, were measured and a plot of $\log k_{ep}^{rel}$ against the vertical ionization potential, shown in Figure 3-20, indicates that electron transfer is a possibility in epoxidations (Traylor and Xu, 1988). Rates of epoxidation by *m*-chloroperbenzoic acid (eq. 82) are shown on the same plot. Whereas this oxidation also follows the ionization potentials for simple alkenes, reaction rates of 1,4-diphenylbutadiene and *cis*-stilbene fall badly off the line, i.e., react much more slowly with MCPBA than predicted by their ionization potentials.

$$\text{RCO}_3\text{H} + \overset{\diagdown}{\underset{\diagup}{C}}=\overset{\diagup}{\underset{\diagdown}{C}} \xrightarrow{k} \overset{O}{\underset{\diagup \diagdown}{C-C}} \tag{82}$$

This contrast definitely points to two different mechanisms for epoxidation (eqs. 83 and 84). In conjugated alkenes, eq. (83) is retarded because of loss of delocalization energy at the transition state. But eq. (84), going to the radical cation, takes advantage of low ionization potentials, and these alkenes fit the $\log k_{ep}^{rel}$ versus IP correlation accurately. These and other observations point to this electron-transfer "rebound" epoxidation for dienes and other alkenes with low ionization potentials.

Oxene, Iron(IV) Cation Radical Reactions 145

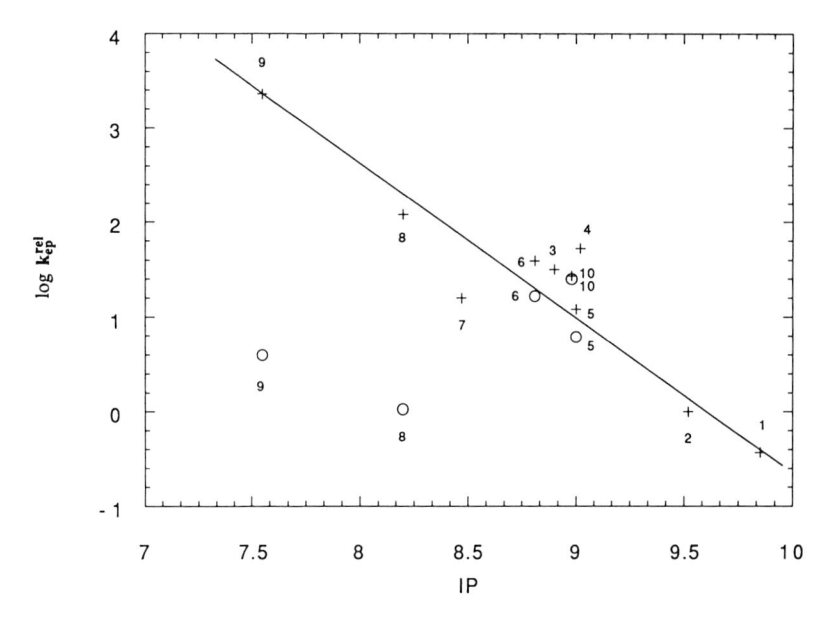

**FIGURE 3-20** Plot of log $k_{ep}^{rel}$ vs. vertical ionization potentials for alkene epoxidation with PFIB catalyzed by Fe$^{III}$DCPP(+); MCPBA epoxidation (O) (Traylor and Xu, 1988). Alkenes: 1 = vinyltrimethylsilane; 2 = 1-octene; 3 = vinylcyclopropane; 4 = allyltrimethylsilane; 5 = cyclohexene; 6 = norbornene; 7 = styrene; 8 = *cis*-stilbene; 9 = 1,4-diphenylbutadiene; 10 = *cis*-cyclooctene.

Reactions of simple alkenes (e.g., cyclohexene or norbornene) are not as easy to interpret. One of the most interesting observations with regard to mechanisms of epoxidation concerns the regiochemistry of epoxidation of norbornene. Whereas epoxidations by MCPBA or dioxiranes produce at least 1,000 times (Traylor et al., 1986b) as much exo- as endo-epoxynorbornane, reactions with some of the oxene species produce exo/endo ratios of less than 60 (eq. 85). In the latter case, the exo/endo ratio varies with the nature of the iron(III) porphyrin but is independent of the nature of the oxidant (iodosylbenzenes, peracids, hydroperoxides) (Traylor et al., 1989c).

$$\text{norbornene} + RCO_3H \text{ or } \overset{|}{\underset{|}{Fe}}{=}O \longrightarrow \text{exo-epoxide} + \text{endo-epoxide} \qquad (85)$$

An explanation for this observed regiochemistry is that the reaction goes by the electron transfer-rebound mechanism, as shown in eq. (86).

$$\text{norbornene} + \overset{|}{\underset{|}{Fe}}{=}O \longrightarrow [\text{radical cation} + OFe] \longrightarrow exo + endo \qquad (86)$$

It is known that norbornyl radicals collapse with other radicals or reactive species, such as chlorine, to give a large amount of endo product. However, it is unlikely that primary alkenes or other alkenes with high ionization potentials undergo the election-transfer rebound mechanism (Mirafzal et al., 1992).

An alternative explanation would be very long C---O bonds at the transition state for direct atom transfer, i.e., a mechanism very close to electron transfer (eq. 87) (Traylor and Miksztal, 1989). As the incipient C—O bonds become longer at the transition state, steric differentiation (exo vs. endo) decreases.

$$\text{norbornene} + \overset{|}{\underset{|}{Fe}}{=}O \longrightarrow [\text{long bond transition state with O-Fe}]^{\ddagger} \longrightarrow exo + endo \qquad (87)$$

The electron transfer, followed by a radical to oxoiron collapse, produces a carbocation intermediate. This is consistent with the rearrangements often observed in hemin-catalyzed oxidations of strained alkenes (eqs. 88 and 89) (Traylor et al., 1986a). However, the recent application of cation-radical clock reactions to this problem (Mirafzal et al., 1992; Tian, 1992) suggests that simple alkenes such as norbornene do not form radical cations as epoxidation intermediates. Further applications of such clock reactions are in order.

## EVIDENCE FROM SUICIDE LABELING

Ortiz de Montellano et al. (Ortiz de Montellano et al., 1980; Ortiz de Montellano and Correia, 1983) showed that cytochrome P-450-catalyzed epoxidations of terminal alkenes and alkynes to N-alkylated porphyrins were accompanied by suicide (i.e., loss of catalytic activity) (eq. 90).

The structure of the N-alkylated product seemed inconsistent with a carbocation intermediate, which should give the regiochemistry in eq. (91).

An identical type of N-alkylation was subsequently discovered with simple hemin catalysts (Tian, 1992), which allowed the N-alkylation process to be studied in greater detail.

Previous studies of the N-alkylation products with cytochrome P-450 (Ortiz de Montellano and Correia, 1983; Ortiz de Montellano, 1986)— and other studies of the products and kinetics of N-alkylation with simple hemins (Mashiko et al., 1985)—based their conclusions on isolated (i.e., stable) N-alkyl hemins. Studies of the absolute rates of epoxidation and N-alkylation by stopped-flow methods, however, revealed three important points: (1) almost all alkenes (including cyclic alkenes and trans internal alkenes) lead to N-alkylation (Mashiko et al., 1985; Artaud et al., 1989); (2) terminal alkenes react predominantly to give secondary amines rather than primary amines (Traylor et al., 1987b; Tian, 1992); (3) and most important, all of the secondary N-alkylated products are unstable and return to produce the original catalyst (Traylor et al., 1986b). Equations (92–94) show this for norbornene and 4,4-dimethyl-1-pentene. The ratios of epoxidation ($k_1$) to N-alkylation ($k_2 + k_{2'}$) are fairly constant, varying from about 50 to 150 over a wide variety of alkenes. Much higher ratios have been reported for certain terminal alkenes (Collman et al., 1986b), but these ratios may be too high since regeneration of the catalyst via $k_3$ was not considered.

$$C_6F_5IO + Fe^+ + \text{[norbornene]} \xrightarrow{k_1}_{k_2} \text{[epoxide]} + \text{[N-alkyl hemin]} \quad (92)$$

$$\text{[N-alkyl hemin]} \xrightarrow{k_3} \text{[epoxide]} + Fe^+ \quad (93)$$

It is interesting that the only product isolated from the decomposition of the norbornene-derived N-alkyl hemin is the exo-epoxide (eq. 93) (Richards and Traylor, unpublished). This means that, at a minimum, the reverse of N-alkylation shares an intermediate with epoxidation.

$$C_6F_5IO + Fe^+ + \text{[alkene]} \xrightarrow{k_1, k_2, k_{2'}} \text{products} \quad (94)$$

From these observations, a general epoxidation process can be postulated, with the caveat that multiple simultaneous pathways are still possible. This is exemplified with 4,4-dimethylpentene in Scheme 3-4.

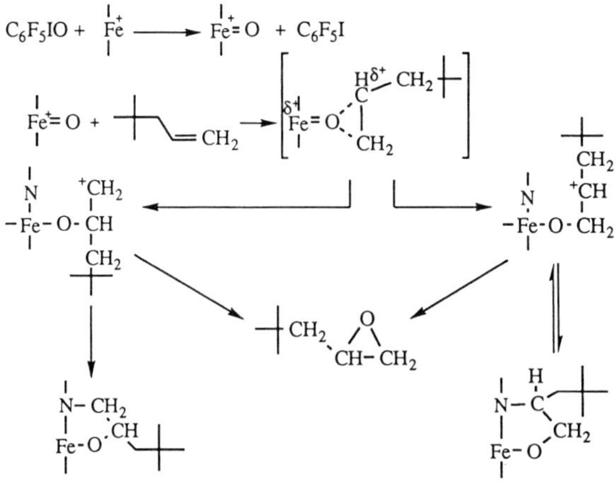

**Scheme 3.4**

Some general observations concerning the stabilities of suicide-labeled (N-alkylated) hemins are consistent with the mechanisms presented above. As the alkene is made more electronegative, the N-alkylhemin becomes more stable. These results—and the fact that the secondary N-alkyl hemin is much more rapidly formed and yet more labile than the primary one—are consistent with heterolytic cleavage of the C—N bond (eq. 95). These mechanisms require that the oxygen atom close the oxirane ring before rotation about the C—C bond. Bond N-alkylation and its reverse appear to be stereospecific (Collman et al., 1990a,b).

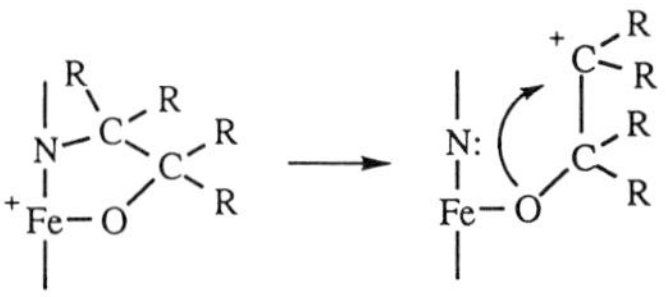

$$\quad (95)$$

## STEREOCHEMISTRY OF EPOXIDATION

Hemin-catalyzed epoxidations under an inert atmosphere are usually completely stereospecific (Groves et al., 1979; Groves and Nemo, 1983). An example is given in eq. (96).

$$F_5C_6IO + TDCPPFe^+Cl^- + \underset{Ph}{\overset{H}{>}}C=C\underset{Ph}{\overset{H}{<}} \longrightarrow \underset{Ph}{\overset{H}{>}}\overset{O}{C-C}\underset{Ph}{\overset{H}{<}} \quad (96)$$

Such results exclude the intervention of free-radical intermediates. However, loss of stereospecificity is often reported, especially when hydroperoxides are the oxidants. Whereas iodosylbenzenes or peracids with hemin catalysts provide stereospecific epoxidation, the use of t-butyl hydroperoxide has been reported to result in complete loss of stereospecificity (eq. 97) (Lee and Bruice, 1985; Mansuy, 1987; Balasubramanian et al., 1989), attributed to epoxidation by t-butyl-peroxyl radicals (Traylor et al., 1989).

$$t\text{-BuOOH} + \text{TPPFeCl} + \underset{Ph}{\overset{H}{>}}C=C\underset{Ph}{\overset{H}{<}} \longrightarrow \underset{Ph}{\overset{H}{>}}\overset{O}{C-C}\underset{H}{\overset{Ph}{<}} + \underset{Ph}{\overset{H}{>}}\overset{O}{C-C}\underset{Ph}{\overset{H}{<}} \quad (97)$$

Side reactions involving peroxyl radicals are indicated in eqs. (62–64) (Mansuy et al., 1984; Battioni et al., 1988). Important competition by hydroperoxides and alkenes for the oxene species leads to low yields of epoxide and loss of stereospecificity (eqs. 98 and 99) (Traylor and Xu, 1987).

$$\overset{|}{\underset{|}{Fe^+}}=O + ROOH \xrightarrow{k_{ab}} \overset{|}{\underset{|}{Fe^+}}=OH + ROO\cdot \quad (98)$$

$$\overset{|}{\underset{|}{Fe^+}}=O + \underset{R}{\overset{\ }{>}}C=C\underset{R}{\overset{\ }{<}} \xrightarrow{k_{ep}} \underset{R}{\overset{\ }{>}}\overset{O}{C-C}\underset{R}{\overset{\ }{<}} \quad (99)$$

Both can be prevented by lowering the $k_{ab}/k_{ep}$ ratio. Fortunately, this is accomplished by the same changes which cause the catalysts to resist oxidative destruction—namely, making the catalyst more electron deficient. For example, the hemin ($F_{28}TPPFe^+Cl^-$) has a very low (<1) value of $k_{ab}/k_{ep}$, whereas tetramesitylporphyrinFe$^+$Cl$^-$ has a high value (~100). With the first catalyst, almost quantitative yields of epoxide are obtained using t-BuOOH or $H_2O_2$ as oxidants (eq. 100) (Traylor et al., 1993). Therefore, useful synthetic epoxidations with hydrogen peroxide are now possible. With the second catalyst, less than 5% yields are produced.

$$F_{28}TPPFe^+Cl^- + H_2O_2 + \text{cyclooctene} \xrightarrow{CH_2Cl_2/MeOH} \text{cyclooctene oxide} \quad (100)$$
$$\text{100\% based on } H_2O_2$$

## HEMIN-CATALYZED OXIDATIONS

### Arenes

One of the first remarkable properties reported for cytochrome P-450 was the *NIH shift* during oxidation of substituted benzenes (Jerina and Daly, 1974). In this oxidation, a deuterium moves from the position of substitution to the adjacent carbon (eq. 101). At first puzzling, this interesting reaction, also observed with a hemin by Chang and Ebina (1981), was shown to involve an epoxidation mechanism (eq. 102) similar to that discussed in the previous section. This shift could occur either along the epoxidation pathway or by opening the epoxide (Ortiz de Montellano, 1986b).

$$\text{(101)}$$

$$\text{(102)}$$

Relative reactivities of arenes toward the high-valent iron intermediate have not been thoroughly studied. Relative rates increase as the ionization potential of the arene decreases, as might be expected from either an electron transfer or direct-attack mechanism (eq. 103).

$$\text{(103)}$$

### Amines

The oxidation of dimethylaniline leads to demethylation in both enzymatic and simple hemin catalysis. This, and the low isotope effect (in the CH$_3$ group), led to the suggestion of electron transfer (eq. 104) (Lindsay Smith and Mortimer, 1985, 1986; Iley et al., 1990).

$$\underset{\text{Me}}{\text{Me}}\text{N}-\text{C}_6\text{H}_5 + \overset{+}{\text{Fe}}=\text{O} \longrightarrow \left[ \underset{\text{H}_3\text{C}}{\overset{+}{\text{N}}} \underset{\text{CH}_3}{\text{CH}_3}-\text{C}_6\text{H}_5 \;\; \text{OFe} \right] \xrightarrow{-\text{H}^+} \underset{\text{H}_3\text{C}}{\text{N}}\underset{}{\text{CH}_2^\cdot}-\text{C}_6\text{H}_5 + \text{HOFe}^+ \longrightarrow \longrightarrow \underset{\text{H}_3\text{C}}{\text{NH}}-\text{C}_6\text{H}_5 \quad (104)$$

The ring opening observed in the enzyme-catalyzed oxidation of cyclopropylamines (Guengerich and MacDonald, 1984) suggests that electron-transfer oxidation occurs in alkylamines as well (eq. 105). All amine oxidations probably occur by this electron-transfer mechanism. It has been suggested that the oxidative dealkylation of alkyl sulfides also proceeds by the same mechanism (Ortiz de Montellano, 1986).

$$\text{cyclopropyl(Me)(NH}_2) + \overset{+}{\text{Fe}}=\text{O} \longrightarrow [\text{cyclopropyl(Me)(NH}_2^{+\cdot}) \; \text{OFe}] \longrightarrow \cdot\text{CH}_2\text{CH}_2\text{C(Me)}=\overset{+}{\text{NH}}_2 \longrightarrow \text{products} \quad (105)$$

### Ethers

The higher oxidation potentials of ethers make electron transfer less attractive. The large CH$_3$CH$_2$(CD$_2$) isotope effect ($k_H/k_D$ = 13–14) observed in the cytochrome P-450-catalyzed O-deethylation of 7-ethoxycoumarin (Miwa et al., 1984) is certainly inconsistent with rate-limiting electron transfer.

However, it is also difficult to understand certain aspects of simple hemin-catalyzed ether oxidations as a hydroxylation reaction involving rate-limiting hydrogen abstraction. Oxidative demethylation of cyclohexylmethyl ether gives more methyl than cyclohexyl oxidation products (eq. 106), whereas the preferred order of hydrogen abstraction is tertiary > secondary > primary.

$$\text{Cy-O-CH}_3 + \overset{+}{\text{Fe}}=\text{O} \longrightarrow \underset{(1)}{\text{Cy}=\text{O}} + \underset{2.5}{\text{Cy-OH}} \quad (106)$$

Currently, the preferred mechanism envisions a rather polarized hydrogen abstraction transition state (eq. 107) for hemin-catalyzed oxidation of ethers. More work is needed.

## Alkanes

The reactions of an oxene ($\overset{|}{\underset{|}{Fe^+}}=O$) with simple alkanes show properties consistent with a hydrogen atom abstraction process (eq. 108).

$$\overset{|}{\underset{|}{Fe^+}}=O + RH \rightarrow \left[\overset{|}{\underset{|}{Fe^+}}-O\overset{H}{\overset{\diagup}{\cdot R}}\right] \rightarrow \overset{|}{\underset{|}{Fe^+}}-\overset{H}{\underset{|}{O}}-R \quad (108)$$

Evidence for the existence of free radicals also comes from the formation of alkyl bromide in the presence of bromotrichloromethane (eq. 109) (Groves et al., 1983b).

$$R\cdot + Cl_3CBr \rightarrow RBr + \cdot CCl_3 \quad (109)$$

However, the formation of a competitive free radical does not account for all of the product since the reaction is at least partially stereospecific. Complete stereospecific hydroxylation of *cis*-decalin has been reported (eq. 110) (Groves et al., 1983b).

(110)

The similarity of stereochemistry from enzyme-catalyzed and simple hemin-catalyzed hydroxylations is shown in eq. (111) (Traylor et al., 1992a). The values in parentheses are the percentage of deuterated (either $D_3$ or $D_4$) exo and endo alcohol products, respectively.

## exo / endo product distribution

Catalysts:
- Br$_8$TDCPPFeCl — (83), (17), (13), (87)
- Cytochrome P-450 — (75), (25), (9), (91)

These results seem to indicate partial retention of configuration with both the liver microsomal enzyme (60% retention for exo) and the model system (47% retention for exo), which is consistent with Groves' (Groves et al., 1978; Groves and Nemo, 1983) proposition that the "rebound" mechanism is predominantly followed by cage collapse to the ROH product (eq. 112).

$$RH + Fe^+{=}O \rightarrow \left[ R \cdot O - {}^+\overset{H}{Fe} \right] \rightarrow ROH + Fe^+ \rightarrow R\cdot + Fe^+OH \quad (112)$$

Additional evidence for a radical cage intermediate was provided by the cytochrome P-450-catalyzed hydroxylation of cyclohexene in which allylic rearrangement was observed (eq. 113) (Groves et al., 1978; Groves and Subramanian, 1984).

(113)

Recently, it has been shown that the primary isotope effect in cyclohexane oxidation ($k_{C_6H_{12}}/k_{C_6D_{12}}$) is very sensitive to hemin structure. As the iron(III) porphyrin changes from the electropositive tetramesitylporphyrin iron(III) Cl, through a series of more halogenated iron(III) porphyrins, to perhalotetraphenyl porphyrin iron(III) Cl, the isotope effect decreases from 9 to less than 3 (Fann, 1992). This strongly suggests a change in the transition state for hydrogen atom abstraction. An appealing explanation is that as the attacking radical is made more electronegative, the transition state adopts more hydride or electron-transfer character (eq. 114).

$$\left[ \begin{array}{c} Y \\ \uparrow \delta^+ \\ Fe=O \cdots H \cdots \underset{\delta^+}{\diagup\!\!\!\diagdown} \\ \downarrow \\ Y \end{array} \right]^{\ddagger} \qquad (114)$$

It is useful to translate all of the mechanistic studies of oxene (Fe$^+$=O)/substrate reactions into a predictive scheme. Scheme 3-5 is an effort to classify mechanisms of oxidation into a predictive model based on CH or OH bond strength, ionization potential of the substrate, and reduction potential of the "oxene" catalyst.

This idea can be viewed as the simple mechanistic continuum:

Hydrogen atom abstraction ---- Electron transfer ---- O atom transfer

The position of the transition state along this continuum depends on both the ionization potential of the substrate and the reduction potential of the "oxene." Highly electronegatively substituted iron porphyrins and substrates with low ionization potentials will tend toward electron transfer. Relative rates of substrate oxidations in Scheme 3-5 increase going down the lists (some have not been measured), with

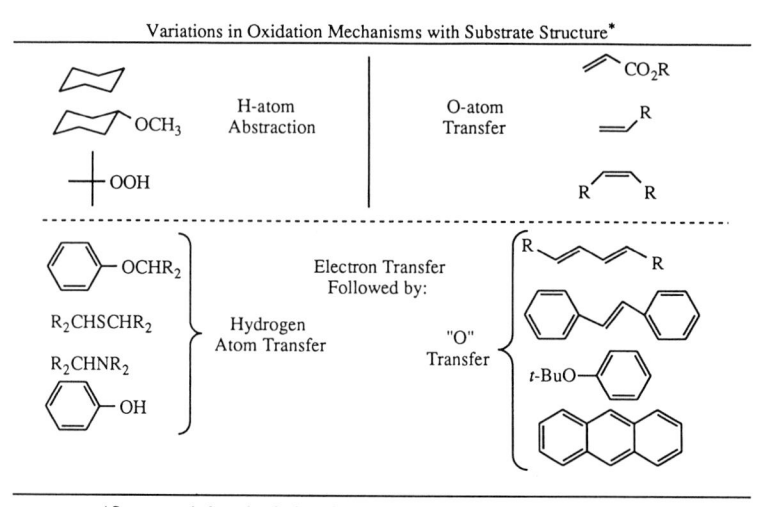

*Structures below the dashed line are thought to react by electron transfer.

**Scheme 3.5**

one exception: *t*-butyl hydroperoxide. The OH bond in *t*-butyl hydroperoxide is unusually weak, leading to facile hydrogen atom abstraction. Phenol also has a weak OH bond and a low ionization potential. It appears to react by electron transfer. This might be expected to change to a hydrogen abstraction mechanism as the phenol is substituted with electronegative groups. This possibility has not been investigated.

## SUMMARY

This brief review has been confined to the mechanistic aspects of dioxygen transport and activation involving heme proteins and related iron porphyrin systems. We have discussed the various controls on the affinities and dynamics of dioxygen binding and the mechanisms by which substrates are catalytically oxidized by the O–O bond in dioxygen, hydrogen peroxide, and hydroperoxides. Other metalloporphyrins and other porphyrin-like metal complexes have been studied in detail and are reviewed elsewhere (Mansuy, 1987; Stubbe and Kozarich, 1987; McGall et al., 1992; Meunier, 1992). Other important biological dioxygen carriers in which µ-oxo diiron (Wieghardt, 1986; Hartman et al., 1987; Wilkins and Wilkins, 1987; Beer et al., 1991; Stubbe, 1991) or di-copper complexes (Reed, 1986; Karlin and Gultneh, 1987, 1989; Latour, 1988; Fenton, 1989; Schindler et al., 1989; Sorrell, 1989; Salvato and Beltrami, 1990; Paul et al., 1991; Karlin et al., 1992) are involved have been studied and elegantly modeled. Reviews are available. Overall reviews of metal-catalyzed oxidations, both biochemical and chemical, have also been summarized (Sheldon and Kochi, 1981; Sheldon, 1985; Koola and Kochi, 1987).

Although major differences are expected as the metal type and its environment are changed, it seems reasonable that many of the principles described here will be applicable to other systems.

## REFERENCES

AASA, R., VANNGARD, T., and DUNFORD, H. B. (1975) EPR Studies on Compound I of Horseradish Peroxidase. *Biochim. Biophys. Acta*, 391, 259–264.

ADAM, W., CURCI, R., and EDWARDS, J. O. (1989) Dioxiranes: A New Class of Powerful Oxidants. *Acc. Chem. Res.*, 22, 205–211.

ADEDIRAN, S. A., and, DUNFORD, H. B. (1983) Structure of Horseradish Peroxidase Compound I. Kinetic Evidence for the Incorporation of One

Oxygen Atom from the Oxidizing Substrate into the Enzyme. *Eur. J. Biochem.*, 132, 147–150.

ALBERDING, N., AUSTIN, R. H., CHAN, S. S., EISENSTEIN, L., FRAUENFELDER, H., GOOD, D., KAUFMANN, K., MARDEN, M., NORDLUND, T. M., REINISCH, L., REYNOLDS, A. H., SORENSEN, L. B., WAGNER, G. C., and YUE, K. T. (1978a) Fast Reactions in Carbon Monoxide Binding to Heme Proteins. *Biophys. J.*, 24, 319–334.

ALBERDING, N., CHAN, S. S., EISENSTEIN, L., FRAUENFELDER, H., GOOD, D., GUNSALUS, I. C., NORDLUND, T. M., PERUTZ, M. F., REYNOLDS, A. H., and SORENSEN, L. B. (1978b) Binding of Carbon Monoxide to Isolated Hemoglobin Chains. *Biochemistry*, 17, 43–51.

ALMOG, J., BALDWIN, J. E., DYER, R. L., HUFF, J., and WILKERSON, C. J. (1974) Reversible Binding of Dioxygen to Mesoporphyrin IX Derivatives at Low Temperatures. *J. Am. Chem. Soc.*, 96, 5600–5601.

ALMOG, J., BALDWIN, J. E., DYER, R. L., and PETERS, M. (1975a) Condensation of Tetraaldehydes with Pyrrole. Direct Synthesis of Capped Porphyrins. *J. Am. Chem. Soc.*, 97, 226–227.

ALMOG, J., BALDWIN, J. E., and HUFF, J. (1975b) Reversible Oxygenation and Autoxidation of a Capped Porphyrin Iron(II) Complex. *J. Am. Chem. Soc.*, 97, 227–228.

ALPERT, B., EL MOHSNI, S., LINDQVIST, L., and TFIBEL, F. (1979) Transient Effects in the Nanosecond Laser Photolysis of Carboxyhemoglobin: "Cage" Recombination and Spectral Evolution of the Protein. *Chem. Phys. Lett.*, 64, 11–16.

ALTMAN, J., LIPKA, J. J., KUNTZ, I., and WASKELL, L. (1989) Identification by Proton Nuclear Magnetic Resonance of the Histidines in Cytochrome $b_5$ Modified by Diethyl Pyrocarbonate. *Biochemistry*, 28, 7516–7523.

AMICONI, G., ANTONINI, E., BRUNORI, M., TORMANECK, H., and HUBER, R. (1972) Functional Properties of Native and Reconstituted Hemoglobins from *Chironomus thummi thummi*. *Eur. J. Biochem.*, 31, 52–58.

AMUNDSEN, A. R., and VASKA, L. (1975) Oxygenation of *meso*-Tetrakis (2,4,6-alkoxyphenyl)-porphinato Complexes of Iron(II): Some Unusual Observations. *Inorg. Chim. Acta*, 14, L49–L51.

ANDERSON, D. L., WESCHLER, C. J., and BASOLO, F. (1974) Reversible Reaction of Simple Ferrous Porphyrins with Molecular Oxygen at Low Temperatures. *J. Am. Chem. Soc.*, 96, 5599–5600.

ANFINRUD, P. A., HAN, C., and HOCHSTRASSER, R. M. (1989) Direct Observations of Ligand Dynamics in Hemoglobin by Subpicosecond Infrared Spectroscopy. *Proc. Natl. Acad. Sci. U.S.A.*, 86, 8387–8391.

ANSARI, A., DIIORIO, E. E., DLOTT, D. D., FRAUENFELDER, H., IBEN, I. E. T., LANGER, P., RODER, H., SAUKE, T. B., and SHYAMSUNDER, E. (1986) Ligand Binding to Heme Proteins: Relevance of Low-temperature Data. *Biochemistry*, 25, 3139–3146.

ANTONINI, E., and BRUNORI, M. (1971) *Hemoglobin and Myoglobin in Their Reactions with Ligands*, North-Holland, Amsterdam.

ANTONINI, E., and GIBSON, Q. H. (1960) Some Observations on the Kinetics of the Reactions with Gases of Natural and Reconstituted Haemoglobins. *Biochem. J.*, 76, 534–538.

ANZENBACHER, P., DAWSON, J. H., and KITAGAWA, T. (1989) Towards a Unified Concept of Oxygen Activation by Heme Enzymes: The Role of the Proximal Ligand. *J. Mol. Struct.*, 214, 149–158.

ARASASINGHAM, R. D., BALCH, A. L., and LATOS-GRAZYNSKI, L. (1987) Identification of Intermediates and Products in the Reaction of Porphyrin Iron(III) Alkyl Complexes with Dioxygen. *J. Am. Chem. Soc.*, 109, 5846–5847.

ARTAUD, I., GREGOIRE, N., and MANSUY, D. (1989) Suicidal Inactivation of Iron-porphyrins during Trans Hex-2-ene Oxidation: First Isolation and Characterization of N-alkyl-porphyrins with a N-CHR-CHR'OH structure. *New J. Chem.*, 13, 581–586.

AUSTIN, R. H., BEESON, K. W., EISENSTEIN, L., FRAUENFELDER, H., and GUNSALUS, I. C. (1975) Dynamics of Ligand Binding to Myoglobin. *Biochemistry*, 14, 5355–5373.

BABCOCK, G. T., VAROTSIS, C., and ZHANG, Y. (1992) Oxygen Activation in Cytochrome Oxidase and in Other Heme Proteins. *Biochim. Biophys. Acta*, 1101, 192–194.

BABCOCK, G. T., and WIKSTRÖM, M. (1992) $O_2$ Activation and the Conservation of Energy in Cell Respiration. *Nature*, 356, 301–309.

BALASUBRAMANIAN, P. N., LINDSAY SMITH, J. R., DAVIES, M. J., KAARET, T. W., and BRUICE, T. C. (1989) Dynamics of Reaction of (*meso*-Tetrakis(2,6-dimethyl-3-sulfonatophenyl)porphinato)-iron(III) Hydrate with *tert*-Butyl Hydroperoxide in Aqueous Solution. 2. Establishment of a Mechanism That Involves Homolytic O—O Bond Breaking and One-electron Oxidation of the Iron(III) Porphyrin. *J. Am. Chem. Soc.*, 111, 1477–1483.

BALASUBRAMANIAN, P. N., SCHMIDT, E. S., and BRUICE, T. C. (1987) Catalase Modeling. 2. Dynamics of Reaction of a Water-soluble and Non µ-Oxo Dimer Forming Manganese(III) Porphyrin with Hydrogen Peroxide. *J. Am. Chem. Soc.*, 109, 7865–7873.

BALCH, A. L. Private communication.

BALCH, A. L., CHAN, Y.-W., CHENG, R.-J., LA MAR, G. N., LATOS-GRAZYNSKI, L., and RENNER, M. W. (1984) Oxygenation Patterns for Iron(II) Porphyrins. Peroxo and Ferryl ($Fe^{IV}O$) Intermediates Detected by $^1H$ Nuclear Magnetic Resonance Spectroscopy During the Oxygenation of (Tetramesitylporphyrin)iron(II). *J. Am. Chem. Soc.*, 106, 7779–7785.

BALCH, A. L., CORNMAN, C. R., LATOS-GRAZYNSKI, L., and RENNER, M. W. (1992) Highly Oxidized Iron Complexes of N-methyltetra-*p*-tolylporphyrin. *J. Am. Chem. Soc.*, 114, 2230–2237.

BALCH, A. L., HART, R. L., LATOS-GRAZYNSKI, L., and TRAYLOR, T. G. (1990) Nuclear Magnetic Resonance Studies of the Formation of Tertiary Alkyl Complexes of Iron(III) Porphyrins and Their Reactions with Dioxygen. *J. Am. Chem. Soc.*, 112, 7382–7388.

BALCH, A. L., LATOS-GRAZYNSKI, L., and RENNER, M. W. (1985) Oxidation of Red Ferryl [Fe$^{IV}$O)$^{2+}$] Porphyrin Complexes to Green Ferryl [Fe$^{IV}$(O)$^{2+}$] porphyin Radical Complexes. *J. Am. Chem. Soc.*, 107, 2983–2985.

BALDWIN, J. E., CAMERON, J. H., CROSSLEY, M. J., DAGLEY, I. J., HALL, S. R., and KLOSE, T. (1984) Synthesis of Iron(II) C$_2$-capped Strapped Porphyrin Complexes and Their Reaction with Dioxygen. *J. Chem. Soc., Dalton Trans.*, 1739–1746.

BALDWIN, J. E., CROSSLEY, M. J., KLOSE, T., O'REAR, E. A., III, and PETERS, M. K. (1982) Syntheses and Oxygenation of Iron(II) Strapped Porphyrin Complexes. *Tetrahedron*, 38, 27–39.

BALDWIN, J. E., KLOSE, T., and PETERS, M. (1976) Syntheses of "Strapped" Porphyrins and the Oxygenation of Their Iron(II) Complexes. *J. Chem. Soc., Chem. Commun.*, 881–883.

BALDWIN, J. M. (1975) Structure and Function of Haemoglobin. *Prog. Biophys. Molec. Biol.*, 29, 225–320.

BANDYOPADHYAY, D., MAGDE, D., TRAYLOR, T. G., and SHARMA, V. S. (1992) Quaternary Structure and Geminate Recombination in Hemoglobin: Flow-flash Studies on $\alpha_2^{CO}\beta_2$ and $\alpha_2\beta_2^{CO}$. *Biophys. J.*, 673–681.

BANDYOPADHYAY, D., WALDA, K. N., MAGDE, D., TRAYLOR, T. G., and SHARMA, V. S. (1990) Quaternary Structure and the Geminate Recombination of Carp Hemoglobin with Methylisocyanide. *Biochem. Biophys. Res. Commun.*, 171, 306–312.

BARTLETT, P. D., and HIATT, R. (1958) A Series of Tertiary Butyl Peresters Showing Concerted Decomposition. *J. Am. Chem. Soc.*, 80, 1398–1405.

BASOLO, F., HOFFMAN, B. M., and IBERS, J. A. (1975) Synthetic Oxygen Carriers of Biological Interest. *Acc. Chem. Res.*, 8, 384–392.

BATTERSBY, A. R., and HAMILTON, A. D. (1980) Synthesis of a Doubly-bridged Oxygen Carrier Which Shows Reduced Affinity for Carbon Monoxide. *J. Chem. Soc., Chem. Commun.*, 117–119.

BATTERSBY, A. R., HARTLEY, S. G., and TURNBULL, M. G., (1978) Synthetic Routes to Singly and Doubly Bridged Porphyrins. *Tetrahedron Lett.* 3169–3172.

BATTIONI, P., RENAUD, J. P., BARTOLI, J. F., REINA-ARTILES, M., FORT, M., and MANSUY, D. (1988) Monooxygenase-like Oxidation of Hydrocarbons by H$_2$O$_2$ Catalyzed by Manganese Porphyrins and Imidazole: Selection of the Best Catalytic System and Nature of the Active Oxygen Species. *J. Am. Chem. Soc.*, 110, 8462–8470.

BEER, R. H., TOLMAN, W. B., BOTT, S. G., and LIPPARD, S. J. (1991) Effects of a Bridging Dicarboxylate Ligand on the Synthesis and Physical Properties of (μ-Oxo)*bis*(μ-carboxylato)diiron(III) Complexes. *Inorg. Chem.*, 30, 2082–2092.

BENNETT, L. E. (1973) Metalloprotein Redox Reactions, in *Prog. Inorg. Chem.*, (S. J. Lippard, Ed.), John Wiley & Sons, New York, 18, pp. 1–176.

BERINSTAIN, A. B., ENGLISH, A. M., HILL, B. C., and SHARMA, D. (1990) Picosecond Flash Photolysis of Carboxy Horseradish Peroxidase: Rapid Geminate Recombination in the Presence of Benzohydroxamic Acid. *J. Am. Chem. Soc.*, 112, 9649–9651.

BILL, E., DING, X. Q., BOMINAAR, E. L., TRAUTWEIN, A. X., WINKLER, H., MANDON, D., WEISS, R., GOLD, A., and JAYARAJ, K. (1990) Evidence for Variable Metal-radical Spin Coupling in Oxoferryl–Porphyrin Cation Radical Complexes. *Eur. J. Biochem.*, 188, 665–672.

BONAVENTURA, J., and WOOD, S. C. (1980) Respiratory Pigments: Overview. *Am. Zool.*, 20, 5–6.

BOSO, B., LANG, G., MCMURRY, T. J., and GROVES, J. T. (1983) Mössbauer Effect Study of Tight Spin Coupling in Oxidized Chloro-5,10,15,20-tetra(mesity)porphyrinato Iron(III). *J. Chem. Phys.*, 79, 1122–1126.

BOSSHARD, H. R., ANNI, H., and YONETANI, T. (1991) Yeast Cytochrome C, Peroxidase, in *Peroxidases in Chemistry and Biology* (J. Everse, K. E. Everse, and M. B. Grisham, Eds.), CRC, Boca Raton, FL, 2, pp, 51–84.

BRINIGAR, W. S., CHANG, C. K., GEIBEL, J., and TRAYLOR, T. G. (1974) Solvent Effects on Reversible Formation and Oxidative Stability of Heme–Oxygen Complexes. *J. Am. Chem. Soc.*, 96, 5597–5599.

BRUICE, T. C. (1986) Chemical Studies Pertaining to the Chemistry of Cytochrome P-450 and the Peroxidases. *Ann. N.Y. Acad. Sci.*, 471, 83–98.

BRUICE, T. C. (1991) Reactions of Hydroperoxides with Metallotetraphenylporphyrins in Aqueous Solutions. *Acc. Chem. Res.*, 24, 243–249.

BRUICE, T. C., BALASUBRAMANIAN, P. N., LEE, R. W., and LINDSAY SMITH, J. R. (1988) The Mechanism of Hydroperoxide O—O Bond Scission on Reaction of Hydroperoxides with Iron(III) Porphyrins. *J. Am. Chem. Soc.*, 110, 7890–7892.

BRUICE, T. C., ZIPPLIES, M. F., and LEE, W. A. (1986) The pH Dependence of the Mechanism of Reaction of Hydrogen Peroxide with a Nonaggregating, Non-μ-oxo Dimer-forming Iron(III) Porphyrin in Water. *Proc. Natl. Acad. Sci. U.S.A.*, 83, 4646–4649.

BRUNORI, M., and CHANCE, B. (Eds.) (1988) Cytochrome Oxidase: Structure, Function and Physiopathology. *Ann. NY Acad. Sci.*, 550, 1–382.

BUDGE, J. R., ELLIS, P. E., JR., JONES, R. D., LINARD, J. E., BASOLO, F., BALDWIN, J. E., and DYER, R. L. (1979) The Iron(II) "Homologous Cap" Porphyrin. A Novel Dioxygen Binder. *J. Am. Chem. Soc.*, 101, 4760–4762.

CALDIN, E. F., and HASINOFF, B. B. (1975) Diffusion-controlled Kinetics in the Reaction of Ferroprotoporphyrin IX with Carbon Monoxide. *J. Chem. Soc., Faraday Trans. I*, 71, 515–527.

CAMPESTRINI, S., ROBERT, A., and MEUNIER, B. (1991) Ozone Epoxidation of Olefins Catalyzed by Highly Robust Manganese and Iron Porphyrin Complexes. *J. Org. Chem.*, 56, 3725–3727.

CAPALDI, R. A. (1990a) Structure and Function of Cytochrome *c* Oxidase. *Ann. Rev. Biochem.*, 59, 569–596.

CAPALDI, R. A. (1990b) Structure and Assembly of Cytochrome *c* Oxidase. *Arch. Biochem. Biophys.*, 280, 252–262.

CAUGHEY, W. S. (1970) Carbon Monoxide Bonding in Hemeproteins. *Ann. N.Y. Acad. Sci.*, 174, 148–153.

CHAN, S. I., and LI, P. M. (1990) Cytochrome *c* Oxidase: Understanding Nature's Design of a Proton Pump. *Biochemistry*, 29, 1–12.

CHANCE, B. (1951) Enzyme Substrate Compounds: Mechanism of Action of Hydroperoxidases, in *The Enzymes*, 2 (Part 1) Academic Press, New York pp 428–434.

CHANCE, B. (1952) The Transition from the Primary to the Secondary Peroxidase–Peroxide Complex. *Arch. Biochem. Biophys.*, 37, 235–237.

CHANCE, M. R., COURTNEY, S. H., CHAVEZ, M. D., ONDRIAS, M. R., and FRIEDMAN, J. M. (1990) $O_2$ and CO Reactions with Heme Proteins: Quantum Yields and Geminate Recombination on Picosecond Time Scales. *Biochemistry*, 29, 5537–5545.

CHANG, C. K., and ABDALMUHDI, I. (1984) Biphenylenediporphyrin: Two Cofacially Ordered Porphyrins with Biphenylene Bridge. *Angew. Chem.*, 96, 154–155.

CHANG, C. K., and EBINA, F. (1981) NIH Shift in Haemin-iodosylbenzene-mediated Hydroxylations. *J. Chem. Soc., Chem. Commun.*, 778–779.

CHANG, C. K., KUO, M.-S., and WANG, C.-B. (1977a) Stacked Double-macrocyclic Ligands II. Synthesis of Cofacial Diporphyrins (1–2). *J. Heterocyclic Chem.*, 14, 943–945.

CHANG, C. K., POWELL, D., and TRAYLOR, T. G. (1977b) Kinetics and Mechanisms of Oxidation of Hemoprotein Model Compounds. *Croatia Chem. Acta*, 49, 295–307.

CHANG, C. K., and TRAYLOR, T. G. (unpublished).

CHANG, C. K., and TRAYLOR, T. G. (1973a) Solution Behavior of a Synthetic Myoglobin Active Site. *J. Am. Chem. Soc.*, 95, 5810–5811.

CHANG, C. K., and TRAYLOR, T. G. (1973b) Synthesis of the Myoglobin Active Site. *Proc. Natl. Acad. Sci. U.S.A.*, 70, 2647–2650.

CHANG, C. K., and TRAYLOR, T. G. (1975a) Reversible Oxygenation of Protoheme–Imidazole Complex in Aqueous Solution. *Biochem. Biophys. Res. Commun.*, 62, 729–735.

CHANG, C. K., and TRAYLOR, T. G. (1975b) Kinetics of Oxygen and Carbon Monoxide Binding to Synthetic Analogs of the Myoglobin and Hemoglobin Active Sites. *Proc. Natl. Acad. Sci, U.S.A.*, 72, 1166–1170.

CHANG, C. K., and WANG, C.-B. (1982) Binuclear Porphyrin Complexes: Heme Models for Oxygen Binding and Reduction, in *Electron Transport and Oxygen Utilization* (C. Ho, Ed.), Elsevier North-Holland, Amsterdam, pp. 237–243.

CHANG, C. K., WARD, B., YOUNG, R., and KONDYLIS, M. P. (1988) Fine Tuning of Heme Reactivity: Hydrogen-bonding and Dipole Interactions Affecting Ligand Binding to Hemoproteins. *J. Macromolecular Sci.—Chem.*, A25(10 & 11), 1307–1326.

CHIN, D.-H., BALCH, A. L., and LA MAR, G. N. (1980a) Formation of Porphyrin Ferryl (FeO$^{2+}$) Complexes Through the Addition of Nitrogen Bases to Peroxo-bridged Iron(III) Porphyrins. *J. Am. Chem. Soc.*, 102, 1446–1448.

CHIN, D.-H., DEL GAUDIO, J., LA MAR, G. N., and BALCH, A. L. (1977) Detection and Characterization of the Long-postulated Fe-OO-Fe Intermediate in the Autoxidation of Ferrous Porphyrins. *J. Am. Chem. Soc.*, 99, 5486–5488.

CHIN, D.-H., LA MAR, G. N., and BALCH, A. L. (1980b) On the Mechanism of Autoxidation of Iron(II) Porphyrins. Detection of a Peroxo-bridged Iron(III) Porphyrin Dimer and the Mechanism of Its Thermal Decomposition to the Oxo-bridged Iron(III) Porphyrin Dimer. *J. Am. Chem. Soc.*, 102, 4344–4350.

CHIN, D.-H., LA MAR, G. N., and BALCH, A. L. (1980c) Role of Ferryl (FeO$^{2+}$) Complexes in Oxygen Atom Transfer Reactions. Mechanism of Iron(II) Porphyrin Catalyzed Oxygenation of Triphenylphosphine. *J. Am. Chem. Soc.*, 102, 5945–5947.

CHU, M. M. L., CASTRO, C. E., and HATHAWAY, G. M. (1978) Oxidation of Low-spin Iron(II) Porphyrins by Molecular Oxygen. An Outer Sphere Mechanism. *Biochemistry*, 17, 481–486.

COLLMAN, J. P. (1977) Synthetic Models for the Oxygen-binding Hemoproteins. *Acc. Chem. Res.*, 10, 265–272.

COLLMAN, J. P., BRAUMAN, J. I., COLLINS, T. J., IVERSON, B. L., LANG, G., PETTMAN, R. B., SESSLER, J. L., and WALTERS, M. A. (1983a) Synthesis and Characterization of the "Pocket" Porphyrins. *J. Am. Chem. Soc.*, 105, 3038–3052.

COLLMAN, J. P., BRAUMAN, J. I., IVERSON, B. L., SESSLER, J. L., MORRIS, R. M., and GIBSON, Q. H. (1983b) O$_2$ and CO Binding to Iron(II) Porphyrins: A Comparison of the "Picket Fence" and "Pocket" Porphyrins. *J. Am. Chem. Soc.*, 105, 3052–3064.

COLLMAN, J. P., BRAUMAN, J. I., COLLINS, T. J., IVERSON, B. L., and SESSLER, J. L. (1981) The "Pocket" Porphyrin: A Hemoprotein Model with Lowered CO Affinity. *J. Am. Chem. Soc.*, 103, 2450–2452.

COLLMAN, J. P., BRAUMAN, J. I., and DOXSEE, K. M. (1979) Carbon Monoxide Binding to Iron Porphyrins. *Proc. Natl. Acad. Sci. U.S.A.*, 76, 6035–6039.

COLLMAN, J. P., BRAUMAN, J. I., DOXSEE, K. M., HALBERT, T. R., BUNNENBERG, E., LINDER, R. E., LAMAR, G. N., Del GAUDIO, J., LANG, G., and SPARTALIAN, K. (1980) Synthesis and Characterization of "Tailed Picket Fence" Porphyrins. *J. Am. Chem. Soc.*, 102, 4182–4192.

COLLMAN, J. P., BRAUMAN, J. I., DOXSEE, K. M., HALBERT, T. R., and SUSLICK, K. S. (1978) Model Compounds for the T State of Hemoglobin. *Proc. Natl. Acad. Sci. U.S.A.*, 75, 564–568.

COLLMAN, J. P., BRAUMAN, J. I., HALBERT, T. R., and SUSLICK, K. S. (1976) Nature of $O_2$ and CO Binding to metalloporphyrins and Heme Proteins. *Proc. Natl. Acad. Sci. U.S.A.*, 73, 3333–3337.

COLLMAN, J. P., ELLIOTT, C. M., HALBERT, T. R., and TOVROG, B. S. (1977) Synthesis and Characterization of "Face-to-face" Porphyrins. *Proc. Natl. Acad. Sci. U.S.A.*, 74, 18–22.

COLLMAN, J. P., GAGNÉ, R. R., HALBERT, T. R., MARCHON, J. C., and REED, C. A. (1973) Reversible Oxygen Adduct Formation in Ferrous Complexes Derived from a Picket Fence Porphyrin. Model for Oxymyoglobin. *J. Am. Chem. Soc.*, 95, 7868–7870.

COLLMAN, J. P., GAGNÉ, R. R., REED, C. A., HALBERT, T. R., LANG, G., and ROBINSON, W. T. (1975) "Picket Fence Porphyrins." Synthetic Models for Oxygen Binding Hemoproteins. *J. Am. Chem. Soc.*, 97, 1427–1439.

COLLMAN, J. P., HAMPTON, P. D., and BRAUMAN, J. I. (1986a) Stereochemical and Mechanistic Studies of the "Suicide" Event in Biomimetic P-450 Olefin Epoxidation. *J. Am. Chem. Soc.*, 108, 7861–7862.

COLLMAN, J. P., HAMPTON, P. D., and BRAUMAN, J. I. (1990a) Suicide Inactivation of Cytochrome P-450 Model Compounds by Terminal Olefins. 1. A Mechanistic Study of Heme N-Alkylation and Epoxidation. *J. Am. Chem. Soc.*, 112, 2977–2986.

COLLMAN, J. P., HAMPTON, P. D., and BRAUMAN, J. I. (1990b) Suicide Inactivation of Cytochrome P-450 Model Compounds by Terminal Olefins. 2. Steric and Electronic Effects in Heme N-Alkylation and Epoxidation. *J. Am. Chem. Soc.*, 112, 2986–2998.

COLLMAN, J. P., KODADEK, T., and BRAUMAN, J. I. (1986b) Oxygenation of Styrene by Cytochrome P-450 Model Systems: A Mechanistic Study. *J. Am. Chem. Soc.*, 108, 2588–2594.

COLLMAN, J. P., KODADEK, T., RAYBUCK, S. A., BRAUMAN, J. I., and PAPAZIAN, L. M. (1985) Mechanism of Oxygen Atom Transfer from High Valent Porphyrins to Olefins: Implications to the Biological Epoxidation of Olefins by Cytochrome P-450. *J. Am. Chem. Soc.*, 107, 4343–4345.

COOPER, C. E. (1990) The Steady-state Kinetics of Cytochrome-c Oxidation by Cytochrome Oxidase. *Biochim. Biophys. Acta*, 1017, 187–203.

CUSANOVICH, M. A., MEYER, T. E., and TOLLIN, G. (1988) c-Type cytochromes: Oxidation-reduction Properties. *Advances in Inorganic Biochemistry* (G. L. Eichhorn and L. G. Marzilli, Eds.), Elsevier, New York 7, pp. 37–91.

DAVID, S., DOLPHIN, D., and JAMES, B. R. (1986) Discriminatory Binding of Carbon Monoxide Versus Dioxygen within Heme Proteins and Model Hemes, in *Frontiers in Bioinorganic Chemistry* (A. Xavier, Ed.) VCH Publishers, New York, pp. 163–182.

DAVID, S., JAMES, B. R., DOLPHIN, D., TRAYLOR, T. G., and LOPEZ, M. A. (1994) Dioxygen and Carbon Monoxide Binding to Apolar Cyclophane Hemes: Durene-capped Hemes. *J. Am. Chem. Soc.*, 116, 6–14.

DAVIES, D. M., and LAWTHER, J. M. (1989) Kinetics and Mechanism of Electron Transfer from Dithionite to Microsomal Cytochrome b5 and to Forms of the Protein Associated with Charged and Neutral Vesicles. *Biochem. J.*, 258, 375–380.

DEISSEROTH, A., and DOUNCE, A. L. (1970) Catalase: Physical and Chemical Properties, Mechanism of Catalysis, and Physiological Role. *Physiol. Rev.*, 50, 319–375.

DE YOUNG, A., PENNELLY, R. R., TAN-WILSON, A. L., and NOBLE, R. W. (1976) Kinetic Studies on the Binding Affinity of Human Hemoglobin for the 4th Carbon Monoxide Molecule, $L_4$. *J. Biol. Chem.*, 251, 6692–6698.

DICKEN, C. M., WOON, T. C., and BRUICE, T. C. (1986) Kinetics and Mechanisms of Oxygen Transfer in the Reaction of p-Cyano-N,N-dimethylaniline N-oxide with Metalloporphyrin Salts. 3. Catalysis by [*meso*-Tetrakis(2,6-dichlorophenyl)porphinato]iron(III) Chloride. *J. Am. Chem. Soc.*, 108, 1636–1643.

DICKERSON, R. E. (1972) The Structure and History of an Ancient Protein. *Sci. Am.*, 226, 58–72.

DICKERSON, R. E. (1980) Cytochrome *c* and the Evolution of Energy Metabolism. *Sci. Am.*, 242, 137–153.

DIEKMANN, H., CHANG, C. K., and TRAYLOR, T. G. (1971) Cyclophane Porphyrin. *J. Am. Chem. Soc.*, 93, 4068–4070.

DIX, T. A., FONTANA, R., PANTHANI, A., and MARNETT, L. J. (1985) Hematin-catalyzed Epoxidation of 7,8-Dihydroxy-7,8-dihydrobenzo[a]pyrene by Polyunsaturated Fatty Acid Hydroperoxides. *J. Biol. Chem.*, 260, 5358–5365.

DIX, T. A., MARNETT, L. J. (1981) Free-radical Epoxidation of 7,8-Dihydroxy-7,8-dihydrobenzo-[a]pyrene by Hematin and Polyunsaturated Fatty Acid Hydroperoxides. *J. Am. Chem. Soc.*, 103, 6744–6746.

DIX, T. A., and MARNETT, L. J. (1983) Hematin-catalyzed Rearrangement of Hydroperoxylinoleic Acid to Epoxy Alcohols Via an Oxygen Rebound. *J. Am. Chem. Soc.*, 105, 7001–7002.

DIX, T. A., and MARNETT, L. J. (1985) Conversion of Linoleic Acid Hydroperoxide to Hydroxy, Keto, Epoxyhydroxy, and Trihydroxy Fatty Acids by Hematin. *J. Biol. Chem.*, 260, 5351–5357.

DLOTT, D. D., FRAUENFELDER, H., LANGER, P., RODER, H., and DIIORIO, E. E. (1983) Nanosecond Flash Photolysis Study of Carbon Monoxide Binding to the β Chain of Hemoglobin Zurich [β63(E7)His→Arg]. *Proc. Natl. Acad. Sci. U.S.A.*, 80, 6239–6243.

DOLPHIN, D., FORMAN, A., BORG, D. C., FAJER, J., and FELTON, R. H. (1971) Compounds I of Catalase and Horseradish Peroxidase: π-Cation Radicals. *Proc. Natl. Acad. Sci. U.S.A.*, 68, 614–618.

DOSTER, W., BOWNE, S. F., FRAUENFELDER, H., REINISCH, L., and SHYAMSUNDER, E. (1987) Recombination of Carbon Monoxide to Ferrous Horseradish Peroxidase Types A and C. *J. Mol. Biol.*, 194, 299–312.

DUDDELL, D. A., MORRIS, R. J., and RICHARDS, J. T. (1979) Ultra-fast Recombination in Nanosecond Laser Photolysis of Carbonylhaemoglobin. *J. Chem. Soc., Chem. Commun.*, 75–76.

DUNFORD, H. B. (1991) Horseradish Peroxidase: Structure and Kinetic Properties, in *Peroxidases in Chemistry and Biology* (J. Everse, K. E. Everse, and M. B. Grisham, Eds.), CRC, Boca Raton, FL, 2, 1–24.

DUNFORD, H. B., and STILLMAN, J. S. (1976) On the Function and Mechanism of Action of Peroxidases. *Coord. Chem. Rev.*, 19, 187–251.

EDWARDS, S. L., XUONG, N. H., HAMLIN, R. C., and KRAUT, J. (1987) Crystal Structure of Cytochrome *c* Peroxidase Compound I. *Biochemistry*, 26, 1503–1511.

EGEBERG, K. D., SPRINGER, B. A., MARTINIS, S. A., SLIGAR, S. G., MORIKIS, D., and CHAMPION, P. M. (1990) Alteration of Sperm Whale Myoglobin Heme Axial Ligation by Site-directed Mutagenesis. *Biochemistry*, 29, 9783–9791.

ELLIS, P. E., JR., LINARD, J. E., SZYMANSKI, T., JONES, R. D., BUDGE, J. R., and BASOLO, F. (1980) Axial Ligation Constants of Iron(II) and Cobalt(II) "Capped" Porphyrins. *J. Am. Chem. Soc.*, 102, 1889–1896.

ERMAN, J. E., VITELLO, L. B., MILLER, M. A., and KRAUT, J. (1992) Active-site Mutations in Cytochrome *c* Peroxidase: A Critical Role for Histidine-52 in the Rate of Formation of Compound. I. *J. Am. Chem. Soc.*, 114, 6592–6593.

EVERSE, J., EVERSE, K. E., and GRISHAM, M. B. (Eds.) (1991) *Peroxidases in Chemistry and Biology.* CRC, Boca Raton, FL.

FANN, W.-P. (1992) Cytochrome P-450 Model Compounds: *Mechanisms of Iron(III) Porphyrin Catalyzed Oxidation Reactions.* Doctoral Dissertation, University of California, San Diego.

FELTON, R. H., OWEN, G. S., DOLPHIN, D., and FAJER, J. (1971) Iron(IV) Porphyrins. *J. Am. Chem. Soc.*, 93, 6332–6334.

FENTON, D. E. (1989) Copper Biosites: The Merits of Models. *Pure Appl. Chem.*, 61, 903–908.

FINZEL, B. C., POULOS, T. L., and KRAUT, J. (1984) Crystal Structure of Yeast Cytochrome *c* Peroxidase Refined at 1.7-Å Resolution. *J. Biol. Chem.*, 259, 13027–13036.

FITA, I., and ROSSMANN, M. G. (1985) The Active Center of Catalase. *J. Mol. Biol.*, 185, 21–37.

FRANZUS, B. BAIRD, W. C., JR., CHAMBERLAIN, N. F., HINES, T., and SNYDER, E. I. (1968) On the Question of 7-*syn* and 7-*anti*-proton Absorptions in the Nuclear Magnetic Resonance Spectra of Norbornenes. *J. Am. Chem. Soc.*, 90, 3721–3724.

FRANZUS, B., BAIRD, W. C., JR., and SURRIDGE, J. H. (1968) Synthesis of *exo,exo*-5,6-Dideuterio-*syn*-7-acetoxynorbornene and *exo,exo*-5,6-Dideuterio-2-norbornene. *J. Org. Chem.*, 33, 1288–1290.

FREW, J. E., and JONES, P. (1984) Structure and Functional Properties of Peroxidases and Catalases, in *Advances in Inorganic and Bioinorganic Mechanisms* (A. G. Sykes, Ed.), Academic Press, New York, 3, pp. 175–212.

FUJII, H., and ICHIKAWA, K. (1992) Preparation and Characterization of an A₁u Oxoiron(IV) Porphyrin π–cation–radical Complex. *Inorg. Chem.*, 31, 1110–1112.

GEIBEL, J., CANNON, J., CAMPBELL, D., and TRAYLOR, T. G. (1978) Model Compounds for R-state and T-state Hemoglobins. *J. Am. Chem. Soc.*, 100, 3575–3585.

GIACOMETTI, G. M., BRUNORI, M., ANTONINI, E., DIIORIO, E. E., and WINTERHALTER, K. H. (1980) The Reaction of Hemoglobin Zürich with Oxygen and Carbon Monoxide. *J. Biol. Chem.*, 255, 6160–6165.

GIBSON, Q. H. (1959) The Kinetics of Reactions Between Haemoglobin and Gases, in *Progress in Biophysics and Biophysical Chemistry* (J. A. V. Butler and B. Katz, Eds.), Pergamon Press, New York 9, pp. 1–53.

GIBSON, Q. H. (1970) The Reaction of Oxygen with Hemoglobin and the Kinetic Basis of the Effect of Salt on Binding of Oxygen. *J. Biol. Chem.*, 245, 3285–3288.

GIBSON, Q. H., and SMITH, M. H. (1965) Rates of Reaction of *Ascaris* Haemoglobins with Ligands. *Proc. R. Soc. (London) Ser. B.*, 163, 206–214.

GOLD, A., JAYARAJ, K., DOPPELT, P., WEISS, R., BILL, E., DONG, X. Q., BOMINAAR, E. L., TRAUTWEIN, A. X., and WINKLER, H. (1989) Oxoferryl (2,6-Dichlorophenyl) Porphyrin Cation Radical: Evidence for Weak Metal–Radical Spin Coupling. *New J. Chem.*, 13, 169–172.

GOLD, A., JAYARAJ, K., DOPPELT, P., WEISS, R., CHOTTARD, G., BILL, E., DING, X., and TRAUTWEIN, A. X. (1988) Oxoferryl Complexes of the Halogenated(porphinato)iron Catalyst: (Tetrakis(2,6-dichlorophenyl)-porphinato) Iron. *J. Am. Chem. Soc.*, 110, 5756–5761.

GOLLY, I., HLAVICA, P., and SCHARTAU, W. (1988) The Functional Role of Cytochrome *b*5 Reincorporated Into Hepatic Microsomal Fractions. *Arch. Biochem. Biophys.*, 260, 232–240.

GOY, M. F. (1991) cGMP: The Wayward Child of the Cyclic Nucleotide Family. *Trends Neurosci.*, 14, 293–299.

GREENE, B. I., HOCHSTRASSER, R. M., WEISMAN, R. B., and EATON, W. A. (1978) Spectroscopic Studies of Oxy- and Carbonmonoxyhemoglobin After Pulsed Optical Excitation. *Proc. Natl. Acad. Sci. U.S.A.*, 75, 5255–5259.

GROVES, J. T. (1985) Key Elements of the Chemistry of Cytochrome P-450. The Oxygen Rebound Mechanism. *J. Chem. Educ.*, 62, 928–931.

GROVES, J. T., and GILBERT, J. A. (1986) Electrochemical Generation of an Iron(IV) Porphyrin. *Inorg. Chem.*, 25, 123–125.

GROVES, J. T., HAUSHALTER, R. C., NAKAMURA, M., NEMO, T. E., and EVANS, B. J. (1981) High-valent Iron–Porphyrin Complexes Related to Peroxidase and Cytochrome P-450. *J. Am. Chem. Soc.*, 103, 2884–2886.

GROVES, J. T., KRUPER, W. J., JR., and HAUSHALTER, R. C. (1980) Hydrocarbon Oxidations with Oxometalloporphinates. Isolation and Reactions of a (Porphinato)manganese(V) Complex. *J. Am. Chem. Soc.*, 102, 6375–6377.

GROVES, J. T., MCCLUSKY, G. A., WHITE, R. E., and COON, M. J. (1978) Aliphatic Hydroxylation by Highly Purified Liver Microsomal Cytochrome P-450. Evidence for a Carbon Radical Intermediate. *Biochem. Biophys. Res. Commun.*, 81, 154–160.

GROVES, J. T., and MYERS, R. S. (1983) Catalytic Asymmetric Epoxidations with Chiral Iron Porphyrins. *J. Am. Chem. Soc.*, 105, 5791–5796.

GROVES, J. T., and NEMO, T. E. (1983a) Epoxidation Reactions Catalyzed by Iron Porphyrins. Oxygen Transfer from Iodosylbenzene. *J. Am. Chem. Soc.*, 105, 5786–5791.

GROVES, J. T., and NEMO, T. E. (1983b) Aliphatic Hydroxylation Catalyzed by Iron Porphyrin Complexes. *J. Am. Chem. Soc.*, 105, 6243–6248.

GROVES, J. T., NEMO, T. E., and MYERS, R. S. (1979) Hydroxylation and Epoxidation Catalyzed by Iron–Porphyrine Complexes. Oxygen Transfer from Iodosylbenzene. *J. Am. Chem. Soc.*, 101, 1032–1033.

GROVES, J. T., and SUBRAMANIAN, D. V. (1984) Hydroxylation by Cytochrome P-450 and Metalloporphyrin Models. Evidence for Allylic Rearrangement. *J. Am. Chem. Soc.*, 106, 2177–2181.

GROVES, J. T., and WATANABE, Y. (1986a) The Mechanism of Olefin Epoxidation by Oxo-iron Porphyrins. Direct Observation of an Intermediate. *J. Am. Chem. Soc.*, 108, 507–508.

GROVES, J. T., and WATANABE, Y. (1986b) Oxygen Activation by Metalloporphyrins Related to Peroxidase and Cytochrome P-450. Direct Observation of the Oxygen–Oxygen Bond Cleavage Step. *J. Am. Chem. Soc.*, 108, 7834–7836.

GROVES, J. T., and WATANABE, Y. (1988) Reactive Iron Porphyrin Derivatives Related to the Catalytic Cycles of Cytochrome P-450 and Peroxidase. Studies of the Mechanism of Oxygen Activation. *J. Am. Chem. Soc.*, 110, 8443–8452.

GUENGERICH, F. P. (1991) Reactions and Significance of Cytochrome P-450 Enzymes. *J. Biol. Chem.*, 266, 10019–10022.

GUENGERICH, F. P., and MACDONALD, T. L. (1984) Chemical Mechanisms of Catalysis by Cytochromes P-450: A Unified View. *Acc. Chem. Res.*, 17, 9–16.

GUENGERICH, F. P., and MACDONALD, T. L. (1990) Mechanisms of Cytochrome P-450 Catalysis. *FASEB J.*, 4, 2453–2459.

GUNTER, M. J., and TURNER, P. (1991) Metalloporphyrins as Models for the Cytochromes P-450. *Coord. Chem. Rev.*, 108, 115–161.

HAMBRIGHT, P. (1975) Dynamic Coordinaton Chemistry of Metalloporphyrins, in *Porphyrins and Metalloporphyrins* (K. M. Smith, Ed.), Elsevier, New York, pp. 236–278.

HARTMAN, J. R., RARDIN, R. L., CHAUDHURI, P., POHL, K., WIEGHARDT, K., NUBER, B., WEISS, J., PAPAEFTHYMIOU, G. C., FRANKEL, R. B., LIPPARD, S. J. (1987) Synthesis and Characterization of (µ-Hydroxo)*bis*(µ-acetato)diiron(II) and (µ-Oxo)bis(µ-acetato)diiron(III)1,4,7-trimethyl-1,4,7-triazacyclononane Complexes as Models for Binuclear Iron Centers in Biology; Properties of the Mixed Valence Diiron(II,III) Species. *J. Am. Chem. Soc.*, 109, 7387–7396.

HASHIMOTO, T., DYER, R. L., CROSSLEY, M. J., BALDWIN, J. E., and BASOLO, F. (1982) Ligand, Oxygen, and Carbon Monoxide Affinities of Iron(II) Modified "Capped' Porphyrins. *J. Am. Chem. Soc.*, 104, 2101–2109.

HASINOFF, B. B. (1974) Kinetic Activation Volumes of the Binding of Oxygen and Carbon Monoxide to Hemoglobin and Myoglobin Studied on a High-pressure Laser Flash Photolysis Apparatus. *Biochemistry*, 13, 3111–3117.

HASINOFF, B. B. (1977) Simultaneous Diffusion and Chemical Activation Control of the Kinetics of the Binding of Carbon Monoxide to Ferroprotoporphyrin IX in Glycerol–Water Mixtures of High Viscosity. *Can. J. Chem.*, 55, 3955–3960.

HASINOFF, B. B. (1981) Diffusion-controlled Reaction Kinetics of the Binding of Carbon Monoxide to the Heme Undecapeptide of Cytochrome *c* (Microperoxidase 11) in High Viscosity Solvents. *Arch. Biochem. Biophys.*, 211, 396–402.

HE, G.-X., and BRUICE, T. C. (1991) Nature of the Epoxidizing Species Generated by Reaction of Alkyl Hydroperoxides with Iron(III) Porphyrins. Oxidation of *cis*-Stilbene and (Z)-1,2-bis(*trans*-2,*trans*-3-Diphenylcyclopropyl)ethene by *t*-BuOOH in the Presence of [*meso*-Tetrakis (2,4,6-trimethylphenyl)porphinato, [*meso*-Tetrakis(2,6-dichlorophenyl) porphinato]-, and [*meso*-Tetrakis(2,6-dibromophenyl)porphinato]iron(III) Chloride. *J. Am. Chem. Soc.*, 113, 2747–2753.

HELMS, A., HEILER, D., and MCLENDON, G. (1992) Electron Transfer in bis-Porphyrin Donor–Acceptor Compounds with Polyphenylene Spacers Shows a Weak Distance Dependence. *J. Am. Chem. Soc.*, 114, 6227–6238.

HENRY, E. R., SOMMER, J. H., HOFRICHTER, J., and EATON, W. A. (1983) Geminate Recombination of Carbon Monoxide to Myoglobin. *J. Mol. Biol.*, 166, 443–451.

HERRON, N., CAMERON, J. H., NEER, G. L., and BUSCH, D. H. (1983) A Totally Synthetic (Nonporphyrin) Iron(II) Dioxygen Carrier That Is Fully Functional Under Ambient Conditions. *J. Am. Chem. Soc.*, 105, 298–301.

HEWSON, W. D., and HAGER, L. P. (1979) Peroxidases, Catalases, and Chloroperoxidase, in *The Porphyrins* (D. Dolphin, Ed.), Academic Press, New York, 7, pp. 295–332.

HILDEBRANDT, A., and ESTABROOK, R. W. (1971) Evidence for the Participation of Cytochrome $b_5$ in Hepatic Microsomal Mixed-function Oxidation Reactions. *Arch. Biochem. Biophys.*, 143, 66–79.

HIOM, J., PAINE, J. B., III, ZAPF, U., and DOLPHIN, D. (1983) The Synthesis of Cofacial Porphyrin Dimers. *Can. J. Chem.*, 61, 2220–2223.

HIRAO, T., OHNO, M., and OHSHIRO, Y. (1990) Mediation of Orthoquinones in the MnTPPCl-catalyzed Epoxidation with Hydrogen Peroxide. *Tetrahedron Lett.*, 31, 6039–6042.

HOARD, J. L. (1966) Stereochemistry of Porphyrins, in *Hemes and Hemoproteins* (B. Chance, R. W. Estabrook, and T. Yonetani, Eds.), Academic Press, New York, pp. 9–24.

HOFMANN, F., DOSTMANN, W., KEILBACH, A., LANDGRAF, W., and RUTH, P. (1992) Structure and Physiological Role of cGMP-dependent Protein Kinase. *Biochim. Biophys. Acta*, 1135, 51–60.

HOLM, R. H. (1987) Metal-centered Oxygen Atom Transfer Reactions. *Chem. Rev.*, 87, 1401–1449.

HUESTIS, W. H., and RAFTERY, M. A. (1975) Conformation and Cooperativity in Hemoglobin. *Biochemistry*, 14, 1886–1892.

HUTCHINSON, J. A., TRAYLOR, T. G., and NOE, L. J. (1982) Picosecond Study of the Photodissociation of a Model Hemoprotein Compared to Hemoglobin. *J. Am. Chem. Soc.*, 104, 3221–3223.

IGNARRO, L. J. (1990) Haem-dependent Activation of Guanylate Cyclase and Cyclic GMP Formation by Endogenous Nitric Oxide: A Unique Transduction Mechanism for Transcellular Signaling. *Pharm. Toxicol.*, 67, 1–7.

ILEY, J., CONSTANTINO, L., NORBERTO, F., and ROSA, E. (1990) Oxidation of the Methyl Groups of N,N-Dimethylbenzamides by a Cytochrome P450 Mono-oxygenase Model System. *Tetrahedron Lett.*, 31, 4921–4922.

IMAMURA, T., RIGGS, A., and GIBSON, Q. H. (1972) Equilibria and Kinetics of Ligand Binding by Leghemoglobin from Soybean Root Nodules. *J. Biol. Chem.*, 247, 521–526.

JAMESON, G. B., MOLINARO, F. S., IBERS, J. A., COLLMAN, J. P., BRAUMAN, J. I., ROSE, E., and SUSLICK, K. S. (1980) Models for the Active Site of Oxygen-binding Hemoproteins. Dioxygen Binding Properties and the Structures of (2-Methylimidazole-*meso*-tetra(α,α,α,α-*o*-pivalamidophenyl)

Porphyrinatoiron(II)-ethanol and Its Dioxygen Adduct. *J. Am. Chem. Soc.*, 102, 3224–3237.

JAMESON, G. B., ROBINSON, W. T., and IBERS, J. A. (1982) Mercaptan-Tail Porphyrins. Synthetic Analogs for the Active Site of Cytochrome P-450, in *Hemoglobin and Oxygen Binding* (C. Ho, Ed.), Elsevier, New York, pp. 37–40.

JAMESON, G. B., RODLEY, G. A., ROBINSON, W. T., GAGNE, R. R., REED, C. A., and COLLMAN, J. P. (1978) Structure of a Dioxygen Adduct of (1-Methylimidazole) - *meso* - tetrakis(α,α,α,α, - *o* - pivalamidophenyl)porphinatoiron(II). An Iron Dioxygen Model for the Heme Component of Oxymyoglobin. *Inorg. Chem.*, 17, 850–857.

JERINA, D. M., and DALY, J. W. (1974) Arene Oxides: A New Aspect of Drug Metabolism. *Science*, 185, 573–582.

JOB, D., and DUNFORD, H. B. (1976) Substituent Effect on the Oxidation of Phenols and Aromatic Amines by Horseradish Peroxidase Compound I. *Eur. J. Biochem.*, 66, 607–614.

JONES, P., MANTLE, D., DAVIES, D. M., and KELLY, H. C. (1977) Hydroperoxidase Activities of Ferrihemes: Heme Analogues of Peroxidase Enzyme Intermediates. *Biochemistry*, 16, 3974–3978.

JONES, P., MANTLE, D., and WILSON, I. (1983) Influence of Ligand Modification on the Kinetics of the Reactions of Iron(III) Porphyrins with Hydrogen Peroxide in Aqueous Solutions. *J. Chem. Soc., Dalton Trans.*, 161–164.

JONGEWARD, K. A., MAGDE, D., TAUBE, D. J., MARSTERS, J. C., TRAYLOR, T. G., and SHARMA, V. S. (1988a) Picosecond and Nanosecond Geminate Recombination of Myoglobin with CO, $O_2$, NO, and Isocyanides. *J. Am. Chem. Soc.*, 110, 380–387.

JONGEWARD, K. A., MAGDE, D., TAUBE, D. J., and TRAYLOR, T. G. (1988b) Picosecond Kinetics of Cytochromes $b_5$ and *c*. *J. Biol. Chem.*, 263, 6027–6030.

JONGEWARD, K. A., MARSTERS, J. C., MITCHELL, M. J., MAGDE, D., and SHARMA, V. S. (1986) Picosecond Geminate Recombination of Nitrosylmyoglobins. *Biochem. Biophys. Res. Commun.*, 140, 962–966.

KARAMAN, R., ALMARSSON, O., BLASKO, A., and BRUICE, T. C. (1992) Design, Synthesis, and Characterization of a "Shopping Basket" Bisporphyrin. The First Examples of Triply Bridged Closely Interspaced Cofacial Porphyrin Dimers. *J. Org. Chem.*, 57, 2169–2173.

KARLIN, K. D., and GULTNEH, Y. (1987) Binding and Activation of Molecular Oxygen by Copper Complexes. *Prog. Inorg. Chem.*, 35, 219–327.

KARLIN, K. D., TYEKLAR, Z., FAROOQ, A., HAKA, M. S., GHOSH, P., CRUSE, R. W., GULTNEH, Y., HAYES, J. C., TOSCANO, P. J., and ZUBIETA, J. (1992) Dioxygen-copper Reactivity and Functional Modeling of Hemocyanins: Reversible Binding of $O_2$ and CO to Dicopper(I) Complexes $[Cu_2^I(L)]^{2+}$

(L = dinucleating ligand) and the Structure of a Bis (Carbonyl) [Cu$_2^I$(L)(CO$_2$)]$^{2+}$. *Inorg. Chem.*, 31, 1436–1451.

KELLY, H. C., DAVIES, D. M., KING, M. J., and JONES, P. (1977) Pre-steady-state Kinetics of Intermediate Formation in the Deuteroferriheme–Hydrogen Peroxide System. *Biochemistry*, 16, 3543–3549.

KIM, K., and IBERS, J. A. (1991) Structure of a Carbon Monoxide Adduct of a "Capped" Porphyrin: Fe (C$_2$-cap)(CO)(1-methylimidazole). *J. Am. Chem. Soc.*, 113, 6077–6081.

KOOLA, J. D., and KOCHI, J. K. (1987) Cobalt-catalyzed Epoxidation of Olefins. Dual Pathways for Oxygen Atom Transfer. *J. Org. Chem.*, 52, 4545–4553.

KREYSEL, M., and VOEGTLE, F. (1992) One-pot Synthesis of a Fourfold Bridged Double-decker Porphyrin. *Synthesis*, 733–734.

KUNZE, K. L., MANGOLD, B. L. K., WHEELER, C., BEILAN, H. S., and ORTIZ DE MONTELLANO, P. R. (1983) The Cytochrome P-450 Active Site. *J. Biol. Chem.*, 258, 4202–4207.

LABEQUE, R., and MARNETT, L. J. (1988) Reaction of Hematin with Allylic Fatty Acid Hydroperoxides: Identification of Products and Implications for Pathways of Hydroperoxide-dependent Epoxidation of 7,8-Dihydroxy-7,8-dihydrobenzo[a]pyrene. *Biochemistry*, 27, 7060–7070.

LABEQUE, R., and MARNETT, L. J. (1989) Homolytic and Heterolytic Scission of Organic Hydroperoxides By (*meso*-Tetraphenylporphinato)iron and Its Relation to Olefin Epoxidation. *J. Am. Chem. Soc.*, 111, 6621–6627.

LA MAR, G. N., and DE ROPP, J. S. (1980) Proton Nuclear Magnetic Resonance Characterization of the Electronic Structure of Horseradish Peroxidase Compound I. *J. Am. Chem. Soc.*, 102, 395–397.

LA MAR, G. N., and WALKER, F. A. (1979) Nuclear Magnetic Resonance of Paramagnetic Metalloporphyrins, in *The Porphyrins* (D. Dolphin, Ed.), Academic Press, New York, IV, pp. 61–157.

LATOS-GRAZYNSKI, L., CHENG, R.-J., LA MAR, G. N., and BALCH, A. L. (1982) Oxygenation Patterns for Substituted *meso*-Tetraphenylporphyrin Complexes of Iron(II). Spectroscopic Detection of Dioxygen Complexes in the Absence of Amines. *J. Am. Chem. Soc.*, 104, 5992–6000.

LATOUR, J. M. (1988) Binuclear Active Sites of Copper Proteins. Stimulants for a New Copper Coordination Chemistry. *Bull. Soc. Chim. Fr.*, 508–523.

LAVALETTE, D., and MOMENTEAU, M. (1982) Transient Oxygenation of Chelated Iron(II) Porphyrins: Improved Kinetics of Carbon Monoxide Replacement by Oxygen. *J. Chem. Soc., Perkin Trans. 2*, 385–388.

LAVALETTE, D., TÉTREAU, C., and MOMENTEAU, M. (1979) Laser Photolysis of Hemochromes. Kinetics of Nitrogenous Bases Binding to Four-coordinated and Five-coordinated Iron(II) Tetraphenylporphyrine. *J. Am. Chem. Soc.*, 101, 5395–5401.

LEE, C. H., and LEE, C. K. (1992) Synthesis of Sterically Hindered Strapped Porphyrins. *Bull. Korean Chem. Soc.*, 13, 352–354.

LEE, W. A., and BRUICE, T. C. (1985) Homolytic and Heterolytic Oxygen–Oxygen Bond Scissions Accompanying Oxygen Transfer to Iron(III) Hydroperoxides. A Mechanistic Criterion for Peroxidase and Cytochrome P-450. *J. Am. Chem. Soc.*, 107, 513–514.

LEE, W. A., GRAETZEL, M., and KALYANASUNDARAM, K. (1984) Anomalous Ortho Effects in Sterically Hindered Porphyrins: Tetrakis(2,6-dimethylphenyl) Porphyrin and Its Sulfonato Derivative. *Chem. Phys. Lett.*, 107, 308–312.

LEE, W. A., YUAN, L.-C., and BRUICE, T. C. (1988) Oxygen Transfer from Percarboxylic Acids and Alkylhydroperoxides to (*meso*-Tetraphenylporphinato)iron(III) and -chromium(III). *J. Am. Chem. Soc.*, 110, 4277–4283.

LEMBERG, R. and BARRETT, J. (1973) *Cytochromes*, Academic Press, London, pp. 122–216.

LINARD, J. E., ELLIS, P. E., JR., BUDGE, J. R., JONES, R. D., and BASOLO, F. (1980) Oxygenation of Iron(II) and Cobalt(II) "Capped" Porphyrins. *J. Am. Chem. Soc.*, 102, 1896–1904.

LINCOLN, T. M., and CORNWELL, T. L. (1991) Towards an Understanding of the Mechanism of Action of Cyclic AMP and Cyclic GMP in Smooth Muscle Relaxation. *Blood Vessels*, 28, 129–137.

LINDSAY SMITH, J. R., BALASUBRAMANIAN, P. N., and BRUICE, T. C. (1988) The Dynamics of Reaction of a Water-soluble and Non-µ-oxo Dimer Forming Iron(III) Porphyrin with *tert*-Butyl Hydroperoxide in Aqueous Solution. 1. Studies Using a Trap for Immediate Oxidation Products. *J. Am. Chem. Soc.*, 110, 7411–7418.

LINDSAY SMITH, J. R., and LOWER, R. (1991) The Mechanism of Reaction Between *tert*-Butyl Hydroperoxide and 5,10,15,20-Tetra(N-methyl-4-pyridyl)porphyrinatoiron(III) Pentachloride in Aqueous Solution. *J. Chem. Soc., Perkin Trans.*, 2, 31–39.

LINDSAY SMITH, J. R., and MORTIMER, D. N. (1985) Oxidative N-Dealkylation of N,N-Dimethylbenzylamines by Metalloporphyrin-catalysed Model Systems for Cytochrome P450 Monooxygenases. *J. Chem. Soc., Chem. Commun.*, 64–65.

LINDSAY SMITH, J. R., and MORTIMER, D. N. (1986) Model Systems for Cytochrome P450-dependent Mono-oxygenases. Part 5. Amine Oxidation. Part 17. Oxidative N-Dealkylation of Tertiary Amines by Metalloporphyrincatalysed Model Systems for Cytochrome P450 Monooxygenases. *J. Chem. Soc., Perkin Trans.*, 2, 1743–1749.

LINDSAY SMITH, J. R., and SLEATH, P. R. (1982) Model Systems for Cytochrome P450 Dependent Monooxygenases. Part 1. Oxidation of Alkenes and Aromatic Compounds by Tetraphenylporphinatoiron(III) Chloride and Iodosylbenzene. *J. Chem. Soc., Perkin Trans.*, 2, 1009–1015.

LINDSAY SMITH, J. R., and SLEATH, P. R. (1983a) Model Systems for Cytochrome P450 Dependent Monooxygenases. Part 2. Kinetic Isotope Effects for the Oxidative Demethylation of Anisole and Anisole-Me-3d by Cytochrome P450 Dependent Monooxygenases and Model Systems. *J. Chem. Soc., Perkin Trans.*, 2, 621–628.

LINDSAY SMITH, J. R., and SLEATH, P. R. (1983b) Model Systems for Cytochrome P450-dependent Monooxygenases. Part 3. The Stereochemistry of Hydroxylation of Cis- and Trans-decahydronaphthalene by Chemical Models for Cytochrome P450 Monooxygenases. *J. Chem. Soc., Perkin Trans.*, 2, 1165–1169.

LONG, E. C., and HECHT, S. M. (1988) Direct Comparison of Oxygen Transfer By Iron Bleomycin and Zinc Bleomycin. *Tetrahedron Lett.*, 28, 6413–6416.

LOPEZ, M. A. (1987) *Dioxygen and Carbon Monoxide Binding to Model Heme Systems.* Doctoral Dissertation, University of California, San Diego.

MCGALL, G. H., RABOW, L. E., ASHELEY, G. W., WU, S. H., KOZARICH, J. W. and STUBBE, J. (1992) New Insight Into the Mechanism of Base Propenal Formation During Bleomycin-mediated DNA Degradation. *J. Am. Chem. Soc.*, 114, 4958–4967.

MAILLARD, P., GUERQUIN-KERN, J.-L, MOMENTEAU, M., and GASPARD, S. (1989) Glycoconjugated Tetrapyrrolic Macrocycles. *J. Am. Chem. Soc.*, 111, 9125–9127.

MAILLARD, P., SCHAEFFER, C., HUEL, C., LHOSTE, J. M., and MOMENTEAU, M. (1988) Both-faces Hindered Porphyrins. Part 5. Synthesis and Characterization of Iron(III) Basket Handle Porphyrins Having Mutiple Secondary Amide Groups Inserted in the Superstructures. *J. Chem. Soc., Perkin Trans.*, 1, 3285–3296.

MALMSTRÖM, B. G. (1990a) Cytochrome *c* Oxidase as a Redox-linked Proton Pump. *Chem. Rev.*, 90, 1247–1260.

MALMSTRÖM, B. G. (1990b) Cytochrome Oxidase: Some Unsolved Problems and Controversial Issues. *Arch. Biochem. Biophys.*, 280, 233–241.

MANSUY, D. (1987) Cytochrome P-450 and Synthetic Models. *Pure App. Chem.*, 59, 759–770.

MANSUY, D. (1990) Biomimetic Catalysts for Selective Oxidation in Organic Chemistry. *Pure App. Chem.*, 62, 741–746.

MANSUY, D., BARTOLI, J.-F., CHOTTARD, J.-C., and LANGE, M. (1980) Metalloporphyrin-catalyzed Hydroxylation of Cyclohexane by Alkyl Hydroperoxides: Pronounced Efficiency of Iron-porphyrins. *Angew. Chem. Int. Ed. Engl.*, 19, 909–910.

MANSUY, D., BATTIONI, P., and BATTIONI, J.-P. (1989) Chemical Model Systems for Drug-metabolizing Cytochrome P-450-dependent Monooxygenases. *Eur. J. Biochem.*, 184, 267–285.

MANSUY, D., BATTIONI, P., and RENAUD, J.-P. (1984) In the Presence of Imidazole, Iron- and Manganese-porphyrins Catalyse the Epoxidation of Alkenes by Alkyl Hydroperoxides. *J. Chem. Soc., Chem. Commun.,* 1255–1257.

MANSUY, D., DEVOCELLE, L., ARTAUD, I., and BATTIONI, J.-P. (1985) Alkene Oxidations by Iodosylbenzene Catalyzed by Iron-porphyrins: Fate of the Catalyst and Formation of *n*-Alkyl-porphyrin Green Pigments from Monosubstitued Alkenes as in Cytochrome P-450 Reactions. *Nouv. J. Chim.,* 9, 711–716.

MARDEN, M. C. (1982) A Coupled Diffusion and Barrier Model for the Recombination Kinetics of Myoglobin with Carbon Monoxide. *Eur. J. Biochem.,* 128, 399–404.

MARDEN, M. C., HAZARD, E. S., III, and GIBSON, Q. H. (1986) Protoheme–Carbonmonoxide Geminate Kinetics. *Biochemistry,* 25, 2786–2792.

MARTIN, J. L., MIGUS, A., POYART, C., LECARPENTIER, Y., ANTONETTI, A., and ORSZAG, A. (1982) Femtosecond Photodissociation and Picosecond Recombination of $O_2$ in Myoglobin: A Plausible Explanation for the Low Quantum Yield in $MbO_2$. *Biochem. Biophys. Res. Commun.,* 107, 803–810.

MARTIN, J. L., MIGUS, A., POYART, C., LECARPENTIER, Y., ASTIER, R., and ANTONETTI, A. (1983) Femtosecond Photolysis of CO-ligated Protoheme and Hemoproteins: Appearance of Deoxy Species with a 350-fsec Time Constant. *Proc. Natl. Acad. Sci. U.S.A.,* 80, 173–177.

MARTIN, J. L., and VOS, M. H. (1992) Femtosecond Biology. *Annu. Rev. Biophys. Biomol. Struct.,* 21, 199–222.

MASHIKO, T., DOLPHIN, D., NAKANO, T., and TRAYLOR, T. G. (1985) N-Alkylporphyrin Formation During the Reactions of Cytochrome P-450 Model Systems. *J. Am. Chem. Soc.,* 107, 3835–3736.

MATHEWS, F. S. (1985) The Structure, Function and Evolution of Cytochromes. *Prog. Biophys. Mol. Biol.,* 45, 1–56.

MEUNIER, B. (1986) Metalloporphyrin-catalyzed Oxygenation of Hydrocarbons. *Bull. Soc. Chim. Fr.,* 4, 578–594.

MEUNIER, B. (1992) Metalloporphyrins as Versatile Catalysts for Oxidation Reactions and Oxidative DNA Cleavage. *Chem. Rev.,* 92, 1411–1456.

MEUNIER, B., GUILMET, E., DE CARVALHO, M.-E., and POILBLANC, R. (1984) Sodium Hypochlorite: A Convenient Oxygen Source for Olefin Epoxidation Catalyzed By (Porphinato)manganese Complexes. *J. Am. Chem. Soc.,* 106, 6668–6676.

MIERS, J. B., POSTLEWAITE, J. C., COWEN, B. R., ROEMIG, G. R., LEE, I.-Y.S., and DLOTT, D. D. (1991) Preexponential-limited Solid State Chemistry: Ultrafast Rebinding of a Heme Ligand Complex in a Glass or Protein Matrix. *J. Chem. Phys.,* 94, 1825–1836.

MIERS, J. B., POSTLEWAITE, J. C., ZYUNG, T. H., CHEN, S., ROEMIG, G. R., WEN, X., DLOTT, D. D., and SZABO, A. (1990) Diffusion Can Explain the Nonexponential Rebinding of Carbon Monoxide to Protoheme. *J. Chem. Phys.,* 93, 8771–8776.

MIMS, M. P., PORRAS, A. G., OLSON, J. S., NOBLE, R. W., and PETERSON, J. A. (1983) Ligand Binding to Heme Proteins. *J. Biol. Chem.*, 258, 14219–14232.

MINCEY, T., and TRAYLOR, T. G. (1979) Anion Complexes of Ferrous Porphyrins. *J. Am. Chem. Soc.*, 101, 765–766.

MIRAFZAL, G. A., KIM, T., LIU, J. P., and BAULD, N. L. (1992) Cation Radical Probes: Development and Application to Metalloporphyrin-catalyzed Epoxidation. *J. Am. Chem. Soc.*, 114, 10968–10969.

MISPELTER, J., MOMENTEAU, M., LAVALETTE, D., and LHOSTE, J.-M. (1983) Hydrogen-bond Stabilization of Oxygen in Hemoprotein Models. *J. Am. Chem. Soc.*, 105, 5165–5166.

MIWA, G. T., WALSH, J. S., and LU, A. Y. H. (1984) Kinetic Isotope Effects on Cytochrome P-450-catalyzed Oxidation Reactions: The Oxidative O-Dealkylation of 7-Ethoxycoumarin. *J. Biol. Chem.*, 259, 3000–3004.

MOFFAT, K., DEATHERAGE, J. F., and SEYBERT, D. W. (1979) A Structural Model for the Kinetic Behavior of Hemoglobin. *Science*, 206, 1035–1042.

MOMENTEAU, M. (1986) Synthesis and Coordination Properties of Superstructured Iron-porphyrins. *Pure App. Chem.*, 58, 1493–1502.

MOMENTEAU, M., and LAVALETTE, D. (1978) Photodissociation of Nitrogenous Bases from Hemochromes and Kinetics of Recombination of Axial Bases. *J. Am. Chem. Soc.*, 100, 4322–4324.

MOMENTEAU, M., and LAVALETTE, D. (1982) Kinetic Evidence for Dioxygen Stabilization in Oxygenated Iron(II)-porphyrins by Distal Polar Interactions. *J. Chem. Soc., Chem. Commun.*, 341–343.

MOMENTEAU, M., LOOCK, B., LAVALETTE, D., TÉTREAU, C., and MISPELTER, J. (1983a) Iron(II) "Hanging Imidazole" Porphyrin: Synthesis and Proximal Ligand Effect on CO and $O_2$ Binding. *J. Chem. Soc., Chem. Commun.*, 962–964.

MOMENTEAU, M., MISPELTER, J., LOOCK, B., and BISAGNI, E. (1983b) Both-faces Hindered Porphyrins. Part 1. Synthesis and Characterization of Basket-handle Porphyrins and Their Iron Complexes. *J. Chem. Soc., Perkin Trans.*, 1, 189–196.

MOORE, G. R., and PETTIGREW, G. W. (1990) *Cytochromes c: Evolutionary, Structural and Physiochemical Aspects*, Springer-Verlag, New York, pp. 1–478.

MOORE, J. N., HANSEN, P. A., and HOCHSTRASSER, R. M. (1987) A New Method for Picosecond Time-resolved Infrared Spectroscopy: Application to CO Photodissociation from Iron Porphyrins. *Chem. Phys. Lett.*, 138, 110–114.

MORIKIS, D., CHAMPION, P. M., SPRINGER, B. A., and SLIGAR, S. G. (1989) Resonance Raman Investigations of Site-directed Mutants of Myoglobin: Effects of Distal Histidine Replacement. *Biochemistry*, 28, 4791–4800.

MORRISON, M., and SCHONBAUM, G. R. (1976) Peroxidase-Catalyzed Halogenation, in *Ann. Rev. Biochemistry*, (E. E. Snell, P. D. Boyer, A. Meister, and C. C. Richardson, Eds.), 45, pp. 861–888.

MURAD, F. (1986) Cyclic Guanosine Monophosphate as a Mediator of Vasodilation. *J. Clin. Invest.*, 78, 1–5.

MURTHY, M. R. N., REID, T. J., III, SICIGNANO, A., TANAKA, N., and ROSSMANN, M. G. (1981) Structure of Beef Liver Catalase. *J. Mol. Biol.*, 152, 465–499.

NAGATA, T. (1992) Synthesis and Characterization of Doubly-strapped Porphyrins. *Bull. Chem. Soc. Jpn.*, 65, 385–391.

NAKAMURA, M. (1989) Dissociation Rates of Axially Coordinated Imidazoles and Formation Constants of Low Spin Ferric Complexes Derived from Tetraphenylporphyrin and Tetramesitylporphyrin. *Inorg. Chim. Acta*, 161, 73–80.

NAKATSU, K., and DIAMOND, J. (1989) Role of cGMP in Relaxation of Vascular and Other Smooth Muscle. *Canadian J. Physiol. Pharm.*, 67, 251–262.

NANTHAKUMAR, A., NASIR, M. S., KARLIN, K. D., RAVI, N., and HUYNH, B. H. (1992) A Cytochrome *c* Oxidase Reactivity Model: Generation of a Peroxo-bridged Iron/Copper Dinuclear Complex. *J. Am. Chem. Soc.*, 114, 6564–6566.

NOBLE, R. W., PARKHURST, L. J., and GIBSON, Q. H. (1970) The Effect of pH on the Reactions of Oxygen and Carbon Monoxide with the Hemoglobin of the Carp, *Cyprinus carpio*. *J. Biol. Chem.*, 245, 6628–6633.

NOE, L. J., EISERT, W. G., and RENTZEPIS, P. M. (1978) Picosecond Photodissociation and Subsequent Recombination Processes in Carbon Monoxide Hemoglobin. *Proc. Natl. Acad. Sci. U.S.A.*, 75, 573–577.

NORDBLUM, G. D., WHITE, R. E., and COON, M. J. (1976) Studies on Hydroperoxide-dependent Substrate Hydroxylation by Purified Liver Microsomal Cytochrome P-450 *Arch. Biochem. Biophys*, 175, 524–533.

NOSHIRO, M., ULLRICH, V., and OMURA, T. (1981) Cytochrome $b_5$ as Electron Donor for Oxy-cytochrome P-450. *Eur. J. Biochem.*, 116, 521–526.

OLSON, J. S., ANDERSEN, M. E., and GIBSON, Q. H. (1971) The Dissociation of the First Oxygen Molecule from Some Mammalian Oxyhemoglobins. *J. Biol. Chem.*, 246, 5919–5923.

OLSON, J. S., MATHEWS, A. J., ROHLFS, R. J., SPRINGER, B. A., EGEBERG, K. D., SLIGAR, S. G., TAME, J., RENAUD, J.-P., and NAGAI, K. (1988) The Role of the Distal Histidine in Myoglobin and Haemoglobin. *Nature*, 336, 265–266.

OLSON, J. S., MCKINNIE, R. E., MIMS, M. P., and WHITE, D. K. (1983) Mechanisms of Ligand Binding to Pentacoordinate Protoheme. *J. Am. Chem. Soc.*, 105, 1522–1527.

ONUCHIC, J. N., BERATAN, D. N., WINKLER, J. R., and GRAY, H. B. (1992)

Pathway Analysis of Protein Electron Transfer Reactions. *Ann. Rev. Biophys. Biomol. Struct.*, 21, 349–377.

ORTIZ DE MONTELLANO, P. R. (Ed.) (1986a) *Cytochrome P-450: Structure, Mechanism and Biochemistry*, Plenum Press, New York.

ORTIZ DE MONTELLANO, P. R. (1986b) Oxygen Activation and Transfer, in *Cytochrome P-450: Structure, Mechanism and Biochemistry*, Plenum Press, New York, pp. 217–271.

ORTIZ DE MONTELLANO, P. R. (1989) Cytochrome P-450 Catalysis: Radical Intermediates and Dehydrogenation Reactions. *Trends Pharm. Sci.*, 10, 354–359.

ORTIZ DE MONTELLANO, P. R. (1992) Catalytic Sites of Hemoprotein Peroxidases. *Ann. Rev. Pharm. Toxicol.*, 32, 89–107.

ORTIZ DE MONTELLANO, P. R., and CORREIA, M. A. (1983) Suicidal Destruction of Cytochrome P-450 During Oxidative Drug Metabolism. *Ann. Rev. Pharmacol. Toxicol.*, 23, 481–503.

ORTIZ DE MONTELLANO, P. R., and KUNZE, K. L. (1981) Cytochrome P-450 Inactivation: Structure of the Prosthetic Heme Adduct with Propyne. *Biochemistry*, 20, 7266–7271.

ORTIZ DE MONTELLANO, P. R., KUNZE, K. L., and MICO, B. A. (1980) Destruction of Cytochrome P-450 by Olefins: N-Alkylation of Prosthetic Heme. *Mol. Pharmacol.*, 18, 602–605.

OSTOVIC, D., and BRUICE, T. C. (1989) Intermediates in the Epoxidation of Alkenes by Cytochrome P-450 Models. 5. Epoxidation of Alkenes Catalyzed by a Sterically Hindered (*meso*-Tetrakis(2,6-dibromophenyl) porphinato)iron(III) Chloride. *J. Am. Chem. Soc.*, 111, 6511–6517.

OSTOVIC, D., and BRUICE, T. C. (1992) Mechanism of Alkene Epoxidation by Iron, Chromium, and Manganese Higher Valent Oxo-metalloporphyrins. *Acc. Chem. Res.*, 25, 314–320.

OSUKA, A., KOBAYASHI, F., NAGATA, T., and MARUYAMA, K. (1990) One-pot Synthesis of Strapped Porphyrins and Face-to-face Dimeric Porphyrins. *Chem. Lett.*, 287–290.

PADBURY, G., SLIGAR, S. G., LABEQUE, R., and MARNETT, L. J. (1988) Ferric Bleomycin Catalyzed Reduction of 10-Hydroperoxy-8,12-octadecadienoic Acid: Evidence for Homolytic O–O Bond Scission. *Biochemistry*, 27, 7846–7852.

PAINE, J. B., III, and DOLPHIN, D. (1978) Synthesis of Covalently-linked Dimeric Porphyrins. *Can. J. Chem.*, 56, 1710–1712.

PARKHURST, L. J. (1979) Hemoglobin and Myoglobin Ligand Kinetics. *Ann. Rev. Phys. Chem.*, 30, 503–546.

PARKHURST, L. J. SIMA, P., and GOSS, D. J. (1980) Kinetics of Oxygen and Carbon Monoxide Binding to Hemoglobins of *glycera* Dibranchiati. *Biochemistry*, 19, 2688–2692.

PAUL, P. P., TYEKLAR, Z., JACOBSON, R. R., and KARLIN, K. D. (1991)

Reactivity Patterns and Comparisons in Three Classes of Synthetic Copper–Dioxygen {Cu$_2$–O$_2$} Complexes: Implication for Structure and Biological Relevance. *J. Am Chem. Soc.*, 113, 5322–5332.

PENNER-HAHN, J. E., EBLE, K. S., MCMURRY, T. J., RENNER, M., BALCH, A. L., GROVES, J. T., DAWSON, J. H., and HODGSON, K. O. (1986) Structural Characterization of Horseradish Peroxidase Using EXAFS Spectroscopy. Evidence for Fe=O Ligation in Compounds I and II. *J. Am. Chem. Soc.*, 108, 7819–7825.

PERUTZ, M. F. (1979) Regulation of Oxygen Affinity in Hemoglobin. *Ann. Rev. Biochem.*, 48, 327–386.

PERUTZ, M. F., and TEN EYCK, L. F. (1971) Stereochemistry of Cooperative Effects in Hemoglobin, *Cold Spring Harbor Symp. Quant. Biol.*, 36, 295–310.

PETERSON, M. W., RIVERS, D. S., and RICHMAN, R. M. (1985) Mechanistic Considerations in the Photodisproportionation of µ-Oxo-bis((tetraphenylporphinato)iron(III)). *J. Am. Chem. Soc.*, 107, 2907–2915.

PETRICH, J. W., POYART, C., and MARTIN, J. L. (1988) Photophysics and Reactivity of Heme Proteins: A Femtosecond Absorption Study of Hemoglobin, Myoglobin, and Protoheme. *Biochemistry*, 27, 4049–4060.

PHILLIPS, S. E. V. (1980) Structure and Refinement of Oxymyoglobin at 1.6 Resolution. *J. Mol. Biol.*, 142, 531–554.

POMPON, D., and COON, M. J. (1984) On the Mechanism of Action of Cytochrome P-450. Oxidation and Reduction of the Ferrous Dioxygen Complex of Liver Microsomal Cytochrome P-450 and Cytochrome $b_5$. *J. Biol. Chem.*, 259, 15377–15385.

PORTELLA, C. F. (1987) Fast Kinetic Methods in the Study of Carbon Monoxide and Imidazoles Binding to Iron(II) Porphyrins. The Effect of Ortho Substituents in Tetraphenylhemes. Doctoral dissertation, University of California, San Diego.

PORTELA, C. F., MAGDE, D., and TRAYLOR, T. G. (1993) The Ortho Effect in Ligation of Iron Tetraphenylporphyrins. *Inorg. Chem.*, 32, 1313–1320.

PORTER, T. D., and COON, M. J. (1991) Cytochrome P-450. Multiplicity of Isoforms, Substrates, and Catalytic and Regulatory Mechanisms. *J. Biol. Chem.*, 266, 12469–12472.

PORTSMOUTH, D., and BEAL, E. A. (1971) Peroxidase Activity of Deuterohemin. *Eur. J. Biochem.*, 19, 479–487.

POSTLEWAITE, J. C., MIERS, J. B., and DLOTT, D. D. (1989) Ultrafast Ligand Rebinding to Protoheme and Heme Octapeptide at Low Temperature. *J. Am Chem. Soc.*, 111, 1248–1255.

POULOS, T. L. Heme Enzyme Crystal Structures, in *Advances in Inorganic Biochemistry* (G. L. Eichhorn and L. G. Marzilli, Eds.), Elsevier, New York, pp. 1–36.

POULOS, T. L., FINZEL, B. C., and HOWARD, A. J. (1987) High-resolution Crystal Structure of Cytochrome P450$_{cam}$. *J. Mol. Biol.*, 195, 687–700.

POULOS, T. L. FREER, S. T., ALDEN, R. A., EDWARDS, S. L., SKOGLAND, U., TAKIO, K., ERIKSSON, B., XUONG, N.-H, YONETANI, and KRAUT, J. (1980) The Crystal Structure of Cytochrome $c$ Peroxidase. *J. Biol. Chem.*, 255, 575–580.

POULOS, T. L., and KRAUT, J. (1980) The Stereochemistry of Peroxidase Catalysis. *J. Biol. Chem.*, 255, 8199–8205.

QUAGLIARIELLO, E., PAPA, S., PALMIER, F., SLATER, E. C., and SILIPRANDI, N. (Eds.) (1975) *Electron Transport Chains and Oxidative Phosphorylation*, Academic Press, New York.

READDY, D., and CHANDRASHEKAR, T. K. (1992) Short-chain Basket Handle Porphyrins: Synthesis and Characterization. *J. Chem. Soc., Dalton Trans.*, 619–625.

REED, C. A. (1986) Hemocyanin Cooperativity: A Copper Coordination Chemistry Perspective. in *Biological Inorganic Copper Chemistry* (K. D. Karlin and J. Zubieta, Eds.), Adenine Press, Guilderland, NY, 1, pp. 61–73.

REISBERG, P. I., and OLSON, J. S. (1980) Kinetic and Cooperative Mechanisms of Ligand Binding to Hemoglobin. *J. Biol. Chem.*, 255, 4159–4169.

RENAUD, J.-P., BATTIONI, P., BARTOLI, J. F., and MANSUY, D. (1985) A Very Efficient System for Alkene Epoxidation by Hydrogen Peroxide: Catalysis by Manganese-porphyrins in the Presence of Imidazole. *J. Chem. Soc., Chem. Commun.*, 888–889.

RICHARDS, J. L., and TRAYLOR, T. G. (unpublished).

RIFKIND, J. M. (1988) Hemoglobin, in *Adv. Inorg. Biochem.* (G. L. Eichhorn and L. G. Marzilli, Eds.), Elsevier, New York, pp. 155–244.

ROBERTS, J. E., HOFFMAN, B. M., RUTTER, R., and HANER, L. P. (1981) Electron-nuclear Double Resonance of Horseradish Peroxidase Compound I. *J. Biol. Chem.*, 256, 2118–2121.

ROHLFS, R. J., MATHEWS, A. J., CARVER, T. E., OLSON, J. S., SPRINGER, B. A., EGEBERG, K. D., and SLIGAR, S. G. (1990) The Effects of Amino Acid Substitution at Position E7 (Residue 64) on the Kinetics of Ligand Binding to Sperm Whale Myoglobin. *J. Biol. Chem.*, 265, 3168–3176.

ROHLFS, R. J., OLSON, J. S., and GIBSON, Q. H. (1988) A Comparison of the Geminate Recombination Kinetics of Several Monomeric Heme Proteins. *J. Biol. Chem.*, 263, 1803–1813.

ROMERO-HERRERA, A. E., GOODMAN, M., DENE, H., BARTNICKI, D. E., and MIZUKAMI, H. (1981) An Exceptional Amino Acid Replacement on the Distal Side of the Iron Atom in Proboscidean Myoglobin. *J. Molec. Evolution*, 17, 140–147.

ROSE, E. J., and HOFFMAN, B. M. (1983) Nitric Oxide Ferrohemes: Kinetics of Formation and Photodissociation Quantum Yields. *J. Am. Chem. Soc.*, 105, 2866–2873.

ROSE, E. J., VENKATASUBRAMANIAN, P. N., SWARTZ, J. C., JONES, R. D.,

BASOLO, F., and HOFFMAN, B. M. (1982) Carbon Monoxide Binding Kinetics in "Capped" Porphyrin Compounds. *Proc. Natl. Acad. Sci. U.S.A.*, 79, 5742–5745.

ROUGÉE, M., and BRAULT, D. (1975) Influence of Trans Weak or Strong Field Ligands Upon the Affinity of Deuteroheme for Carbon Monoxide. Monoimidazoleheme as a Reference for Unconstrained Five-coordinate Hemoproteins. *Biochemistry*, 14, 4100–4106.

ROUGHTON, F. J. W. (1954) The Equilibrium Between Carbon Monoxide and Sheep Haemoglobin at Very High Percentage Saturations. *J. Physiol. (London)*, 126, 359–383.

SAFO, M. K., GUPTA, G. P., WALKER, F. A., and SCHEIDT, W. R. (1991) Models of the Cytochromes *b*. Control of Axial Ligand Orientation with a "Hindered" Porphyrin System. *J. Am. Chem. Soc.*, 113, 5497–5510.

SAFO, M. K., GUPTA, G. P., WATSON, C. T., SIMONIS, U., WALKER, F. A., and SCHEIDT, W. R. (1992) Models of the Cytochromes *b*. Low-spin Bis-ligated (Porphinato)iron(III) Complexes with "Unusual" Molecular Structures and NMR, EPR, and Mössbauer Spectra. *J. Am. Chem. Soc.*, 114, 7066–7075.

SALEMME, F. R. (1977) Structure and Function of Cytochromes *c*. *Ann. Rev. Biochem.*, 46, 299–329.

SALVATO, B., and BELTRAMINI, M. (1990) Hemocyanins: Molecular Architecture, Structure and Reactivity of the Binuclear Copper Active Site. *Life Chem. Rep.*, 8, 1–47.

SAUNDERS, B. C., HOLMES-SIEDEL, A. G., and STARK, B. P. (Eds.) (1964) *Peroxidases*, Butterworths, London.

SAWICKI, C. A., and GIBSON, Q. H. (1977) Properties of the T State of Human Oxyhemoglobin Studied by Laser Photolysis. *J. Biol. Chem.*, 252, 7538–7547.

SCHAPPACHER, M., WEISS, R., MONTIEL-MONTOYA, R., TRAUTWEIN, A., and TABARD, A. (1985) Formation of an Iron(IV)-oxo "Picket Fence" Porphyrin Derivative Via Reduction of the Ferrous Dioxygen Adduct Reaction with Carbon Dioxide. *J. Am. Chem. Soc.*, 107, 3736–3738.

SCHINDLER, S., KUHNE, T., and ELIAS, H. (1989) Preparation and Kinetic Investigation of Model Complexes for Hemocyanin. *Proc. Conf. Coord. Chem.*, 12th, 313–314.

SCHMEISSER, M., DAHMEN, K., and SARTORI, P. (1967) Perfluoroacyloxy Compounds of Positive Iodine. *Chem. Ber.*, 100, 1633–1637.

SCHONBAUM, G. R., and CHANCE, B. (1976) Catalase, in *The Enzymes*, 3rd ed. (P. D. Boyer, Ed.), Academic Press, New York, 13, pp. 363–408.

SCHONBAUM, G. R., and LO, S. (1972) Interaction of Peroxidases with Aromatic Peracids and Alkyl Peroxides. *J. Biol. Chem.*, 247, 3353–3560.

SCOTT, R. A. (1989) X-ray Absorption Spectroscopic Investigations of

Cytochrome *c* Oxidase Structure and Function. *Ann. Rev. Biophys. Biophys. Chem.*, 18, 137–158.

SESSLER, J. L., JOHNSON, M. R., LIN, T.-Y., and CREAGER, S. E. (1988) Quinone-substituted Monometalated Porphyrin Dimers: Models for Photoinduced Charge Separation at Fixed Orientation and Energy. *J. Am. Chem. Soc.*, 110, 3659–3661.

SHAANAN, B. (1983) Structure of Human Oxyhaemoglobin at 2.1 Å Resolution. *J. Mol. Biol.* 171, 31–59.

SHAANAN, B. (1988) The Iron–Oxygen Bonding in Human Oxyhaemoglobin. *Nature (London)*, 296, 683–684.

SHANK, C. V., IPPEN, E. P., and BERSOHN, R. (1976) Time-resolved Spectroscopy of Hemoglobin and Its Complexes with Subpicosecond Optical Pulses. *Science*, 193, 50–51.

SHANNON, P., and BRUICE, T. C. (1981) A Novel P-450 Model System for the N-dealkylation Reaction. *J. Am. Chem. Soc.*, 103, 4580–4582.

SHARMA, V. S., GEIBEL, J. F., and RANNEY, H. M. (1978) "Tension" on Heme by the Proximal Base and Ligand Reactivity: Conclusions Drawn from Model Compounds for the Reaction of Hemoglobin. *Proc. Natl. Acad. Sci. U.S.A.*, 75, 3747–3750.

SHARMA, V. S., SCHMIDT, M. R., and RANNEY, H. M. (1976) Dissociation of CO from Carboxy-hemoglobin. *J. Biol. Chem.*, 251, 4267–4272.

SHELDON, R., (1985) Catalytic Oxidations in Organic Synthesis. *Bull. Soc. Chim. Belg.*, 94, 651–670.

SHELDON, R., and KOCHI, J. K. (1981) *Metal-Catalyzed Oxidations of Organic Compounds*. Academic Press, New York.

SORRELL, T. N. (1986) Binuclear Copper Complexes: Synthetic Models for the Active Site of Type III Copper Proteins, in *Biological Inorganic Copper Chemistry* (K. D. Karlin, and J. Zubieta, Eds.), Adenine Press, Guilderland, NY, 2, pp. 41–55.

SORRELL, T. N. (1989) Synthetic Models for Binuclear Copper Proteins. *Tetrahedron*, 45, 3–68.

SPRINGER, B. A., EGEBERG, K. D., SLIGAR, S. G., ROHLFS, R. J., MATHEWS, A. J., and OLSON, J. S. (1989) Discrimination Between Oxygen and Carbon Monoxide and Inhibition of Autooxidation by Myoglobin. *J. Biol. Chem.*, 264, 3057–3060.

St. GEORGE, R. C. C., and PAULING, L. (1951) The Combining Power of Hemoglobin for Alkyl Isocyanides, and the Nature of the Heme–Heme Interactions in Hemoglobin. *Science*, 114, 629–634.

STAYTON, P. S., and SLIGAR, S. G. (1990) Cytochrome P-450cam Binding Surface Defined by Site-directed Mutagenesis and Electrostatic Modeling. *Biochemistry*, 29, 7381–7386.

STEINMEIER, R. C., and PARKHURST, L. J. (1975) Kinetic Studies on the Five

Principal Components of Normal Adult Human Hemoglobin. *Biochemistry*, 14, 1564–1572.

STERN, M. K., and GROVES, J. T. (1992) Oxygen Transfer Reactions of Oxo-Manganese Porphyrins, in *Manganese Redox Enzymes* (V. L. Pecoraro, Ed.), VCH, New York, pp. 233–259.

STEARNS, R. A., and ORTIZ DE MONTELLANO, P. R. (1985) Cytochrome P-450 Catalyzed Oxidation of Quadricyclane. Evidence for a Radical Cation Intermediate. *J. Am. Chem. Soc.*, 107, 4081–4082.

STRYER, L. (1988) *Biochemistry* (3rd ed.), W. H. Freeman, New York, pp. 143–176.

STUBBE, J. (1991) Dinuclear Non-heme Iron Centers: Structure and Function. *Curr. Opin. Struct. Biol.* 1, 788–795.

STUBBE, J., and KOZARICH, J. W. (1987) Mechanisms of Bleomycin-induced DNA Degradation. *Chem. Rev.*, 87, 1107–1136.

SUSLICK, K. S., and FOX, M. M. (1983) A Bis-pocket Porphyrin. *J. Am. Chem. Soc.*, 105, 3507–3510.

SUSLICK, K. S., FOX, M. M. and REINERT, T. J. (1984) Influences on CO and $O_2$ Binding to Iron(II) Porphyrins. *J. Am. Chem. Soc.*, 106, 4522–4525.

TABUSHI, I. (1988) Reductive Dioxygen Activation by Use of Artificial P-450 Systems. *Coord. Chem. Rev.*, 86, 1–42.

TAKANO, T., KALLAI, O. B., SWANSON, R., and DICKERSON, R. E. (1973) The Structure of Ferrocytochrome *c* at 2.45 Å Resolution. *J. Biol Chem.*, 248, 5234–5255.

TAUBE, D. J., PROJAHN, H.-D., VAN ELDIK, R., MAGDE, D., and TRAYLOR, T. G. (1990) Mechanism of Ligand Binding to Hemes and Hemoproteins. A High-pressure Study. *J. Am. Chem. Soc.*, 112, 6880–6886.

THEORELL, H. (1941) Crystalline Peroxidase. *Enzymologia*, 10, 250–252.

THEORELL, H., EHRENBERG, A., and CHANCE, B. (1952) Electronic Structure of the Peroxidase–Peroxide Complexes. *Arch. Biochem. Biophys.*, 37, 237–239.

THOMPSON, J. A., and WAND, M. D. (1985) Interaction of Cytochrome P-450 with a Hydroperoxide Derived from Butylated Hydroxytoluene. *J. Biol. Chem.*, 260, 10637–10644.

TIAN, Z.-Q. (1992) *Mechanisms of Iron Porphyrin Catalyzed Oxidation of Alkenes and the Concomitant N-alkylhemin Formation*, Doctoral dissertation, University of California, San Diego.

TRAYLOR, T. G. (1981) Synthetic Model Compounds for Hemoproteins. *Acc. Chem. Res.*, 14, 102–109.

TRAYLOR, T. G. (1989) Synthesis of Biological Active Sites, in *New Aspects of Organic Chemistry* I (Z. Yoshida, T. Shiba, and Y. Oshiro, Eds.), Kodansha, Tokyo, pp. 509–527.

TRAYLOR, T. G., and BERZINIS, A. P. (1980) Binding of $O_2$ and CO to Hemes and Hemoproteins. *Proc. Natl. Acad. Sci. U.S.A.*, 77, 3171–3175.

TRAYLOR, T. G., BERZINIS, A., CAMPBELL, D., CANNON, J., LEE, W., MCKINNON, D., MINCEY, T., and WHITE, D. K. (1979a) Factors Controlling Hemoprotein Reactivity as Studied with Synthetic Model Compounds, in *Biochemical and Clinical Aspects of Oxygen* (W. S. Caughey, Ed.), Academic Press, New York, pp. 455–476.

TRAYLOR, T. G., CAMPBELL, D., and TSUCHIYA, S. (1979b) Cyclophane Porphyrin. 2. Models for Steric Hindrance to CO Ligation in Hemoproteins. *J. Am. Chem. Soc.*, 101, 4748–4749.

TRAYLOR, T. G., BERZINIS, A. P., CANNON, J. B., CAMPBELL, D. H., GEIBEL, J. F., MINCEY, T., TSUCHIYA, S., and WHITE, D. K. (1980a) The Chemical Basis of Variations in Hemoglobin Reactivity, in *Advances in Chemistry Series* (D. Dolphin, C. McKenna, Y. Murakami, and I. Tabushi, Eds.), American Chemical Society, Washington, DC, pp. 219–233.

TRAYLOR, T. G., CAMPBELL, D., TSUCHIYA, S., MITCHELL, M., and STYNES, D. V. (1980b) Cyclophane Hemes. 3. Magnitudes of Distal Side Steric Effects in Hemes and Hemoproteins. *J. Am. Chem. Soc.*, 102, 5939–5941.

TRAYLOR, T. G., CAMPBELL, D. H., TSUCHIYA, S., STYNES, D. V., and MITCHELL, M. J. (1982) Steric Effects in Hemoprotein Reactivities, in *Hemoglobin and Oxygen Binding* (C, Ho, Ed.), Elsevier North-Holland, Amsterdam, pp. 425–433.

TRAYLOR, T. G., CHANG, C. K., GEIBEL, J., BERZINIS, A., MINCEY, T., and CANNON, J. (1979c) Syntheses and NMR Characterization of Chelated Heme Models of Hemoproteins. *J. Am. Chem. Soc.*, 101, 6716–6731.

TRAYLOR, T. G., and CICCONE, J. P. (1989) Mechanism of Reactions of Hydrogen Peroxide and Hydroperoxides with Iron(III) Porphyrins. Effects of Hydroperoxide Structure on Kinetics. *J. Am. Chem. Soc.*, 111, 8413–8420.

TRAYLOR, P. S., DOLPHIN, D., and TRAYLOR, T. G. (1984) Sterically Protected Hemins with Electronegative Substituents: Efficient Catalysts for Hydroxylation and Epoxidation. *J. Chem. Soc., Chem. Commun.*, 279–280.

TRAYLOR, T. G., FANN, W.-P., and BANDYOPADHYAY, D. (1989) A Common Heterolytic Mechanism for Reactions of Iodosobenzenes, Peracids, Hydroperoxides, and Hydrogen Peroxide with Iron(III) Porphyrins. *J. Am. Chem., Soc.*, 111, 8009–8010.

TRAYLOR, T. G., HILL, K. W., FANN, W.-P., TSUCHIYA, S., and DUNLAP, B. E. (1992a) Aliphatic Hydroxylation Catalyzed by Iron(III) Porphyrins. *J. Am. Chem. Soc.*, 114, 1308–1312.

TRAYLOR, T. G., IAMAMOTO, Y., and NAKANO, T. (1986a) Mechanisms of Hemin-catalyzed Oxidations: Rearrangements During the Epoxidation of *trans*-Cyclooctene. *J. Am. Chem. Soc.*, 108, 3529–3531.

TRAYLOR, T. G., KOGA, N., and DEARDRUFF, L. A. (1985a) Structural Differentiation of CO and $O_2$ Binding to Iron Porphyrins: Polar Pocket Effects. *J. Am. Chem. Soc.*, 107, 6504–6510.

TRAYLOR, T. G., KOGA, N., DEARDURFF, L. A., SWEPSTON, P. N., and IBERS, J. A. (1984a) 1,3-Adamantane-3,13-porphyrin-6,6-cyclophane: Crystal Structure of the Free Base and Steric Effects on Ligation of the Iron(II) Complex. *J. Am. Chem. Soc.*, 106, 5132–5143.

TRAYLOR, T. G., LEE, W. A., and STYNES, D. V. (1984b) Model Compound Studies Related to Peroxidases. Mechanisms of Reactions of Hemins with Peracids. *J. Am. Chem. Soc.*, 106, 755–764.

TRAYLOR, T. G., KIM, C., RICHARDS, J., XU, F., and PERRIN, C. L. (1995) Reactions of Iron(III) Porphyrins with Oxidants: Structure-Reactivity Studies. *J. Am. Chem. Soc.* 117, 3468–3474.

TRAYLOR, T. G., LUO, J., SIMON, J. A., and FORD, P. C. (1992b) Pressure-induced Change from Activation to Diffusion Control in Fast Reactions of Carbon Monoxide with Hemes. *J. Am. Chem. Soc.*, 114, 4340–4345.

TRAYLOR, T. G., MAGDE, D., LUO, J., WALDA, K. N., BANDYOPADHYAY, D., WU, G.-Z., and SHARMA, V. S. (1992d) Mechanisms of Cage Reactions: Kinetics of Combination and Diffusion after Picosecond Photolysis of Iron(II) Porphyrin Ligated Systems. *J. Am. Chem. Soc.*, 114, 9011–9017.

TRAYLOR, T. G., MAGDE, D., TAUBE, D. J., and JONGEWARD, K. (1987a) Geminate Recombination of Iron(II) Porphyrin with Methyl, *tert*-Butyl, and Tosylmethyl Isocyanide and 1-Methylimidazole. *J. Am. Chem. Soc.*, 109, 5864–5865.

TRAYLOR, T. G., MAGDE, D., TAUBE, D. J., JONGEWARD, K. A., BANDYOPADHYAY, D., LUO, J., and WALDA, K. N. (1992c) Geminate Recombination of Carbon Monoxide Complexes of Hemes and Heme Proteins. *J. Am. Chem. Soc.*, 114, 417–429.

TRAYLOR, T. G., MARSTERS, J. C., JR., NAKANO, T., and DUNLAP, B. E. (1985c) Kinetics of Iron(III) Porphyrin Catalyzed Epoxidations. *J. Am. Chem. Soc.*, 107, 5537–5539.

TRAYLOR, T. G., and MIKSZTAL, A. R. (1987) Mechanisms of Hemin-catalyzed Epoxidations: Electron Transfer from Alkenes. *J. Am. Chem. Soc.*, 109, 2770–2774.

TRAYLOR, T. G., MITCHELL, M. J., TSUCHIYA, S., CAMPBELL, D. H., STYNES, D. V., and KOGA, N. (1981a) Cyclophane Hemes. 4. Steric Effects on Dioxygen and Carbon Monoxide Binding to Hemes and Heme Proteins. *J. Am. Chem. Soc.*, 103, 5234–5236.

TRAYLOR, T. G., NAKANO, T., DUNLAP, B. E., TRAYLOR, P. S., and DOLPHIN, D. (1986b) Mechanisms of Hemin-catalyzed Alkene Epoxidation. The Effect of Catalyst on the Regiochemistry of Epoxidation. *J. Am. Chem. Soc.*, 108, 2782–2784.

TRAYLOR, T. G., NAKANO, T., MIKSZTAL, A. R., and DUNLAP, B. E. (1987b) Transient Formation of N-Alkylhemins during Hemin-catalyzed Epoxidation of Norbornene. Evidence Concerning the Mechanism of Epoxidation. *J. Am. Chem. Soc.*, 109, 3625–3632.

TRAYLOR, T. G., TAUBE, D. J., JONGEWARD, K. A., and MAGDE, D. (1990) Steric Effects on Geminate Recombinations. *J. Am. Chem. Soc.*, 112, 6875-6880.

TRAYLOR, T. G., and TRAYLOR, P. S. (1982) Considerations for the Design of Useful Synthetic Oxygen Carriers. *Ann. Rev. Biophys. Bioeng.*, 11, 105-127.

TRAYLOR, T. G., TSUCHIYA, S., BYUN, Y.-S., and KIM, C. (1993) High Yield Epoxidations with Hydrogen Peroxide and *Tert*-Butyl Hydroperoxide Catalyzed by Iron(III) Porphyrins: Heterolytic Cleavage of Hydroperoxides. *J. Am. Chem. Soc.*, 115, 2775-2781.

TRAYLOR, T. G., TSUCHIYA, S., CAMPBELL, D., MITCHELL, M., STYNES, D., and KOGA, N. (1985b) Anthracene Heme Cyclophanes. Steric Effects in CO, $O_2$, and RNC Binding. *J. Am. Chem. Soc.*, 107, 604-614.

TRAYLOR, T. G., WHITE, D. K., CAMPBELL, D. H., and BERZINIS, A. P. (1981b) Electronic Effects on the Binding of Dioxygen and Carbon Monoxide to Hemes. *J. Am. Chem. Soc.*, 103, 4932-4936.

TRAYLOR, T. G., and XU, F. (1987) A Biomimetic Model for Catalase: the Mechanisms of Reaction of Hydrogen Peroxide and Hydroperoxides with Iron(III) Porphyrins. *J. Am. Chem. Soc.*, 109, 6201-6202.

TRAYLOR, T. G., and XU, F. (1988) Model Reactions Related to Cytochrome P-450. Effects of Alkene Structure on the Rates of Epoxide Formation. *J. Am. Chem. Soc.*, 110, 1953-1958.

TRAYLOR, T. G., and XU, F. (1990) Mechanisms of Reactions of Iron(III) Porphyrins with Hydrogen Peroxide and Hydroperoxides: Solvent and Solvent Isotope Effects. *J. Am. Chem. Soc.*, 112, 178-186.

TSUCHIYA, S., and SENO, M. (1989) Novel Synthetic Method of Phenol from Benzene Catalyzed by Perfluorinated Hemin. *Chem. Lett.*, 263-266.

TUCKER, P. W., PHILLIPS, S. E. V., PERUTZ, M. F., HOUTCHENS, R., and CAUGHEY, W. S. (1978) Structure of Hemoglobins Zürich[His E7(63)β→Arg] and Sydney[Val E11(67)β→Ala] and Role of the Distal Residues in Ligand Binding. *Proc. Natl. Acad. Sci. U.S.A.*, 75, 1076-1080.

TYEKLAR, Z., and KARLIN, K. D. (1989) Copper Dioxygen Chemistry: A Bioinorganic Challenge. *Acc. Chem. Res.*, 22, 241-248.

VAINSHTEIN, B. K., MELIK-ADAMYAN, W. R., BARYNIN, V. V., VAGIN, A. A., GREBENKO, A. I., BORISOV, V. V., BARTELS, K. S., FITA, I., and ROSSMANN, M. G. (1986) Three-dimensional Structure of Catalase from Penicillium Vitale at 2.0 Å Resolution. *J. Mol. Biol.*, 188, 49-61.

VOLBEDA, A., and HOL, W. G. J. (1986) The Structure of the Copper-containing Oxygen-transporting Hemocyanins from Arthropods, in *Frontiers in Bioinorganic Chemistry*. (A. V. Xavier, Ed.), pp. 584-593.

WAGNER, G. C., and KASSNER, R. J. (1974) Spectroscopic Properties of Protoheme Complexes Undergoing Reversible Oxygenation. *J. Am. Chem. Soc.*, 96, 5593-5595.

WALDMAN, S. A., and MURAD, F. (1988) Biochemical Mechanisms Underlying

Vascular Smooth Muscle Relaxation: The Guanylate Cyclase-cyclic GMP System. *J. Cardiovas. Pharm.*, 12(Suppl. 5), S115–S118.

WALKER, F. A., and SIMONIS, U. (1991) Models of the Cytochromes *b*. 8. Two-dimensional Nuclear Overhauser and Exchange Spectroscopy Studies of Paramagnetic "Cavity" Type (Tetra-phenylporphinato)iron(III) Complexes of Planar Ligands. *J. Am. Chem. Soc.*, 113, 8652–8657.

WALTER, U. (1989) Physiological Role of cGMP and cGMP-dependent Protein Kinase in the Cardiovascular System. *Rev. Physiol. Biochem. Pharm.*, 41–88.

WANG, J. H. (1958) Hemoglobin Studies. II. A Synthetic Material with Hemoglobin-like Property. *J. Am. Chem. Soc.*, 80, 3168–3169.

WANG, J. H. (1962) Hemoglobin and Myoglobin, in *Oxygenases* (O. Hayaishi, Ed.), Academic Press, New York, pp. 469–516.

WANG, J.-S., BAEK, H. K., and VAN WART, H. E. (1991) High-valent Intermediates in the Reaction of N Alpha–acetyl Microperoxidase-8 with Hydrogen Peroxide: Models for Compounds 0, I and II of Horseradish Peroxidase. *Biochem. Biophys. Res. Commun.*, 179, 1320–1324.

WARD, B., WANG, C., and CHANG, C. K. (1981) Nonbonding Steric Effect on CO and $O_2$ Binding to Hemes. Kinetics of Ligand Binding in Iron-copper Cofacial Diporphyrins and Strapped Hemes. *J. Am. Chem. Soc.*, 103, 5236–5238.

WARME, P. K., and HAGER, L. P. (1970) Heme Sulfuric Anhydrides. II. Properties of Heme Models Prepared from Mesoheme Sulfuric Anhydrides. *Biochemistry*, 9, 1606–1614.

WESCHLER, C. J., ANDERSON, D. L., and BASOLO, F. (1975) Kinetics and Thermodynamics of Oxygen and Carbon Monoxide Binding to Simple Ferrous Porphyrins at Low Temperatures. *J. Am. Chem. Soc.*, 97, 6707–6713.

WESTRICK, J. A., PETERS, K. S., ROPP, J. D., and SLIGAR, S. G. (1990) Role of the Arginine-45 Salt Bridge in Ligand Dissociation from Sperm Whale Carboxymyoglobin as Probed by Photoacoustic Calorimetry. *Biochemistry*, 29, 6741–6746.

WHITE, D. K., CANNON, J. B., and TRAYLOR, T. G. (1979) A Kinetic Model for R- and T-state Hemoglobin. Flash Photolysis of Heme–Imidazole–Carbon Monoxide Mixtures. *J. Am. Chem. Soc.*, 101, 2443–2454.

WHITE, P. W. (1990) Mechanistic Studies and Selective Catalysis with Cytochrome P-450 Model Systems. *Bioorg. Chem.*, 18, 440–456.

WHITE, R. E., and COON, M. J. (1980) Oxygen Activation by Cytochrome P-450. *Ann. Rev. Biochem.*, 49, 315–356.

WHITFORD, D., GAO, Y., PIELAK, G. J., WILLIAMS, R. J., MCLENDON, G. L., and SHERMAN, F. (1991) The Role of the Internal Hydrogen Bond Network in First-order Protein Electron Transfer Between *Saccharomyces cerevisiae* Iso-1-cytochrome *c* and Bovine Microsomal Cytochrome $b_5$. *Eur. J. Biochem.*, 200, 359–367.

WIEGHARDT, K. (1986) Accurate Synthetic Models for the Diiron Centers in Hemerythrin, in *Frontiers Bioinorg. Chem.* (A. V. Xavier Ed.), VCH, New York, pp. 246–255.

WIJESEKERA, T. P., PAINE, J. B., III, and DOLPHIN, D. (1988) Improved Synthesis of Covalently Strapped Porphyrins. Application to Highly Deformed Porphyrin Synthesis. *J. Org. Chem.*, 53, 1345–1352.

WIKSTRÖM, M. (1989) Identification of the Electron Transfers in Cytochrome Oxidase That Are Coupled to Proton-pumping. *Nature*, 338, 776–778.

WIKSTRÖM, M., KRAB, K., and SARASTE, M. (Eds.) (1981) *Cytochrome Oxidase: A Synthesis*, Academic Press, New York.

WILKINS, P. C., and WILKINS, R. G. (1987) The Coordination Chemistry of the Binuclear Iron Site in Hemerythrin. *Coord. Chem. Rev.*, 79, 195–214.

WINKLER, J. R., and GRAY, H. B. (1992) Electron Transfer in Ruthenium-modified Proteins. *Chem. Rev.*, 92, 369–377.

WITTENBERG, J. B., APPLEBY, C. A., and WITTENBERG, B. A. (1972) The Kinetics of the Reactions of Leghemoglobin with Oxygen and Carbon Monoxide. *J. Biol. Chem.*, 247, 527–531.

WITTENBERG, J. B., NOBLE, R. W., WITTENBERG, B. A., ANTONINI, E., BRUNORI, M., and WYMAN, J. (1967) Studies on the Equilibria and Kinetics of the Reactions of Peroxidase with Ligands. *J. Biol. Chem.*, 242, 626–634.

YAMAGUCHI, K., WATANABE, Y., and MORISHIMA, I. (1992) Push Effect on Heterolytic O–O Bond Cleavage of Peroxoiron(III)porphyrin Adducts. *Inorg. Chem.*, 31, 156–157.

YUMIBE, N. P., and THOMPSON, J. A. (1988) Fate of Free Radicals Generated During One-electron Reductions of 4-Alkyl-1,4-Peroxyquinols by Cytochrome P-450. *Chem. Res. Toxicol.*, 1, 385–390.

ZIPPLIES, M. F., LEE, W. A., and BRUICE, T. C. (1986) Influence of Hydrogen Ion Activity and General Acid-base Catalysis on the Rate of Decomposition of Hydrogen Peroxide by a Novel Nonaggregating Water-soluble Iron(III) Tetraphenylporphyrin Derivative. *J. Am. Chem. Soc.*, 108, 4433–4445.

# 4
# Dioxygen Reactivity in Copper Proteins and Complexes

STEPHEN FOX AND KENNETH D. KARLIN

### INTRODUCTION

Copper ion is an essential trace element found in living systems, and its importance resides in its role as a protein or enzyme active-site constituent. Thus, copper proteins occur widely in nature, performing a diverse array of functions (Adman, 1991; Beinert, 1991; Kitajima, 1992; Solomon et al., 1992; Karlin and Tyeklár, 1993). All of these involve oxidation-reduction (i.e., "redox") activity, since a number of oxidation states are known for copper ion (Hathaway, 1987). Ligand donors for copper which are typically available in protein matrices include the side-chain imidazole group of histidine, the phenol oxygen donor of tyrosine, or the sulfur atom of the thiol group from cysteine. With combinations of these, the $Cu^I$ and $Cu^{II}$ oxidation states are readily accessible and interconvertible under physiological conditions, using available oxidants (e.g., molecular oxygen [dioxygen; $O_2$]) or reducing agents such as glutathione ($\gamma$-Glu-Cys-Gly) a sulfhydryl-containing tripeptide which is abundant ($\sim$1 mM) in biological media, or ascorbic acid (i.e., vitamin C).

Protein active-site copper ions perform the functions of electron transfer (e.g., as electron carriers in photosynthetic organisms or in respiratory pathways in certain bacteria), reversible $O_2$ binding and transport, mono- or dioxygenation of organic substrates (*oxygenases*, incorporating one or both atoms of $O_2$), or oxidation/dehydrogenation of substrates accompanied by $O_2$ reduction to either hydrogen peroxide or water (*oxidases*). Comparable functions carried out by heme and nonheme iron enzymes are reviewed in Chapters 3 and 5, respectively. An important copper/zinc-containing superoxide dismutase (SOD; $2O_2^- + 2H^+ \rightarrow H_2O_2 + O_2$) is well characterized (Tainer et al., 1983; Bertini et al., 1990), and recent evidence suggests that a point mutation in this SOD gene may cause a degenerative disease of motor neurons, namely, the inherited form of amyotrophic lateral sclerosis (ALS), often called *Lou Gehrig's disease* (Beckman et al., 1993; McNamara and Fridovich, 1993). Other redox-active copper proteins are found in certain denitrifying bacteria, effecting the reduction of nitrogen-oxide species such as nitrite (e.g., $NO_2^- \rightarrow NO$) (Godden et al., 1991; Brittain et al., 1992) and nitrous oxide (e.g., $N_2O \rightarrow N_2$) (Farrar et al., 1991). These substrates act as electron acceptors in certain anaerobic organisms, and their biochemistry constitutes a vital part of the global nitrogen cycle, while the presence of such $NO_x$ species is also of environmental concern. In association with mechanisms to regulate the levels of copper in cells, i.e., copper homeostasis, copper-thiolate proteins (*metallothioneins*) are also found in biological systems. These are cysteine-rich, and they tightly bind and store small clusters of four to eight copper(I) ions; alternatively, transcriptional activator proteins such as ACE1 bind copper ion, thus turning on metallothionein biosynthesis (O'Halloran, 1993).

The focus of this review will be on $O_2$-activation mechanisms involving copper metalloproteins. Here our intention is to emphasize reactions involving the binding of dioxygen to copper centers and either its subsequent oxygen atom transfer to substrate or its reduction reactions. Both reactions involve O–O bond cleavage, mediated at least in part by metal–ion interactions. Appropriate chemical model systems will also be mentioned, since biomimetic modeling approaches have previously provided and continue to constitute a vital component in elucidating basic metal-based chemistry, potentially leading to insights in biological $O_2$-activation mechanisms or even to the development of reagents or catalysts for use in oxidation processes of practical use. Biological organic substrates which are oxidized or oxygenated will be described, and where they are known, protein active-site structures and reaction pathways will be delineated. An assortment of structures and copper-ligating situations are observed, and even the number of copper ions (i.e., the nuclearity) involved in the active-site chemistry varies. An emerging theme is the occurrence of active-site protein derived *cofactors*, organics located nearby or even coordinated with the copper ion, which become intimately involved in the reaction chemistry. For the purpose of overviewing and highlighting, the major classes of copper proteins involved in dioxygen processing, their origins, and their primary functions are provided in Table 4-1.

## REVERSIBLE BINDING OF DIOXYGEN TO COPPER ION

There currently exists a significant understanding of the nature of the reversible binding of $O_2$ to hemoglobin and myoglobin, the activation of $O_2$ by heme proteins such as cytochrome P-450 monooxygenase, and corresponding chemistry deduced from studies on model compounds (see Chapter 3). Much less is known about mechanisms of $O_2$ binding and activation by nonheme iron (see Chapter 5) and copper enzymes (Table 4-1). However, a common denominator is that $O_2$ activation most often requires prior $O_2$ binding at the metal center. Thus, a knowledge of copper ion–dioxygen interactions, resulting structures, and reactivity patterns is an essential prerequisite for elucidating $O_2$ activation mechanistic principles. Accordingly, the reversible binding of dioxygen to the $O_2$ carrier protein hemocyanin will be discussed. Several well-characterized synthetic model systems are described, along with relevant reactivity patterns, where known. The close relationship between binding of $O_2$ and subsequent activation is well illustrated for at least one copper monooxygenase, tyrosinase (see below); this enzyme mediates *o*-hydroxylations of phenolic substrates via reaction of a $Cu_2-O_2$ adduct, where the latter structure is essentially identical to that found in hemocyanin.

### Biological $O_2$ Storage and Transport: Hemocyanins

Invertebrate molluscan and arthropodal hemocyanins (Hcs) are very large (MW $4.5-90 \times 10^5$) $O_2$-transporting proteins. These multisubunit proteins exhibit highly cooperative behavior. Molluscan hemocyanins contain 10–20 subunits where the functional unit has a molecular weight of ~55,000. Arthropodal hemocyanins are hexamers or multi-hexamers with larger subunits (~75,000). In both classes of carriers, $O_2$ binding occurs at a dinuclear copper center, i.e., two copper ions in close proximity comprise the active site. In spite of clear differences in sequence homologies between the molluscan and arthropodal forms, analogies in at least half of the dicopper sites and spectroscopic similarities indicate a closely related active-site structure and a dioxygen-binding mode (Sorrell, 1989; Volbeda and Hol, 1989; Solomon et al., 1992).

**TABLE 4-1** Main Classes of Copper Proteins Involved in Dioxygen Processing

| Protein | Source | Biological Function |
|---|---|---|
| *Dioxygen Carrier* | | |
| Hemocyanin (Hc) | Molluscs and arthropods | Hemolymph $O_2$ carrier |
| *Copper Oxygenases* | | |
| Tyrosinase (Tyr) | Fungal, mammal | Tyrosine oxidation |
| Dopamine β-monooxygenase (DβM) | Adrenal, brain | Dopamine → norepinephrine |
| Peptidylglycine α-amidating monooxygenase (PAM) | Pituitary, heart | Oxidative N-dealkylation |
| Phenylalanine hydroxylase (PAH) | *Chromobacterium violaceum* | Phenylalanine → tyrosine |
| Methane monooxygenase (MMO) | Methanogenic bacteria | Methane → methanol |
| Ammonia monooxygenase (AMO) | Bacteria | $NH_3 \rightarrow NH_2OH$ |
| *Copper Dioxygenases* | | |
| Quercetinase | Fungal | Quercetin oxidative cleavage |
| *Copper Oxidases* | | |
| "Blue" multicopper Oxidases | | |
| Laccase | Tree, fungal | Phenol and diamine oxidation |
| Ascorbate oxidase | Plants | Oxidation of L-ascorbate |
| Ceruloplasmin | Human, animal serum | Weak oxidase activity |
| "Nonblue" Oxidases | | |
| Amine oxidase | Most animals | Elastin and collagen formation |
| Galactose oxidase | Molds | Galactose oxidation |
| Cytochrome *c* oxidase | Mitochondria | Terminal oxidase ($O_2 \rightarrow H_2O$) and proton pump |
| Phenoxazinone synthase | *Streptomyces* | Phenoxazinone formation |
| *Other* | | |
| Superoxide dismutase | Red blood cells, animals | $O_2^-$ detoxification |

The unoxygenated reduced forms of hemocyanins are colorless, reflecting a $3d^{10}$ dicopper(I) formulation. While chemical and X-ray absorption spectroscopic studies had shed considerable light on the nature of the deoxy-Hc dicopper binding site, the actual nature of the site was unclear until recently. There now exist two X-ray crystal structures, the first on the spiny lobster Hc, *Panulirus interruptus* (Volbeda and Hol, 1989) and a recent one of the horseshoe crab *Limulus II* protein (Hazes et al., 1993). The structures exhibit rather different active-site characteristics, and since the spiny lobster HC was crystallized at low pH and possesses rather odd copper(II) ligation, the latter *Limulus II* structure is probably representative. It indicates that the two $Cu^I$ ions are 4.6 Å apart, each found in a trigonal-planar coordination environment with Cu-$N^{His}$ bond distances of ~2.0 Å. Intersubunit $O_2$-binding cooperative effects are probably initiated and transmitted as a result of movement of histidine residue(s) and copper ion(s) on $O_2$ coordination (see below).

Binding of $O_2$ causes the appearance of an intense blue color, and resonance Raman studies of this oxy-Hc form show that the dioxygen is bound as peroxide (formally $O_2^{2-}$) with $v_{O-O}$ ~750 cm$^{-1}$, indicating an oxidative addition of $O_2$ to give a peroxodicopper(II) oxy-form (Solomon et al., 1992). This oxy site is electron paramagnetic resonance (EPR) silent and diamagnetic, reflecting strong magnetic coupling between the two $Cu^{II}$ centers. A very recent X-ray structure (Magnus et al., 1994) reveals a side-on $\mu$-$\eta^2$:$\eta^2$-peroxo ligation with Cu . . . Cu = 3.6 Å. The electronic spectrum of oxy-Hc is distinctive and dominated by $O_2^{2-} \rightarrow Cu^{II}$ ligand-to-metal charge transfer transitions at $\lambda_{max}$ of

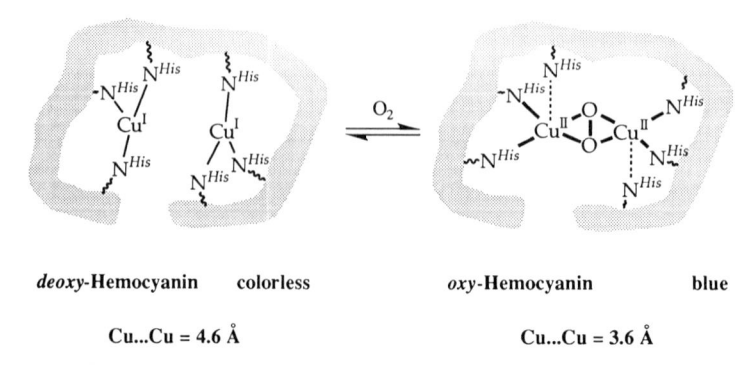

*deoxy*-Hemocyanin    colorless            *oxy*-Hemocyanin         blue

Cu...Cu = 4.6 Å                              Cu...Cu = 3.6 Å

**FIGURE 4-1** Schematic representation of the active site structures of deoxy (unoxygenated) and oxy (oxygenated) *Limulus II* (horseshoe crab) hemocyanin.

345 ($\varepsilon = 20{,}000\,M^{-1}\,cm^{-1}$) and 570 ($\varepsilon = 1{,}000\,M^{-1}\,cm^{-1}$) nm, with an additional circular dichroic feature at 485 nm (Solomon et al., 1992).

## Synthetic Copper–Dioxygen Complexes

The dicopper nature of the oxy-hemocyanin active site, along with its interesting spectroscopic and magnetic properties, has attracted considerable attention from the inorganic chemical community in the last dozen years. In part, it has been of considerable interest to mimic dioxygen-binding behavior by copper ion in synthetic systems and to elucidate basic patterns involved in copper(I)/dioxygen chemistry. Such efforts have led to a number of advances, including the X-ray structural characterization of two copper–dioxygen complexes.

Utilizing sterically encumbering tris(pyrazolyl)borate ligands, Kitajima and coworkers provided a breakthrough in synthetic modeling chemistry (Kitajima, 1992; Kitajima et al., 1992; Kitajima and Moro-Oka, 1994). They synthesized and characterized the compound {Cu[HB(3,5-$i$Pr$_2$pz)$_3$]}$_2$(O$_2$) (3), (HB(3,5-$i$Pr$_2$pz)$_3$ = hydrotris(3,5-diisopropyl-pyrazolyl)borate anion), with $\mu$-$\eta^2$:$\eta^2$-peroxo-dicopper(II) ligation (Cu...Cu = 3.56 Å) and physical properties (e.g., $\lambda_{max}$ = 349, $\varepsilon = 21{,}000\,M^{-1}cm^{-1}$; $\nu_{O-O}$ = 741 cm$^{-1}$) which are essentially identical to that found in oxy-hemocyanin. This compound can be generated at reduced temperatures in solution by O$_2$ reaction with 1 or via an acid-base reaction of hydrogen peroxide with dicopper(II) species 2 (Figure 4-2). In fact, the discovery of 3 preceded the confirmation that the protein structure also possessed this unusual mode of binding of dioxygen.

The other structurally characterized Cu$_2$–O$_2$ complex comes from our own laboratories. [{(TMPA)Cu}$_2$(O$_2$)]$^{2+}$ (6, TMPA = tris[2-pyridylmethyl]amine) forms reversibly by reaction of the mononuclear Cu$^I$ complex [(TMPA)Cu(RCN)]$^+$ (4, R = Me or Et) with O$_2$ at 193 K in EtCN (Figure 4-3) (Tyeklár et al., 1993). A detailed kinetics/thermodynamics study (Karlin et al., 1993b) reveals that a Cu/O$_2$ = 1:1 adduct [(TMPA)Cu(O$_2$)]$^+$ (5) forms as a spectroscopically detectable ($\lambda_{max}$ = 410 nm, $\varepsilon = 4000\,M^{-1}cm^{-1}$) initial transient adduct, with rate constant $k_1 = 2 \times 10^4\,M^{-1}s^{-1}$ and $k_1/k_{-1} = K_1 = 1.9 \times 10^3\,M^{-1}$ at 183 K, while values extrapolated to 298 K are $k_1 \sim 10^8\,M^{-1}s^{-1}$ and $K_{formation}$ ~0.3 M$^{-1}$. The final stable product 6 forms with $K_{formation} = 4.3 \times 10^{11}\,M^{-2}$ at 183 K, but its room temperature stability is precluded by a highly unfavorable entropy term. The X-ray structure (Tyeklár et al., 1993) of the stable final product 6 shows a $trans$-$\mu$-1,2-O$_2^{2-}$ dicopper(II) coordination with a Cu...Cu' distance of 4.359 Å; the O—O' bond

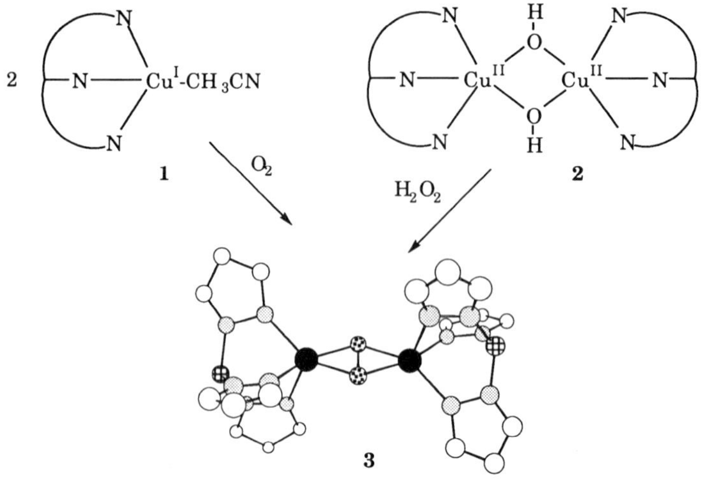


$\{Cu[HB(3,5\text{-}iPr_2pz)_3]\}_2(O_2)$

**FIGURE 4-2** $\{Cu[HB(3,5\text{-}iPr_2pz)_3]\}_2(O_2)$ (**3**), a copper–dioxygen complex which possesses structural features and physical properties nearly identical to those of the $O_2$-carrier protein oxy-hemocyanin.

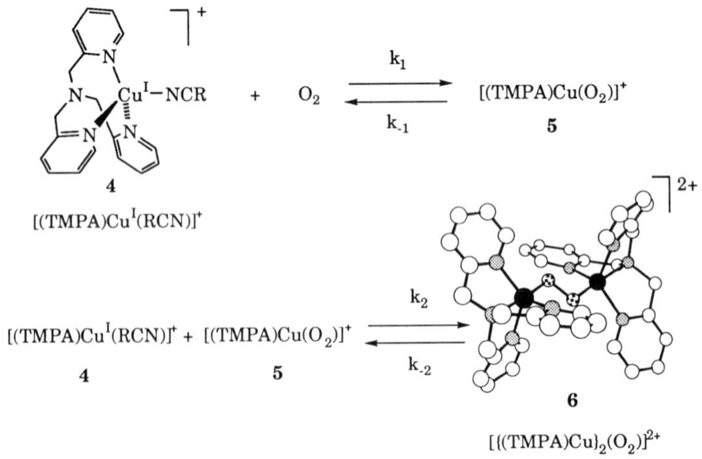


**FIGURE 4-3** The peroxo–dicopper(II) complex $[\{(TMPA)Cu\}_2(O_2)]^{2+}$ (**6**) forms reversibly by reaction of $O_2$ with mononuclear species $[(TMPA)Cu^I(RCN)]^+$ (**4**).

length is 1.432 Å. Relevant spectroscopic data for this intensely purple species are charge-transfer transitions at $\lambda_{max} = 440$ ($\varepsilon = 2,000\,M^{-1}\,cm^{-1}$), 525 (11,500), 590 (sh, 7,600) and a $d$-$d$ band at 1,035 (180) nm (Baldwin et al., 1991). Resonance Raman studies showed an intraperoxide (O—O) stretch (832 cm$^{-1}$) and a Cu—O stretch (561 cm$^{-1}$) (Baldwin et al., 1991).

The insight that binding of dioxygen to a copper ion center can occur via a coordination mode other than that found in the protein $O_2$ carrier hemocyanin is important. In fact, a variety of binding modes to one or two copper ions have been observed (Karlin and Gultneh, 1987; Kitajima, 1992; Karlin, 1993; Kitajima and Moro-Oka, 1994), and their subsequent reactivity also differs as a function of the detailed $Cu_2$—$O_2$ structure. Thus, the $trans$-$\mu$-1,2-$O_2^{2-}$–dicopper(II) complex **6** reacts as a nucleophile or base in reactions with substrates such as carbon dioxide or protons, while Kitajima's $\mu$-$\eta^2$:$\eta^2$-peroxo dicopper(II) complex **3** and other types of $Cu_2$—$O_2$ complexes with this side-on bridging peroxo group behave somewhat differently, as indicated below.

## COPPER OXYGENASES

### Tyrosinase and Aromatic Hydroxylation Chemistry

*Protein studies and proposed mechanism.* In 1955, Mason and coworkers carried out an $^{18}O_2$ labeling experiment confirming the incorporation of an oxygen atom from dioxygen and establishing this previously well-known enzyme as the first to be characterized as a monooxygenase (Mason et al., 1955). Tyrosinase occurs widely in bacteria, fungi, plants, and mammals, where it functions in melanin pigment formation and in the well-known browning reaction observed in fruits and vegetables. Tyrosinases carry out the oxygenation of monophenols, affording $o$-quinones, and these reactions are considered to proceed via the intermediacy of the corresponding $o$-catechols. In addition to this so-called cresolase activity (see discussion below), tyrosinases efficiently oxidize $o$-catechol substrates to $o$-quinones (catecholase activity).

Chemical and spectroscopic evidence (Wilcox et al., 1985; Solomon et al., 1992) indicates that tyrosinases have a coupled dinuclear active site which is essentially identical to that found in hemocyanins; a stable oxy form (oxy-Tyr; $\lambda_{max}$ 350 nm, $\varepsilon = 26,000\,M^{-1}cm^{-1}$) can be generated either from reaction of the reduced dicopper(I) form with $O_2$ or by exposure of oxidized dicopper(II) met-Tyr to hydrogen peroxide. An important difference between hemocyanin and tyrosinase is that the active site in the latter protein is highly accessible to exogenous ligands. Amino-acid sequence (Beltramini et al., 1990) comparisons support a simplistic view in which hemocyanins consist of a tyrosinase-like protein, with an additional protein sequence making up a domain which shields the active site, preventing binding of large ligand substrates. Thus, hemocyanins possess only very limited tyrosinase activity.

With the close similarity of the active-site spectroscopy and biophysical chemistry of $\{Cu[HB(3,5\text{-}iPr_2pz)_3]\}_2(O_2)$ (3), hemocyanin, and tyrosinase establishing oxy-Tyr as possessing a $\mu\text{-}\eta^2:\eta^2$-peroxo dicopper(II) structure, Solomon and coworkers (Ross and Solomon, 1991; Solomon et al., 1992) have proposed a mechanism of $O_2$ activation

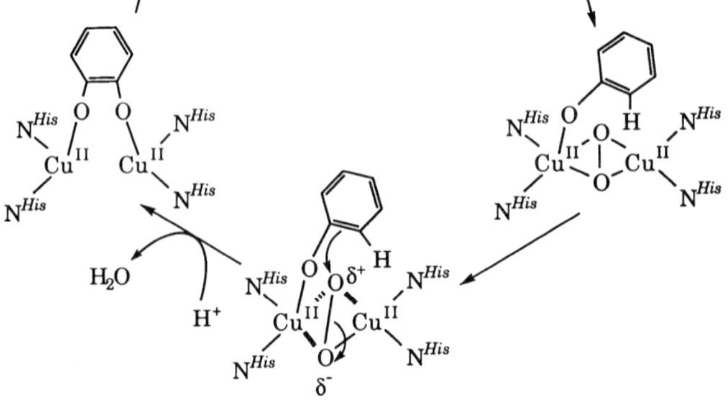

**FIGURE 4-4** Tyrosinase (Tyr) mechanism proposed for the o-hydroxylation of substrate phenols by oxy-Tyr, involving electrophilic attack by peroxide, and subsequent two-electron oxidation to give o-quinones. Adapted from Solomon et al. (1992).

for Tyr cresolase activity (Figure 4-4). Binding of a phenol substrate would lead to a structural rearrangement of the copper (i.e., from tetragonal to trigonal bipyramidal), as deduced from studies on competitive inhibitors; this interaction is stabilized by a strong substrate–protein interaction. The result is an unsymmetrically coordinated peroxide which is heterolytically polarized (Figure 4-4), and thus activated for electrophilic attack on the phenol ring. Theoretical calculations show side-on bridged binding of peroxide to be less negative, i.e., more electrophilic, than end-on bound peroxide. From this analysis, the O—O bond is also weakened and activated for cleav-

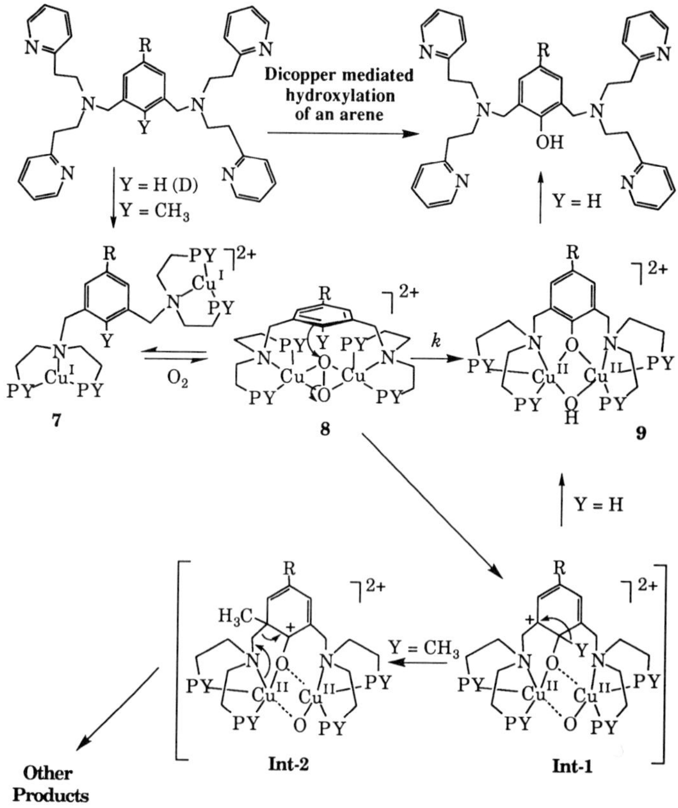

**FIGURE 4-5** A copper monooxygenase model system involving activation of dioxygen via interaction of $O_2$ with a dicopper(I) center, resulting in the hydroxylation of the xylyl portion of the dinucleating ligand. PY = 2-pyridyl. See text for further details.

age due to the presence of some σ* character in the Cu–(O$_2$)–Cu highest occupied molecular orbital (HOMO) (Ross and Solomon, 1991). Subsequent reaction (Figure 4-4) would lead to a catecholate-dicopper(II) intermediate, a species known in coordination chemistry (Karlin et al., 1985), followed by an intramolecular redox reaction giving product quinone and regenerated dicopper(I) deoxy-tyrosinase.

*A model system with C—H activation by an electrophilic Cu$_2$—O$_2$ moiety.* A chemical model system which provides an example of one of the very few cases of activation of dioxygen for hydroxylation of a C—H bond involves a dinuclear copper center, thus potentially bearing on the chemistry of tyrosinase (Karlin, 1993). Dicopper(I) complex **7** reacts rapidly with O$_2$ (1 atm) quantitatively, producing **9** (Figure 4-5). Isotopic labeling experiments prove that the oxygen atom in the phenolic product is derived from dioxygen. Given the observed stoichiometry of reaction (i.e., Cu:O$_2$ = 2:1), the process **7** + O$_2$ → **9** is reminiscent of reactions catalyzed by biological monooxygenases (Hayaishi, 1974).

Detailed kinetic/mechanistic studies (Nasir et al., 1992; Karlin et al., 1994; Karlin and Tyeklár, 1994) reveal that a Cu$_2$—O$_2$ intermediate **8** forms *reversibly* on reaction of **7** with O$_2$; **8** further irreversibly converts to hydroxylated product **9**, from which phenol product can be extracted. Spectroscopic studies on **8** and analogs with methylene chain spacers [—(CH$_2$)$_n$—, n = 3, 4, 5] replacing the xylyl group, i.e., reversible dioxygen carriers [Cu$_2^I$(Nn)]$^{2+}$ (**10**; see the drawing below, PY = 2-pyridyl) and O$_2$ adducts [Cu$_2$(Nn)(O$_2$)]$^{2+}$ (**11**) (Karlin et al., 1992; Karlin and Tyeklár, 1994), suggest that species **11** and **8** possess a side-on μ-η$^2$:η$^2$-peroxo dicopper(II) structure, as indicated in Figure 4-5. In fact, their distinctive UV-Vis properties also greatly resemble those of oxy-Hc and oxy-Tyr.

**10**
[Cu$_2^I$(Nn)]$^{2+}$
  
**11**
[Cu$_2$(Nn)(O$_2$)]$^{2+}$

A number of lines of evidence indicate that the reaction **8** → **9** takes place via a mechanism best described as electrophilic attack on the substrate arene ring by the Cu$_2$—O$_2$ group (Karlin et al., 1993a, 1994; Karlin and Tyeklár, 1994):

1. Kinetic studies on complexes with R-xylyl substituents opposite to the position that is hydroxylated (Figure 4-5) show that electron-donating groups increase the rate of hydroxylation (i.e., k).
2. When a deuterium atom is substituted for the hydrogen on the 2-xylyl carbon which is hydroxylated, no measurable isotope effect is observed, consistent with an electrophilic attack by the $Cu_2-O_2$ moiety on the arene π-system in the rate-limiting step, rather than direct attack on the C—H(D) bond.
3. Electrophilic attack and O—O bond cleavage (stabilized by Cu—O—Cu forming interactions) are suggested to lead to nonobservable intermediate cyclohexadienyl cations **Int-1** and **Int-2** (Figure 4-5). Experiments in which a methyl group is placed in the 2-xylyl position (Y=CH$_3$) lead to products that are consistent with the occurrence of a methyl 1,2-migration; after workup of the reaction mixture, the observed products are HN[CH$_2$CH$_2$-PY]$_2$, formaldehyde (derived from a benzylic methylene group), and a phenol, all obtained by transformation from **Int-2**. This is the so-called NIH shift, well established in certain iron hydroxylases (see Chapters 3 and 5) and shown to occur here in a copper ion-based system (Nasir et al., 1992).
4. Independent studies comparing the reactivity of $[Cu_2(Nn)(O_2)]^{2+}$ (**11**) and $[\{(TMPA)Cu\}_2(O_2)]^{2+}$ (**6**) reveal that the peroxo group in **11** is nonbasic or electrophilic when compared directly to the behavior of **6** toward the same reagents, where a basic or nucleophilic character is observed (Figure 4-6) (Paul et al., 1991). Thus, at least within the framework of studies of these particular molecules, it appears that the side-on μ-η$^2$:η$^2$-peroxo ligand bound in a dicopper(II) structure can act as an electrophile; this finding is consistent with the theoretical calculations carried out on such $Cu_2-O_2$ species (see above).

In summary, this xylyl hydroxylation model system reveals how a dicopper center can activate $O_2$ for hydrocarbon oxidation under mild conditions. As such, it may be of interest in exploiting the structural and mechanistic principles garnered for use in practical applications. It also serves as a *functional mimic* for copper hydroxylases. Similar to an enzyme active site, the peroxo group in **8** is generated in a highly favorable proximity to the xylyl ligand substrate, and facile hydroxylation of a C—H bond occurs. Observations in this system also support the basic notions expressed in Solomon's mechanism for tyrosinase (see above), i.e., that of electrophilic attack on a hydrocarbon substrate by a $Cu_2-O_2$ moiety possessing a μ-η$^2$:η$^2$-peroxo ligand.

*Other model studies and an alternative tyrosinase mechanism.* While the μ-η$^2$:η$^2$-peroxodicopper(II) complexes of the type **8** (Figure 4-5) and $[Cu_2(Nn)(O_2)]^{2+}$(**11**) react toward a variety of substrates in a

**Basic/Nucleophilic Peroxide**

```
      O—O     PPh₃      O₂ + Cuᴵ-PPh₃ complex
Cu       Cu   ─────→
                H⁺       {Cuᴵᴵ₂(⁻OOH)}  ──H⁺──→ H₂O₂
         6    ─────→
              RC(O)⁺
              ─────→   {Cuᴵᴵ₂(RC(O)O₂⁻)}
              CO₂
              ─────→   {Cuᴵᴵ₂(O₂CO₂²⁻)} ──────→ {Cuᴵᴵ₂(CO₃²⁻)}
              PhOH
              ─────→   PhO⁻ + {Cuᴵᴵ₂(⁻OOH)}
```

**Non-Basic/Electrophilic Peroxide**

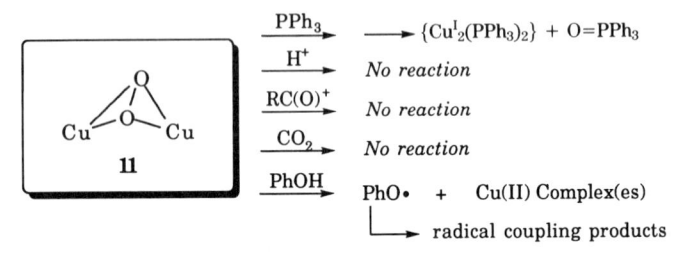

```
        O
       /|\          PPh₃
      / O \         ─────→  {Cuᴵ₂(PPh₃)₂} + O=PPh₃
   Cu─────Cu        H⁺
       11           ─────→  No reaction
                    RC(O)⁺
                    ─────→  No reaction
                    CO₂
                    ─────→  No reaction
                    PhOH
                    ─────→  PhO• + Cu(II) Complex(es)
                            └──→ radical coupling products
```

**FIGURE 4-6** A summary of reactivity studies performed on **6**, with *trans*-μ-1,2-peroxodicopper(II) coordination, compared to those observed for **11**, with a μ-η²:η²-peroxodicopper(II) structure.

manner which is basically electrophilic, Kitajima's complex {Cu[HB(3,5-*i*Pr₂pz)₃]}₂(O₂) (**3**) (Figure 4-2) behaves somewhat differently, and the reactivity patterns his group observed with the bound O₂ ligand in **3** have led him to suggest an alternative mechanism for O₂ activation in tyrosinase (see below) (Kitajima et al., 1990; Kitajima, 1992; Kitajima and Moro-Oka, 1994).

{Cu[HB(3,5-*i*Pr₂pz)₃]}₂(O₂) (**3**) itself undergoes a thermal decomposition reaction which is first order in the peroxo complex. In the absence of any substrate, this results in the production of the μ-oxo–dicopper(II) complex {Cu[HB(3,5-*i*Pr₂pz)₃]}₂(O) (**12**). The interpretation offered is that an O—O bond homolysis occurs, producing a transient, **Int-3**, described as a Cuᴵᴵ/oxygen radical; an alternative description is a Cuᴵᴵᴵ-oxo species, but no physical-spectroscopic data are available. **Int-3** may then react with the mononuclear copper(I) complex {Cuᴵ[HB(3,5-*i*Pr₂pz)₃]} (**1**) that is present in solution to afford the μ-oxo product **12** (Figure 4-7). The radical behavior of the reaction can be seen if substrates are present. In the presence of cyclohexene and excess dioxygen, a classical radical chain reaction occurs (Walling, 1995) and 2-cyclohexene-1-one is the product; if no excess O₂ exists,

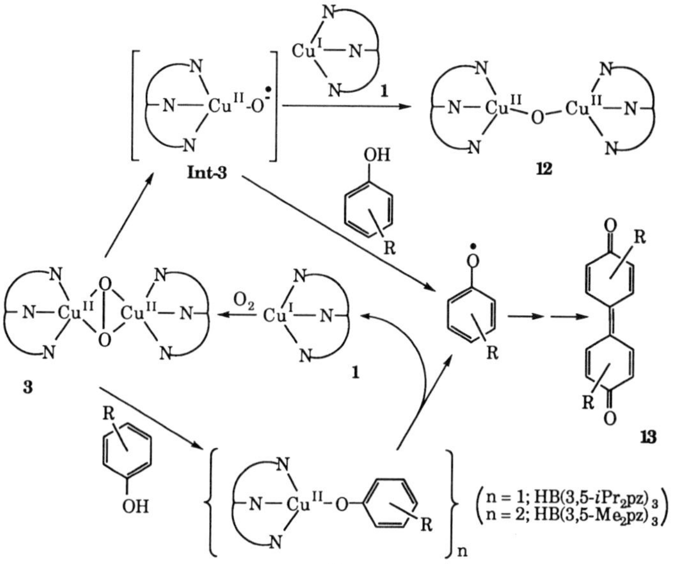

**FIGURE 4-7** Reactivity of $\mu\text{-}\eta^2:\eta^2$-peroxo complex **3** and its ensuing decomposition product, **Int-3**, with phenols.

2-cyclohexenyl-1-chloride is obtained via chlorine abstraction from dichloromethane solvent. Particularly revealing observations are made with phenol substrates. The kinetics reveal two reaction pathways (Figure 4-7). The first is the O—O homolysis producing the $Cu^{II}$—O· intermediate **Int-3**, which initiates production of phenoxyl radicals (via H· abstraction), ultimately leading to the usual types of oxidatively coupled diphenoquinone products **13**, especially when using sterically hindered phenol substrates. Importantly, another pathway involves an acid-base reaction whereby hydrogen peroxide is displaced and $Cu^{II}$-phenoxide products **14** form. These react further with phenol substrates to give the same diphenoquinones as the major products. Here the interpretation given is that Cu—O bond homolysis occurs, giving $Cu^{I}$ and phenoxyl radicals. The latter again couple to give diphenoquinone products, while the $\{Cu^{I}[HB(3,5\text{-}iPr_2pz)_3]\}$ (**1**) produced can react again with the $O_2$ present, giving catalytic turnover in these phenol oxidative coupling reactions. Support for the proposed reaction pathways is claimed from additional observations: (1) acidic phenols such as $p$-fluoro-phenol afford isolable and structurally characterized $Cu^{II}$-phenoxide products **14**, and (2) $Cu^{II}(^-OOtBu)(HB(3,5\text{-}iPr_2pz)_3)$, an isolable mononuclear $Cu^{II}$-alkylperoxo complex, exhibits reactivity where the OO*t*Bu group either behaves as a nucleophilic oxidant (i.e., since it oxidizes tetrahydrothiophene S-oxide but not tetrahydrothiophene) or Cu—O bond homolysis occurs and radical reactions typical of ·OO*t*Bu occur (Kitajima et al., 1993).

The chemistry of $\{Cu[HB(3,5-iPr_2pz)_3]\}_2(O_2)$ (**3**), with its $\mu\text{-}\eta^2\text{:}\eta^2$-peroxodicopper(II) structure and physical properties closely matching those of the oxy-hemocyanin and oxy-tyrosinase active sites, leads Kitajima to another perspective on the proposed mechanism for tyrosinase action. It might initially appear that phenol reactions with **3** are not good tyrosinase mimics, since diphenoquinones (i.e., radical coupling) and not benzoquinones (i.e., *o*-oxygenated) products form. However, it is important to realize that when hindered phenols are reacted with tyrosinase, diphenoquinones are in fact produced, as in the model study (Pandey et al., 1990; Kitajima, 1992). Thus, Kitajima suggests the tyrosinase mechanism depicted in Figure 4-8. Here a single phenol substrate approaches the peroxo dicopper(II) active site, and in an acid-base reaction (akin to his model system; see above) it coordinates as phenoxide to one copper ion, and with a proton transfer gives a hydroperoxo-$Cu^{II}$ intermediate species **Int-4**. Since Kitajima observed that $Cu^{II}$—OOR species can cleave homolytically (see above), he suggests that peroxo-cyclohexadienone dicopper(I) intermediate species **Int-5** forms via coupling of a HOO· radical to a phenoxyl

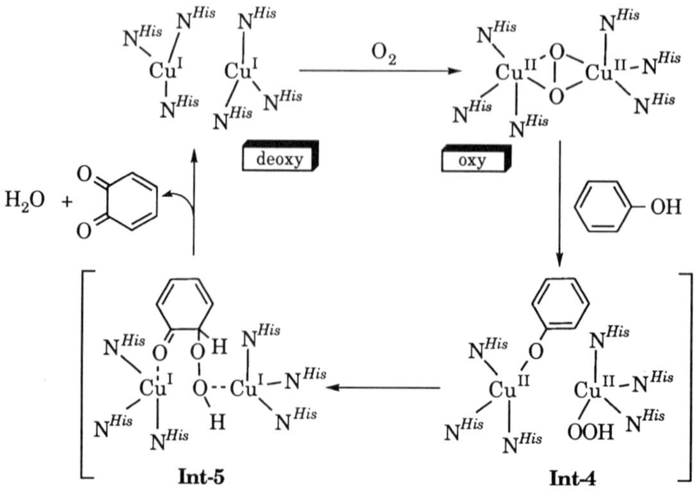

**FIGURE 4-8** Proposed mechanism for tyrosinase. Adapted from Kitajima (1992).

radical; the *ortho* position of the phenoxo group coordinated to $Cu^{II}$ is susceptible to radical coupling chemistry, as seen by the possible contributing resonance structures for this moiety (Kitajima, 1992), and it has already been mentioned that $Cu^{II}(^-OOtBu)(HB(3,5-iPr_2pz)_3)$ can undergo radical reactions. Another view is that cyclohexadienone intermediate **Int-5** forms in a synchronous process via attack of the bound peroxo (or hydroperoxo) group in oxy-Tyr, initiated by phenol binding to the enzyme. The final step in the Kitajima mechanism is O—O cleavage and release of quinone product and water, which restores dicopper(I) deoxy-tyrosinase, as shown.

$$Cu^{II}-O-C_6H_5 \longleftrightarrow Cu^{I}-O=C_6H_5^\bullet \longleftrightarrow Cu^{I}-O=C_6H_5 \bullet$$

A very recent chemical study (Sayre and Nadkarni, 1994) lends support to aspects of this mechanism, for which a noteworthy feature is that no *o*-hydroxylated catechol intermediate is produced in the reaction sequence because phenol substrate is directly oxygenated to *o*-quinone product. While a cresolase activity (see above) has most often been ascribed to tyrosinase, in fact there is no direct evidence for this. In carefully reexamining mechanistic aspects of two chemical model systems that appear to exhibit monooxygenase activity (Karlin and Tyeklár, 1994) and Sayre and Nadkarni (1994) demonstrated that $Cu^I$–phenolate complexes react with $O_2$ to produce *o*-quinones directly. For example, when sodium 4-carbomethoxyphenolate is reacted with the dicopper(I) complex **16** and $O_2$ in acetonitrile solvent, a good yield of a coupled catechol **17** is obtained; this product is derived from Michael addition of the starting phenol to the inherently unstable 4-carbomethoxy-1,2-benzoquinone **Int-6** (Figure 4-9). Independently it was shown that 4-carbomethoxy-1,2-catechol is inert to oxidation under the reaction conditions employed, and a study employing $^{18}O_2$ gas indicated that the Michael product shown contains only one labeled oxygen atom. The conclusion from these and other observations is that the products observed are quinones or compounds derived from their subsequent reactions, and *no catechol intermediate could be formed in these systems.*

Based on these observations, the authors suggest $O_2$-activation processes in these chemical systems which involve 6-peroxy-2,4-cyclohexadienone metal salt intermediates, which have some chemical precedent (Barbaro et al., 1992; Sayre and Nadkarni, 1994); (See also Chapter 5). These transform to *o*-quinones, as indicated below. Thus,

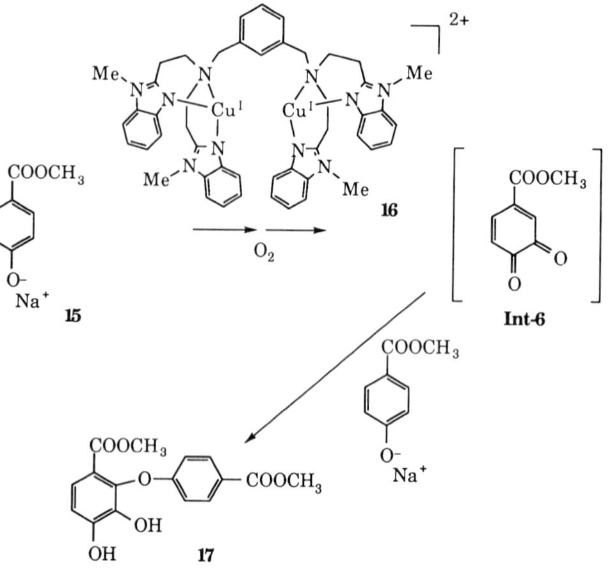

**FIGURE 4-9** A tyrosinase model system, indicating that direct phenol → quinone conversion occurs. See text for further explanation.

the mechanism described here greatly resembles that suggested by Kitajima for tyrosinase and supports the notion of direct phenol to quinone conversion.

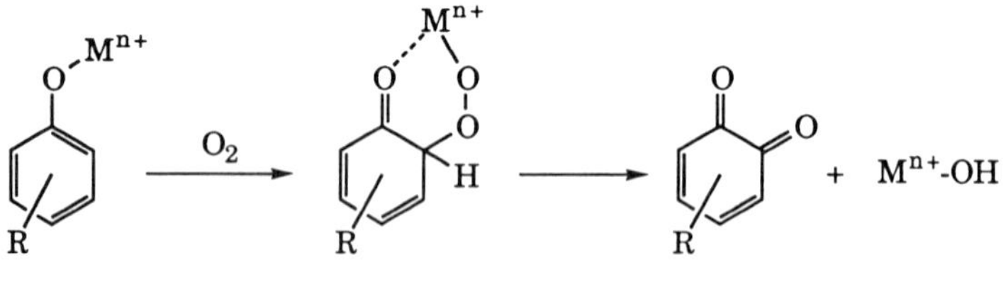

*Summary.* Studies on the enzyme tyrosinase and related $O_2$-carrier hemocyanin have spurred considerable interest in the development of copper(I)–$O_2$ coordination chemistry and reactivity studies. Reactions of discrete well-characterized $Cu_2$–$O_2$ copper–dioxygen complexes such as **3, 6, 7,** and **11** provide insights into reactivity patterns, indicating that the detailed nature (i.e., type of complex and coordination mode) of the bound peroxo group can determine whether one observes nucleophilic or electrophilic chemistry. However, an increased understanding is clearly necessary. The $\mu$-$\eta^2$:$\eta^2$-peroxodicopper(II) species is theoretically predicted to be an electrophile, and it can attack the C—H bond of an arene in the example of the xylyl hydroxylation model system **7 → 9** or perhaps in tyrosinase. However, in Kitajima's complex **3**, somewhat modified behavior is seen, and a localized radical coupling or even nucleophilic peroxo behavior can explain the chemistry observed in reactions of **3** and also perhaps of tyrosinase. These differences may in the end be explained by differences that the synthetic complexes have in their access to substrates which could (or might not) come into close contact with the potentially reactive peroxo ligand. Or perhaps it is simply the special nature of phenolic substrates, which are electron-rich arenes that may be susceptible to electrophilic peroxides, or which also are particularly amenable to radical chemistry, depending on the particular circumstances of reaction.

Another point worth mentioning is that the tyrosinase and chemical mechanisms suggested by the Kitajima and Sayre chemistry, i.e., with peroxo-cyclohexadienone intermediates (see above), is distinct from those usually associated with well-studied monooxygenases such as cytochrome P-450 (see Chapter 3). Here O—O cleavage is traditionally thought to occur prior to attack on substrate, leading to a high-valent (porphyrin)Fe=O ferryl intermediate which effects the oxygenation reactions. Thus, Figure 4-8 depicts a somewhat nontraditional mechanism in which dioxygen incorporation into substrate occurs first and O—O cleavage follows. Note that the mechanism suggested for the xylyl hydroxylation model system (Figure 4-5) involves synchronous substrate attack and O—O cleavage. In fact, additional recent evidence indicates that metal-peroxides can and do react directly with substrates; this subject is discussed in Chapter 1.

## Dopamine β-Monooxygenase (DβM)

This mammalian enzyme, located in the adrenal medulla, catalyzes the ascorbate-dependent, stereospecific benzylic hydroxylation of phenylethylamines such as dopamine to the neurotransmitter norepinephrine (Stewart and Klinman, 1988). With appropriate substrates, it has also been shown that the enzyme can effect olefin oxygenation and oxidative N-dealkylation reactions. Interestingly, the activity of this enzyme has been found to be reduced in the brains of patients with Alzheimer's disease (Cowburn et al., 1990).

A stoichiometry of two coppers per catalytic unit is required, but "normal" UV-Vis and EPR spectroscopic properties between paramagnetic $Cu^{II}$ centers in the fully oxidized protein preclude Cu...Cu electronic interactions, at least in this oxidation state. Thus, dinuclear copper–$O_2$–substrate reactions seem unlikely. Brenner and Klinman (1989) suggest that the two copper ions have different roles: one is an electron acceptor with a nearby ascorbate binding site, while the other carries out $O_2$ and substrate binding and reaction. Blackburn and coworkers (1990) have recently demonstrated the nonequivalence of the Cu sites, since they found that only one of two $Cu^I$ ions binds CO in competition with $O_2$.

A variety of spectroscopic techniques suggest that the copper sites in the DβM protein complex each possess two or three histidine ligands (Stewart and Klinman, 1988). Results from an Extended X-ray Absorption Fine Structure (EXAFS) spectroscopic study (Blackburn et al., 1991) suggest that substantial reorganization occurs on reduction of the $Cu^{II}$ enzyme to $Cu^I$. This study also resolves a previous disagreement on EXAFS spectrocopy of this protein, confirming that a sulfur ligand, probably methionine, is present.

The currently proposed mechanism of action (Stewart and Klinman, 1988; Tian et al., 1994) (Figure 4-10) initially requires a fully reduced active enzyme. Following $O_2$ and phenylethylamine substrate binding, electron transfer to dioxygen generates a mononuclear copper-hydroperoxide (e.g., $Cu^{II}$—OOH), this conclusion being reached because a proton (presumably supplied by an active-site amino acid group) is required for catalysis. Recent intricate $^{18}O$ kinetic isotope experiments (Tian et al., 1994) make the novel suggestion that the activation step involves reductive cleavage of the O—O bond prior to attack of substrate. This result is rationalized (Figure 4-10) by the proposed involvement of a nearby tyrosine residue, leading to the production of water, a tyrosyl radical, and a copper-oxo-radical oxidant. Hydrogen abstraction leads to substrate benzylic radical formation and H-atom transfer back to the tyrosyl radical, followed by copper-oxygen "rebound" to couple the radicals then generates a copper(II)-alkoxide intermediate. The hydroxylated product is eliminated on protonation and electron transfer (from the other copper site), regenerating the reduced active site.

While a number of chelated copper ion complexes exhibit $Cu^I_n/O_2$

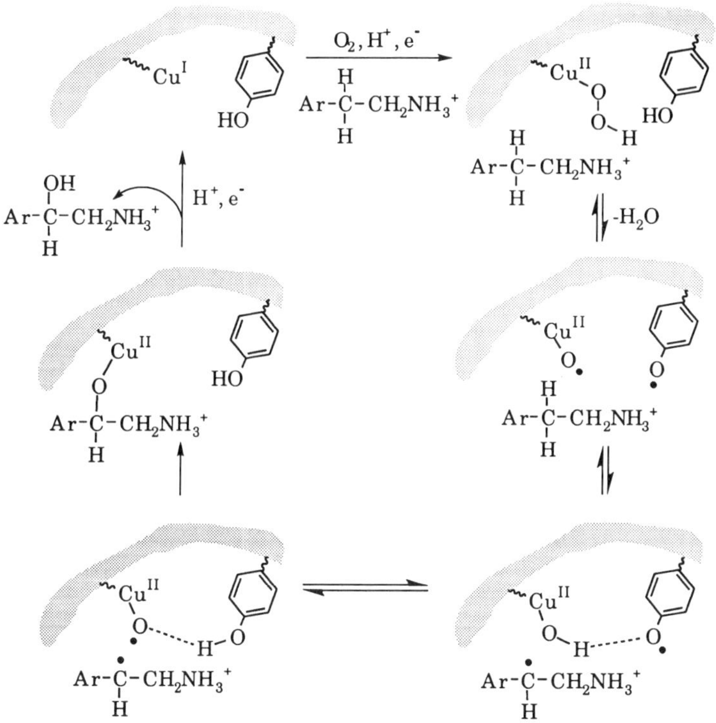

**FIGURE 4-10** Proposed mechanism for phenethylamine hydroxylation at the copper ion center in DβM. Adapted from Tian et al. (1994).

reactivity where hydrocarbon groups are oxygenated (Karlin and Tyeklár, 1994), mechanistic insights are lacking and good DβM mimics do not exist. There are a number of reports of alkane oxidations in which alcohols or ketones are produced by reacting simple copper salts (e.g., $Cu(OH)_2$ or $Cu(ClO_4)_2$) with substrate, utilizing either hydrogen peroxide (Geletii et al., 1988; Barton et al., 1992) or molecular oxygen (Barton et al., 1992; Murahashi et al., 1993) as oxidant. Considerable effort is needed to help elucidate mechanisms of copper ion catalyzed alkane oxygenations; it will be of great interest to learn if that proposed for DβM (i.e., Figure 4-10) can be confirmed, i.e., involving a critical active-site neighboring group (e.g., a tyrosine phenol) to induce dioxygen activation. Hydroxylation of C—H bonds is also relevant to the action of phenylalanine hydroxylase and copper-dependent methane monooxygenase, as mentioned below.

## Peptidylglycine α-Amidating Monooxygenase (PAM)

A large number of physiological secretory polypeptide hormones and neuropeptides (e.g., gastrin, calcitonin, vasopressin) have carboxy terminal amides, —C(O)NH$_2$. These are known to be generated enzymatically from glycine-extended prohormone peptides, which become active when an enzymatic posttranslational oxidative cleavage reaction occurs. Sharing a considerable protein sequence homology as well as copper ion, ascorbate and dioxygen dependence with DβM, PAM effects this oxidative N-dealkylation process (Bradbury and Smyth, 1991; Eipper et al., 1993). PAM can also support monooxygenation of alternate substrates, i.e., in sulfoxidation, amine N-dealkylation and O-dealkylation reactions (Katopodis and May, 1990).

The enzyme is actually bifunctional, carrying out two reactions in sequence (Figure 4-11). The first is the (Cu-dependent) stereospecific hydroxylation reaction carried out by peptidyl α-hydroxylating monooxygenase (PHM), which generates an isolable hydroxyglycine intermediate. This is followed by lyase activity catalyzed by peptidyl amidoglycolate lyase (PAL) to give glyoxylate and carboxamide products. A recent paper provides the first evidence that PAM is a true monooxygenase, since an 18-O label was retained in an isolable hydroxyglycyl peptide intermediate when the reaction was carried out using $^{18}O_2$ (Merkler et al., 1992). The mechanism of hydroxylation could in principle proceed either (1) by oxidative dehydrogenation to give an N-acylimine intermediate which would react by hydrolysis, giving products, or (2) by direct hydroxylation at the α-carbon of the C-terminal glycine residue followed by rearrangement and product elimination. In

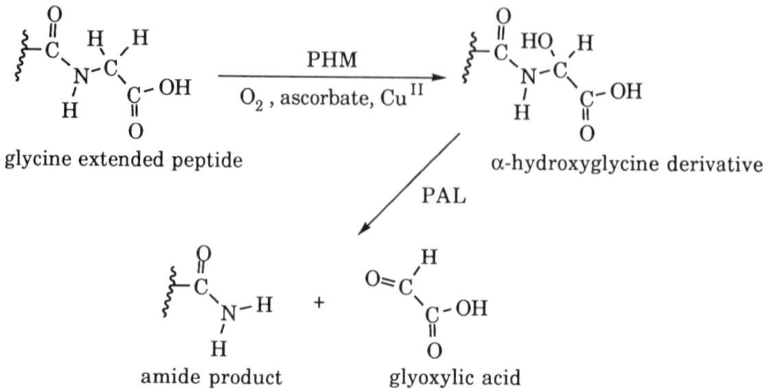

**FIGURE 4-11** Action of PAM, with its two sequential hydroxylation (PHM) and lyase (PAL) functions.

experiments employing various mixtures of 18-O isotopically substituted $O_2$ and/or $H_2O$ with model peptide substrates, only products derived from labeled $^{18}O_2$ were observed, supporting the direct hydroxylation mechanism (Merkler et al., 1992). It would seem, however, that these results do not completely exclude the possibility that an active-site copper–dioxygen adduct, or product derived from it, could effect an initial dehydrogenation reaction, followed by "rebound" (without O-atom exchange from solvent water) of OH(H) from a proximate Cu—OH(H) derived from $O_2$.

$$\begin{array}{c}
\text{O} \quad \boxed{N\text{-acylimine}} \\
\text{~~C-N=CH-COOH} \\
\end{array}$$

$$\text{~~C-NH-CH}_2\text{-COOH} \xrightarrow{} \text{~~C-NH}_2 + \overset{^{18}O}{\underset{H}{C}}\text{-COOH}$$

$$^{18}O_2 \searrow \quad \text{~~C-NH-CH-COOH} \\ H^{18}O$$

The importance of the chemistry involving oxidative N-dealkylation reactions suggests the need for chemical studies bearing on questions regarding how copper(I)–dioxygen chemistry can effect such transformations. A few model systems for PHM have been described (Karlin and Tyeklár, 1994). Using electrochemical techniques or strong chemical oxidants such as persulfate or $IrCl_6^{2-}$, it has been shown that various peptide–Cu$^{II}$ complexes with C-terminal glycine groups can be made to produce carboxamide products; Cu$^{III}$ and initial one-electron N-oxidation intermediates were suggested to form (Reddy et al., 1990). In another system (Capdevielle and Maumy, 1991), oxidant mixtures consisting of Cu$^0$/O$_2$/pyridine or Cu$^{II}$/H$_2$O$_2$/pyridine were shown to convert N-salicyloyl-glycine (**18**) into N-salicyloyl-α-hydroxyglycine (**19**) and salicylamide (20). The nature of the possible active copper-based oxygenating agent was not probed, although speculations included $Cu^{III}$ oxidant intermediates.

Future studies of copper complex reactions using $O_2$ in oxidative N-dealkylation reactions will be of great interest in elucidating the chemistry relevant to the active site reactions of PAM.

## Other Copper Monooxygenases

Several other copper monooxygenases are listed in Table 4-1, but will only be briefly mentioned here. These are phenylalanine hydroxylase, methane monooxygenase, and ammonia monooxygenase; the first two involve oxygenation of C—H bonds.

*Phenylalanine hydroxylase.* In mammalian and many bacterial systems, the biosynthesis of tyrosine proceeds by the aromatic hydroxylation of phenylalanine via a nonheme iron, (tetrahydro)pterin-cofactor and dioxygen-dependent enzyme, phenylalanine hydroxylase (PAH). It is thought that the role of the metal involves iron(II) ion stabilization and/or activation of a peroxypterin intermediate (see Chapter 5). In 1986, it was reported (Pember et al., 1986) that bacterial *Chromobacterium violaceum* PAH was copper and not iron dependent, and a number of subsequent biochemical, spectroscopic and kinetic studies were used to describe the binding of $Cu^{II}$ to pterin or protein ligands (Blackburn et al., 1992), or the possible role of a copper-dioxygen reaction intermediate (Pember et al., 1989). A mechanism of action similar to that of the iron-dependent enzymes was suggested. However, a recent report from the same research group (Carr and Benkovic, 1993) suggests that in fact this PAH is not at all metal dependent. It will be interesting to see if these findings hold up.

*Methane monooxygenase.* Methanotrophic bacteria utilize methane produced in anaerobic sediments as their sole carbon source. The first step in this process is carried out by methane monooxygenases (MMOs), which fix this hydrocarbon substrate by oxygenation to methanol. A soluble form of MMO is a multicomponent enzyme in which a dinuclear iron center effects the methane hydroxylation on its reaction with $O_2$; the X-ray crystal structure of this nonheme iron hydroxylase has recently been reported (Rosenzweig et al., 1993). The enzyme exhibits broad substrate specificity. This has led to the use of methanotrophic bacteria for applications including bioremediation of oil-contaminated land and oxidative removal of trichloroethylene from drinking water.

However, there also exists a particulate, i.e., membrane-bound MMO enzyme form which in fact may be the sole highly active MMO biosynthesized when sufficient or stimulated levels of copper ion are present. This *particulate*-methane monooxygenase (*p*-MMO) has proved difficult to purify, but recent efforts employing EPR and magnetic susceptibility measurements have added support to the supposition that *p*-MMO is a copper-containing enzyme system (Chan et al., 1993; Nguyen et al., 1994). These data suggest that the active site consists of a novel trinuclear (i.e., Cu$_3$), ferromagnetically interacting copper(II) ion cluster, which on reduction (to Cu$^I_3$) and exposure to O$_2$/substrate can effect the CH$_4$ → CH$_3$OH conversion. It would be very exciting if such trinuclear clusters indeed turn out to be new types of enzyme catalysts for monooxygenation of C—H bonds.

*Ammonia monooxygenase.* It should be mentioned that *p*-MMO is thought to be homologous to copper-dependent ammonia monooxygenase (AMO; Table 4-1), an enzyme found in bacteria which obtain all their energy for growth by oxidation of ammonia to nitrite. The first step in this sequence is the conversion of NH$_3$ to hydroxylamine (Ensign et al., 1993; McTavish et al., 1993).

## The Dioxygenase Quercetinase

Dioxygenases incorporate *both* atoms of O$_2$ into a substrate without the need for a coreductant such as ascorbate. From that perspective, an understanding of their mechanisms of action should be of great interest in the application of such transformations to efficient laboratory or commerical oxidative processes (Sheldon, 1993). Unfortunately, far less is generally known about this type of oxygenation chemistry.

Quercetin-2,3-dioxygenase is established as a copper dioxygenase in which quercetin (3′,4′,5,7,-tetrahydroxyflavonol, **21**, R = OH) is converted to the corresponding depside (e.g., phenolic carboxylic acid ester **22**) and carbon monoxide (Ingraham and Meyer, 1985). The pale green enzyme in *Aspergillus flavus* contains copper(II) and has a molecular weight of 110,000. Labeling studies show that the oxygen atom in the CO product is not derived from $O_2$. The formation of a copper chelate with the 3-hydroxy and 4-carbonyl groups of the substrate has been postulated.

A number of biomimetic chemical studies have led to useful insights regarding the mechanism. With enolic 1,2-cyclohexanediones (**23**) as model substrates, addition of cupric ions and dioxygen affords dioxygenation products, the 1,5-keto acid (**24**) and carbon monoxide (Utaka et al., 1984) (Figure 4-12). The mechanism was suggested to proceed via $Cu^{II}$ chelation to the deprotonated α-hydroxyketone, which would be in resonance with a $Cu^I$-substrate-radical **Int-7**. Oxygenation affords a peroxy radical (**Int-8**) which could attack the carbonyl group, generating a transient endoperoxide (**Int-9**), leading to observed products. Side products obtained provided support for the suggested peroxy and endoperoxy intermediates.

Speier and coworkers (Speier, 1993) have crystallographically characterized copper complexes with model flavonols, obtaining structural evidence for the hydroxy-keto site of copper(II) coordination for such substrates. Moreover, EPR spectroscopic studies demonstrate that ligand-based organic radicals (i.e., susceptible to attack by $O_2$) can be stabilized in such systems. Both stoichiometric and even catalytic biomimetic transformations could be effected. For example, heating $[Cu^{II}(Fla)(tmeda)]^+$ (**25**; Fla = **21**, R = H; tmeda = tetramethyleneethylenediamine) with $O_2$ in dimethylformamide afforded high yields of O-benzoylsalicylato–copper(II) complex **26** and CO.

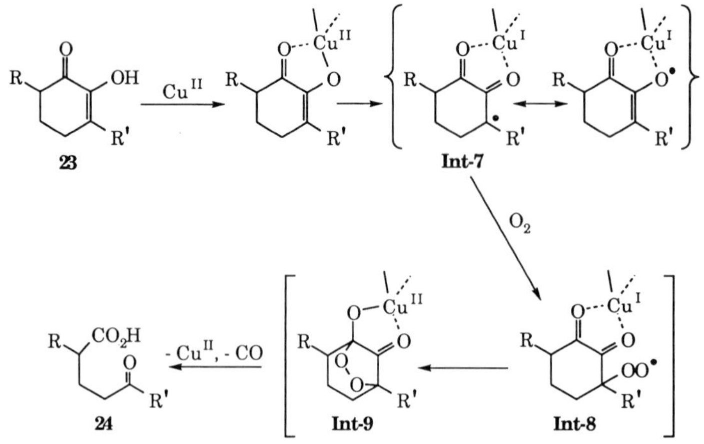

**FIGURE 4-12** Model system for the dioxygenase quercetinase using 1,2-cyclohexanedione substrates. See text for further explanation.

[Cu$^{II}$(Fla)(tmeda)]$^+$ + O$_2$ $\xrightarrow[\text{DMF}]{100 - 120\ °C}$ **26** + CO

**25**

Based on these dioxygenase model systems, it appears that O$_2$ activation involves cupric ion-induced substrate radical formation followed by direct attack by dioxygen.

## COPPER OXIDASES

As a class, copper oxidases effect either the multiple one- (e.g., ascorbate, phenols, cytochrome *c*) or two-electron (e.g., amines, galactose) oxidation of substrates, mediated by reaction of dioxygen at single copper ions or clusters. In the process, O$_2$ is either converted to its two-electron reduced derivative, hydrogen peroxide, or its four-electron product, water. Thus, while oxidases do not incorporate dioxygen-derived O atoms into biological substrates, O—O cleavage reactions do occur; therefore, mechanisms of such processes are of interest within the general context of dioxygen activation by copper ion. To these ends, a brief survey of copper oxidases will be given, emphasizing the nature of active-site chemistry which may occur.

### "Blue" Multicopper Oxidases

Ascorbate oxidase, laccase, and ceruloplasmin couple the one-electron oxidation of substrates such as ascorbate, diamines, monophenols and diphenols, aminophenols (and Fe$^{II}$ for ceruloplasmin) to the reduction of dioxygen to water, i.e., O$_2$ + 4e$^-$ + 4H$^+$ → 2H$_2$O (Messerschmidt and Huber, 1990; Solomon et al., 1992). Laccases are distributed in plants and fungi, and certain species possess some commercial interest, since they are suggested to play a role in polymerization of natural phenols and/or in the biodegradation of wood-containing aromatic polymeric lignins (Kawai et al., 1988). For fully oxidized proteins, the copper(II) sites were historically given spectroscopic designations type 1 [T1; strong ($\varepsilon \sim 3000\,\text{M}^{-1}\,\text{cm}^{-1}$) blue absorption in the ~600-nm region, high redox potential and low Cu hyperfine $A_\parallel$ value in its EPR

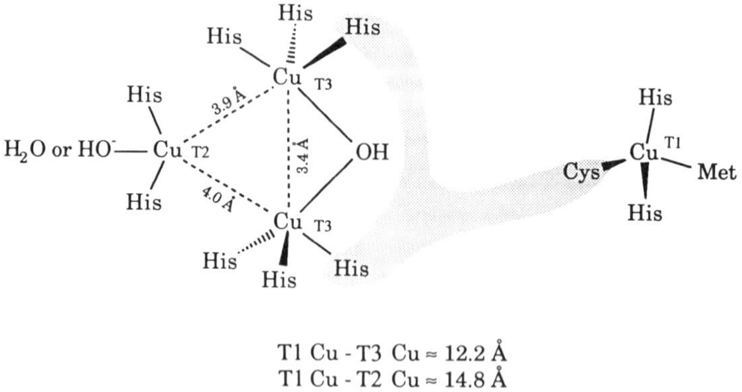

214    *Dioxygen Reactivity in Copper Proteins and Complexes*

**FIGURE 4-13** The four-copper-containing active site of ascorbate oxidase.

spectrum], type 2 [T2; normal weak visible absorption and EPR spectroscopic behavior], and type-3 [T3; dinuclear antiferromagnetically coupled $Cu^{II}$ pair with strong $\lambda_{max}$ = 330-nm absorption (Malkin and Malmström, 1970]. However, in the last decade, elegant spectroscopic studies (Solomon et al., 1992) on laccases have indicated that a trinuclear copper ion cluster is present, a finding supported by recent X-ray structural studies on ascorbate oxidase (Figure 4-13) (Messerschmidt et al., 1992; Messerschmidt, 1993).

The apparent mode of action of this enzyme starts with acceptance of electrons by the T1 site, with a nearby binding site for ascorbate substrate. Electrons then "feed" from this copper through a cysteine-histidine protein electron-transfer pathway to the trinuclear center, where dioxygen reduction occurs. Recent laccase protein spectroscopic studies (Clark and Solomon, 1992; Solomon et al., 1992) on $O_2$-derived reaction intermediates suggest that the four-electron reduction of $O_2$ may proceed via two two-electron steps. The first is suggested to be a hydroperoxo (i.e., $^-OOH$) bridged mixed-valent dicopper species **27**. Subsequent one-electron transfer from the T1 $Cu^I$ and the still reduced

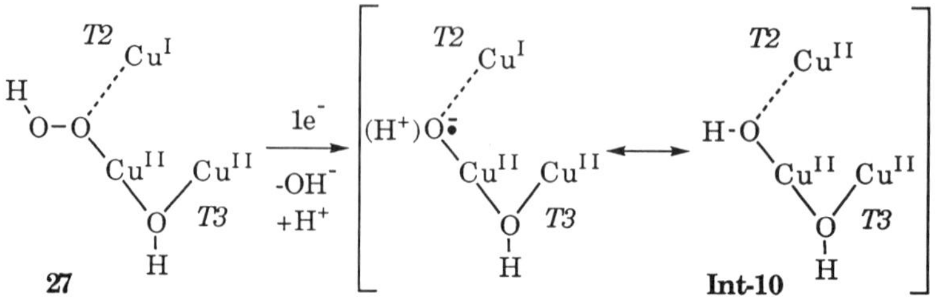

T2 Cu$^I$ give another detectable intermediate which is best described as a hydroxide-bridged Cu$^{II}$ trinuclear complex **Int-10**.

Further protein and synthetic modeling chemical studies are needed to deduce the mechanisms possible for binding, reduction, and possibly activation of O$_2$ (for substrate oxygenation) at copper clusters like those seen in these oxidases; it is intriguing to consider that the chemistry described here may be relevant to such enzymes as copper-dependent MMO (see above). A critical feature of O$_2$ activation is access of substrates to metal-bound dioxygen species, such as described for tyrosinase (see above) and known for the di-iron core in soluble MMO (Rosenzweig et al., 1993) or the heme-iron center in cytochrome P-450 monooxygenases (see Chapter 3).

## "Nonblue" Oxidases

These include galactose oxidase and amine oxidases (e.g., plasma amine oxidase, diamine oxidase, lysyl oxidase), which produce hydrogen peroxide by the two-electron reduction of O$_2$. While the O—O bond is not broken in reaction cycles involving these enzymes, both contain novel active-site protein-derived cofactors which are intimately involved in the chemistry, a theme which is emerging in O$_2$-activating systems (see the discussion of DβM above). Certain other features of the protein functions will also be discussed.

*Galactose oxidase.* This enzyme is an extracellular enzyme secreted by the fungus *Dactylium dendroides* (MW = 68,000), catalyzing the stereospecific two-electron oxidation of D-galactose and other primary alcohols (Ingraham and Meyer, 1985).

It has been known for some time that the catalysis is carried out at an active site containing only a single copper ion. It has thus been a puzzle how this could effect a two-electron process; copper ion usually undergoes only single-electron Cu$^{II}$/Cu$^I$ redox changes. In the past few years, however, insightful biochemical-spectroscopic studies have indicated that a stabilized ligand-protein radical-cation is involved (Whittaker, 1993). A recent X-ray structure (Ito et al., 1991) on the protein has revealed the nature of the highly unusual active site, with a labile water ligand which could be the site of substrate binding, plus a "built-in" protein amino acid-derived cofactor which is apparently able to stabilize a radical cation species. Here a thioether bond links a nearby cysteine to the *ortho*-position of a copper-coordinated tyrosine ligand, and the latter also undergoes a π-stacking interaction with a tryptophan residue (Figure 4-14).

The reaction to oxidize alcohols can be thought of as proceeding by oxidation of a reduced protein and copper(I) center by dioxygen, in a two-electron process, to give oxidized copper(II) and the protein-ligand radical cation $P^{+\cdot}$. The "stored" oxidizing equivalents can then oxidize a coordinated alcohol substrate, producing aldehyde and restoring the reduced enzyme center. Proton transfers from substrate to dioxygen (as product peroxide) are suggested to be mediated via protonation-deprotonation of the coordinated tyrosinate/tyrosine ligands, which are more likely to be protonated and not coordinated when the copper ion is reduced, or possibly because one is a weaker axial ligand (Whittaker, 1993).

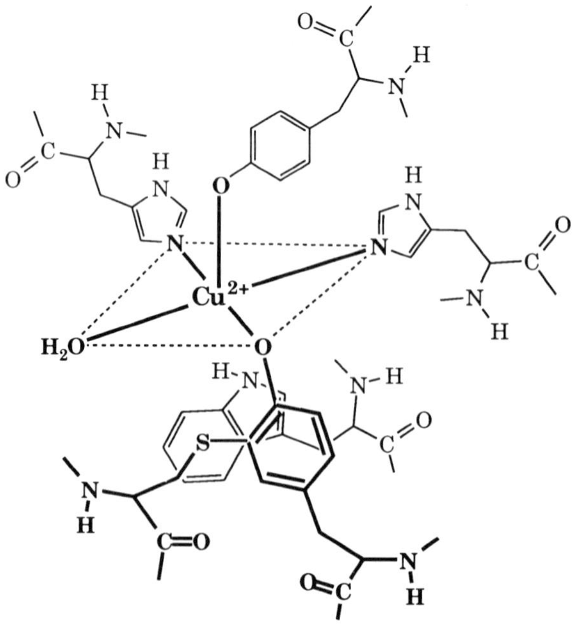

**FIGURE 4-14** Galactose oxidase active site with its unusual modified tyrosine amino acid ligand for copper.

$$(P)Cu^I + O_2 + 2H^+ \rightarrow (P^{+\cdot})Cu^{II} + H_2O_2$$
$$RCH_2OH + (P^{+\cdot})Cu^{II} \rightarrow RC(O)H + (P)Cu^I + 2H^+$$

*Amine oxidases.* Amine oxidases catalyze reactions involving functions such as the crosslinking of collagen and elastin and the regulation of blood plasma biogenic amines (Ingraham and Meyer, 1985; Dooley et al., 1993). As with galactose oxidase, these enzymes also utilize a novel cofactor to help carry out the oxidative deamination of amines giving aldehydes, with hydrogen peroxide again being the reduced dioxygen by-product. No crystal structure is yet available, but spectroscopic studies have shown that in the copper(II)–enzyme form, the ligation is quite typical, with three equatorial histidine ligands suggested (structure **A**). The cofactor has always been known to contain an active carbonyl group capable of reaction with substrate amines or analogs. Recently, however, it was demonstrated that this is a species derived from tyrosine, 6-hydroxy dopa (topa), or, more directly, from its oxidized counterpart topa quinone (structure **B**) (Janes et al., 1990).

The basic mechanism (Hartmann and Klinman, 1991) involves substrate reaction with the topa quinone/enzyme/$Cu^{II}$ to form a Schiff base (Figure 4-15). An internal redox reaction followed by hydrolysis gives aldehyde product and the reduced aminoquinol form of the cofactor, still with $Cu^{II}$. This, however, is in equilibrium with an imino-semiquinone radical-cation $Cu^I$ species; direct evidence for this intermediate has recently been obtained, using electron spin echo envelope modulation (ESEEM) and continuous wave-electron nuclear double resonance (CW-ENDOR) spectroscopies (Warncke et al., 1994). As a reduced copper ion, this is now susceptible to reaction with dioxygen (Dooley et al., 1993). The result is that hydrogen peroxide and ammonia products are produced, along with a regenerated, fully oxidized topa quinone/enzyme/$Cu^{II}$ active site.

While very novel chemistry is involved in the reactions described

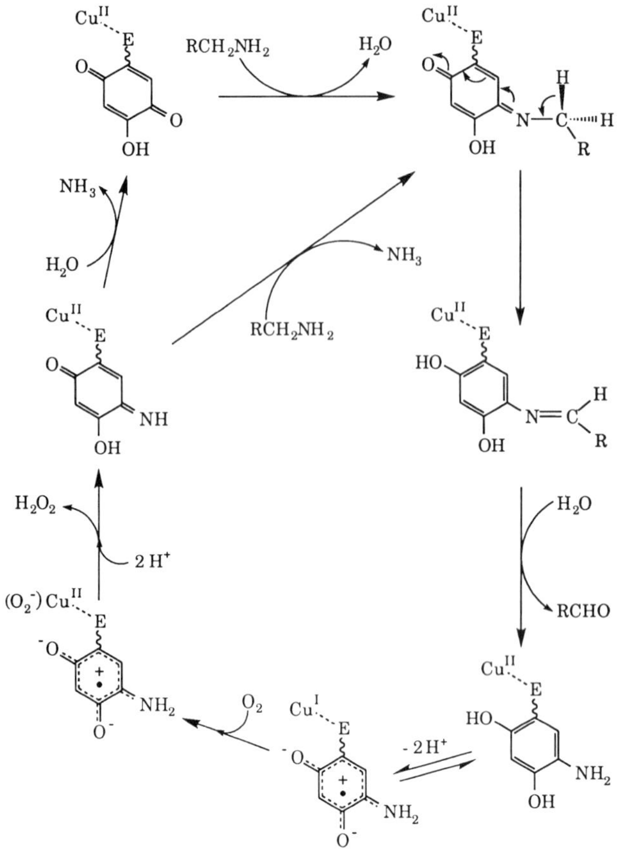

**FIGURE 4-15** Proposed mechanism for copper amine oxidases. Adapted from Hartmann and Klinman (1991) and Dooley et al. (1993). See text for further explanation.

above, strictly speaking there is no element of dioxygen activation. However, an interesting and important point arises when considering how the active site topa or topaquinone is biosynthesized. Klinman and colleagues (Mu et al., 1992) recently carried out spectroscopic and molecular biological analyses which proved that tyrosine is the precursor to topa quinone in mammals. This leaves two reasonable possibilities in considering the means for the biogenesis of topa quinone; both involve oxygen activation chemistry. The overall pathway (Figure 4-16) most likely involves the initial *ortho*-hydroxylation of tyrosine to dopa, followed by two-electron oxidation to dopa quinone. This should

tyrosine → dopa → dopa quinone → topa → topa quinone

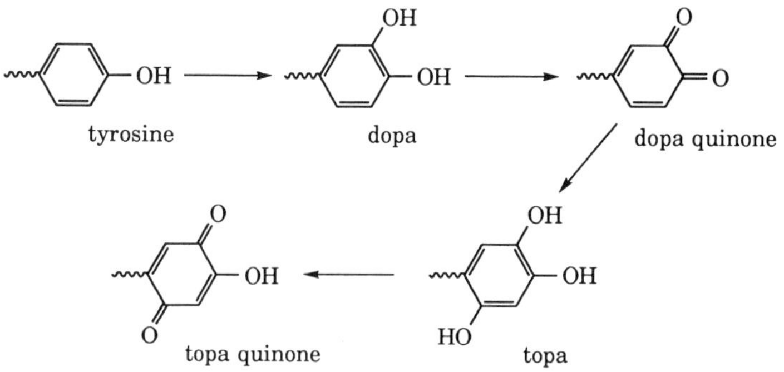

**FIGURE 4-16** Probable pathway for the biogenesis of the active site cofactor topa quinone, from tyrosine. This may occur enzymatically or via self-processing. See text.

readily undergo a hydrolysis reaction initiated at the 6-position, due to the α,β-unsaturated ketone structure, producing topa. Further oxidation under aerobic conditions would lead to topa quinone. The initial step, the oxygenation to give dopa, or the direct conversion of tyrosine to dopa quinone (see above), could in principle be carried out enzymatically by iron/pterin-dependent tyrosine hydroxylases or copper tyrosinases. For various biochemical reasons, this seems unlikely (Mu et al., 1992). So, it is possible that there is a new class of oxidative enzymes that carry out the functions described in Figure 4-16.

On the other hand, a self-processing situation may occur (Mu et al., 1992), in which a copper ion coordinates to its histidine ligands in the native active site, where it is known to be in close proximity to the C-2 position of the topa quinone. If hydrogen peroxide were to find its way to this copper ion, a single metal-promoted hydroxylation of tyrosine to dopa or dopa quinone might take place. Note that this suggestion is very similar to the situation described by the tyrosinase mechanism depicted in Figure 4-8, except that here the protein, instead of a second copper ion, holds the tyrosine substrate in appropriate proximity to the $Cu^{II}$-hydroperoxide. It will be a challenge to unravel the biochemical mechanism by which this posttranslational oxidative transformation (i.e., tyrosine → dopa quinone) occurs. Furthermore, elucidation of the basic chemical reactivity of mononuclear copper hydroperoxo complex species is needed.

### Cytochrome *c* Oxidase

As the terminal respiratory membrane proteins which effect the exergonic reaction, cytochrome *c* oxidase (Babcock and Wikström, 1992;

Malmström, 1993) catalyzes the four-electron reduction of dioxygen to water:

$$O_2 + 4H^+ + 4e^- \text{ (from cytochrome } c) \rightarrow 2H_2O$$

The energy obtained in this process is utilized for subsequent adenosine triphosphate (ATP) synthesis via the pH gradient generated by the simultaneous coupled proton pumping across the membrane. Since it is now known that an intricate mixed-metal copper ion and heme center cleaves the O—O bond in dioxygen, the means by which this occurs is of considerable interest in elucidating the possibilities for copper ion mechanisms of $O_2$ activation.

Extensive biochemical studies on mammalian mitochondrial cytochrome $c$ oxidase, as well as on simpler but structurally similar bacterial forms, provide a basic picture of the arrangement of the metal centers in these proteins (Figure 4-17). Electron flow proceeds from the cytochrome $c$ substrate (i.e., electron donor) to the so-called $Cu_A$ site

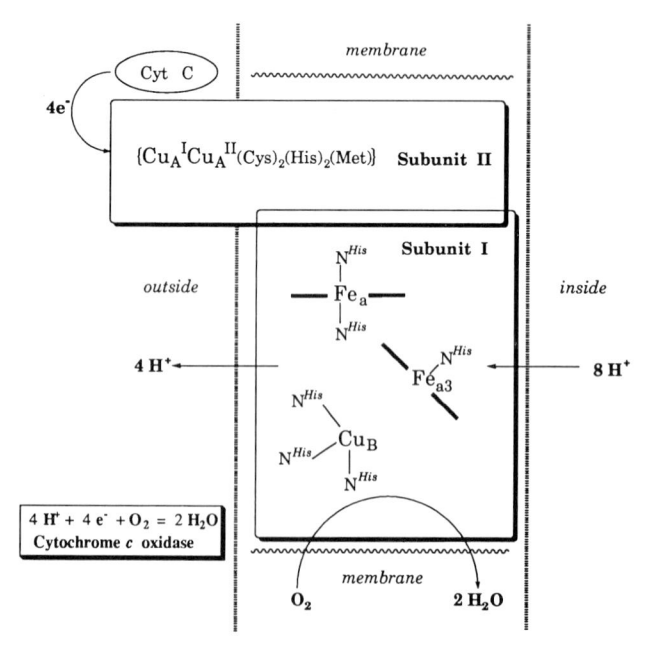

FIGURE 4-17 Schematic representation of the metal ion prosthetic groups in cytochrome $c$ oxidases. The $O_2$-binding and reduction occurs at the heme $a_3$–$Cu_B$ dinuclear center.

residing in subunit II; recent studies indicate that this is a novel dinuclear mixed-valent $Cu^I$—$Cu^{II}$ center possessing cysteine and histidine donor ligands (Kelly et al., 1993). Certain bacterial ubiquinol oxidases lack $Cu_A$, and these enzymes utilize quinol as the alternate electron donor. These reducing equivalents are passed to the "catalytic machinery" residing in subunit I, first to heme *a* (or *b*) and then to the dinuclear heme $a_3$-iron (or heme *o*)-copper ($Cu_B$) unit. This latter heterobimetallic center, known to possess a heme $a_3$ proximal histidine and at least several similar ligands for copper, effects the $O_2$-binding and O—O cleavage reactions.

Recent time-resolved kinetic and spectroscopic investigations have provided considerable insight to the nature of reaction and identity of intermediates (Figure 4-18) (Varotsis et al., 1993), although there are alternative views concerning the possible reaction mechanism (Rousseau et al., 1992; Ogura et al., 1993). Dioxygen initially interacts with the

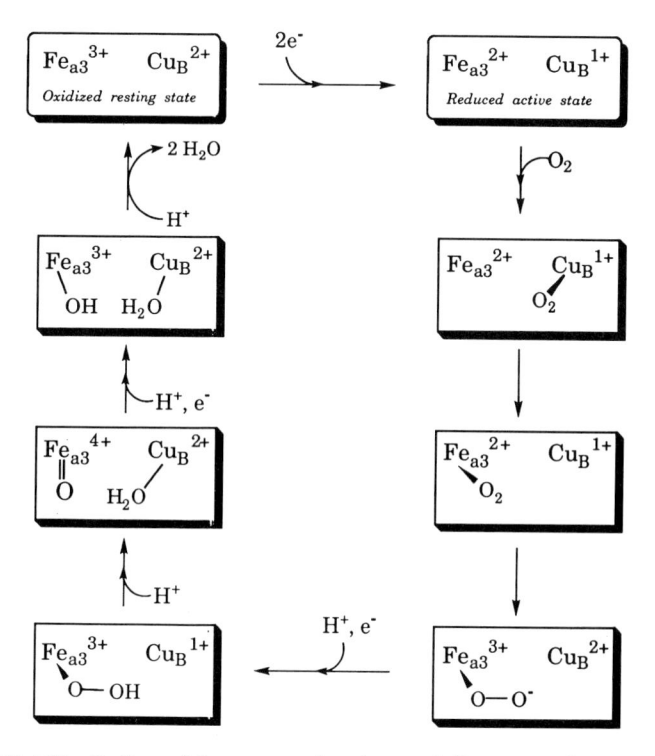

**FIGURE 4-18** Outline of the proposed pathway of dioxygen reduction to water at the dinuclear heme $a_3$–$Cu_B$ center of cytochrome *c* oxidases.

Cu$^I$ of the active form of the enzyme, i.e., with a fully reduced heme $a_3$-Cu$_B$ center. A transient Cu—O$_2$ adduct forms, whereupon the dioxygen switches ligation to give an Fe($a_3$)-O$_2$ species, electronically analogous to oxyhemoglobin or oxymyoglobin. This is followed by production of an Fe$^{III}$-peroxide or hydroperoxide; whether Cu$_B$ ligation takes place is less clear. Cleavage of the O—O bond leads to a distinctive spectroscopically observed ferryl species (i.e., Fe$^{IV}$=O), and further reduction and protonation complete the reaction.

Many questions and details remain unresolved, but the basic picture described above (Figure 4-18) resembles many of the descriptions of O$_2$ activation already provided for other copper-containing enzymes (see above). A two-electron reduced dioxygen species, i.e., as peroxo or hydroperoxo, needs to be generated at the copper (and/or iron) ion center. In copper monooxygenases, interaction with an organic substrate (which is apparently not possible in cytochrome c oxidases) leads to oxygenation chemistry. In general, further addition of one-electron to the aforementioned peroxo or hydroperoxo copper intermediate leads to O—O bond cleavage, whereupon substrate reactivity might again take place (if possible); alternatively, further and complete reduction occurs, giving water. A more detailed understanding of the exact ligation and coordination binding mode of reduced dioxygen species is desirable. In addition, the timing and location of protonation events need to be deduced, since these relate to the activation process for O—O bond cleavage and for the proton pumping cytochrome c oxidase enzyme function. These H$^+$ transfers may involve the ligands bound to the metals and/or nearby protein residues, in addition to the reduced dioxygen moieties.

## Phenoxazinone Synthase

This oxidase catalyzes the overall six-electron oxidative coupling of 2-aminophenols to form 2-aminophenoxazinone, the final step in the

R = (L)-Thr-(D)-Val-(L)-Pro-Sar-(N-Me-L)-Val

bacterial biosynthesis of the antineoplastic agent actinomycin. The structures of the active sites are not yet well characterized, although recent spectroscopic studies (Freeman et al., 1993) reveal the presence of four or five copper atoms per protein monomer, with one site being of the "blue" type 1 electron transfer variety, while the other, type 2, copper ion centers possess different functions. The nature of the product suggests that it forms via radical initiated oxidative coupling reaction chemistry.

## CONCLUSIONS

It should be clear from this summary that a variety of mechanisms for dioxygen binding, and for activation and/or reduction, are observed for copper proteins as well as for synthetically derived inorganic model complexes. As with dioxygen activation for iron-containing systems, a basic critical feature is the availability of a substrate proximate to a copper-bound, reduced dioxygen species (e.g., peroxo or hydroperoxo); the latter is formed via $O_2$ reaction with a reduced copper(I) ion center. As currently understood, subsequent reactions with substrate appear to follow two basic mechanistic pathways: (1) concerted attack by copper (hydro)peroxo species, i.e., in tyrosinase or (2) O—O bond cleavage to give radical intermediates, as in DβM. Dioxygenase chemistry is basically different, involving copper(II)-promoted substrate activation, leading to direct $O_2$ attack on stabilized substrate radicals.

The active sites of the copper biological centers may contain single, dinuclear, or trinuclear copper ion cluster moieties. While there has been recent progress in elucidating the chemistry of dioxygen binding and activation at dinuclear copper centers (i.e., hemocyanin, tyrosinase, and model complexes), very little is known about what happens at single or trinuclear metal sites. Yet, this latter type of chemistry is clearly associated with enzymes such as DβM, PHM, and p-MMO, as well as various copper oxidases. The employment of active-site (normal or modified) amino acid residues which intimately participate in the redox and $O_2$ activation process has also been clearly established; more examples are likely to be found.

Presumably, there will be continued extensive research involving all of these oxygenase and oxidase proteins, as well as related chemical modeling studies. Fascinating new dioxygen-binding modes and reactivity patterns are likely to be uncovered, leading to new insights into the nature of copper ion-mediated $O_2$ activation.

*Note Added in Proof:* Two new model dioxygen complexes, in the form of Cu/O$_2$ = 1:1 copper(II)-superoxo species, have recently appeared (Harata et al., 1994) and copper/O$_2$ mediated ligand hydroxylation at a benzylic position modeling dopamine β-monooxygenase (DβM) was recently reported (Itoh et al., 1995). A recently published study suggests that neither bacterial nor mammalian phenylalanine hydroxylases (PAH) require a metal (Cu or Fe) for O$_2$ activation (Carr et al., 1995). Further recent evidence supports the self-processing of tyrosine to yield topa quinone, as opposed to an enzymatic mechanism in copper amine oxidases from both yeast (Cai and Klinman 1994) and bacterial sources (Matsuzaki et al., 1995). A new model for heme-copper oxidase proton pumping involving metal–ligand and O$_2$-derived intermediates has recently appeared (Morgan et al., 1994), and X-ray structures from an *E. coli* amine oxidase have recently been obtained (Knowles, P. F., private communication; submitted for publication).

## Acknowledgment

We would like to thank the National Institutes of Health (GM 28962) for support of K.D.K.'s ongoing research program in copper/dioxygen bioinorganic chemistry.

## REFERENCES

ADMAN, E. T. (1991) Copper Protein Structures. *Adv. Protein Chem.*, 42, 145–197.

BABCOCK, G. T., and WIKSTRÖM, M. (1992) Oxygen Activation and the Conservation of Energy in Cell Respiration. *Nature*, 356, 301–309.

BALDWIN, M. J., ROSS, P. K., PATE, J. E., TYEKLÁR, Z., KARLIN, K. D., and SOLOMON, E. I. (1991) Spectroscopic and Theoretical Studies of an End-on Peroxide-bridged Coupled Binuclear Copper(II) Model Complex of Relevance to the Active Sites in Hemocyanin and Tyrosinase. *J. Am. Chem. Soc.*, 113, 8671–8679.

BARBARO, P., BIANCHINI, C., FREDIANI, P., MELI, A., and VIZZA, F. (1992) Chemoselective Oxidation of 3,5-Di-*tert*-butylcatechol by Molecular Oxygen. Catalysis by an Iridium(III) Catecholate Through Its Dioxygen Adduct. *Inorg. Chem.*, 31, 1523–1529.

BARTON, D. H. R., BEVIERE, S. D., CHAVASIRI, W., CSUHAI, E., and DOLLER, D. (1992) The Functionalization of Saturated Hydrocarbons. Part XXI+. The Fe(III)-catalyzed and the Cu(II)-catalyzed Oxidation of Saturated Hydrocarbons by Hydrogen Peroxide: A Comparative Study. *Tetrahedron*, 48, 2895–2910.

BECKMAN, J. S., CARSON, M., SMITH, D. C., and KOPPENOL, W. H. (1993) ALS, SOD, and Peroxynitrite. *Nature*, 364, 524.

BEINERT, H. (1991) Copper in Biological Systems. A Report from the 6th Manziana Conference, September 23–27, 1990. *J. Inorg. Biochem.*, 44, 173–218.

BELTRAMINI, M., SALVATO, B., SANTAMARIA, M., and LERCH, K. (1990) The Reaction of CN⁻ with the Binuclear Copper Site of *Neurospora* Tyrosinase: Its Relevance for a Comparison Between Tyrosinase and Hemocyanin Active Sites. *Biochim. Biophys. Acta*, 1040, 365–372.

BERTINI, I., BANCI, L., PICCIOLI, M., and LUCHINAT, C. (1990) Spectroscopic Studies on $Cu_2Zn_2$-SOD: A Continuous Advancement of Investigation Tools. *Coord. Chem. Rev.*, 100, 67–103.

BLACKBURN, N. J., HASNAIN, S. S., PETTINGILL, T. M., and STRANGE, R. W. (1991) Copper K-extended X-ray Absorption Fine Structure Studies of Oxidized and Reduced Dopamine β-Hydroxylase. *J. Biol. Chem.*, 266, 23120–23127.

BLACKBURN, N. J., PETTINGILL, T. M., SEAGRAVES, K. S., and SHIGETA, R. T. (1990) Characterization of a Carbon Monoxide Complex of Reduced Dopamine β-Hydroxylase. *J. Biol. Chem.*, 265, 15383–15386.

BLACKBURN, N. J., STRANGE, R. W., CARR, R. T., and BENKOVIC, S. J. (1992) X-ray Absorption Studies of the Cu-dependent Phenylalanine Hydroxylase from *Chromobacterium violaceum*. Comparison of the Copper Coordination in Oxidized and Dithionite-reduced Enzymes. *Biochemistry*, 31, 5298–5303.

BRADBURY, A. F., and SMYTH, D. G. (1991) Peptide Amidation. *Trends Biochem. Sci.*, 16, 112–115.

BRENNER, M. C., and KLINMAN, J. P. (1989) Correlation of Copper Valency with Product Formation in Single Turnovers of Dopamine β-Monooxygenase. *Biochemistry*, 28, 2124–2130.

BRITTAIN, T., BLACKMORE, R., GREENWOOD, C., and THOMSON, A. J. (1992) Bacterial Nitrite-reducing Enzymes. *Eur. J. Biochem.*, 209, 793–802.

CAI, D., and KLINMAN, J. P. (1994) Evidence for a Self-catalytic Mechanism of 2,4,5-Trihydroxyphenylalanine Quinone Biogenesis in Yeast Copper Amine Oxidase. *J. Biol. Chem.*, 269, 32039–32042.

CAPDEVIELLE, P., and MAUMY, M. (1991) Copper-mediated α-Hydroxylation of N-salicyloyl-glycine. A Model for Peptidyl-Glycine α-Amidating Monooxygenase (PAM). *Tetrahedron Lett.*, 32, 3831–3834.

CARR, R. T., and BENKOVIC, S. J. (1993) An Examination of the Copper Requirement of Phenylalanine Hydroxylase from *Chromobacterium violaceum*. *Biochemistry*, 32, 14132–14138.

CARR, R. T., BALASUBRAMANIAN, S., HAWKINS, P. C. D., and BENKOVIC, S. J. (1995) Mechanism of Metal-Independent Hydroxylation by *Chromobacterium violaceum* Phenylalanine Hydroxylase. *Biochemistry*, 34, 7525–7532.

CHAN, S. I., NGUYEN, H.-H. T., SHIEMKE, A. K., and LIDSTROM, M. E. (1993) The Copper Ions in the Membrane-Associated Methane Monooxygenase, in *Bioinorganic Chemistry of Copper* (K. D. Karlin and Z. Tyeklár, Eds.), Chapman & Hall, New York, pp. 184–195.

CLARK, P. A., and SOLOMON, E. I. (1992) Magnetic Circular Dichroism Spectroscopic Definition of the Intermediate Produced in the Reduction of Dioxygen to Water by Native Laccase. *J. Am. Chem. Soc.*, 114, 1108–1110.

COWBURN, J. D., FARRAR, G., and BLAIR, J. A. (1990) Alzheimer's Disease— Some Biochemical Clues. *Chem. Brit.*, 1169–1173.

DOOLEY, D. M., BROWN, D. E., CLAGUE, A. W., KEMSLEY, J. N., MCCAHON, C. D., MCGUIRL, M. A., TUROWSKI, P. N., MCINTIRE, W. S., FARRAR, J. A., and THOMSON, A. J. (1993) Structure and Reactivity of Copper-containing Amine Oxidases, in *Bioinorganic Chemistry of Copper* (K. D. Karlin and Z. Tyeklár, Eds.), Chapman & Hall, New York, pp. 459–470.

EIPPER, B. A., MILGRAM, S. L., HUSTEN, E. J., YUN, H.-Y., and MAINS, R. W. (1993) Peptidylglycine α-Amidating Monooxygenase: A Multifunctional Protein with Catalytic, Processing and Routing Domains. *Protein Sci.*, 2, 489–497.

ENSIGN, S. A., HYMAN, M. R., and ARP, D. J. (1993) *In vitro* Activation of Ammonia Monooxygenase from *Nitrosomonas europaea* by Copper. *J. Bacteriol.*, 175, 1971–1980.

FARRAR, J. A., THOMSON, A. J., CHEESMAN, M. R., DOOLEY, D. M., and ZUMFT, W. G. (1991) A Model of the Copper Centres of Nitrous Oxide Reductase (*Pseudomonas stutzeri*). *FEBS Letters*, 294, 11–15.

FREEMAN, J. C., NAYAR, P. G., BEGLEY, T. P., and VILLAFRANCA, J. J. (1993) Stoichiometry and Spectroscopic Identity of Copper Centers in Phenoxazinone Synthase: A New Addition to the Blue Copper Oxidase Family. *Biochemistry*, 32, 4826–4830.

FUJISAWA, K., TANAKA, M., MOROKA, Y., and KITAJIMA, N. (1995) A Monomenic Side-On Superoxocopper(II) Complex: $Cu(O_2)$ (HB(3-$t$Bu-$i$Prpz)$_3$). *J. Am. Chem. Soc.*, 116, 12079–12080.

GELETII, Y. V., LAVRUSHKO, V. V., and LUBIMOVA, G. V. (1988) Oxidation of Saturated Hydrocarbons by Hydrogen Peroxide in Pyridine Solution Catalyzed by Copper or Iron Perchlorates. *J. Chem. Soc. Chem. Commun.*, 936–937.

GODDEN, J. W., TURLEY, S., TELLER, D. C., ADMAN, E. T., LIU, M. Y., PAYNE, W. J., and LEGALL, J. (1991) The 2.3 Ångstrom X-ray Structure of Nitrite Reductase from *Achromobacter cycloclastes*. *Science*, 253, 438–442.

HARATA, M., JITSUKAWA, K., MASUDA, H., and EINEGA, H. (1994) A Structurally Characterized Mononuclear Copper(II)-Superoxo Complex. *J. Am. Chem. Soc.*, 116, 10817–10818.

HARTMANN, C., and KLINMAN, J. P. (1991) Structure-function Studies of Substrate Oxidation by Bovine Serum Amine Oxidase: Relationship to Cofactor Structure and Mechanism. *Biochemistry*, 30, 4605–4611.

HATHAWAY, B. J. (1987) Copper, in *Comprehensive Coordination Chemistry* (G. Wilkinson, Ed.), Pergamon, New York, 5, pp. 533–774.

HAYAISHI, O. (Ed.) (1974) *Molecular Mechanisms of Oxygen Activation*, Academic Press, New York.

HAZES, B., MAGNUS, K. A., BONAVENTURA, C., BONAVENTURA, J., DAUTER, Z., KALK, K. H., and HOL, W. G. J. (1993) Crystal Structure of Deoxygenated *Limulus polyphemus* Subunit II Hemocyanin at 2.18 Å Resolution: Clues for a Mechanism for Allosteric Regulation. *Protein Sci.*, 2, 597–619.

INGRAHAM, L. I., and MEYER, D. L. (1985) *Biochemistry of Dioxygen*. Vol. 4. *Biochemistry of the Elements* (E. Frieden, Ed.), Plenum, New York.

ITO, N., PHILLIPS, S. E. V., STEVENS, C., OGEL, Z. B., MCPHERSON, M. J., KEEN, J. N., YADAV, K. D. S., and KNOWLES, P. F. (1991) Novel Thioether Bond Revealed by a 1.7 Å Crystal Structure of Galactose Oxidase. *Nature*, 350, 87–90.

ITOH, S., KONDO, T., KOMATSU, M., OHSHIRO, Y., LI, C., KANEHISA, N., KAI, Y., and FUKUZUMI, S. (1995) Functional Model of Dopamine β-Hydroxylase. Quantitative Ligand Hydroxylation at the Benzylic Position of a Copper Complex by Dioxygen. *J. Am. Chem. Soc.*, 117, 4714–4715.

JANES, S. M., MU, D., WEMMER, D., SMITH, A. J., KAUR, S., MALTBY, D., BURLINGAME, A. L., and KLINMAN, J. P. (1990) A New Redox Cofactor in Eukaryotic Enzymes: 6-Hydroxydopa at the Active Site of Bovine Serum Amine Oxidase. *Science*, 248, 981–987.

KARLIN, K. D. (1993) Metalloenzymes, Structural Motifs, and Inorganic Models. *Science*, 261, 701–708.

KARLIN, K. D., and GULTNEH, Y. (1987) Binding and Activation of Molecular Oxygen by Copper Complexes. *Prog. Inorg. Chem.*, 35, 219–327.

KARLIN, K. D., GULTNEH, Y., NICHOLSON, T., and ZUBIETA, J. (1985) Catecholate Coordination to Copper: Structural Characterization of a Tetrachloro-*o*-catecholate-bridged Dicopper(II) Complex as a Model for Intermediates in Copper-catalyzed Oxidation of Catechols. *Inorg. Chem.*, 24, 3725–3727.

KARLIN, K. D., NASIR, M. S., COHEN, B. I., CRUSE, R. W., KADERLI, S., and ZUBERBÜHLER, A. D. (1994) Reversible Dioxygen Binding and Aromatic Hydroxylation in $O_2$-Reactions with Substituted Xylyl Dinuclear Copper(I) Complexes: Syntheses and Low-temperature Kinetic/Thermodynamic and Spectroscopic Investigations of a Copper Monooxygenase Model System. *J. Am. Chem. Soc.*, 116, 1324–1336.

KARLIN, K. D., and TYEKLÁR, Z. (Eds.) (1993) *Bioinorganic Chemistry of Copper*, Chapman & Hall, New York.

KARLIN, K. D., and TYEKLÁR, Z. (1994) Functional Biomimics for Copper Proteins Involved in Reversible $O_2$-Binding, Substrate Oxidation/Oxygenation and Nitrite Reduction, in *Adv. Inorg. Biochem.*, (G. L. Eichorn and L. G. Marzilli, Eds.), Prentice-Hall, Englewood Cliffs, NJ, pp. 123–172.

KARLIN, K. D., TYEKLÁR, Z., FAROOQ, A., HAKA, M. S., GHOSH, P., CRUSE, R. W., GULTNEH, Y., HAYES, J. C., and ZUBIETA, J. (1992) Functional Modeling of Hemocyanins. Reversible Binding of $O_2$ and CO to Dicopper(I) Complexes $[Cu(I)_2(L)]^{2+}$ (L = Dinucleating Ligand) and the Structures of Two Bis(Carbonyl) Adducts, $[Cu(I)_2(L)(CO)_2]^{2+}$. *Inorg. Chem.*, 31, 1436–1451.

KARLIN, K. D., TYEKLÁR, Z., and ZUBERBÜHLER, A. D. (1993a) Formation, Structure, and Reactivity of Copper Dioxygen Complexes, in *Bioinorganic Catalysis* (J. Reedijk, Ed.), Marcel Dekker, New York, pp. 261–315.

KARLIN, K. D., WEI, N., JUNG, B., KADERLI, S., NIKLAUS, P., and ZUBERBÜHLER, A. D. (1993b) Kinetics and Thermodynamics of Formation of Copper–Dioxygen Adducts: Oxygenation of Mononuclear Cu(I) Complexes Containing Tripodal Tetradentate Ligands. *J. Am. Chem. Soc.*, 115, 9506–9514.

KATOPODIS, A. G., and MAY, S. W. (1990) Novel Substrates and Inhibitors of Peptidylglycine α-Amidating Monooxygenase. *Biochemistry*, 29, 4541–4548.

KAWAI, S., UMEZAWA, T., and HIGUCHI, T. (1988) Degradation Mechanisms of Phenolic β-1 Lignin Substructure Model Compounds by Laccase of *Coriolus versicolor*. *Arch. Biochem. Biophys.*, 262, 99–110.

KELLY, M., LAPPALAINEN, P., TALBO, G., HALTIA, T., VAN DER OOST, J., and SARASTE, M. (1993) Two Cysteines, Two Histidines, and One Methionine are Ligands of a Binuclear Purple Copper Center. *J. Biol. Chem.*, 268, 16781–16787.

KITAJIMA, N. (1992) Synthetic Approach to the Structure and Function of Copper Proteins. *Adv. Inorg. Chem.*, 39, 1–77.

KITAJIMA, N., FUJISAWA, K., FUJIMOTO, C., MORO-OKA, Y., HASHIMOTO, S., KITAGAWA, T., TORIUMI, K., TASUMI, K., and NAKAMURA, A. (1992) A New Model for Dioxygen Binding in Hemocyanin. Synthesis, Characterization and Molecular Structure of the $\mu\text{-}\eta^2:\eta^2$ Peroxo Dinuclear Copper(II) Complexes, $[Cu(HB(3,5\text{-}R_2pz)_3]_2(O_2)$ (R = i-Pr and Ph). *J. Am. Chem. Soc.*, 114, 1277–1291.

KITAJIMA, N., KATAYAMA, T., FUJISAWA, K., IWATA, Y., and MORO-OKA, Y. (1993) Synthesis, Molecular Structure and Reactivity of (Alkylperoxo)-copper(II) Complex. *J. Am. Chem. Soc.*, 115, 7872–7873.

KITAJIMA, N., KODA, T., IWATA, Y., and MORO-OKA, Y. (1990) Reaction Aspects of a μ-Peroxo Binuclear Copper(II) Complex. *J. Am. Chem. Soc.*, 112, 8833–8839.

KITAJIMA, N., and MORO-OKA, Y. (1994) Copper Dioxygen Complexes. Inorganic and Bioinorganic Perspectives. *Chem. Rev.*, 94, 737–757.

MAGNUS, K. A., HAZES, B., TON-THAT, H., BONAVENTURA, C., BONAVENTURA, J., and HOL, W. G. J. (1994) Crystallographic Analysis of Oxygenated and Deoxygenated States of Arthropod Hemocyanin Shows Unusual Differences. *Proteins: Struct., Funct., Genet.*, 19, 302–309.

MALKIN, R., and MALMSTRÖM, B. G. (1970) The State and Function of Copper in Biological Systems. *Adv. Enzymol.*, 33, 177–244.

MALMSTRÖM, B. G. (1993) Vectorial Chemistry in Bioenergetics: Cytochrome *c* Oxidase as a Redox-linked Proton Pump. *Acc. Chem. Res.*, 26, 332–338.

MASON, H. S., FOWLKS, W. L., and PETERSON, E. (1955) Oxygen Transfer and Electron Transport by the Phenolase Complex. *J. Am. Chem. Soc.*, 77, 2914–2915.

MATSUZAKI, R., SUZUKI, S., YAMAGUCHI, K., FUKUI, T., and TANIZAWA, K. (1995) Spectroscopic Studies on the Mechanism of the Topa Quinone Generation on Bacterial Monoamine Oxidase. *Biochemistry*, 34, 4524–4530.

MCNAMARA, J. O., and FRIDOVICH, I. (1993) Did Radicals Strike Lou Gehrig? *Nature*, 362, 20–21.

MCTAVISH, H., FUCHS, J. A., and HOOPER, A. B. (1993) Sequence of the Gene Coding for Ammonia Monooxygenase in *Nitrosomonas europaea*. *J. Bacteriol.*, 175, 2436–2444.

MERKLER, D. J., KULATHILA, R., CONSALVO, A. P., YOUNG, S. D., and ASH, D. E. (1992) $^{18}$O Isotopic $^{13}$C NMR Shift as Proof That Bifunctional Peptidylglycine α-Amidating Enzyme Is a Monooxygenase. *Biochemistry*, 31, 7282–7288.

MESSERSCHMIDT, A. (1993) Ascorbate Oxidase Structure and Chemistry, in *Bioinorganic Chemistry of Copper* (K. D. Karlin and Z. Tyeklár, Eds.), Chapman & Hall, New York, pp. 471–484.

MESSERSCHMIDT, A., and HUBER, R. (1990) The Blue Oxidases, Ascorbate Oxidase, Laccase and Ceruloplasmin. *Eur. J. Biochem.*, 187, 341–352.

MESSERSCHMIDT, A., LADENSTEIN, R., HUBER, R., BOLOGNESI, M., AVIGLIANO, L., PETRUZZELLI, R., ROSSI, A., and FINAZZI-AGRO, A. (1992) Refined Crystal Structure of Ascorbate Oxidase at 1.9 Å Resolution. *J. Mol. Biol.*, 224, 179–205.

MORGAN, J. E., VERKHOVSKY, M. L., and WIKSTRÖM, M. (1994) The Histidine Cycle: A New Model for Proton Translocation in the Respiratory Heme-Copper Oxidases. *J. Bioeng. Biomembr.*, 26, 599–608.

MU, D., JANES, S. M., SMITH, A. J., BROWN, D. E., DOOLEY, D. M., and KLINMAN, J. P. (1992) Tyrosine Codon Corresponds to Topa Quinone at the Active Site of Copper Amine Oxidases. *J. Biol. Chem.*, 267, 7979–7982.

MURAHASHI, S.-I., ODA, Y., NAOTA, T., and KOMIYA, N. (1993) Aerobic Oxidations of Alkanes and Alkenes in the Presence of Aldehydes Catalyzed by Copper Salts. *J. Chem. Soc. Chem. Commun.*, 139–140.

NASIR, M. S., COHEN, B. I., and KARLIN, K. D. (1992) Mechanism of Aromatic Hydroxylation in a Copper Monooxygenase Model System. 1,2-Methyl Migrations and the NIH Shift in Copper Chemistry. *J. Am. Chem. Soc.*, 114, 2482–2494.

NGUYEN, H. T., SHIEMKE, A. K., JACOBS, S. J., HALES, B. J., LIDSTYOM, M. E., and CHAN, S. I. (1994) Nature of the Copper Ions in the Membranes Containing the Particulate Methane Monooxygenase from *Methylococcus capsilatus* (Bath). *J. Biol. Chem.*, 269, 14995–15005.

O'HALLORAN, T. V. (1993) Transition Metals in Control of Gene Expression. *Science*, 261, 715–725.

OGURA, T., TAKAHASHI, S., HIROTA, S., SHINZAWA-ITOH, K., YOSHIKAWA, S., APPELMAN, E. H., and KITAGAWA, T. (1993) Time-resolved Resonance Raman Elucidation of the Pathway for Dioxygen Reduction by Cytochrome *c* Oxidase. *J. Am. Chem. Soc.*, 115, 8527–8536.

PANDEY, G., MURALIKRISHNA, C., and BHALERAO, U. T. (1990) Mushroom Tyrosinase Catalyzed Coupling of Hindered Phenols: A Novel Approach for the Synthesis of Diphenoquinones and Bisphenols. *Tetrahedron Lett.*, 31, 3771–3774.

PAUL, P. P., TYEKLÁR, Z., JACOBSON, R. R., and KARLIN, K. D. (1991) Reactivity Patterns and Comparisons in Three Classes of Synthetic Copper–Dioxygen {Cu$_2$–O$_2$} Complexes: Implication for Structure and Biological Relevance. *J. Am. Chem. Soc.*, 113, 5322–5332.

PEMBER, S. O., JOHNSON, K. A., VILLAFRANCA, J. J., and BENKOVIC, S. J. (1989) Mechanistic Studies on Phenylalanine Hydroxylase from *Chromobacterium violaceum*. Evidence for the Formation of an Enzyme–Oxygen Complex. *Biochemistry*, 28, 2124–2130.

PEMBER, S. O., VILLAFRANCA, J. J., and BENKOVIC, S. J. (1986) Phenylalanine Hydroxylase from *Chromobacterium violaceum* is a Copper-Containing Monooxygenase. Kinetics of the Reductive Activation of the Enzyme. *Biochemistry*, 25, 6611–6619.

REDDY, K. V., JIN, S.-J., ARORA, P. K., SFEIR, D. S., MALONEY, S. C. F., URBACH, F. L., and SAYRE, L. M. (1990) Copper-mediated Oxidative C-terminal N-dealkylation of Peptide-derived Ligands. A Possible Model for Enzymatic Generation of Desglycine Peptide Amides. *J. Am. Chem. Soc.*, 112, 2332–2340.

ROSENZWEIG, A. C., FREDERICK, C. A., LIPPARD, S. J., and NORDLUND, P. (1993) Crystal Structure of a Bacterial Non-haem Iron Hydroxylase that Catalyses the Biological Oxidation of Methane. *Nature*, 366, 537–543.

ROSS, P. K., and SOLOMON, E. I. (1991) An Electronic Structural Comparison of Copper–Peroxide Complexes of Relevance to Hemocyanin and Tyrosinase Active Sites. *J. Am. Chem. Soc.*, 113, 3246–3259.

ROUSSEAU, D. L., HAN, S., SONG, S., and CHING, Y.-C. (1992) Catalytic Mechanism of Cytochrome *c* Oxidase. *J. Raman Spectros.*, 23, 551–556.

SAYRE, L. M., and NADKARNI, D. (1994) Direct Conversion of Phenols to *o*-Quinones by Copper(I)–Dioxygen. Questions Regarding the Monophenolase Activity of Tyrosinase Mimics. *J. Am. Chem. Soc.*, 116, 3157–3158.

SHELDON, R. A. (1993) History of Oxygen Activation 1773–1993, in *The Activation of Dioxygen and Homogeneous Catalytic Oxidation* (D. H. R. Barton, A. E. Martell, and D. T. Sawyer, Eds.), Plenum, New York, pp. 9–30.

SOLOMON, E. I., BALDWIN, M. J., and LOWERY, M. D. (1992) Electronic Structures of Active Sites in Copper Proteins: Contributions to Reactivity. *Chem. Rev.*, 92, 521–542.

SORRELL, T. N. (1989) Synthetic Models for Binuclear Copper Proteins. *Tetrahedron*, 45, 3–68.

SPEIER, G. (1993) Copper Dioxygenation Chemistry Relevant to Quercetin Dioxygenase, in *Bioinorganic Chemistry of Copper* (K. D. Karlin and Z. Tyeklár, Eds.), Chapman & Hall, New York, pp. 382–394.

STEWART, L. C., and KLINMAN, J. P. (1988) Dopamine β-hydroxylase of Adrenal Chromaffin Granules: Structure and Function. *Annu. Rev. Biochem.*, 57, 551–592.

TAINER, J. A., GETZOFF, E. D., RICHARDSON, J. S., and RICHARDSON, D. C. (1983) Structure and Mechanism of Copper, Zinc Superoxide Dismutase. *Nature*, 306, 284–287.

TIAN, G., BERRY, J. A., and KLINMAN, J. P. (1994) Oxygen-18 Kinetic Isotope Effects in the Dopamine β-Monooxygenase Reaction: Evidence for a New Chemical Mechanism in Non-heme Metallomonooxygenases. *Biochemistry*, 33, 226–234.

TYEKLÁR, Z., JACOBSON, R. R., WEI, N., MURTHY, N. N., ZUBIETA, J., and KARLIN, K. D. (1993) Reversible Reaction of $O_2$ (and CO) with a Copper(I) Complex: X-ray Structures of Relevant Mononuclear Cu(I) Precursor Adducts and the *trans*-(μ-1,2-Peroxo)–Dicopper(II) Product. *J. Am. Chem. Soc.*, 115, 2677–2689.

UTAKA, M., HOJO, M., FUJII, Y., and TAKEDA, A. (1984) Cupric Ion-catalyzed Dioxygenation of 1,2-Cyclohexanediones. A Nonenzymatic Analog for Quercetinase Dioxygenation. *Chem. Lett.*, 635–638.

VAROTSIS, C., ZHANG, Y., APPELMAN, E. H., and BABCOCK, G. T. (1993) Resolution of the Reaction Sequence During the Reduction of $O_2$ by Cytochrome Oxidase. *Proc. Natl. Acad. Sci. U.S.A.*, 90, 237–241.

VOLBEDA, A., and HOL, W. G. J. (1989) Crystal Structure of Hexameric Haemocyanin from *Panulirus interruptus* Refined at 3.2 Å Resolution. *J. Mol. Biol.*, 209, 249–279.

WALLING, C. (1995) Autoxidation, in *Active Oxygen in Chemistry* (C. S. Foote, J. S. Valentine, A. Greenberg, and J. F. Liebman, Eds.), Chapman & Hall, New York, pp. 24–65.

WARNCKE, K., BABCOCK, G. T., DOOLEY, D. M., MCGUIRL, M. A., and MCCRACKEN, J. (1994) Structure of the Topa-semiquinone Catalytic Intermediate of Amine Oxidase as Revealed by Magnetic Interactions with Exchangeable $^2$H and $^1$H Nuclei. *J. Am. Chem. Soc.*, 116, 4028–4037.

WHITTAKER, J. M. (1993) Active Site Interactions in Galactose Oxidase, in *Bioinorganic Chemistry of Copper* (K. D. Karlin and Z. Tyeklár, Eds.), Chapman & Hall, New York, pp. 447–458.

WILCOX, D. E., PORRAS, A. G., HWANG, Y. T., LERCH, K., WINKLER, M. E., and SOLOMON, E. I. (1985) Substrate Analogue Binding to the Coupled Binuclear Copper Active Site in Tyrosinase. *J. Am. Chem. Soc.*, 107, 4015–4027.

# 5
# Oxygen Activation at Nonheme Iron Centers

LAWRENCE QUE, JR.

Progress in understanding the mechanism of dioxygen activation by nonheme iron enzymes has lagged behind that of their heme counterparts, probably because of their relative spectroscopic inaccessibility. However, developments in the past decade have resulted in greater efforts in this subfield and provided insights into how such enzymes function. As a group, these enzymes catalyze a diverse array of metabolically important reactions which involve a number of interesting chemical transformations. The chapter will focus on the mechanistic principles that tie together these seemingly unrelated reactions.

## A PARADIGM FOR OXYGEN ACTIVATION AT A METAL CENTER

Cytochrome P-450 catalyzes a variety of metabolically important oxidative transformations, including the hydroxylation of unactivated alkanes (McMurry and Groves, 1986). The mechanism proposed for cytochrome P-450, perhaps the most studied monooxygenase (Ortiz de Montellano, 1986), serves as a paradigm for oxygen activation mechanisms associated with nonheme iron enzymes. The principal features of its mechanism are summarized in Figure 5-1. Thus dioxygen binds at a ferrous center and is reduced by two electrons to form a ferric peroxide complex; subsequent O—O bond heterolysis converts it to a formally oxoiron(V) porphyrin species, which is believed to be re-

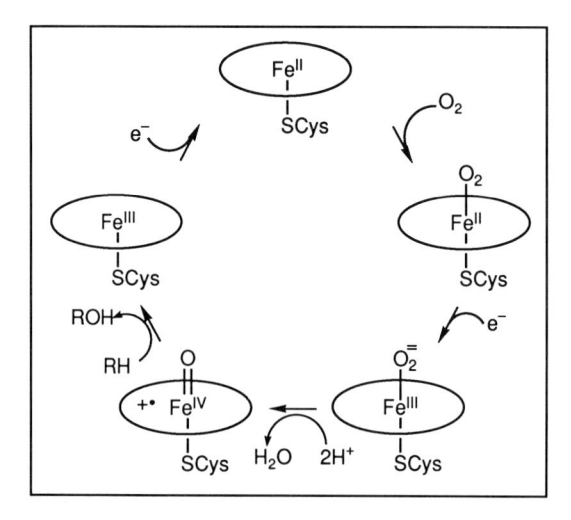

FIGURE 5-1  Catalytic cycle of cytochrome P-450.

sponsible for the oxygen transfer chemistry. Although such a species has not been directly observed in the catalytic cycle of the enzyme, evidence for its likely involvement comes from two observations. First, the reaction of horseradish peroxidase with $H_2O_2$ produces Compound I, which is spectroscopically characterized as an oxoiron(IV) porphyrin cation radical complex (Dawson, 1988). Secondly, ferricytochrome P-450 and synthetic $Fe^{III}$ porphyrin complexes are known to react with oxygen atom donors such as peroxides, PhIO, and $ClO^-$ to generate a species capable of performing the same reactions as the native enzyme (Figure 5-2). This reaction bypasses the $Fe^{II}$ oxidation state in the P-450 cycle and is known as the *peroxide shunt* (Groves, 1985).

Oxygen activation proceeds in at least two steps: the binding and reduction of $O_2$ by an electron-rich species to form an intermediate at the peroxide oxidation level and the cleavage of the peroxide O—O bond concomitant with or followed by utilization of the oxidizing equivalents generated to effect transformation of the substrate.

## BLEOMYCIN

Among nonheme iron centers, the Fe-bleomycin system is perhaps the system that most closely corresponds in mechanism to the cytochrome

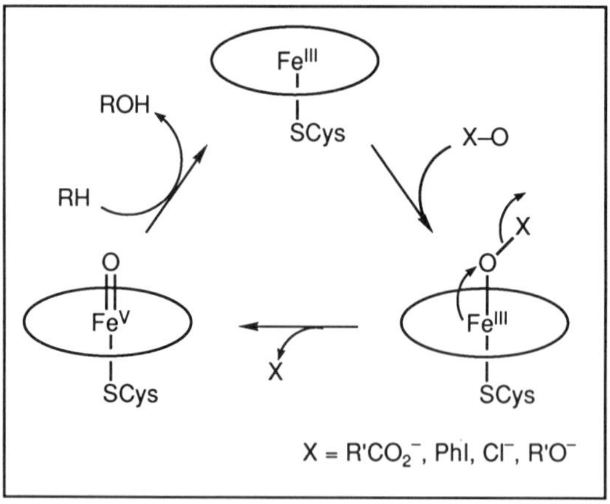

**FIGURE 5-2** Peroxide shunt pathway for cytochrome P-450.

P-450 paradigm. Bleomycins (BLMs) are a family of glycopeptide-derived antibiotics that have the ability to bind and degrade DNA, an ability that is believed to be responsible for their antitumor activity (Stubbe and Kozarich, 1987). $Fe^{II}$ and $O_2$ are required for the DNA cleavage. On the basis of spectroscopic studies, the proposed structure of the $Fe^{II}BLM$ complex is shown in Figure 5-3. The pyrimidine, the secondary amine, the imidazole, and the amidate provide a square planar $N_4$ environment akin to that of a porphyrin, with the primary amine acting as the axial ligand. The involvement of the amide nitrogen in

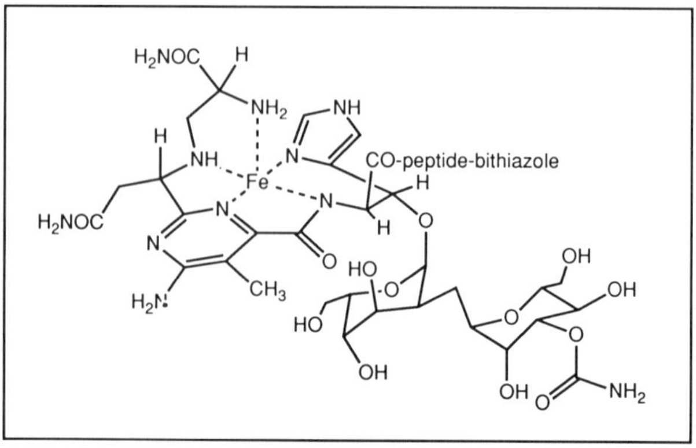

**FIGURE 5-3** Proposed structure of $Fe^{II}BLM$.

iron coordination is perhaps a novel aspect of the BLM coordination chemistry. It is implicated as a ligand by the NMR solution structure of the Fe$^{II}$BLM–CO complex (Oppenheimer et al., 1979), and model complexes suggest that such ligation is reasonable (Brown and Mascharak, 1988; Brown et al., 1989). The amidate is a very basic moiety and the only anionic ligand in the proposed coordination sphere; its coordination serves to stabilize Fe$^{III}$ relative to Fe$^{II}$ and lowers the Fe$^{III/II}$ redox potential to a range suitable for O$_2$ binding. Like ferrous hemes, the high-spin Fe$^{II}$BLM complex binds CO, NO, RNC, and, of course, O$_2$ (Burger et al., 1981, 1983). The proposed reaction cycle for Fe$^{II}$BLM is shown in Figure 5-4. On O$_2$ binding, the iron ion is converted to a diamagnetic [and therefore electron paramagnetic resonance (EPR) silent], low-spin center (Burger et al., 1983). Though the diamagnetism of the oxy complex could be construed as the result of a low-spin ferrous center with a bound dioxygen that is spin paired, the Mössbauer parameters favor the formulation as a low-spin ferric center antiferromagnetically coupled to a bound superoxide, similar to the electronic descriptions of oxygenated heme complexes.

One-electron reduction of the oxy complex affords "activated BLM," which should be at the oxidation level of a ferric–peroxide complex or its equivalent (Burger et al., 1981). This electron can be provided by various reductants, including another Fe$^{II}$BLM molecule. By analogy to the peroxide shunt pathway in the cytochrome P-450 cycle, "activated BLM" can also be formed from Fe$^{III}$BLM and H$_2$O$_2$ or EtOOH (Burger et al., 1981). Similar DNA cleavage products are produced by either preparation route, indicating that the same activated BLM moiety is formed in either case.

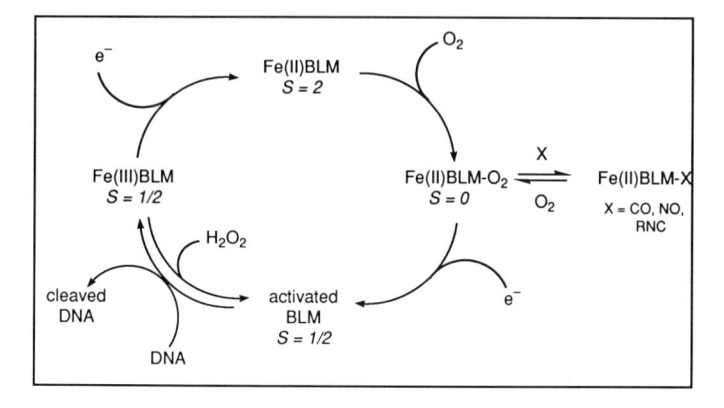

**FIGURE 5-4** Proposed reaction cycle of FeBLM.

Chemical titrations show that activated BLM contains two oxidizing equivalents above the $Fe^{III}$ oxidation state (Burger et al., 1985). In terms of formal oxidation states, activated BLM is at the same oxidation level as heme peroxidase Compounds I and the "reactive species" in the cytochrome P-450 cycle. But are their electronic descriptions analogous? Activated BLM hydroxylates naphthalene (at C-1) and 4-deuterioanisole [at C-4 with a concomitant NIH shift (see Guroff et al., 1967)], epoxidizes olefins, and demethylates $N,N$-dimethylaniline, reactions also catalyzed by cytochrome P-450 (Murugesan and Hecht, 1985; Heimbrook et al., 1987; Natrajan et al., 1990). Furthermore, $Fe^{III}$BLM can act as an oxygen atom transfer agent, using donors such as PhIO and periodate (Murugesan and Hecht, 1985). Thus, by analogy to the high-valent chemistry of cytochrome P-450 and its models, the involvement of a formally $Fe^V{=}O$ species in bleomycin chemistry is strongly implicated.

However, activated BLM exhibits an $S = 1/2$ EPR spectrum with $g$ values at 2.26, 2.17, and 1.94 (Burger et al., 1981). Isotope substitution experiments show that $^{57}Fe$ causes a 22 G splitting of the $g = 1.94$ signal, and the use of $^{17}O_2$ results in the broadening of the $g = 2.17$ signal, indicating that the species involves both iron and oxygen. (The persistence of the $^{17}O$ effect in $H_2^{16}O$ solvent also suggests that exchange of label with solvent is slow.) However, the EPR signals are typical of low-spin $Fe^{III}$, not of $Fe^{IV}$ or $Fe^V$, a conclusion which is corroborated by the Mössbauer parameters (Burger et al., 1983). Furthermore, the visible spectrum of activated BLM appears indistinguishable from that of $Fe^{III}$BLM (Burger et al., 1981). The spectroscopic data thus appear to be inconsistent with an $Fe^V{=}O$ formulation.

An attractive alternative formulation for activated BLM which accounts for the EPR and Mössbauer data, the redox titration data, and the retention of the $^{17}O$ label is a low-spin ferric–peroxide complex. The two oxidizing equivalents needed for the oxidation chemistry would thus be localized on the dioxygen moiety instead of on the metal center; such a species would be analogous to $[Fe(TPP)O_2]^-$ (McCandlish et al., 1980) and $[Fe(EDTAH)O_2]^{2-}$ (Ahmad et al., 1988). Such a formulation would be consistent with the observations that ethylhydroperoxide (EtOOH) generates the characteristic EPR spectrum of activated BLM (Burger et al., 1981) but iodosylbenzene (PhIO) does not (Padbury et al., 1988).

Evidence for the peroxide formulation has recently been obtained from electrospray ionization mass spectral studies (Sam et al., 1994). In this scheme, "activated BLM" either directly effects the cytochrome P450-like transformations or serves as the precursor to a very short-lived iron-oxo species that carries out these transformations. This scheme is also consistent with the lack of solvent label incorporation into either the spectroscopically characterized activated BLM or the product epoxides (Heimbrook et al., 1987). The relatively long-lived activated BLM, being an iron-peroxo species, cannot exchange label with solvent, while the transient iron-oxo species, if present, is too short-lived to exchange with solvent. The details of how activated bleomycin interacts with DNA to effect strand cleavage have been investigated in the elegant studies of Stubbe and Kozarich (Stubbe and Kozarich, 1987; McCall et al., 1992).

## DIIRON ENZYMES: METHANE MONOOXYGENASE AND RIBONUCLEOTIDE REDUCTASE

Methane monooxygenase (MMO) and ribonucleotide reductase (RNR) are diiron enzymes that activate dioxygen to hydroxylate methane to methanol and to generate a protein-bound tyrosyl radical, respectively. MMO is isolated from methanotrophs that utilize methane as their energy and carbon source (Dalton, 1980). RNR is a key enzyme in DNA biosynthesis (Reichard and Ehrenberg, 1983; Sjöberg and Gräslund, 1983; Stubbe, 1990) found in mammals, viruses, and *Escherichia coli*. These two enzymes belong to an emerging subclass of proteins with diiron active sites called the *diiron-oxo proteins* (Sanders-Loehr, 1989; Que and True, 1990), which also includes hemerythrin, purple acid phosphatase, and rubrerythrin. Of these, only hemerythrin, MMO, and RNR are known to interact with $O_2$ as part of their biological function; this section will thus focus only on these three proteins.

### Hemerythrin, the Prototypical Diiron-Oxo Protein

Hemerythrin, the dioxygen carrier for certain marine invertebrates, is prototypical of this group (Que and True, 1990). X-ray crystallography (Stenkamp et al., 1985; Sheriff et al., 1987; Holmes and Stenkamp, 1991; Holmes et al., 1991) and extensive spectroscopic and biomimetic investigations (Garbett et al., 1969; Okamura et al., 1969; Dawson et al., 1972; Sanders-Loehr et al., 1980, 1989; Clark and Webb, 1981; Shiemke et al., 1984; Stenkamp et al., 1985; Zhang et al., 1988; Reem et al., 1989; Kurtz, 1990) have established the structures of deoxyhemerythrin and oxyhemerythrin (Figure 5-5), the novel features being the tribridged diiron site and the terminal hydroperoxide hydrogen bonded to the oxo bridge. The structural and physical properties of the tribridged diferric site of oxyHr and of the closely related metHr are quite well

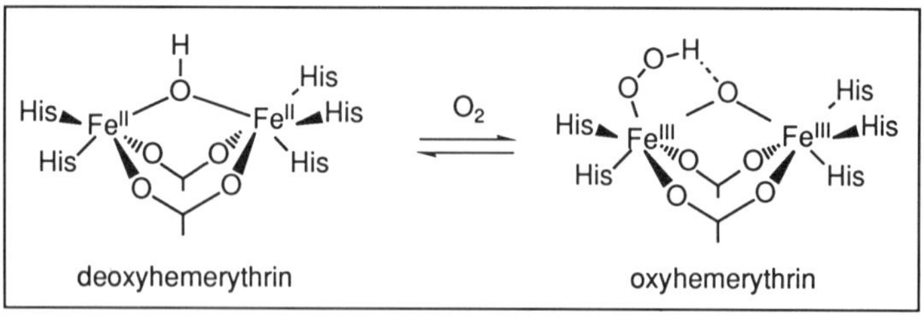

FIGURE 5-5  Active-site structures of deoxyhemerythrin and oxyhemerythrin.

established, having been modeled by a number of synthetic complexes (Kurtz, 1990). The oxo bridge in the diferric form defines many of its distinctive properties; it gives rise to short Fe-μ-oxo bonds (ca. 1.8 Å) (Zhang et al., 1988), strong antiferromagnetic coupling (J = −120 ± 10 cm$^{-1}$ for H = −2$JS_1 \cdot S_2$) (Dawson et al., 1972), and large Mössbauer quadrupole splittings ($\Delta E_Q \geq$ 1.3 mm/s) (Okamura et al., 1969; Clark and Webb, 1981). The carboxylate bridges bend the Fe—O—Fe unit to an angle of 120–130°, resulting in an Fe-Fe separation of 3.2(1) Å and giving rise to a series of characteristic visible features (ε = 100–1000 M$^{-1}$cm$^{-1}$) (Garbett et al., 1969; Sanders-Loehr et al., 1980; Reem et al., 1989), and Raman vibrations (Shiemke et al., 1984; Sanders-Loehr et al., 1989) associated with a bent Fe—O—Fe unit.

OxyHr is characterized by an intense visible band near 500 nm, which is associated with a peroxide-to-Fe$^{III}$ charge transfer transition (Shiemke et al., 1986). Resonance Raman studies confirm this assignment with the observation of $v_{Fe-OO}$ and $v_{O-O}$ features at 503 and 844 cm$^{-1}$, respectively, both of which downshift appropriately on $^{18}$O substitution (Shiemke et al., 1986; Sanders-Loehr et al., 1989). That the bound dioxygen is present as a hydroperoxide—that is, hydrogen bonded to the oxo bridge, as shown in Figure 5-5—is indicated by the downshift observed for $v_{Fe-OO}$ in D$_2$O and the upshift of the symmetric $v_{Fe-O-Fe}$ feature at 492 cm$^{-1}$ (Shiemke et al., 1986; Sanders-Loehr et al., 1989).

By comparison, the tribridged core of deoxyHr is somewhat more expanded due to the presence of the μ-OH bridge and the longer Fe$^{II}$-ligand bonds (Stenkamp et al., 1985; Zhang et al., 1988), a fact corroborated by [Fe$_2$(OH)(OAc)$_2$(Me$_3$TACN)$_2$]$^+$, which is crystallographically characterized to have a (μ-hydroxo)bis(μ-carboxylato) diferrous core (Figure 5-6) (Chaudhuri et al., 1985; Hartman et al., 1987). The lengthening of the Fe—μ—O bond engenders weak antiferromagnetic coupling between the ferrous ions, with -J estimated

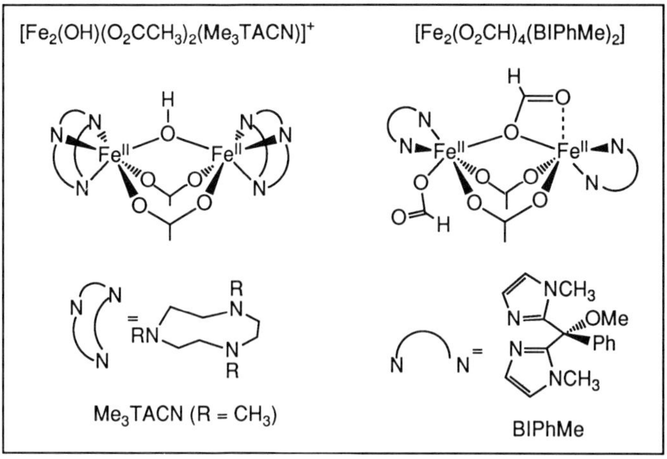

**FIGURE 5-6** Models for deoxyhemerythrin.

to be 15 cm$^{-1}$ on the basis of NMR measurements (Maroney et al., 1986) and between 15 and 36 cm$^{-1}$ from MCD studies (Reem and Solomon, 1987). The J value of −13 cm$^{-1}$ for the synthetic complex is thus in agreement with the protein estimates. Another synthetic diferrous complex (Tolman et al., 1989, 1991) models a different aspect of the deoxyHr active site. [Fe$_2$(BIPhMe)$_2$(O$_2$CH)$_4$] possesses a (μ-formato-*O*)bis(μ-formato-*O,O'*)diiron core, with the bidentate BIPhMe ligand completing the coordination sphere on one iron and a BIPhMe and a monodentate formate as terminal ligands on the other (Figure 5-6); thus this complex has a six-coordinate and a five-coordinate iron center, as found in deoxyHr. However, neither of these synthetic complexes binds O$_2$.

Dioxygen binding to deoxyHr, forming oxyHr, thus entails a transfer of two electrons from the diferrous center to the bound dioxygen, affording a peroxide moiety coordinated to a diferric center. Unlike the mononuclear iron sites of cytochrome P-450 and BLM, the delivery of the second electron is built in by the presence of a second iron center.

### Active-Site Structure

Based on similarities of spectroscopic and magnetic properties, the diiron centers of MMO and RNR are proposed to have structures related to that found for hemerythrin. The RNR from *E. coli* consists of two protein components, R1 and R2. The R1 protein contains the substrate and effector binding sites and the conserved thiols that provide the electrons for the deoxygenation process (Thelander, 1974; von Döbeln and Reichard, 1976). The R2 protein consists of two identical polypeptide chains (Thelander, 1973), each of which has a diiron cluster (Lynch et al., 1989) and a Tyr at position 122 (Larsson and Sjöberg, 1986) that can be oxidized to its radical form (Atkin et al., 1973). The diiron center in this enzyme does not serve a catalytic function, but instead acts in a regulatory role to generate the catalytically essential tyrosyl radical during the course of the cell's life cycle (Fontecave et al., 1989).

The crystal structure of RNR $R_2$ (Figure 5-7) shows a (μ-oxo)diiron(III) unit (Nordlund et al., 1990), as expected from the earlier magnetic (Petersson et al., 1980), Mössbauer (Atkin et al., 1973), resonance Raman (Sjöberg et al., 1982), and EXAFS data (Scarrow et al., 1987). The iron centers are further bridged by a single carboxylate, which bends the Fe—O—Fe unit to an angle of 130° and affords an Fe—Fe separation of 3.22 Å. Such dibridged cores have been demonstrated in synthetic complexes to be difficult to distinguish spectroscopically from the tribridged cores (Norman et al., 1990b). The major distinction from hemerythrin is the predominantly oxygen-ligand environment. There is only one His residue per iron center, and there are three terminal carboxylates, each coordinated in a distinct manner: one chelated,

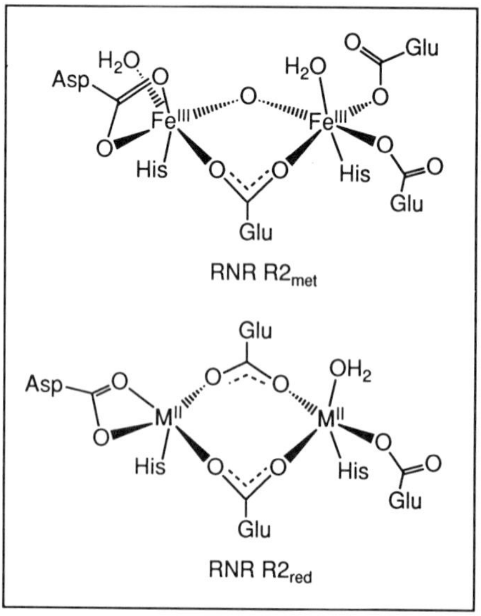

**FIGURE 5-7** Structures of the diiron cluster in the $R_2$ protein of RNR.

the second monodentate *syn*, and the third monodentate *anti*. The carboxylate-rich ligand environment decreases the Lewis acidity of the ferric centers relative to that of diferric (or met) hemerythrin and increases the energy gap between the ligand and the metal d orbitals (Reem et al., 1989). The differences in the ligand environment between metHr and RNR R2 give rise to the generally blue-shifted absorption features of the RNR R2 visible spectrum which have dominant oxo-to-Fe$^{III}$ charge transfer bands (Norman et al., 1990a; Ménage and Que, 1991). Tyr122 is found to be 5 Å away from the dinuclear cluster and appears not to affect any of the properties of the metal center.

The diferrous form of R2, R2$_{red}$, can be obtained by reduction of native R2 with dithionite and appropriate mediators or by anaerobic reconstitution of apoR2 with Fe$^{II}$ (Lynch et al., 1989; Sahlin et al., 1989). It is very air sensitive and readily forms native R2 on exposure to O$_2$. The crystal structure of the corresponding Mn$^{II}$R2 has been reported and may represent the structure of R2$_{red}$ as well (Figure 5-7) (Atta et al., 1992). In this form, no single atom bridge appears to be present and Glu 238, terminal monodentate in the diferric protein, now bridges the dinuclear unit. The metal centers are coordinatively unsaturated, consistent with the (NIR–CD) data on R2$_{red}$ (McCormick et al., 1991), thereby affording sites for O$_2$ binding.

MMO is an enzyme complex consisting of three components: a hydroxylase, a reductase, and a third component required for efficient coupling of the reductive and oxidative parts of the overall reaction (Woodland and Dalton, 1984; Fox et al., 1989, 1991). The diiron active site is localized on the hydroxylase component and receives two electrons from NADH via the reductase to activate dioxygen. Single-turnover experiments have demonstrated that the 2e$^-$-reduced hydroxylase by itself can effect the oxidative transformations catalyzed by the enzyme complex (Fox et al., 1989, 1991).

The crystal structure of the hydroxylase component of MMO, MMOH, has been solved and shows an active site that resembles that of RNR R2 (Figure 5-8) (Rosenzweig et al., 1993; Rosenzweig and Lippard, 1994). However unlike for RNR R2, the diferric unit in MMO is hydroxo-bridged as indicated by the absence of any visible absorption features, weak antiferromagnetic coupling derived from Mössbauer and integer spin EPR evidence, and the longer Fe—Fe distance of 3.44 Å (Fox et al., 1993; DeWitt et al., 1991; Rosenzweig et al., 1993).

The diferrous form exhibits an integer spin EPR signal with $g = 16$ (Fox et al., 1988; Hendrich et al., 1990), thereby providing a convenient spectroscopic probe for an oxidation state that is otherwise difficult to monitor. Indeed, this signal has been used to show that it is the

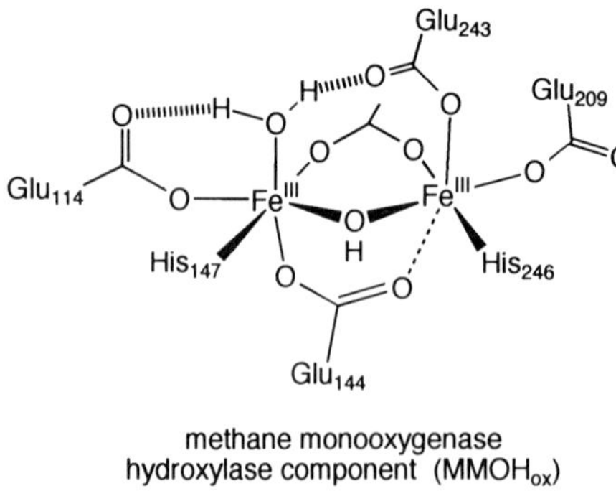

methane monooxygenase
hydroxylase component (MMOH$_{ox}$)

**FIGURE 5-8** Crystal structure of the hydroxylase component of methane monooxygenase (MMOH$_{ox}$).

diferrous state, not the mixed valence state, that reacts with dioxygen (Fox et al., 1989). Analysis of the EPR signal indicates that it can only arise from ferromagnetic coupling of the two $S = 2$ Fe$^{II}$ centers. A similar signal ($g = 18$) has been observed for the azide adduct of deoxyhemerythrin and is proposed to result from the protonation of the hydroxo bridge to an aqua bridge (Reem and Solomon, 1987; Hendrich et al., 1991). Examples of diferrous complexes that exhibit integer spin signals include [Fe$_2$(BPMP)($\mu$-O$_2$CC$_2$H$_5$)$_2$]$^+$ (Borovik et al., 1990) and [Fe$_2$(salmp)$_2$]$^{2-}$ (Snyder et al., 1989); both have phenoxo bridges with Fe-$\mu$-O-Fe angles of <110°. Such acute angles very likely give rise to ferromagnetic coupling between the ferrous centers. In contrast, synthetic complexes with hydroxo or alkoxo bridges have Fe-$\mu$-O-Fe angles of 110° and afford EPR-silent, antiferromagnetically coupled complexes. More work is needed to understand the structural basis for this phenomenon.

The striking differences in terminal ligands between hemerythrin, on the one hand, and RNR R2 and MMO, on the other, are intriguing; these may reflect the different functions of the three proteins, with nitrogen-rich ligation promoting reversible oxygen binding in the case of hemerythrin and oxygen-rich ligation favoring oxygen activation in the cases of RNR R2 and MMO.

## Oxygen Activation Mechanism

A common mechanism for oxygen activation by both enzymes is proposed in Figure 5-9, starting with the premise that a diferrous unit

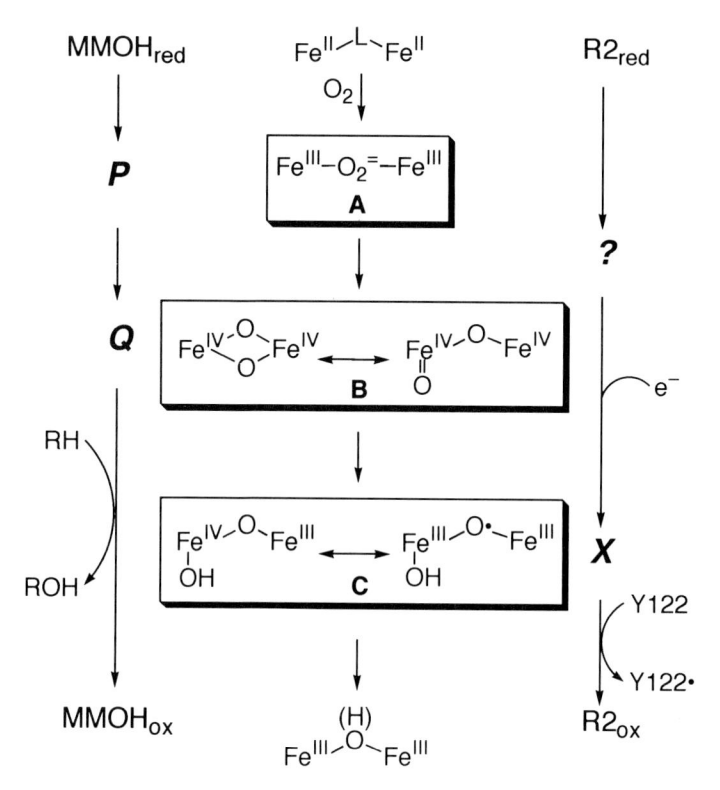

**FIGURE 5-9** Common oxygen activation mechanism proposed for ribonucleotide reductase and methane monooxygenase

is the oxygen-binding species. It is based on observations of the various workers in the field and incorporates many elements found in heme peroxidase and cytochrome P-450 chemistry. The diferrous unit provides the two electrons necessary to reduce dioxygen, yielding diferric peroxide complex **A**. This intermediate then loses water and converts into high-valent species **B** and **C** that are diiron analogs of heme peroxidase Compounds I and II (Dawson, 1988). It may be argued that the second iron in the Compound I analog can serve the function of the porphyrin in storing and oxidizing equivalent. These high-valent species then effect the critical substrate oxidation steps in the enzyme mechanisms.

Support for this proposal in the R2 and MMO cycles is found in experiments demonstrating the availability of a peroxide shunt pathway. Diferric RNR R2 reacts with $H_2O_2$ (Sahlin et al., 1990) or oxygen atom donors (Fontecave et al., 1990), such as iodosobenzoate and peracids, to produce the tyrosyl radical, albeit less efficiently than the $O_2$ reaction. Similarly, $MMO_{ox}$ can hydroxylate substrates with $H_2O_2$ (Andersson et al., 1991). As with cytochrome P-450, these experiments strengthen arguments for the intermediacy of the diferric peroxide and high-valent iron-oxo species in the $O_2$ activation mechanisms of these enzymes. The formation of the tyrosyl radical in R2 would then occur by the reaction of Tyr122 with either high-valent intermediate, analogous to the phenol oxidations effected by horseradish peroxidase Compounds I and II, while alkane hydroxylation by MMO would proceed by hydrogen abstraction and oxygen rebound by analogy to the cytochrome P-450 mechanism.

To what extent is there spectroscopic evidence for any of these proposed intermediates? Figure 5-9 summarizes the present state of knowledge regarding intermediates in the MMOH and R2 redox cycles derived from elegant stopped flow and rapid freeze-quench studies of the diiron(II) forms of MMOH and R2 with dioxygen (Lee et al., 1993a, 1993b; Liu et al., 1994; Ravi et al., 1994; Bollinger et al., 1994a, 1994b). For MMO, two intermediates have been observed, **P** and **Q**. Since both are kinetically competent to hydroxylate methane, they both still have the two oxidizing equivalents needed for the reaction. Compound **P** exhibits a Mössbauer quadrupole doublet with parameters ($\delta$ = 0.66 mm/s; $\Delta E_Q$ = 1.51 mm/s) proposed to arise from an antiferromagnetically coupled dinuclear high spin ferric species (Liu et al., 1994), though these parameters differ somewhat from those of a majority of diferric complexes. This suggests that the oxidizing equivalents may be stored on the dioxygen moiety as a peroxide. Compound **P** would thus correspond to intermediate A in Figure 9. Compound **Q**, on the other hand, exhibits Mössbauer parameters ($\delta$ = 0.17 mm/s; $\Delta E_Q$ = 0.53 mm/s) and is associated with a diamagnetic ground state. These unusual properties are proposed to arise from an antiferromagnetically coupled diiron(IV) species (Lee et al., 1993a), suggesting that **Q** corresponds to intermediate B in Figure 5-9.

The reaction of $R2_{red}$ with $O_2$ gives rise to an intermediate with a paramagnetic cluster **X**, which has been characterized by EPR and Mössbauer spectroscopy (Bollinger et al., 1991b; Ravi et al., 1994). Species **X** corresponds in oxidation state to intermediate C in Figure 5-9, analogous to heme peroxidase Compound II. Since species X decays concomitant with tyrosyl radical formation, its lifetime may be increased in the absence of Tyr122. Indeed the use of the mutant Y122F protein allows X to be generated in nearly quantitative yield. Being formally $Fe^{III}Fe^{IV}$, this intermediate is expected to exhibit a half integer spin EPR signal. In fact, a broad isotropic signal near $g = 2$ is observed, typical of a $S = 1/2$ state, indicating that antiferromagnetic coupling of the metal centers occurs. The signal is split by $^{57}Fe$ hyperfine coupling, implicating at least one of the iron centers with this signal. Amazingly, the Mössbauer spectrum of this species shows no evidence for an $Fe^{IV}$ center but instead two distinct high spin $Fe^{III}$ ions in a 1:1 ratio with respective ($\delta$, $\Delta E_Q$, $A/g_n\beta n$) values of (0.55 mm/s, −1.0 mm/s, 52.5 T) and (0.36 mm/s, −1.0 mm/s, 24.0 T). The different $A$ values for the two ferric centers indicate that the unpaired spin is delocalized onto both iron centers, but to different extents. Thus this intermediate must be formulated not as $Fe^{III}$-X-$Fe^{IV}$ but as its valence tautomer $Fe^{III}$-X··-$Fe^{III}$ with its oxidizing equivalent stored at a ligand (proposed in the scheme as a bridging ligand). The precise description for species X remains to be elaborated.

The kinetic scheme shows that species X is formed concomitant with the tyrosyl radical when substoichiometric $Fe^{II}$ is used in the assembly reaction. These two EPR species may be sufficiently close in space as to engender a dipolar coupling interaction. Such a dipolar interaction is manifested by EPR signal line broadening and observed only in the substoichiometric experiments, thereby providing further validation for the proposed kinetic scheme.

As further support for a common oxygen activation mechanism for the two enzymes, it has been observed that the F208Y mutant of R2 may function as a hydroxylase (Ormö et al., 1992; Åberg et al., 1993; Ling et al., 1994). F208 is a residue near the diiron site; when mutated to tyrosine, this residue appears to be hydroxylated to DOPA in the course of the oxidation of the diferrous protein. The protein isolated from F208Y mutant cells has an intense blue color which is associated with a visible absorption maximum near 700 nm, corresponding to a catecholate-to-$Fe^{III}$ charge transfer band. The replacement of F208 by the more electron-rich tyrosine residue may provide a suitable target for the active oxygen species in the R2 active site and give rise to the hydroxylation chemistry implied by the spectroscopic results. This is an exciting observation and serves to emphasize the strong similarities of the oxygen chemistry that occurs in the R2 and MMO active sites.

## Reactivity Models

Insights into the oxidative mechanisms of RNR R2 and MMO have also been gleaned from synthetic complexes that mimic the reactivity observed and permit the characterization of proposed reaction intermediates. A bis(μ-carboxylato)diferrous complex [Fe$_2$(TPA)$_2$(OAc)$_2$] (BPh$_4$)$_2$ (Ménage et al., 1992) which models the putative diiron core of R2$_{red}$ has been crystallographically characterized (Figure 5-10A). Exposure of this complex to O$_2$ elicits an immediate reaction affording a (μ-

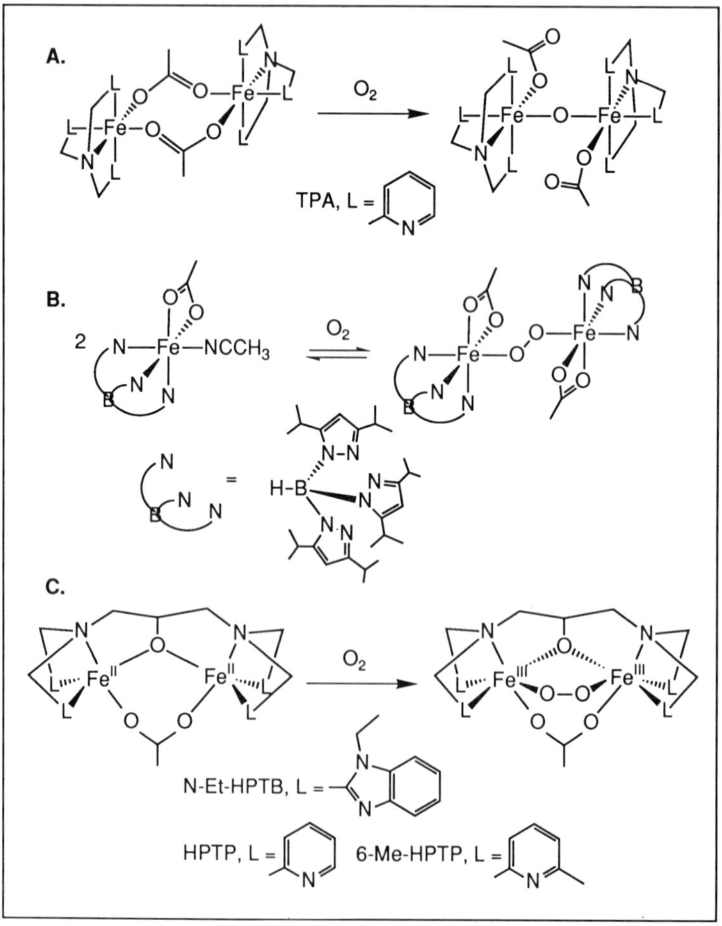

FIGURE 5-10 Model chemistry related to RNR R2.

oxo)diferric complex wherein the bridging carboxylates of the precursor have become terminal monodentate ligands. This carboxylate shift (Rardin et al., 1991) is precisely that ascribed to Glu 238 in the oxidation of R2$_{red}$ (see Figure 5-7). Unfortunately, no oxygenated intermediate could be detected even at low temperatures.

Models capable of dioxygen binding have been reported; in all four cases, adduct formation appears to be facilitated by the presence of available coordination sites. [Fe{HB(3,5-i-Pr$_2$-pz)$_3$}(OBz)(CH$_3$CN)] (Kitajima et al., 1990), a six-coordinate mononuclear complex, binds O$_2$ reversibly in toluene at $-20\,°C$, presumably via the loss of CH$_3$CN (Figure 5-10B). The adduct exhibits a visible absorption maximum at 679 nm and is proposed to be a (μ-peroxo)diiron(III) complex on the basis of its 2:1 Fe:O$_2$ stoichiometry and the Raman $\nu_{O-O}$ at 876 cm$^{-1}$. The other complexes, [Fe$_2$(N-Et-HPTB)(OBz)]$^{2+}$, [Fe$_2$(HPTP)OBz]$^{2+}$, and [Fe$_2$(6-Me-HPTP)OBz(H$_2$O)]$^{2+}$, are dinuclear, with two iron centers bridged by an alkoxide and a carboxylate (Ménage et al., 1990; Hayashi et al., 1992; Dong et al., 1993), each center having a vacant site or one with an easily displaced solvent molecule. [Fe(N-Et-HPTB)OBz]$^{2+}$ binds O$_2$ irreversibly in CH$_2$Cl$_2$ at $-60\,°C$ to form an adduct that exhibits a visible absorption maximum at 588 nm ($\varepsilon$ = 1500 M$^{-1}$cm$^{-1}$) and a Raman $\nu_{O-O}$ at 900 cm$^{-1}$. The carboxylate remains coordinated in the adduct as the $\lambda_{max}$ blue shifts to 570 nm, as expected, when the more basic propionate replaces benzoate. The accumulated spectroscopic evidence suggests that O$_2$ binds as peroxide in a symmetric μ-1,2 mode (Figure 5-10C). A similar complex is formed for [Fe$_2$(6-Me-HPTP)OBz(H$_2$O)]$^{2+}$, but the presence of the methyl groups on the pendant pyridines renders O$_2$ binding reversible (Hayashi et al., 1992).

The proposed mechanism in Figure 5-9 implicates a nonheme iron-oxo species derived from the heterolytic cleavage of the intermediate peroxide. Of great interest are questions regarding the electronic structure of such a species relative to its heme counterpart and the stability of such a species in the absence of a porphyrin ligand. Treatment of [Fe$_2$O(TPA)$_2$(OH)(H$_2$O)](ClO$_4$)$_3$ with H$_2$O$_2$ in CH$_3$CN at $-40\,°C$ affords a metastable high valent nonheme iron-oxo intermediate (Leising et al., 1991; Dong et al., 1995). Like [Fe$_2$O(TPA)$_2$(OAc)](ClO$_4$)$_3$ (Leising et al., 1993), [Fe$_2$O(TPA)$_2$(OH)(H$_2$O)](ClO$_4$)$_3$ catalyzes the hydroxylation of cyclohexane by t-BuOOH (Leising et al., 1991a). With H$_2$O$_2$, the reaction is much slower, but a transient green color appears during the course of this reaction. On cooling to $-40\,°C$, the green intermediate ($\lambda_{max}$ 614 nm) can be maintained for over 2 hours. This species reacts with Ph$_3$P, losing its color immediately and quantitatively forming the corresponding phosphine oxide.

The green intermediate exhibits a resonance Raman feature at 666 cm$^{-1}$ which shifts to 638 cm$^{-1}$ when H$_2$$^{18}$O is added to the reaction mixture and is assigned to the $\nu_{Fe-O}$ of an iron–oxo complex (Leising et al., 1991a). Furthermore, it exhibits an S = 3/2 EPR spectrum with $g$ values at 4.5, 3.9, and 2 and Mössbauer parameters consistent with Fe$^{IV}$. On this basis, the green intermediate was originally formulated as [(TPA$^{+\cdot}$)Fe$^{IV}$ = O], but the recent isolation of the corresponding 5-Me-TPA derivative as a solid has allowed a better characterization of this intermediate (Dong et al., 1995). The green intermediate is now formulated as [Fe$_2$(μ-O)$_2$(TPA)$_2$]$^{3+}$, a species in the same oxidation state as species C in Figure 5-9. The existence of this nonheme species demonstrates that ligands without the extended π system of the porphyrin can access these high-valent oxidation states. Efforts to characterize the relevance of this intermediate to the mechanism of oxygen activation by diiron active sites are ongoing.

## ENZYMES WITH MONONUCLEAR IRON SITES

Unlike the diiron enzymes, in which the electrons required for O$_2$ reduction to the peroxide level can be stored within the diiron unit, the mononuclear nonheme iron enzymes need a slightly different mechanism. For cytochrome P-450, the second electron required to generate the ferric peroxide intermediate is delivered by an iron-sulfur protein. The analogous nonheme iron enzymes are represented by putidamonooxin (Bernhardt and Kuthan, 1981; Bill et al., 1985; Twilfer et al., 1985) and phthalate dioxygenase (Batie et al., 1987; Gassner et al., 1993). Both enzymes contain a mononuclear nonheme iron center, where O$_2$ binding is proposed to occur, and a Rieske-type Fe$_2$S$_2$ cluster which delivers the second electron. The Rieske-type Fe$_2$S$_2$ center differs from Fe$_2$S$_2$ clusters found in ferredoxins in having two terminal cysteines on one iron and two terminal histidines on the other (Gurbiel et al., 1991). Putidamonooxin functions as an O-demethylase in bacterial metabolism, (reaction a, Figure 5-11) but has also been shown to convert an alkene into a diol (reaction b, Figure 5-11) (Wende et al., 1982). Phthalate dioxygenase catalyzes the *cis*-dihydroxylation of the phthalate C4–C5 double bond in the initial step of bacterial phthalate catabolism (reaction c, Figure 5-11). The O-demethylation reaction should be similar in mechanism to N-demethylations by cytochrome P-450; however, little is known about the mechanism of the dihydroxylation reactions. This is an area that clearly deserves attention.

Other mononuclear iron enzymes use cofactors or substrates to

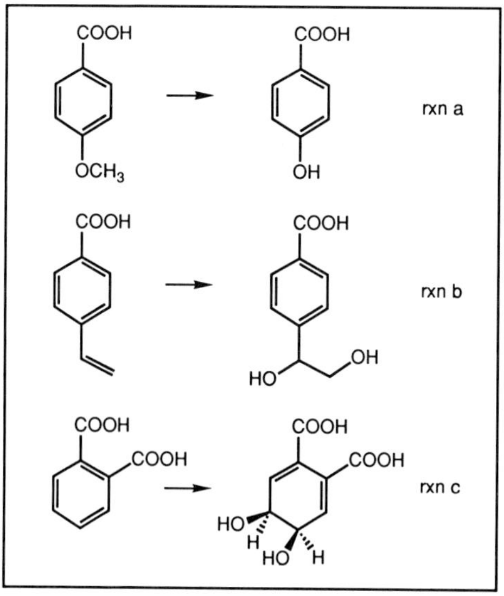

FIGURE 5-11  Reactions catalyzed by putidamonooxin and phthalate dioxygenase.

supply the electrons needed to reduce dioxygen to the peroxide oxidation level. These are discussed below.

## Cofactor as Electron Source

An alternative strategy for dioxygen activation is to have an organic cofactor provide the two electrons needed to reduce $O_2$ to the peroxide oxidation level. For phenylalanine and tyrosine hydroxylase, a tetrahydropterin cofactor is oxidized to dihydropterin in the course of the ring hydroxylation reaction via a hydroperoxypterin moiety (Lazarus et al., 1982; Dix and Benkovic, 1988). For prolyl hydroxylase, an α-ketoacid cofactor is oxidatively decarboxylated to generate a species at the oxidation level of a peracid (Abbott and Udenfriend, 1974; Townsend and Basak, 1991). These peroxo moieties can then interact with the metal center to facilitate cleavage of the O—O bond and generate the oxidant responsible for the hydroxylations. The bioinorganic mechanisms of these cofactor-dependent enzymes are not well understood at present because of the limited supply of protein available for spectroscopic study; however, the application of molecular biology methods is likely to alleviate this shortage in the near future.

The pterin-dependent hydroxylases are involved in the biosynthesis of neurotransmitters. Despite the limited quantities available, several aspects of the iron chemistry have been explored. It appears certain that $Fe^{II}$ is the catalytically important oxidation state (Wallick et al., 1984), but the $Fe^{III}$ state may play a role in feedback regulation. Under certain circumstances, tyrosine hydroxylase is isolated complexed with noradrenaline or epinephrine (Andersson et al., 1988, 1992), which are metabolites of DOPA and act as feedback inhibitors of the enzyme. The complexation of these catecholamines engenders a characteristic blue-green color due to catecholate-to-iron(III) charge transfer transitions. A similar chromophore is generated when catechols are added to phenylalanine hydroxylase in the $Fe^{III}$ state (Cox et al., 1988). The energies of these catecholate (LMCT) bands suggest an iron site of relatively high Lewis acidity with a maximum of one anionic ligand, probably carboxylate. Crystals of the rat liver phenylalanine hydroxylase have very recently been found to diffract to at least 2.4 Å (Celikel et al., 1991), so a protein structure for this subclass of enzyme may be anticipated.

The mechanism presumably entails the formation of an intermediate hydroperoxypterin species (Figure 5-12), the O—O bond of which is cleaved by the $Fe^{II}$ center to generate a species capable of hydroxylating the aromatic ring (Figure 5-12). There is some evidence suggesting that phenylalanine hydroxylase cleaves peroxides primarily by a homolytic mechanism (Benkovic et al., 1986); however, the hydroxyl radical that would result from homolysis would not be expected to engender the NIH shift (hydrogen shift from C-4 to C-3 in the product tyrosine; Figure 5-13) that is characteristic of this enzyme (Guroff et al., 1967). The NIH shift is proposed to result from the attack of an $Fe^{IV}=O$ species derived from heterolysis of the 4a-OOH adduct. The ability of the enzyme to epoxidize 2,5-dihydroxyphenylalanine (Dix and Benkovic, 1988) is also more consistent with a heterolytic mechanism.

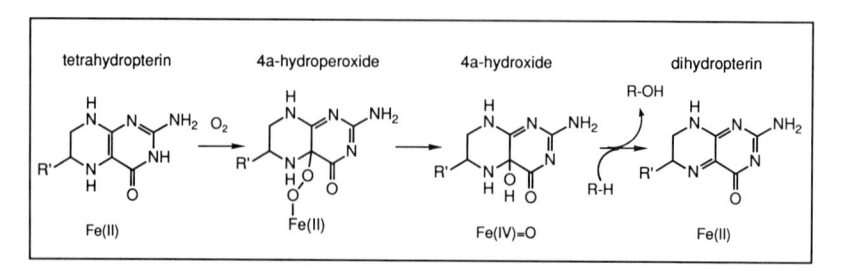

**FIGURE 5-12** Proposed mechanism of action for the pterin-dependent iron hydroxylases.

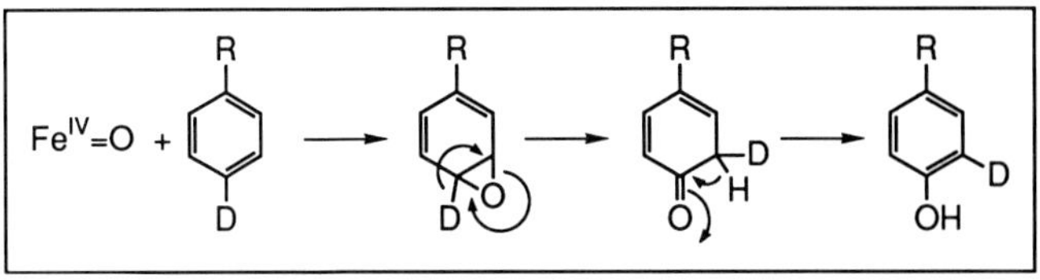

FIGURE 5-13 The NIH shift in the hydroxylation of phenylalanine.

Sorting out these alternative possibilities will require further enzyme studies and biomimetic efforts.

The α-ketoacid-dependent enzymes (Abbott and Udenfriend, 1974; Townsend and Basak, 1991) catalyze the functionalization of an unactivated C—H bond concomitant with the oxidative decarboxylation of a ketoacid in a variety of metabolic reactions (Figure 5-14). For the

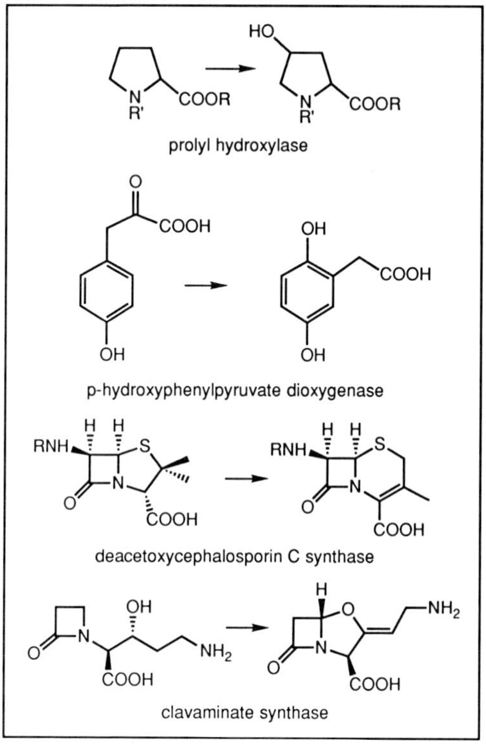

FIGURE 5-14 Some reactions catalyzed by α-ketoacid-dependent enzymes.

hydroxylation reactions, one atom of dioxygen is incorporated into the product, while the other ends up on the carboxylate derived from the ketoacid, i.e.,

$$R\text{-}H + R'COCOOH + O_2 \rightarrow R\text{-}OH + R'COOH + CO_2$$

Examples of such enzymes include those that hydroxylate prolyl and lysyl residues in procollagen (Abbott and Udenfriend, 1974) and aspartyl residues in vitamin K-dependent proteins (Stenflo et al., 1989); such enzymes require α-ketoglutarate as cofactor. However, for *p*-hydroxyphenylpyruvate dioxygenase, the substrate also possesses the keto acid function in an intramolecular variant of the reaction (second reaction of Figure 5-14); in addition, the transformation also entails a 1,2-alkyl shift. α-Ketoacid-dependent enzymes also catalyze reactions wherein a C—H bond becomes functionalized by a heteroatom in the same molecule (reactions 3 and 4 in Figure 5-14) (Townsend and Basak, 1991). In these cases, dioxygen is not incorporated into product; thus deacetoxycephalosporin C synthase and clavaminate synthase are technically oxidases, not oxygenases.

In all of these reactions, $Fe^{II}$ appears to be required (Lindstedt and Rundgren, 1982; Günzler et al., 1986), but little is known of its coordination environment or its role in catalysis. The iron sites of the α-ketoacid-dependent oxygenases are fertile ground for spectroscopic study. Only the iron site in *p*-hydroxyphenylpyruvate dioxygenase has some spectroscopic characterization. As isolated, the enzyme has a deep blue color ($\lambda_{max}$ 595 nm) (Lindstedt and Rundgren, 1982), which is associated with a tyrosinate-to-$Fe^{III}$ charge transfer transition (Bradley et al., 1986). Reduction of the iron center bleaches the color and activates the enzyme. For mechanistic studies, it will be useful to learn about the other endogenous ligands, as well as how the ketoacid cofactor interacts with the metal center.

Some insight into the latter has been provided by the characterization of the first synthetic $Fe^{II}$–ketoacid complexes (Chiou and Que, 1995) Using the tetradentate tripodal ligands TLA and TPA, Chiou and Que synthesized ternary Fe(L)(benzoylformate) complexes. The TLA complex possesses a ketoacid ligand coordinated via a carboxylate and the keto oxygen, while the TPA complex has a monodentate benzoylformate coordinated via the carboxylate. Both react with $O_2$ to afford nearly quantitative yields of benzoate, demonstrating that oxidative decarboxylation occurs. The oxidizing equivalents generated can be trapped only by 2,4-di-*t*-butylphenol thus far. A plausible reaction sequence is shown in Figure 5-15. The model system may be useful for ascertaining

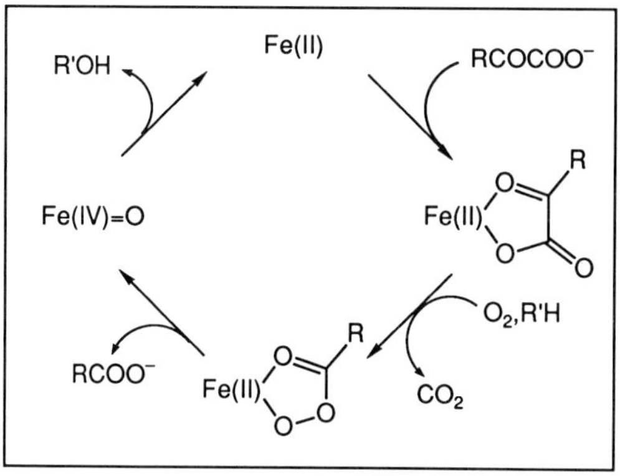

**FIGURE 5-15** Proposed mechanism for the α-ketoacid-dependent hydroxylases.

the involvement of a peracid and an $Fe^{IV}=O$ species. Support for the latter iron-oxo species comes from the sulfoxidation of 3-[(4-hydroxyphenyl)thio]pyruvate when treated with *p*-hydroxyphenylpyruvate dioxygenase (Pascal et al., 1985).

## Substrate as Electron Source

Another subset of mononuclear iron enzymes utilizes the substrate to provide the two electrons needed to reduce $O_2$ to peroxide. The enzymes are exemplified by the catechol dioxygenases and isopenicillin N synthase (IPNS). Substrates in these cases undergo $4e^-$ oxidations, affording ring-opened products in the case of the catechol dioxygenases (Figure 5-16) and converting C—H bonds into C—N and C—S bonds to assemble the penicillin nucleus in the case of IPNS (see Figure 5-21).

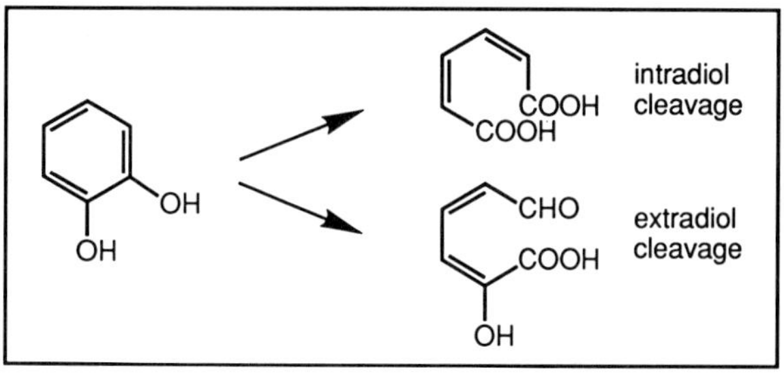

**FIGURE 5-16** Modes of catechol cleavage.

*Catechol dioxygenases.* The catechol dioxygenases serve as part of nature's strategy for degrading aromatic molecules in the environment (Gibson, 1984); they are found in soil bacteria and act in the last step of transforming aromatic precursors into aliphatic products. The intradiol-cleaving enzymes utilize Fe$^{III}$, while the extradiol-cleaving enzymes utilize Fe$^{II}$ (and Mn$^{II}$ in one case) (Que, 1989; Lipscomb and Orville, 1992).

The intradiol-cleaving enzymes are represented by catechol 1,2-dioxygenase and protocatechuate 3,4-dioxygenase (PCD), both of which are well characterized. The crystal structure of the PCD from *Pseudomonas aeruginosa* (Ohlendorf et al., 1988) reveals a trigonal bipyramidal iron site with Tyr 147β and His 162β as the axial ligands and Tyr 108β and His 160β as the equatorial ligands (Figure 5-17). The third equatorial ligand is assigned to a solvent molecule. The iron active site sits at the end of a cleft with a hydrophobic channel that seems likely to be the substrate-binding pocket. The four endogenous ligands appear to be the only bases present in the active-site pocket.

The iron coordination environment defined by the crystal structure corresponds remarkably well to the active site proposed earlier on the basis of spectroscopic studies (Pyrz et al., 1985). The enzymes exhibit a distinct burgundy red color ($\lambda_{max} \sim 460$ nm) that is associated with tyrosinate-to-Fe$^{III}$ charge transfer transitions. The presence of two distinct tyrosine ligands is indicated by the appearance of two tyrosine $\nu_{CO}$ bands at 1254 and 1266 cm$^{-1}$ in the resonance Raman spectrum, each with a different excitation profile (Que et al., 1980; Que and Epstein, 1981; Siu et al., 1992). The latter $\nu_{CO}$ upshifts to 1272 cm$^{-1}$ on

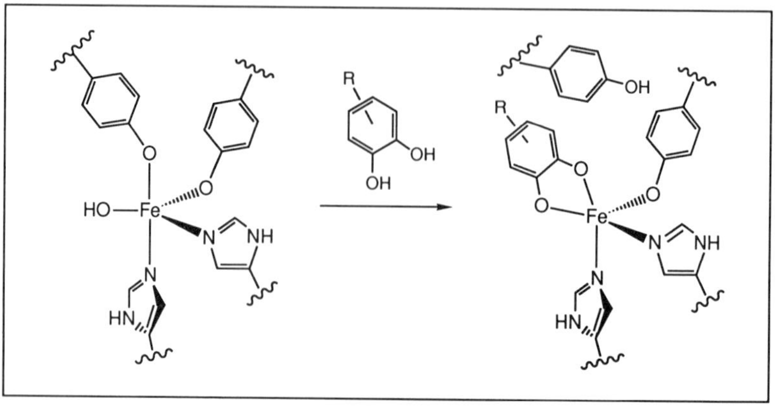

**FIGURE 5-17** The active site of PCD and its enzyme–substrate complex.

H$_2$O/D$_2$O exhange and to ~1290 cm$^{-1}$ on inhibitor binding (Siu et al., 1992). This sensitivity to exogenous ligands has led to its assignment as the more solvent-accessible axial Tyr 146.

The Raman spectrum of PCD also provides evidence for histidine ligation with the observation of a low-energy $v_{Fe-N(Im)}$ feature at 276.5 cm$^{-1}$, which shifts to 274 cm$^{-1}$ in D$_2$O (Felton et al., 1982; Siu et al., 1992). Histidine ligation is also suggested by the presence of second- and third-shell EXAFS features ascribable to imidazole ring atoms at 3 and 4.3 Å from the metal center (Felton et al., 1982). The participation of solvent water is indicated by the line broadening found in the EPR spectrum of the native enzyme from *Brevibacerium fuscum* when dissolved in H$_2$$^{17}$O (Whittaker and Lipscomb, 1984). Subsequent analysis of the first-shell EXAFS data suggests that the solvent water is bound as hydroxide (True et al., 1990). This conclusion is mechanistically significant, as it provides another active-site base to deprotonate incoming substrates or neutral inhibitors such as phenols.

The prime mechanistic question of bioinorganic interest is the role of the iron center. It was clear at the start that the native enzyme had a high-spin ferric center (Fujisawa et al., 1972) and that the enzyme mechanism involved initial substrate binding followed by O$_2$ attack (Nozaki, 1974). An attractive mechanism postulated early in these studies suggested reduction of the Fe$^{III}$ by substrate, followed by dioxygen binding to the Fe$^{II}$. However, later EPR and Mössbauer studies showed that the iron retained its high-spin ferric oxidation state even after substrate bound (Que et al., 1976; Whittaker et al., 1984). Stopped-flow kinetic studies have revealed the involvement of three intermediates subsequent to O$_2$ attack on the ES complex; these intermediates retain their visible chromophores, indicating that the metal center remains Fe$^{III}$ in these species (Bull et al., 1981; Walsh et al., 1983). (Reduction of native enzyme by dithionite results in a colorless solution, which regains its burgundy color on exposure to O$_2$.) These results have led to the postulation of a substrate activation mechanism in which the coordination of catechol to the ferric center activates the catechol for direct attack by O$_2$ (Que et al., 1977). The nature of this substrate activation has been clarified by more recent biomimetic studies (Que et al., 1987; Cox and Que, 1988; Jang et al., 1991), and the mechanism shown in Figure 5-18 has been adapted to incorporate these more recent observations.

The PCD enzyme–substrate complex is clearly a high-spin ferric complex (Que et al., 1976; Whittaker et al., 1984) with a chelated catecholate on the basis of EPR, NMR, and Raman data (Felton et al., 1978; Que and Heistand, 1979; Lauffer and Que, 1982; Orville and

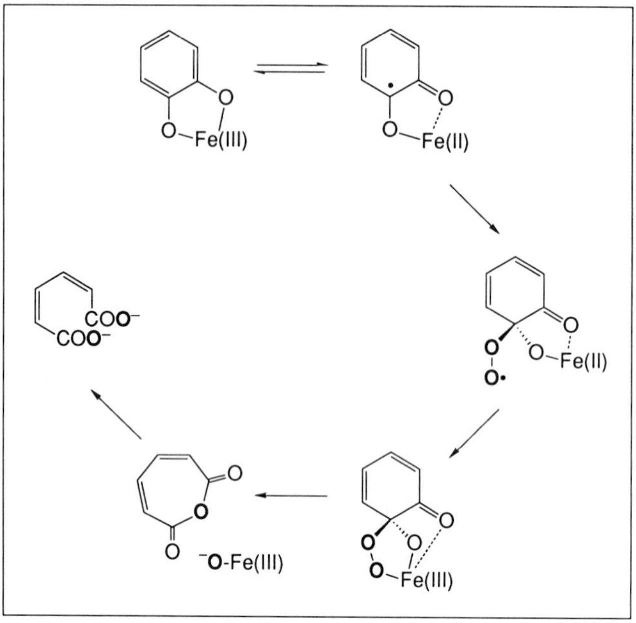

**FIGURE 5-18** Proposed substrate activation mechanism for the intradiol-cleaving catechol dioxygenases.

Lipscomb, 1989), suggesting that both protons are removed when the complex forms. From a mechanistic perspective, the catecholate dianion would certainly be more susceptible to oxidation. Interestingly, the X-ray absorption data on the PCD ES complex indicate that the iron remains 5-coordinate (True et al., 1990). When interpreted together, the spectroscopic data suggest that substrate binding results in the displacement of two ligands found in the native enzyme site. The need to absorb the two protons released from substrate on binding and the absence of bases in the PCD active site save for the iron ligands make the proposed ligand displacement a plausible scenario. Furthermore, the PCD crystal structure shows the substrate-binding plane to be more congruent with the trigonal axis of the bipyramid. Thus, as shown in Figure 5-17, substrate approaches the iron active site along a plane perpendicular to the equatorial plane of the bipyramid and protonates the bound hydroxide and one of the axial ligands. Though the axial His may seem more easily displaced, recent Raman data indicate the loss of the axial tyrosine (T. E. Elgren, unpublished observations). Such an outcome is also preferred on the basis of biomimetic studies (Que et al., 1987; Cox and Que, 1988; Jang et al., 1991)

Que and coworkers synthesized a series of [Fe(L)DBC] complexes in which L is a tetradentate tripodal ligand based on trimethylamine with pendant pyridine, carboxylate, or phenolate ligands (Que et al., 1987; Cox and Que, 1988; Jang et al., 1991). Three of these complexes have been structurally characterized and shown to have a six-coordinate environment with a chelated DBC ligand. All the [Fe(L)DBC] complexes react with $O_2$ to yield the intradiol cleavage product, approaching near-quantitative conversion for the more Lewis acidic iron centers. Furthermore, the rate of reaction accelerates three orders of magnitude from L = nitrilotriacetate to L = tris(2-pyridylmethyl)amine. These results have been rationalized by the argument that increased Lewis acidity of the iron center enhances the covalency of the iron-catecholate bond, and imparts greater semiquinone character to the bound DBC. The enhanced semiquinone character of the bound DBC renders it more prone to oxygen attack and accelerates the rate of oxidative cleavage.

In further support for this hypothesis, Bianchini et al. (Bianchini et al., 1990; Barbaro et al., 1991, 1992) have recently reported that $Rh^{III}$(triphos) and $Ir^{III}$(triphos) catecholate complexes reversibly form adducts with $O_2$. The crystal structure of the Ir adduct reveals that the bidentate catecholate has been transformed into a tridentate peroxy ligand (Figure 5-19), as indicated in the proposed enzyme mechanism. The Ir adduct decomposes to form quinone and $H_2O_2$, while the Rh adduct yields quinone and small amounts of cleavage products.

The extradiol-cleaving catechol dioxygenases represent the more common cleavage pathway, but less is known about these enzymes

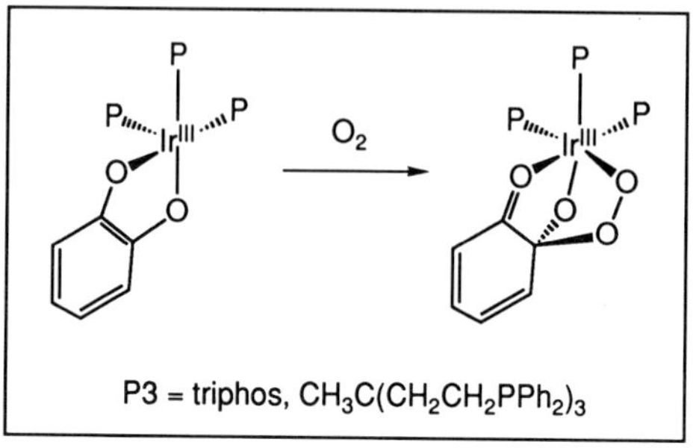

**FIGURE 5-19** Model for the peroxy intermediate in the oxidative cleavage of catechols.

because of the difficulty of probing these $Fe^{II}$ sites spectroscopically. CD and MCD studies on catechol 2,3-dioxygenase suggest that the $Fe^{II}$ site is square pyramidal in both the native state and the enzyme–substrate complex (Mabrouk et al., 1991). From EPR studies, substrate appears to bind in a bidentate fashion (Arciero et al., 1985; Arciero and Lipscomb, 1986; Mabrouk et al., 1991) with the sixth site available for $O_2$ binding. Although there is currently no direct evidence for the identity of the endogenous ligands, the preference of the enzymes for the $Fe^{II}$ oxidation state would seem to eliminate phenolate and thiolate as likely active site ligands: the presence of such ligands would stabilize the $Fe^{III}$ oxidation state and render these sites susceptible to air oxidation. This leaves imidazole and/or carboxylate as the more plausible candidates.

A mechanism based on these observations has been proposed that is analogous to that of the intradiol cleavage mechanism (Figure 5-20) (Lipscomb and Orville, 1992). Substrate chelation should shift the $Fe^{III/II}$ potential sufficiently more negatively to make the iron center bind $O_2$. Subsequently, the bound $O_2$ is postulated to attack a carbon adjacent to

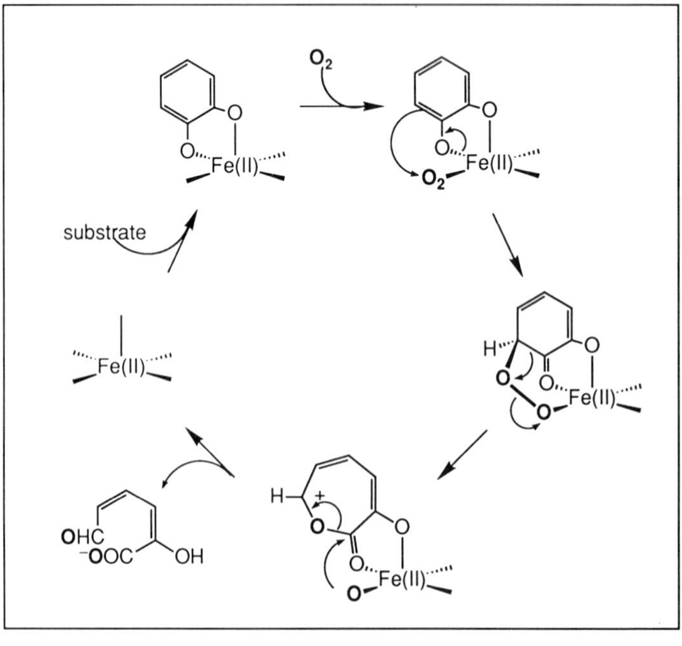

**FIGURE 5-20** Proposed mechanism for the extradiol cleavage of catechol.

one of the carbons bearing a hydroxy group, ensuring the formation of a peroxy intermediate that will decompose in only the desired extradiol fashion. (Attack of $O_2$ on the carbon bearing the hydroxyl group would very likely lead to some product derived from intradiol cleavage.) The precise details of how the peroxy intermediate may form remain to be explored.

*Isopenicillin N synthase.* Isopenicillin N synthase (IPNS) is found in β-lactam antibiotic-producing microorganisms, such as *Cephalosporium*, *Penicillium*, and *Streptomyces*, and catalyzes the formation of isopenicillin N from δ-(L-α-aminoadipoyl)-L-cysteinyl-D-valine (ACV) (Figure 5-21) (Baldwin and Abraham, 1988; Baldwin and Bradley, 1990). Thus the enzyme is technically not an oxygenase, but oxygen activation must nevertheless take place to drive the two ring-forming steps of the reaction.

Important insights into the reaction mechanism have been derived from a substantial effort by Baldwin and coworkers using isotopically labeled substrates and a host of substrate analogs (Baldwin and Abraham, 1988; Baldwin and Bradley, 1990). Based on these studies, a plausible mechanism is proposed in Figure 5-22, which involves thiolate coordination to the iron center, followed by $O_2$ binding and reduction. The reduction of $O_2$ occurs in two stages concomitant with the sequential formation of the β-lactam and the thiazolidine rings. In the first stage, $O_2$ is reduced to peroxide and the Cys thiol is oxidized to thioaldehyde (or some equivalent moiety), followed by nucleophilic attack of the valinyl peptide nitrogen on the thiocarbonyl carbon to

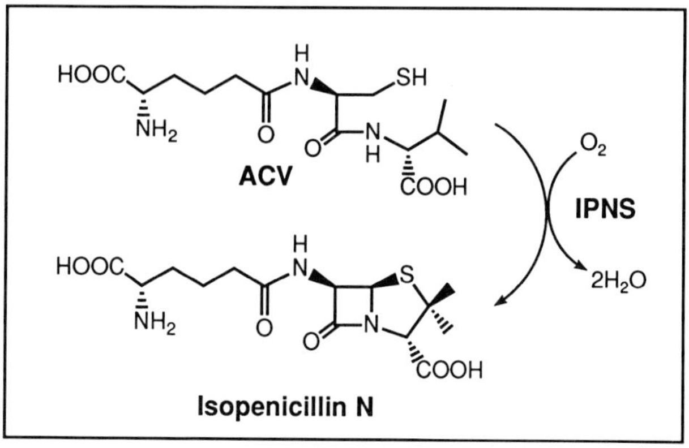

**FIGURE 5-21** Reaction catalyzed by IPNS.

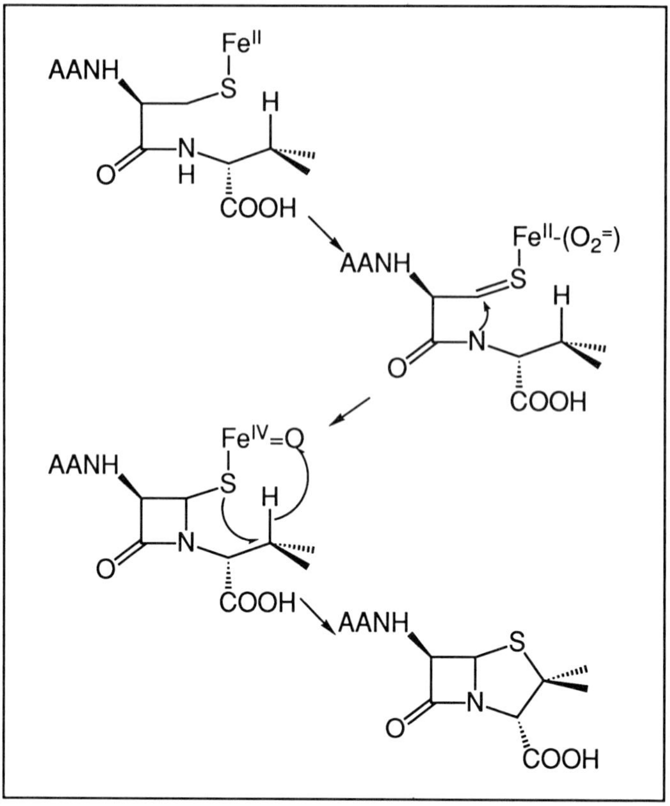

**FIGURE 5-22** Proposed mechanism for IPNS action.

form the β-lactam and regenerating the thiolate. Appropriate modification of the Cys residue can divert the oxidizing equivalents without transforming the Val residue, supporting the notion that the β-lactam ring forms first (Baldwin et al., 1988). For example, substitution of Cys in ACV with β,β-difluorohomocysteine affords the corresponding thiocarboxylic acid, the carboxylate oxygen arising from $O_2$ (Baldwin et al., 1991).

The second stage involves abstraction of the valinyl C3-H by an $Fe^{IV}=O$ species derived from the peroxide heterolysis at the $Fe^{II}$ center and subsequent transfer of the coordinated Cys sulfur to the carbon radical to form the thiazolidine ring. The participation of a carbon radical species is clearly indicated by the rearrangement products observed when β-cyclopropylalanine and allylglycine replace valine in

ACV (Baldwin et al., 1987, 1989). A key intermediate in thiazolidine ring formation is the thiolate coordinated to an iron$^{IV}$-oxo species (Figure 5-22). Reaction of the Val β-H with this species generates a caged radical that collapses to form the thiazolidine ring in a modified rebound mechanism wherein the thiolate sulfur, not the hydroxide, migrates from the metal center to the carbon radical.

Until recently, little was known regarding the nature and role of the iron center. NMR studies of Fe$^{II}$IPNS and Co$^{II}$IPNS have identified four endogenous ligands to the Fe$^{II}$ center: 3 His and 1 Asp (Ming et al., 1990, 1991), analogous to the ligand set found for iron superoxide dismutase (FeSOD) (Stoddard et al., 1990). Like FeSOD, Fe$^{II}$IPNS does not bind O$_2$; however, the coordination of the ACV thiolate to the metal center appears to trigger O$_2$ binding by presumably lowering the Fe$^{III/II}$ redox potential. Three independent observations support the notion that substrate binds via the cysteinyl sulfur. Firstly, the Fe$^{II}$IPNS–ACV–NO displays a pink color that is not observed for Fe$^{II}$IPNS–NO, Fe$^{II}$IPNS–ASerV–NO, or any other Fe$^{II}$–NO complex thus far studied (Chen et al., 1989). Secondly, Cu$^{II}$IPNS develops an intense band at 390 nm on addition of ACV, which has been observed for other Type II copper proteins with S$^-$ ligation and is assigned to RS$^-$-to Cu$^{II}$ charge transfer. EXAFS analyses of the Fe$^{II}$IPNS–ACV and Fe$^{II}$IPNS–ACV–NO complexes indicate a requirement for a sulfur scatterer at ca. 2.3 Å (Fujishima et al., 1991; Scott et al., 1992; Randall et al., 1993).

Support for the proposed thiolate rebound mechanism (Figure 5-22) in the formation of the thiazolidine ring has recently been found in the alkane functionalization chemistry of the [Fe(TPA)X$_2$]$^+$ complexes (Leising et al., 1991b). The stoichiometric formation of cyclohexyl halide was associated with the transfer of halide bound to an iron-oxo center to a caged alkyl radical, demonstrating that oxidative ligand transfer from a high-valent iron-oxo center is a feasible step in the reaction mechanism.

## PERSPECTIVE

In this chapter, I have attempted to assemble the mechanistic ideas for a number of nonheme iron enzymes that activate dioxygen. Whenever available, experimental evidence to support the proposed mechanisms has been included in the discussion; for enzymes that are less well understood, speculative mechanisms have been proposed to stimulate efforts in these areas. The various mechanisms discussed here appear to have a common thread; only the details differ from enzyme to enzyme. As our efforts to understand these enzymes develop and mature, it will be interesting to discover how the various active sites control the metal environment and tailor the iron chemistry to effect the myriad transformations these enzymes catalyze.

## Acknowledgments

My research program on nonheme iron oxygen-activating enzymes has been generously supported by the National Institutes of Health (GM-33162, GM-38767) and the National Science Foundation (DMB-9104669).

## ABBREVIATIONS USED

ACV, δ-(L-α-aminoadipoyl)-L-cysteinyl-D-valine
BIPhMe, 2,2'-Bis(1-methylimidazolyl)phenylmethoxymethane
BLM, bleomycin
BPMP-H, 2,6-bis[bis(2-pyridylmethyl)aminomethyl]-4-methylphenol
DBC · $H_2$, 3,5-di-*tert*-butylcatechol
DOPA, 3,4-dihydroxyphenylalanine
EPR, electron paramagnetic resonance
EXAFS, extended X-ray absorption fine structure
HPTB-H, $N,N,N',N'$-tetrakis(2-benzimidazolylmethyl)-1,3-diamino-2-propanol
HPTP-H, $N,N,N',N'$-tetrakis(2-pyridylmethyl)-1,3-diamino-2-propanol
IPNS, isopenicillin N synthase
LMCT, ligand-to-metal charge transfer
MCD, magnetic circular dichroism
$Me_3$TACN, 1,4,7-trimethyl-1,4,7-triazacyclononane
MMO, methane monooxygenase
NIR, near infrared
NMR, nuclear magnetic resonance
NTA-$H_3$, $N,N$-bis(carboxymethyl)glycine
OAc, acetate
OBz, benzoate
PCD, protocatechuate 3,4-dioxygenase
pz, pyrazole
RNR, ribonucleotide reductase
TACN, 1,4,7-triazacyclononane

TLA, tris[6-methyl-2-pyridyl)methyl]amine
TPA, tris(2-pyridylmethyl)amine
TPP, *meso*-tetraphenylporphin
triphos, tris(diphenylphosphinomethyl)methane

## REFERENCES

ABBOTT, M. T., and UDENFRIEND, S. (1974) α-Ketoglutarate-coupled Dioxygenases, in *Molecular Mechanisms of Oxygen Activation* (O. Hayaishi, Ed.), Academic Press, New York, pp. 167–214.

ÅBERG, A., ORMÖ, M., NORDLUND, P., and SJÖBERG, B. M. (1993). Autocatalytic Generation of Dopa in the Engineered Protein R2 F208Y from *Escherichia coli* Ribonucleotide Reductase and Crystal Structure of the Dopa-208 Protein. *Biochemistry*, 32, 9845–9850.

AHMAD, S., MCCALLUM, J. D., SHIEMKE, A. K., APPELMAN, E. H., LOEHR, T. M., and SANDERS-LOEHR, J. (1988) Raman Spectroscopic Evidence for Side-on Binding of Peroxide Ion to $Fe^{III}$(edta). *Inorg. Chem.*, 27, 2230–2233.

ANDERSSON, K. K., COX, D. D., QUE, L., JR., PETERSSON, L., FLATMARK, T., and HAAVIK, J. (1988) Resonance Raman Studies on the Blue Greencolored Bovine Adrenal Tyrosine 3-Monooxygenase (Tyrosine Hydroxylase). Evidence That the Feedback Inhibitors Adrenaline and Nonadrenaline Are Coordinated to Iron. *J. Biol. Chem.*, 263, 18621–18626.

ANDERSSON, K. K., FROLAND, W. A., LEE., S. -K., and LIPSCOMB, J. D. (1991) Dioxygen Independent Oxygenation of Hydrocarbons by Methane Monooxygenase Hydroxylase Component. *New J. Chem.*, 15, 411–415.

ANDERSSON, K. K., VASSORT, C., BRENNAN, B. A., QUE, L., JR., and WALLICK, D. E. (1992) Purification and Characterization of the Blue-green Rat Phaeochromocytoma (PC12) Tyrosine Hydroxylase with a Dopamine-Fe(III) Complex. *Biochem. J.*, 284, 687–695.

ARCIERO, D. M., and LIPSCOMB, J. D. (1986) Binding of $^{17}O$-labeled Substrate and Inhibitors to Protocatechuate 4,5-Dioxygenase–Nitrosyl Complex. *J. Biol. Chem.*, 261, 2170–2178.

ARCIERO, D. M., ORVILLE, A. M., and LIPSCOMB, J. D. (1985) [$^{17}O$] Water and Nitric Oxide Binding by Protocatechuate 4,5-Dioxygenase and Catechol 2,3-Dioxygenase. *J. Biol. Chem.*, 260, 14035–14044.

ATKIN, C., THELANDER, L., REICHARD, P., and LANG, G. (1973) Iron and Free Radical in Ribonucleotide Reductase. *J. Biol. Chem.*, 248, 7464–7472.

ATTA, M., NORDLUND, P., ÅBERG, A., EKLUND, H., and FONTECAVE, M. (1992) Substitution of Manganese for Iron in Ribonucleotide Reductase from *Escherichia coli*. Spectroscopic and Crystallographic Characterization. *J. Biol. Chem.*, 26, 20682–20688.

BALDWIN, J. E., and ABRAHAM, E. (1988) The Biosynthesis of Penicillins and Cephalosporins. *Natur. Prod. Rep.*, 5, 129–145.

BALDWIN, J. E., ADLINGTON, R. M., BRADLEY, M., PITT, A. R., and TURNER, N. J. (1989) Evidence for Epoxide Formation from Isopenicillin N Synthase. *J. Chem. Soc. Chem. Commun.*, 978–981.

BALDWIN, J. E., ADLINGTON, R. M., DOMAYNE-HAYMAN, B. P., KNIGHT, G., and TING, H.-H. (1987) Use of the Cyclopropylcarbinyl Test to Detect a Radical-like Intermediate in Penicillin Biosynthesis. *J. Chem. Soc. Chem. Commun.*, 1661–1663.

BALDWIN, J. E., and BRADLEY, M. (1990) Isopenicillin N Synthase: Mechanistic Studies. *Chem. Rev.*, 90, 1079–1088.

BALDWIN, J. E., LYNCH, G. P., and SCHOFIELD, C. J. (1991) Isopenicillin N Synthase: A New Mode of Reactivity. *J. Chem. Soc. Chem. Commun.*, 736–738.

BALDWIN, J. E., NORRIS, W. J., FREEMAN, R. T., BRADLEY, M., ADLINGTON, R. M., LONG-FOX, S., and SCHOFIELD, C. J. (1988) γ-Lactam Formation from Tripeptides with Isopenicillin N Synthase. *J. Chem. Soc. Chem. Commun.*, 1128–1130.

BARBARO, P., BIANCHINI, C., LINN, K., MEALLI, C., MELI, A., VIZZA, F., and ZANELLO, P. (1992) Dioxygen Uptake and Transfer by Co(III), Rh(III) and Ir(III) Catecholate Complexes. *Inorg. Chim. Acta*, 198–200, 31–56.

BARBARO, P., BIANCHINI, C., MEALLI, C., and MELI, A. (1991) Synthetic Models for Catechol 1,2-Dioxygenases. Interception of a Metal Catecholate–Dioxygen Adduct. *J. Am. Chem. Soc.*, 113, 3181–3183.

BATIE, C. J., LAHAI, E., and BALLOU, D. P. (1987) Purification and Characterization of Phthalate Oxygenase and Phthalate Oxygenase Reductase from *Pseudomonas cepacia*. *J. Biol. Chem.*, 262, 1510–1518.

BENKOVIC, S. J., BLOOM, L. M., BOLLAG, G., DIX, T. A., GAFFNEY, B., and PEMBER, S. (1986) The Mechanism of Action of Phenylalanine Hydroxylase. *Ann. N.Y. Acad. Sci.*, 471, 226–232.

BERNHARDT, F. H., and KUTHAN, H. (1981) Dioxygen Activation by Putidamonooxin: The Oxygen Species Formed and Released Under Uncoupling Conditions. *Eur. J. Biochem.*, 120, 547–555.

BIANCHINI, C., FREDIANI, P., LASCHI, F., MELI, A., VIZZA, F., and ZANELLO, P. (1990) A Novel Oxygen-carrying and Activating System of Rhodium(III). Oxidation and Oxygenation Reactions of 3,5-Di-*tert*-butylcatechol Catalyzed by a Rhodium(III) Catecholate through Its ($\eta^1$-Superoxo)($\eta^2$-semiquinonato)rhodium(III) Complex. *Inorg. Chem.*, 29, 3402–3409.

BILL, E., BERNHARDT, F.-H., TRAUTWEIN, A. X., and WINKLER, H. (1985) Mössbauer Investigation of the Cofactor Iron of Putidamonooxin. *Eur. J. Biochem.*, 147, 177–182.

BOLLINGER, J. M., EDMONDSON, D. E., HUYNH, B. H., FILLEY, J., NORTON, J., and STUBBE, J. (1991a) Mechanism of Assembly of the Tyrosyl Radical–

Dinuclear Iron Cluster Cofactor of Ribonucleotide Reductase. *Science (Washington, D. C.)*, 253, 292–298.

BOLLINGER, J. M., STUBBE, J., HUYNH, B. H., and EDMONDSON, D. E. (1991b) Novel Diferric Radical Intermediate Responsible for Tyrosyl Radical Formation in Assembly of the Cofactor of Ribonucleotide Reductase. *J. Am. Chem. Soc.*, 113, 6289–6291.

BOLLINGER, J. M., JR., TONG, W. H., RAVI, N., HUYNH, B. H., EDMONSON, D. E., and STUBBE, J. (1994a). Mechanism of Assembly of the Tyrosyl Radical–Diiron(III) Cofactor of *E. coli* Ribonucleotide Reductase. 2. Kinetics of the Excess $Fe^{2+}$ Reaction by Optical, EPR, and Mössbauer Spectroscopies. *J. Am. Chem. Soc.*, 116, 8015–8023.

BOLLINGER, J. M., JR., TONG, W. H., RAVI, N., HUYNH, B. H., EDMONSON, D. E. and STUBBE, J. (1994b). Mechanism of Assembly of the Tyrosyl Radical–Diiron(III) Cofactor of *E. coli* Ribonucleotide Reductase. 3. Kinetics of the Limiting $Fe^{2+}$ Reaction by Optical, EPR, and Mössbauer Spectroscopies. *J. Am. Chem. Soc.*, 116, 8024–8032.

BOROVIK, A. S., HENDRICH, M. P., HOLMAN, T. R., QUE, L., JR., PAPAEFTHYMIOU, V., and MÜNCK, E. (1990) Models for Diferrous Forms of Iron-oxo Proteins. Structure and Properties of $[Fe_2BPMP(O_2CR)_2]BPh_4$ Complexes. *J. Am. Chem. Soc.*, 112, 6031–6038.

BRADLEY, F. C., LINDSTEDT, S., LIPSCOMB, J. D., QUE, L., JR., ROE, A. L., and RUNDGREN, M. (1986) 4-Hydroxyphenylpyruvate Dioxgenase is an Iron-tyrosinate Protein. *J. Biol. Chem.*, 261, 11693–11696.

BROWN, S. J., HUDSON, S. E., STEPHAN, D. W., and MASCHARAK, P. K. (1989) Syntheses, Structures, and Spectral Properties of a Synthetic Analogue of Copper(II)-bleomycin and an Intermediate in the Process of Its Formation. *Inorg. Chem.*, 28, 468–477.

BROWN, S. J., and MASCHARAK, P. K. (1988) Characterization of a Crystalline Synthetic Analogue of Copper(II)-bleomycin. *J. Am. Chem. Soc.*, 110, 1996.

BULL, C., BALLOU, D. P., and OTSUKA, S. (1981) The Reaction of Oxygen with Protocatechuate 3,4-Dioxygenase from *Pseudomonas putida*. *J. Biol. Chem.*, 256, 12681–12686.

BURGER, R. M., BLANCHARD, J. S., HORWITZ, S. B., and PEISACH, J. (1985) The Redox State of Activated Bleomycin. *J. Biol. Chem.*, 260, 15406–15409.

BURGER, R. M., KENT, T. A., HORWITZ, S. B., MÜNCK, E., and PEISACH, J. (1983) Mössbauer Study of Iron Bleomycin and Its Activation Intermediates. *J. Biol. Chem.*, 258, 1559–1564.

BURGER, R. M., PEISACH, J., and HORWITZ, S. B. (1981) Activated Bleomycin. *J. Biol. Chem.*, 256, 11636–11644.

CELIKEL, R., DAVIS, M. D., DAI, X., KAUFMAN, S., and XUONG, N. (1991) Crystallization and Preliminary X-ray Analysis of Phenylalanine Hydroxylase from Rat Liver. *J. Mol. Biol.*, 218, 495–498.

CHAUDHURI, P., WIEGHARDT, K., NUBER, B., and WEISS, J. (1985) $[L_2Fe_2^{II}(\mu$-

CH$_3$CO$_2$)$_2$](ClO$_4$) · H$_2$O, a Model Compound of the Diiron Centers in Deoxyhemerythrin. *Angew. Chem. Int. Ed. Engl.*, 24, 778–779.

CHEN, V. J., FROLIK, C. A., ORVILLE, A. M., HARPEL, M. R., LIPSCOMB, J. D., SURERUS, K. K., and MÜNCK, E. (1989) Spectroscopic Studies of Isopenicillin N Synthase. *J. Biol. Chem.*, 264, 21677–21681.

CHIOU, Y. -M., and QUE, L., JR. (1995) Models for α-Keto Acid-dependent Non-heme Iron Enzymes. Structures and Reactivity of [Fe$^{II}$(L)(O$_2$CCOPh)] (ClO$_4$),pyridyl)methyl]amine](benzoylformate)](ClO$_4$). *J. Am. Chem. Soc.*, 117, 3999–4013.

CLARK, P. E., and WEBB, J. (1981) Mössbauer Spectroscopic Studies of Hemerythrin from *Phascolosoma lurco*. *Biochemistry*, 20, 4628–4632.

COX, D. D., BENKOVIC, S. J., BLOOM, L. M., BRADLEY, F. C., NELSON, M. J., QUE, L., JR., and WALLICK, D. E. (1988) Catecholate LMCT Bands as Probes for the Active Sites of Nonheme Iron Oxygenases. *J. Am. Chem. Soc.*, 110, 2026–2032.

COX, D. D., and QUE, L., JR. (1988) Functional Models for Catechol 1,2-Dioxygenase. The Role of the Iron(III) Center. *J. Am. Chem. Soc.*, 110, 8085–8092.

DALTON, H. (1980) Oxidation of Hydrocarbons by Methane Monooxygenases from a Variety of Microbes. *Adv. Appl. Microbiol.*, 26, 71–87.

DAWSON, J. H. (1988) Probing Structure–Function Relations in Heme-Containing Oxygenases and Peroxidases. *Science*, 240, 433–439.

DAWSON, J. W., GRAY, H. B., HOENIG, H. E., ROSSMAN, G. R., SCHREDDER, J. M., and WANG, R.-H. (1972) A Magnetic Susceptibility Study of Hemerythrin Using an Ultrasensitive Magnetometer. *Biochemistry*, 11, 461–465.

DEWITT, J. G., BENTSEN, J. G., ROSENZWEIG, A. C., HEDMAN, B., GREEN, J., PILKINGTON, S., PAPAEFTHYMIOU, G. C., DALTON, H., HODGSON, K. O., and LIPPARD, S. J. (1991) X-ray Absorption, Mössbauer, and EPR Studies of the Dinuclear Iron Center in the Hydroxylase Component of Methane Monooxygenase. *J. Am. Chem. Soc.*, 113, 9219–9235.

DIX, T. A., and BENKOVIC, S. J. (1988) Mechanism of Oxygen Activation by Pteridine-Dependent Monooxygenases. *Accts. Chem. Res.*, 21, 101–107.

DONG, Y., FUJII, H., HENDRICH, M. P., LEISING, R. A., PAN, G., RANDALL, C. R., WILKINSON, E. C., ZANG, Y., QUE, L., JR., FOX, B. G., KAUFFMANN, K., and MÜNCK, E. (1995) A High Valent Nonheme Iron Intermediate. Structure and Properties of [Fe$_2$(μ-O)$_2$(5-Me-TPA)$_2$](ClO$_4$)$_3$. *J. Am. Chem. Soc.*, 117, 2778–2792.

DONG, Y., MÉNAGE, S., BRENNAN, B. A., ELGREN, T. E., JANG, H. G., PEARCE, L. L., and QUE, L., JR. (1993) Dioxygen Binding to Diferrous Centers. Models for Diiron-oxo Proteins. *J. Am. Chem. Soc.*, 115, 1851–1859.

FELTON, R. H., BARROW, W. L., MAY, S. W., SOWELL, A. L., GOEL, S., BUNKER, G., and STERN, E. A. (1982) EXAFS and Raman Evidence for

Histidine Binding at the Active Site of Protocatechuate 3,4-Dioxygenase. *J. Am. Chem. Soc.*, 22, 6132–6134.

FELTON, R. H., CHEUNG, L. D., PHILLIPS, R. S., and MAY, S. W. (1978) A Resonance Raman Study of Substrate and Inhibitor Binding to Protocatechuate-3,4-Dioxygenase. *Biochem. Biophys. Res. Commun.*, 85, 844–850.

FONTECAVE, M., ELIASSON, R., and REICHARD, P. (1989) Enzymatic Regulation of the Radical Content of the Small Subunit of *E. coli* Ribonucleotide Reductase Involving Reduction of its Redox Centers. *J. Biol. Chem.*, 264, 9164–9170.

FONTECAVE, M., GEREZ, C., ATTA, M., and JEUNET, A. (1990) High Valent Iron Oxo Intermediates Might Be Involved During Activation of Ribonucleotide Reductase: Single Oxygen Atom Donors Generate the Tyrosyl Radical. *Biochem. Biophys. Res. Commun.*, 168, 659–664.

FOX, B. G., FROLAND, W. A., DEGE, J., and LIPSCOMB, J. D. (1989) Methane Monooxygenase from *Methylosinus trichosporium* OB3b. Purification and Properties of a Three-component System with High Specific Activity from a Type II Methanotroph. *J. Biol. Chem.*, 264, 10023–10033.

FOX, B. G., HENDRICH, M. P., SURERUS, K. K., ANDERSSON, K. K., FROLAND, W. A., LIBSCOMB, J. D., and MÜNCK, E. (1993) Mössbauer, EPR, and ENDOR Studies of the Hydroxylase and Reductase Components of Methane Monooxygenase from *Methylosinus trichosporium* OB3b. *J. Am. Chem. Soc.*, 115, 3688–3701.

FOX, B. G., LIU, Y., DEGE, J. E., and LIPSCOMB, J. D. (1991) Complex Formation Between the Protein Components of Methane Monooxygenase from *Methylosinus trichosporium* OB3B. Identification of Sites of Component Interaction. *J. Biol. Chem.*, 266, 540–550.

FOX, B. G., SURERUS, K. K., MÜNCK, E., and LIPSCOMB, J. D. (1988) Evidence for a μ-Oxo-bridged Binuclear Iron Cluster in the Hydroxylase Component of Methane Monooxygenase. Mössbauer and EPR Studies. *J. Biol. Chem.*, 263, 10553–10556.

FUJISAWA, H., UYEDA, M., KOJIMA, Y., NOZAKI, M., and HAYAISHI, O. (1972) Protocatechuate 3,4-Dioxygenase. II. Electron Spin Resonance and Spectral Studies on Interaction of Substrate and Enzyme. *J. Biol. Chem.*, 247, 4414–4421.

FUJISHIMA, Y., SCHOFIELD, C. J., BALDWIN, J. E., CHARNOCK, J. M., and GARNER, C. D. (1991) Recent Physical and Mechanistic Studies on Isopenicillin N Synthase. *J. Inorg. Biochem.*, 43, 564.

GARBETT, K., DARNALL, D. W., KLOTZ, I. M., and WILLIAMS, R. J. P. (1969) Spectroscopy and Structure of Hemerythrin. *Arch. Biochem. Biophys.*, 103, 419–434.

GASSNER, G. T., BALLOU, D. P., LANDRUM, G. A., and WHITTAKER, J. W. (1993) Magnetic Circular Dichroism Studies on the Mononuclear Ferrous

Active Site of Phthalate Dioxygenase from *Pseudomonas cepacia* Show a Change of Ligation State on Substrate Binding. *Biochemistry*, 32, 4820–4825.

GIBSON, D. T. (Ed.) (1984) *Microbial Degradation of Organic Molecules*, Marcel Dekker, New York.

GROVES, J. T. (1985) Key Elements of the Chemistry of Cytochrome P-450. *J. Chem. Ed.*, 62, 928–931.

GÜNZLER, V., MAJAMOA, K., HANAUSKI-ABEL, H. M., and KIVIRIKKO, K. I. (1986) Catalytically Active Ferrous Ions Are Not Released from Prolyl 4-Hydroxylase Under Turnover Conditions. *Biochim. Biophys. Acta*, 873, 38–44.

GURBIEL, R. J., OHNISHI, T., ROBERTSON, D. E., DALDAL, F., and HOFFMAN, B. M. (1991) Q-band ENDOR Spectra of the Rieske Protein from *Rhodobacter capulatus* Ubiquinol-cytochrome *c* Oxidoreductase Show Two Histidines Coordinated to the [2Fe-2S] Cluster. *Biochemistry*, 30, 11579–11584.

GUROFF, G., DALY, J. W., JERINA, D. M., RENSON, J., WITKOP, B., and UDENFRIEND, S. (1967) Hydroxylation-induced Migration: The NIH Shift. *Science*, 157, 1524–1530.

HARTMAN, J. R., RARDIN, R. L., CHAUDHURI, P., POHL, K., WIEGHARDT, K., NUBER, B., WEISS, J., PAPAEFTHYMIOU, G. C., FRANKEL, R. B., and LIPPARD, S. J. (1987) Synthesis and Characterization of (μ-Hydroxo)bis(μ-acetato)diiron(II) and (μ-Oxo)bis(μ-acetato)diiron(III) 1,4,7-Trimethyl-1,4,7-triazacyclononane Complexes as Models for Binuclear Iron Centers in Biology; Properties of the Mixed Valence Diiron(II,III) Species. *J. Am. Chem. Soc.*, 109, 7387–7396.

HAYASHI, Y., SUZUKI, M., UEHARA, A., MIZUTANI, Y., and KITAGAWA, T. (1992) (μ-Alkoxo)diiron(II,II) Complexes of $N,N,N',N'$-tetrakis(2-(6-methylpyridyl)methyl)-1,3-diaminopropane-2-olate and the Reversible Formation of the $O_2$-Adducts. *Chem. Lett.*, 91–94.

HEIMBROOK, D. C., CARR, S. A., MENTZER, M. A., LONG, E. C., and HECHT, S. M. (1987) Mechanism of Oxygenation of *cis*-Stilbene by Iron Bleomycin. *Inorg. Chem.*, 26, 3835–3836.

HENDRICH, M. P., MÜNCK, E., FOX, B. G., and LIPSCOMB, J. K. (1990) Integer-spin EPR Studies of the Fully Reduced Methane Monooxygenase Hydroxylase Component. *J. Am. Chem. Soc.*, 112, 5861–5865.

HENDRICH, M. P., PEARCE, L. L., QUE, L., JR., CHASTEEN, N. D., and DAY, E. P. (1991) Multifield Saturation Magnetization and Multifrequency EPR Measurements of Deoxyhemerythrin Azide. A Unified Picture. *J. Am. Chem. Soc.*, 113, 3039–3044.

HOLMES, M. A., and STENKAMP, R. E. (1991) Structures of Met and Azidomet Hemerythrin at 1.66 Å Resolution. *J. Mol. Biol.*, 220, 723–737.

HOLMES, M. A., TRONG, I. L., TURLEY, S., SIEKER, L. C., and STENKAMP, R. E. (1991) Structures of Deoxy and Oxy Hemerythrin at 2.0 Å Resolution. *J. Mol. Biol.*, 218, 583–593.

JANG, H. G., COX, D. D., and QUE, L., JR. (1991) A Highly Reactive Functional Model for the Catechol Dioxygenases. Structure and Properties of [Fe(TPA)DBC]BPh$_4$. *J. Am. Chem. Soc.*, 13, 9200–9204.

KITAJIMA, N., FUKUI, H., MORO-OKA, Y., MIZUTANI, Y., and KITAGAWA, T. (1990) Synthetic Model for Dioxygen Binding Sites of Non-heme Iron Proteins. X-ray Structure of Fe(OBz)(MeCN)(HB(3,5-iPr$_2$pz)$_3$) and Resonance Raman Evidence for Reversible Formation of a Peroxo Adduct. *J. Am. Chem. Soc.*, 112, 6402–6403.

KURTZ, D. M., JR. (1990) Oxo- and Hydroxo-bridged Diiron Complexes: A Chemical Perspective on a Biological Unit. *Chem. Rev.*, 90, 585–606.

LARSSON, A., and SJÖBERG, B.-M. (1986) Identification of the Stable Free Radical Tyrosine Residue in Ribonucleotide Reductase. *EMBO J.*, 5, 2037–2040.

LAUFFER, R. B., and QUE, L., JR. (1982) $^1$H NMR and $^2$H NMR Studies of the Catechol Dioxygenases. *J. Am. Chem. Soc.*, 104, 7324–7325.

LAZARUS, R. A., DEBROSSE, C. W., and BENKOVIC, S. J. (1982) Phenylalanine Hydroxylase: Structural Determination of the Tetrahydropterin Intermediates by $^{13}$C NMR Spectroscopy. *J. Am. Chem. Soc.*, 104, 6869–6871.

LEE, S.-K., FOX, B. G., FROLAND, W. A., LIPSCOMB, J. D., and MÜNCK, E. (1993a). A Transient Intermediate of the Methane Monooxygenase Catalytic Cycle Containing an Fe$^{IV}$Fe$^{IV}$ Cluster. *J. Am. Chem. Soc.*, 115, 6450–6451.

LEE, S.-K., NESHEIM, J. C., and LIPSCOMB, J. D. (1993b). Transient Intermediates of the Methane Monooxygenase Catalytic Cycle. *J. Biol. Chem.*, 268, 21569–21577.

LEISING, R. A., BRENNAN, B. A., QUE, L., JR., FOX, B. G., and MÜNCK, E. (1991a) Models for Non-heme Iron Oxygenases: A High-valent Iron-oxo Intermediate. *J. Am. Chem. Soc.*, 113, 3988–3990.

LEISING, R. A., ZANG, Y., and QUE, L., JR. (1991b) Oxidative Ligand Transfer to Alkanes: A Model for Iron-mediated C—X Bond Formation in β-Lactam Antibiotic Biosynthesis. *J. Am. Chem. Soc.*, 113, 8555–8557.

LEISING, R. A., KIM, J., PEREZ, M. A., and QUE, L., JR. (1993). Alkane Functionalization at (μ-Oxo)diiron(III) Centers. *J. Am. Chem. Soc.*, 115, 9524–9530.

LINDSTEDT, S., and RUNDGREN, M. (1982) Blue Color, Metal Content, and Substrate Binding in 4-Hydroxyphenylpyruvate Dioxygenase from *Pseudomonas* sp. Strain P. J. 874. *J. Biol. Chem.*, 257, 11922–11931.

LING, J., SAHLIN, M. et al. (1994). Dioxygen is the Source of the μ-Oxo Bridge in Iron Ribonucleotide Reductase. *J. Biol. Chem.*, 269, 5595–5601.

LIPSCOMB, J. D., and ORVILLE, A. M. (1992) Mechanistic Aspects of Dihydroxybenzoate Dioxygenases. *Metal Ions Biol. Syst.*, 28, 243–298.

LIU, K. E., WANG, D., HUYNH, B. H., EDMONDSON, D. E., SALIFOGLOU, A., and LIPPARD, S. J. (1994). Spectroscopic Detection of Intermediates in the Reaction of Dioxygen with the Reduced Methane Monooxygenase

Hydroxylase from *Methylococcus capsulatus* (Bath). *J. Am. Chem. Soc.*, 116, 7465–7466.

LYNCH, J. B., JUAREZ-GARCIA, C., MÜNCK, E., and QUE, L., JR. (1989) Mössbauer and EPR Studies of the Binuclear Iron Center in Ribonucleotide Reductase from *E. coli*. A New Iron-to-Protein Stoichiometry. *J. Biol. Chem.*, 264, 8091–8096.

MABROUK, P. A., ORVILLE, A. M., LIPSCOMB, J. D., and SOLOMON, E. I. (1991) Variable-temperature Variable-field Magnetic Circular Dichroism Studies of the Fe(II) Active Site in Metapyrocatechase: Implications for the Molecular Mechanism of Extradiol Dioxygenases. *J. Am. Chem. Soc.*, 113, 4053–4061.

MARONEY, M. J., KURTZ, D. M., JR., NOCEK, J. M., PEARCE, L. L., and QUE, L., JR. (1986) $^1$H NMR Probes of the Binuclear Iron Cluster in Hemerythrin. *J. Am. Chem. Soc.*, 108, 6871–6879.

MCCALL, G. H., RABOW, L. E., ASHLEY, G. W., WU, S. H., KOZARICH, J. W., and STUBBE, J. (1992) New Insights into the Mechanism of Base Propenal Formation during Bleomycin-Mediated DNA Degradation. *J. Am. Chem. Soc.*, 114, 4958–4967.

MCCANDLISH, E., MIKSZTAL, A. R., NAPPA, M., SPRENGER, A. G., VALENTINE, S. J., STONG, J. D., and SPIRO, T. G. (1980) Reactions of Superoxide with Iron Porphyrins in Aprotic Solvents. A High Spin Ferric Porphyrin Peroxo Complex. *J. Am. Chem. Soc.*, 102, 4268–4271.

MCCORMICK, J. M., REEM, R. C., FOROUGHI, J., BOLLINGER, J. M., JENSEN, G. M., STEPHENS, P. J., STUBBE, J., and SOLOMON, E. I. (1991) Excited State Spectral Features of the Radical Reduced, Native and Fully Reduced Forms of the Coupled Binuclear Non-heme Iron Center in Ribonucleotide Reductase: Active Site Differences Relative to Hemerythrin. *New J. Chem.*, 15, 439–444.

MCMURRY, T. J., and GROVES, J. T. (1986) Metalloporphyrin Models for Cytochrome P-450, in *Cytochrome P-450. Structure, Mechanism, and Biochemistry* (P. R. Ortiz de Montellano, Ed.), Plenum Press, New York, pp. 1–28.

MÉNAGE, S., BRENNAN, B. A., JUAREZ-GARCIA, C., MÜNCK, E., and QUE, L., JR. (1990) Models for Iron-oxo Proteins: Dioxygen Binding to a Diferrous Complex. *J. Am. Chem. Soc.*, 112, 6423–6425.

MÉNAGE, S., and QUE, L., JR. (1991) (µ-Oxo)(µ-carboxylato)diiron(III) Complexes. Effects of the Terminal Ligands. *New. J. Chem.*, 15, 431–438.

MÉNAGE, S., ZANG, Y., HENDRICH, M. P., and QUE, L., JR. (1992) Structure and Reactivity of a Bis(µ-acetato-$O,O'$)diiron(II) Complex, [Fe$_2$(O$_2$CCH$_3$)$_2$(TPA)$_2$](BPh$_4$)$_2$. A Model for the Diferrous Core of Ribonucleotide Reductase. *J. Am. Chem. Soc.*, 114, 7786–7792.

MING, L.-J., QUE, L., JR., KRIAUCIUNAS, A., FROLIK, C. A., and CHEN, V. J. (1990) Coordination Chemistry of the Metal Binding Site of Isopenicillin N Synthase. *Inorg. Chem.*, 29, 1111–1112.

MING, L.-J., QUE, L., JR., KRIAUCIUNAS, A., FROLIK, C. A., and CHEN, V. J. (1991) NMR Studies of the Active Site of Isopenicillin N Synthase, a Nonheme Iron(II) Enzyme. *Biochemistry*, 30, 11653–11659.

MURUGESAN, N., and HECHT, S. M. (1985) Bleomycin as an Oxene Transferase. Catalytic Oxygen Transfer to Olefins. *J. Am. Chem. Soc.*, 107 493–500.

NATRAJAN, A., HECHT S. M., VAN DER MATREL, G. A., and VAN BOOM, J. H. (1990) Activation of Fe(III)·Bleomycin by 10-Hydroperoxy-8,12-octadecadienoic Acid. *J. Am. Chem. Soc.*, 112, 4532–4538.

NORDLUND, P., SJÖBERG, B.-M., and EKLUND, H. (1990) Three-dimensional Structure of the Free Radical Protein of Ribonucleotide Reductase. *Nature*, 345, 393–398.

NORMAN, R. E., HOLZ, R. C., MENAGE, S., O'CONNOR, C. J., and ZHANG, J. H., QUE, L., JR. (1990a) Structures and Properties of Dibridged (μ-Oxo)diiron Complexes. Effects of the Fe-O-Fe Angle. *Inorg. Chem.*, 29, 4629–4637.

NORMAN, R. E., YAN, S., QUE, L., JR., SANDERS-LOEHR, J., BACKES, G., LING, J., ZHANG, J. H., and O'CONNOR, C. J. (1990b) (μ-Oxo)(μ-carboxylato)diiron(III) Complexes with Distinct Iron Sites. Consequences of the Inequivalence and Its Relevance to Dinuclear Iron-oxo Proteins. *J. Am. Chem. Soc.*, 112, 1554–1562.

NOZAKI, M. (1974) Nonheme Iron Dioxygenases, in *Molecular Mechanisms of Oxygen Activation* (O. Hayaishi, Ed.), Academic Press, New York, pp. 135–165.

OHLENDORF, D. H., LIPSCOMB, J. D., and WEBER, P. C. (1988) Structure and Assembly of Protocatechuate 3,4-Dioxygenase. *Nature (London)*, 336, 403–405.

OKAMURA, M. Y., KLOTZ, I. M., JOHNSON, C. E., WINTER, M. R. C., and WILLIAMS, R. J. P. (1969) The State of Iron in Hemerythrin. A Mössbauer Study. *Biochemistry*, 8, 1951–1958.

OPPENHEIMER, N. J., RODRIGUEZ, L. O., and HECHT, S. M. (1979) Structural Studies of the "Active Complex" of Bleomycin: Assignment of Ligands to the Ferrous Ion in a Ferrous–Bleomycin–Carbon Monoxide Complex. *Proc. Natl. Acad. Sci. U.S.A.*, 76, 5616–5620.

ORMÖ, M., DEMARÉ, F., REGNSTRÖM, K., ABERG, A., SAHLIN, M., LING, J., LOEHR, T., SANDERS-LOEHR, J., and SJÖBERG, B.-M. (1992) Engineering of the Iron Site in Ribonucleotide Reductase to a Self-hydroxylating Monooxygenase. *J. Biol. Chem.*, 267, 8711–8714.

ORTIZ DE MONTELLANO, P. R. (Ed.) (1986) *Cytochrome P-450. Structure, Mechanism and Biochemistry*, Plenum Press, New York.

ORVILLE, A. M., and LIPSCOMB, J. D. (1989) Binding of Isotopically Labeled Substrates, Inhibitors, and Cyanide by Protocatechuate 3,4-Dioxygenase. *J. Biol. Chem.*, 264, 8791–8801.

PADBURY, G., SLIGAR, S. G., LATEQUE, R., and MARNET, L. J. (1988) Ferric

Bleomycin Catalyzed Reduction of 10-Hydroperoxy-8,12-octadecadienoic Acid: Evidence for Homolytic O—O Bond Scission. *Biochemistry*, 27, 7846–7852.

PASCAL, R. A., JR., OLIVER, M. A., and CHEN, Y.-C. J. (1985) Alternate Substrates and Inhibitors of Bacterial 4-Hydroxophenylpyruvate Dioxygenase. *Biochemistry*, 24, 3158–3165.

PETERSSON, L., GRÄSLUND, A., EHRENBERG, A., SJÖBERG, B.-M., and REICHARD, P. (1980) The Iron Center in Ribonucleotide Reductase from *E. coli*. *J. Biol. Chem.*, 255, 6706–6712.

PYRZ, J. W., ROE, A. L., STERN, L. J., and QUE, L., JR. (1985) Model Studies of Iron-tyrosinate Proteins. *J. Am. Chem. Soc.*, 107, 614–620.

QUE, L., JR. (1989) The Catechol Dioxygenases, in *Iron Carriers and Iron Proteins* (T. M. LOEHR, Ed.), VCH, New York, pp. 467–524.

QUE, L., JR., and EPSTEIN, R. M. (1981) Resonance Raman Studies of Protocatechuate 3,4-Dioxygenase–Inhibitor Complexes. *Biochemistry*, 20, 2545.

QUE, L., JR., and HEISTAND, R. H., II. (1979) Resonance Raman Studies of Pyrocatechase. *J. Am. Chem. Soc.*, 101, 2219.

QUE, L., JR., HEISTAND, R. H., II, MAYER, R., and ROE, A. L. (1980) Resonance Raman Studies of Pyrocatechase–Inhibitor Complexes. *Biochemistry*, 19, 2258.

QUE, L., JR., KOLANCZYK, R. C., and WHITE, L. S. (1987) Functional Models for Catechol 1,2-Dioxygenase. Structure, Reactivity, and Mechanism. *J. Am. Chem. Soc.*, 109, 5373–5380.

QUE, L., JR., LIPSCOMB, J. D., MÜNCK, E., and WOOD, J. M. (1977) Protocatechuate 3,4-Dioxygenase: Inhibitor Studies and Mechanistic Implications. *Biochim. Biophys. Acta*, 485, 60–74.

QUE, L., JR., LIPSCOMB, J. D., ZIMMERMANN, R., MÜNCK, E., ORME-JOHNSON, N. R., and ORME-JOHNSON, W. H. (1976) EPR and Mössbauer Studies of Protocatechuate 3,4-Dioxygenase from *Pseudomonas aeruginosa*. *Biochim. Biophys. Acta*, 452, 320–334.

QUE, L., JR., and TRUE, A. E. (1990) Dinuclear Iron- and Manganese-oxo Sites in Biology. *Prog. Inorg. Chem.*, 38, 97–200.

RANDALL, C. R., ZANG, Y., TRUE, A. E., QUE, L., JR., CHARNOCK, J. M., GARNER, C. D., FUJISHIMA, Y., SCHOFIELD, C. J., and BALDWIN, J. E. (1993) X-ray Absorption Studies of the Ferrous Active Site of Isopenicillin N Synthase and Related Model Complexes. *Biochemistry*, 32, 6664–6673.

RARDIN, R. L., TOLMAN, W. B., and LIPPARD, S. J. (1991) Monodentate Carboxylate Complexes and the Carboxylate Shift: Implications for Polymetalloprotein Structure and Function. *New J. Chem.*, 15, 417–430.

RAVI, N., J. M. BOLLINGER, J., HUYNH, B. H., EDMONSON, D. E. and STUBBE, J. (1994). Mechanism of Assembly of the Tyrosyl Radical—Diiron(III) Cofactor of *E. coli* Ribonucleotide Reductase. 1. Mössbauer Characterization of the Diferric Radical Precursor. *J. Am. Chem. Soc.*, 116, 8007–8014.

REEM R. C., MCCORMICK, J. M., RICHARDSON, D. E., DEVLIN, F. J., STEPHENS, P. J., MUSSELMAN, R. L., and SOLOMON, E. I. (1989) Spectroscopic Studies of the Coupled Binuclear Ferric Active Site in Methemerythrins and Oxyhemerythrin: The Electronic Structure of Each Iron Center and the Iron–oxo and Iron–Peroxide Bonds. *J. Am. Chem. Soc.*, 111, 4688–4704.

REEM, R. C., and SOLOMON, E. I. (1987) Spectroscopic Studies of the Binuclear Ferrous Active Site of Deoxyhemerythrin: Coordination Number and Probable Bridging Ligands for the Native and Ligand Bound Forms. *J. Am. Chem. Soc.*, 109, 1216–1226.

REICHARD, P., and EHRENBERG, A. (1983) Ribonucleotide Reductase—A Radical Enzyme. *Science (Washington, D.C.)*, 221, 514–519.

ROSENZWEIG, A. C., FREDERICK, C. A., LIPPARD, S. J., and NORDLUND, P. (1993) Crystal Structure of a Bacterical Non-haem Iron Hydroxylase that Catalyses the Biological Oxidation of Methane. *Nature*, 366, 537–543.

ROSENZWEIG, A. C., and LIPPARD, S. J. (1994) Determining the Structure of a Hydroxylase Enzyme That Catalyzes the Conversion of Methane to Methanol in Methanotrophic Bacteria. *Accts. Chem. Res.*, 27, 229–236.

SAHLIN, M., GRÄSLUND, A., PETERSSON, L., EHRENBERG, A., and SJÖBERG, B.-M. (1989) Reduced Forms of the Iron-containing Small Subunit of Ribonucleotide Reductase from *Escherichia coli*. *Biochemistry*, 28, 2618–2625.

SAHLIN, M., SJÖBERG, B.-M., BACKES, G., LOEHR, T. M., and SANDERS-LOEHR, J. (1990) Activation of the Iron-containing $B_2$ Protein of Ribonucleotide Reductase by Hydrogen Peroxide. *Biochem. Biophys. Res. Commun.*, 167, 813–818.

SAM, J. W., TANG, X.-J., and PEISACH, J. (1994) Electrospray Mass Spectrometry of Iron Bleomycin: Demonstration that Activated Bleomycin is a Ferric Peroxide Complex. *J. Am. Chem. Soc.*, 116, 5250–5256.

SANDERS-LOEHR, J. (1989) Binuclear Iron Proteins, in *Iron Carriers and Iron Proteins* (T. M. Loehr, Ed.), VCH, New York, pp. 373–466.

SANDERS-LOEHR, J., LOEHR, T. M., MAUK, A. G., and GRAY, H. B. (1980) An Electronic Spectroscopic Study of Iron Coordination in Hemerythrin. *J. Am. Chem. Soc.*, 102, 6992–6996.

SANDERS-LOEHR, J., WHEELER, W. D., SHIEMKE, A. K., AVERILL, B. A., and LOEHR, T. M. (1989) Electronic and Raman Spectroscopic Properties of Oxo-bridged Dinuclear Iron Centers in Proteins and Model Compounds. *J. Am. Chem. Soc.*, 111, 8084–8093.

SCARROW, R. C., MARONEY, M. J., PALMER, S. M., ROE, A. L., QUE, L., JR., SALOWE, S. P., and STUBBE, J. (1987) EXAFS Studies of Binuclear Iron Proteins: Hemerythrin and Ribonucleotide Reductase. *J. Am. Chem. Soc.*, 109, 7857–7864.

SCOTT, R. A., WANG, S., EIDSNESS, M. K., KRIAUCIUNAS, A., FROLIK, C. A., and CHEN, V. J. (1992) X-ray Absorption Spectroscopic Studies of the High-spin Iron(II) Active Site of Isopenicillin N Synthase: Evidence for Fe–S Interaction in the Enzyme–Substrate Complex. *Biochemistry*, 31, 4596–4601.

SHERIFF, S., HENRICKSON, W. A., and SMITH, J. L. (1987) Structure of Myohemerythrin in the Azidomet State at 1.7/1.3 Å Resolution. *J. Mol. Biol.*, 197, 273–296.

SHIEMKE, A. J., LOEHR, T. M., and SANDERS-LOEHR, J. (1984) Resonance Raman Study of the μ-Oxo-bridged Binuclear Iron Center in Oxyhemerythrin. *J. Am. Chem. Soc.*, 106, 4951–4956.

SHIEMKE, A. K., LOEHR, T. M., and SANDERS-LOEHR, J. (1986) Resonance Raman Study of Oxyhemerythrin and Hydroxomethemerythrin. Evidence for Hydrogen Bonding of Ligands to the Fe–O–Fe Center. *J. Am. Chem. Soc.*, 108, 2437–2443.

SIU, C.-T., ORVILLE, A. M., LIPSCOMB, J. D., OHLENDORF, D. H., and QUE, L., JR. (1992) Resonance Raman Studies of the Protocatechuate 3,4-Dioxygenase from *Brevibacterium fuscum*. *Biochemistry*, 31, 10443–10448.

SJÖBERG, B.-M., and GRÄSLUND, A. (1983) Ribonucleotide Reductase. *Adv. Inorg. Biochem.*, 5, 87–110.

SJÖBERG, B.-M., LOEHR, T. M., and SANDERS-LOEHR, J. (1982) Raman Spectral Evidence for μ-Oxo Bridge in the Binuclear Iron Center of Ribonucleotide Reductase. *Biochemistry*, 21, 96–102.

SNYDER, B. S., PATTERSON, G. S., ABRAHAMSON, A. J., and HOLM, R. H. (1989) Binuclear Iron System Ferromagnetic in Three Oxidation States: Synthesis, Structures, and Electronic Aspects of Molecular with a $Fe_2(OR)_2$ Bridge Unit Containing Fe(III,III), Fe(III,II), and Fe(II,II). *J. Am. Chem. Soc.*, 111, 5214–5223.

STENFLO, J., HOLME, E., LINDSTEDT, S., CHANDRAMOULI, N., HUANG, L. H. T., TAM, J. P., and MERRIFIELD, R. B. (1989) Hydroxylation of Aspartic Acid in Domains Homologous to the Epidermal Growth Factor Precursor. Precursor is Catalyzed by a 2-Oxo Glutarate-dependent Dioxygenase. *Proc. Natl Acad. Sci. U.S.A.*, 86, 444–447.

STENKAMP, R. E., SIEKER, L. C., JENSEN, L. H., McCALLUM, J. D., and SANDERS-LOEHR, J. (1985) Active Site Structures of Deoxyhemerythrin and Oxyhemerythrin. *Proc. Natl. Acad. Sci. U.S.A.*, 82, 713–716.

STODDARD, B. L., HOWELL, P. L., RINGE, D., and PETSKO, G. N. (1990) The 2.1-Å Resolution Structure of Iron Superoxide Dismutase from *Pseudomonas ovalis*. *Biochemistry*, 1990, 8885–8893.

STUBBE, J. (1990) Ribonucleotide Reductases: Amazing and Confusing. *J. Biol. Chem.*, 265, 5329–5332.

STUBBE, J., and KOZARICH, J. W. (1987) Mechanisms of Bleomycin-induced DNA Degradation. *Chem. Rev.*, 87, 1107–1136.

THELANDER, L. (1973) Physicochemical Characterization of Ribonucleoside Diphosphate Reductase from *E. coli*. *J. Biol. Chem.*, 248, 4591–4601.

THELANDER, L. (1974) Reaction Mechanism of Ribonucleoside Diphosphate Reductase from *Escherichia coli*. *J. Biol. Chem.*, 249, 4858–4862.

TOLMAN, W. B., BINO, A., and LIPPARD, S. J. (1989) Self-assembly and

Dioxygen Reactivity of an Asymmetric, Triply Bridged Diiron(II) Complex with Imidazole Ligands and an Open Coordination Site. *J. Am. Chem. Soc.*, 111, 8522–8523.

TOLMAN, W. B., LIU, S., BENTSEN, J. G., and LIPPARD, S. J. (1991) Models of the Reduced Forms of Polyiron-oxo Proteins: An Asymmetric, Triply Carboxylate Bridged Diiron(II) Complex and Its Reaction with Dioxygen. *J. Am. Chem. Soc.*, 113, 152–164.

TOWNSEND, C. A., and BASAK, A. (1991) Experiments and Speculations on the Role of Oxidative Cyclization Chemistry in Natural Product Biosynthesis. *Tetrahedron*, 47, 2591–2602.

TRUE, A. E., ORVILLE, A. M., PEARCE, L. L., LIPSCOMB, J. C., and QUE, L., JR. (1990) An EXAFS Study of the Interaction of Substrate with the Ferric Active Site of Protocatechuate 3,4-Dioxygenase. *Biochemistry*, 29, 10847–10854.

TWILFER, H., BERNHARDT, F.-H., and GERSONDE, K. (1985) Dioxygen-activating Iron Center in Putidamonooxin. Electron Spin Resonance Investigation of the Nitrosylated Putidamonooxin. *Eur. J. Biochem.*, 147, 171–176.

VON DÖBELN, U., and REICHARD, P. (1976) Binding of Substrates to *Escherichia coli* Ribonucleotide Reductase. *J. Biol. Chem.*, 251, 3616–3622.

WALLICK, D. E., BLOOM, L. M., GAFFNEY, B. J., and BENKOVIC, S. J. (1984) Reductive Activation of Phenylalanine Hydroxylase and Its Effect on the Redox State of the Non-heme Iron. *Biochemistry*, 23, 1295–1302.

WALSH, T. A., BALLOU, D. P., MAYER, R., and QUE, L., JR. (1983) Rapid Reaction Studies on the Oxygenation Reactions of Catechol Dioxygenases. *J. Biol. Chem.*, 258, 14422–14427.

WENDE, P., PFLEGER, K., and BERNHARDT, F.-H. (1982) Dioxygen Activation by Putidamonooxin: Substrate-modulated Reaction of Activated Dioxygen. *Biochem. Biophys. Res. Commun.*, 104, 527–532.

WHITTAKER, J. W., and LIPSCOMB, J. D. (1984) $^{17}$O-water and Cyanide Ligation by the Active Site Iron of Protocatechuate 3,4-Dioxygenase. Evidence for Displaceable Ligands in the Native Enzyme and in Complexes with Inhibitors or Transition State Analogs. *J. Biol. Chem.*, 259, 4487–4495.

WHITTAKER, J. W., LIPSCOMB, J. D., KENT, T. A., and MÜNCK, E. (1984) *Brevibacterium fuscum* Protocatechuate 3,4-Dioxygenase. Purification, Crystallization, and Characterization. *J. Biol. Chem.*, 259, 4466–4475.

WOODLAND, M. P., and DALTON, H. (1984) Purification and Characterization of Component A of the Methane Monooxygenase from *Methylococcus capsulatus* (Bath). *J. Biol. Chem.*, 259, 53–59.

ZHANG, K., STERN, E. A., ELLIS, F., SANDERS-LOEHR, J., and SHIEMKE, A. K. (1988) The Active Site of Hemerythrin as Determined by X-ray Absorption Fine Structure. *Biochemistry*, 27, 7470–7479.

# 6
# The Mechanism of Lipoxygenases

MARK J. NELSON AND STEVEN P. SEITZ

## INTRODUCTION

Lipoxygenases catalyze the biosynthesis of fatty acid hydroperoxides from polyunsaturated fatty acids and fatty acid esters. In mammals the fatty acid hydroperoxides are substrates for the pathways that lead to leukotrienes and lipoxins, potent messengers that are involved in the inflammatory response (Samuelsson et al., 1987; Wasserman et al., 1991). Consequently, lipoxygenase inhibitors have been a major goal of the pharmaceutical industry as potential drugs against, e.g., arthritis and asthma (Batt, 1992; McMillan and Walker, 1992). The fatty acid hydroperoxides themselves may play a role in a variety of phenomena, including cell maturation and the development of atherosclerosis (see Chapter 10) (Schewe and Kühn, 1991). In plants the role of lipoxygenase is less well understood; here the fatty acid hydroperoxide is a substrate for pathways that lead to production of species such as jasmonic and traumatic acids that appear to be involved in events as diverse as development and growth regulation, wound response, and pest resistance (Gardner, 1991; Hildebrand and Grayburn, 1991; Siedow, 1991). The most familiar role of these fatty acid hydroperoxides, however, is as substrates for the production of *cis*-3-hexenol and *trans*-2-hexenal, the species responsible for the odor of new-mown grass.

The key feature of the natural substrates of lipoxygenases, whatever the source, is a (1Z, 4Z)-diene (e.g., C-9 through C-13 of linoleic acid, **2**). The primary substrates appear to be arachidonic acid **1** for mammalian lipoxygenases and linoleic acid **2** for plant lipoxygenases.

Introduction 277

Dioxygen is added to one end of the diene, yielding a 1-hydroperoxy-(2E, 4Z)-diene (3). Under optimal conditions, the reaction is regio- and stereoselective for most lipoxygenases, but the preferred site of dioxygenation varies, depending on the source of the enzyme. For example, distinct lipoxygenases have been isolated that will dioxygenate arachidonic acid selectively at the 5, 8, 11, 12, and 15 positions (Corey, 1987).

Lipoxygenases also catalyze peroxidase chemistry, albeit at much slower rates than the dioxygenase reaction. For example, soybean lipoxygenase will use its 13-hydroperoxy-9,11-octadecadienoic acid product 3 as a substrate, yielding 11-hydroxy-12,13-epoxy-9-octadecenoic acid 4, as well as β-scission products such as the 13-aldehyde 5 (Garssen et al., 1976; Corey and Mehrotra, 1983; Chamulitrat and Mason, 1990; Iwahashi et al., 1991a, 1991b). This reaction may be of physiological importance (e.g., in the production of leukotriene A$_4$), but it is outside the scope of this review.

We assume that the mammalian and plant lipoxygenases function by similar, if not identical, chemical mechanisms. Several arguments support this contention: (1) Both types of enzymes use the same type of substrate (fatty acids containing a 1,4-diene) to make the same type of product (allylic hydroperoxides). (2) All of the lipoxygenases that have been analyzed for metals contain one iron per polypeptide (Veldink and Vliegenthart, 1984; Matsuda et al., 1991; Percival, 1991), and it has been shown that the iron is intimately involved in the catalysis (De Groot et al., 1975; Pistorius et al., 1976). (3) There is substantial homology in the amino acid sequences of the mammalian enzymes and (separately) among the plant enzymes (Shibata et al., 1987; Dixon et al., 1988; Sigal et al., 1988; Funk et al., 1990). Comparison of the sequences of enzymes from the two kingdoms yields substantially more homology than expected from coincidence. Indeed, there are regions of very high similarity conserved among all the known lipoxygenases. One such motif, called the *histidine-rich region*, is part of the active site. It includes the binding site for the nonheme iron prosthetic group (Shibata et al., 1988; Feiters et al., 1990; Steczko and Axelrod, 1992), as well as the site of covalent modification of soybean lipoxygenase by phenylhydrazine, an inhibitor that is oxidized by the active-site ferric ion to yield a highly reactive radical species that labels active-site amino acids (Ramachandran et al., 1992). Although most of the mechanistic data available have been obtained using the plant (in particular, soybean) lipoxygenases, the resulting mechanistic hypotheses presumably apply to both types of enzyme.

Dioxygen in the triplet electronic state, the ground state, does not react directly with alkenes at a significant rate. Consequently, the enzyme must activate either the dioxygen or the fatty acid or both to catalyze the reaction. In the next section, we shall examine the chemical precedents for synthesis of allylic hydroperoxides from dioxygen and alkenes in order to lay a foundation for understanding how lipoxygenase might carry out this reaction. Then we will examine studies of the structure and reactions of the enzyme to see if a consistent picture of the mechanism emerges.

## CHEMICAL PRECEDENTS

The transformation catalyzed by lipoxygenases is mimicked by a number of chemical systems. As stated above, the enzyme may function through activation of the fatty acid, dioxygen, or both. For this reason, we will consider the formation of allylic hydroperoxides from substrates in a variety of states, including the parent alkene and its derived radicals, carbanions, and carbenium ions. In addition, we will attempt not to bias the discussion with a presumption of the form of dioxygen involved in the enzymatic reaction.

## Autoxidation

The autoxidation of fatty acids is an important process in the degradation of natural oils and may be involved in biological aging (Porter, 1986; Dussault, 1995; Walling, 1995). The reaction is thought to be a radical chain process that starts with the removal of H-11 (of, e.g., linoleic acid) to give the delocalized radical 7. The relative stability of this pentadienyl radical explains the ready participation of 1,4-dienes in this reaction. Although delocalization of the double bonds lowers the barrier to rotation, the evidence suggests that the initially formed pentadienyl radical has the s-*trans* conformation, retaining the stereochemistry of the alkenes present in linoleic acid. Consistent with the delocalized nature of 7, dioxygen reacts equally at the termini of the radical giving both 9- and 13-peroxyl radicals. Hydrogen atom abstraction by the peroxyl radicals completes initial product formation and generates the pentadienyl radical necessary for chain propagation. The final products depend on the precise reaction conditions but generally do include the *trans, trans* isomers of the dienes (Porter et al., 1981). This loss of the stereochemical integrity of the alkenes is thought to occur via loss of dioxygen from the peroxyl radical generating isomeric pentadienyl radicals such as **10**.

It is important to note that no activation of dioxygen is required in this reaction. If the lipoxygenase reaction utilizes an analogous mechanism, the role of the enzyme would be to activate the fatty acid and to control the regio- and stereospecificity of the dioxygen addition reaction.

## Singlet Dioxygen

Conversion of ground-state dioxygen to the $^1\Delta_g$ state results in a large increase of reactivity toward organic compounds (Foote and Clennan, 1995). Singlet dioxygen is most commonly generated by dye-sensitized photoirradiation, although chemical routes are also known (e.g., oxidation of $H_2O_2$ by HOCl) (Murray, 1979). 1,4-Dienes are excellent participants in the singlet dioxygen–ene reaction; for example, photooxidation of methyl linoleate generates a mixture of four isomeric hydroperoxyoctadecadienoates (9-$\Delta^{10,12}$, 10-$\Delta^{8,12}$, 12-$\Delta^{9,13}$, and 13-$\Delta^{9,11}$) in 44% yield (Gollnick and Kuhn, 1979).

**Isomeric Products**
( e.g. (12E, 10Z)-9-hydroperoxide)

The singlet dioxygen–ene reaction has been a fertile area for mechanistic study. In addition to a concerted mechanism, formation of perepoxides **11**, dioxetanes **13**, diradicals **14**, zwitterions **15**, and exciplexes **12** as intermediates has been considered. Most of the evidence does point to the formation of some intermediate, with the perepoxide **11** (Orfanopoulos et al., 1990) and exciplex **12** (Gorman et al., 1988) structures being the leading contenders.

Perhaps the most compelling evidence for the existence of an intermediate is the stereochemical dependence of the kinetic isotope effect of the reaction of 2,3-dimethylbutene (Grdina et al., 1979). When the competing $CH_3$ and $CD_3$ groups are *trans*, as in **16**, no isotope effect is observed ($k_H/k_D = 1.04$); however, the isomeric compound **17** having

11    12    13

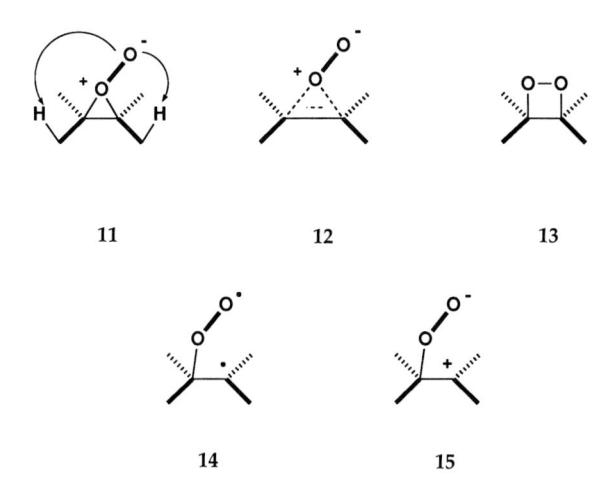

14    15

the competing CH$_3$ and CD$_3$ groups *cis* exhibits a significant isotope effect ($k_H/k_D$ = 1.38). In this case, the oxygen of the perepoxide is poised over one side of the alkene and may attack either of the *cis*-related methyl groups, as shown in structure **11**.

16    17

Theoretical studies have given conflicting views about the nature of the intermediate of this reaction. The first calculations by Dewar at the MINDO/3 level suggested that the perepoxide could be formed exothermically (Dewar and Thiel, 1977). In contrast, GVB-CI calculations favor a diradical intermediate for the reaction (Harding and Goddard, 1980), while an unrestricted Hartree-Fock (STO-3G) study suggests a concerted pathway (Yamaguchi et al., 1981a, 1981b).

To utilize such a mechanism, the enzyme must be able to generate free singlet dioxygen, a species some 23 kcal/mole above the triplet ground state. Further, it would again need to find a way to enforce the regio- and stereochemistry of the reaction. Even though singlet dioxygen is highly reactive, the enzyme still could influence the stereochemical outcome of the reaction by shielding portions of the substrate, as illustrated by the following experiment (Dussault and Hayden, 1992):

## Carbanionic Routes

Simple carbanions react with triplet dioxygen in a radical process to give hydroperoxides. Grignard and organolithium reagents give useful yields of products, particularly with tertiary substrates (Swern, 1979). A novel variant of this scheme has been proposed for the lipoxygenase reaction by Corey (1987). He suggested that the initial abstraction of a proton by the enzyme is facilitated by complexation of the alkene to be oxygenated by $Fe^{3+}$. This process yields the σ-organoiron species 21 as an intermediate. The reaction is then completed by insertion of dioxygen into the C—Fe bond. A conceivable side reaction is formation of fatty acid radicals from **21** by homolysis of the relatively weak Fe—C bond.

Several facets of the organoiron mechanism have been shown to be feasible by model chemistry. The $Fe^{3+}$ species **22** may be formed *in situ* from the reaction of 1-phenylprop-2-enyltributyltin and anhydrous ferric bromide (Corey and Walker, 1987). This organoiron compound reacts rapidly with triplet dioxygen at −78 °C to give phenylvinylketone in 70% yield. Phenylvinylketone is proposed to arise from insertion of dioxygen into the Fe—C bond, followed by a metalloene reaction generating **23**. The ferric–peroxide complex decomposes heterolytically to give the observed product:

Ph—CH=CH—CH₂—FeBr₂ →(O₂) Ph—CH(OOFeBr₂)—CH=CH₂ → Ph—C(O)—CH=CH₂

        **22**                                 **23**                                 **24**

Additional information on the reactivity of Fe—C bonds toward triplet dioxygen has been obtained in the porphyrin series (Arasasingham et al., 1989, 1990). The ethyl complex of [tetra-*p*-tolylporphyrinato (TTP)]Fe$^{III}$ **25** undergoes direct insertion of dioxygen to give the peroxide complex **26**, which was characterized spectroscopically at low temperature. Warming of the peroxide complex yields acetaldehyde as the major organic product.

(TTP)Fe(III)Et —(O₂)→ (TTP)Fe(III)-OOEt → (TTP)Fe(III)-OH + CH₃CHO

    **25**                             **26**                         **27**

Although studies of the insertion of dioxygen into Fe—C bonds thus far have concentrated on σ–organoiron complexes, other coordination states ($\eta^3, \eta^5, \eta^3-\sigma$) are potentially available to iron–diene complexes.

## Nucleophilic Dioxygen Species

A number of hydroperoxide synthetic schemes involve reaction of reduced dioxygen species with electrophilic organic molecules. The most widely applicable method consists of the reaction of hydrogen peroxide or its salts with an organic halide (Hiatt, 1971), in which substrates prone to $S_N1$ reactions may form alkylhydroperoxides under solvolytic conditions. On the other hand, displacement of sulfonate esters by superoxide in an $S_N2$ reaction is also very facile (Lee-Ruff, 1977). Although the ultimate products of the superoxide reaction are usually alcohols, the intermediacy of compounds at the peroxide oxidation level is suggested by examples such as the conversion of dimesylate **28** to the cyclic peroxide **29** (Corey et al., 1975).

$$\underset{\underset{28}{\text{OMs OMs}}}{\bigvee\bigvee}\text{Ph} \quad \xrightarrow[35\%]{KO_2} \quad \underset{29}{\bigvee\underset{O-O}{\bigvee}}\text{Ph}$$

For such a mechanism to obtain with lipoxygenase, the enzyme would be required to activate the substrate at the double bond to be oxygenated, providing a site for nucleophilic attack by peroxide or superoxide. Production of the reduced dioxygen species presumably would be initiated by the active-site iron.

## LIPOXYGENASE-CATALYZED SYNTHESIS OF ALLYLIC HYDROPEROXIDES

### The Structure of the Active Site

Lipoxygenases are members of the class of mononuclear nonheme iron enzymes; that is, they contain a single iron per protein, coordinated primarily to amino acid side chains. The iron is in the ferrous oxidation state in the enzyme as it is isolated from soybeans (Slappendel et al., 1982b; Cheesbrough and Axelrod, 1983). Although still controversial (Wiseman et al., 1988; Wang et al., 1993), it appears that the iron must be oxidized to the ferric form for the enzyme to be active (Schilstra et al., 1992, 1993), and the only oxidant known to yield active ferric soybean lipoxygenase is the fatty acid hydroperoxide product of the dioxygenation reaction. Presumably the lability of 1,4-dienes toward autoxidation leads to contamination of substrate solutions with a small amount of the hydroperoxide to generate the active enzyme.

Very recently, two moderate-resolution (2.6–2.7 Å) crystal structures of ferrous soybean lipoxygenase have been reported (Boyington et al., 1993; Minor et al., 1993). Unfortunately, they are not completely in agreement as to the coordination environment of the iron. Both structures reveal three histidine imidazoles coordinated to the iron, as predicted by site-directed mutagenesis of the "histidine-rich" region of the protein (Nguyen et al., 1991; Steczko and Axelrod, 1992; Steczko et al., 1992; Zhang et al., 1992), as well as by various spectroscopic experiments (see below). A fourth ligand is provided by the carboxylate of the C-terminal isoleucine. The two structures differ as to the existence of a fifth ligand, with Minor et al. proposing the side-chain carboxamide of asparagine 694 as an O-donor ligand to the iron.

Using the structure of the ferrous site as a basis and combining the results of a wide variety of spectroscopic techniques, including X-ray absorption (Navaratnam et al., 1988; Feiters et al., 1990; Van der Heijdt et al., 1992), magnetic circular dichroism (Whittaker and Solomon, 1988; Zhang et al., 1991), electron paramagnetic resonance (Nelson, 1988b), and Mössbauer spectroscopies (Dunham et al., 1990; Funk et al., 1990), a model (**30**) of the structure of the ligand field of the iron in the ferric, active enzyme may be derived. All of the spectroscopic data predict six-coordination for the metal ion in this form. The "?" ligand in **30** could be the asparagine carboxamide seen proximate to the iron in the crystal structure of the ferrous enzyme, or water, for example. Recent extended X-ray absorption fine structure (EXAFS) results are consistent with three imidazoles and three oxygen-donor ligands (e.g., water, hydroxide, carboxylate, carboxamide) (Van der Heijdt et al., 1992), and electron paramagnetic resonance (EPR) data suggest that at least one of these is water or hydroxide (Nelson, 1988b). X-ray absorption spectra of the ferric enzyme support the presence of a hydroxide (Scarrow et al., 1994). Either a bound water (easily generated from the hydroxide) or a carboxamido group would provide a readily displaceable ligand; indeed, the iron in ferric lipoxygenase does form complexes with a variety of exogenous ligands (Nelson, 1988a; Zhang et al., 1991; Gaffney et al., 1993). The ability to react with substrates, intermediates, and products at the metal ion is a critical feature of the mechanisms of other nonheme iron enzymes, in particular the catechol dioxygenases and isopenicillin N synthase (Que et al., 1987; Harpel and Lipscomb, 1990; Orville et al., 1992).

Lipoxygenases are unusual nonheme iron proteins in that they contain neither cysteine nor tyrosine as a ligand to the metal. The largely neutral ligand field effectively stabilizes the ferrous oxidation state, increasing the reduction potential of the ferric ion. Indeed, alkyl hydroxylamines (Clapp et al., 1985), N-alkyl hydroxamic acids, phenols (Nelson et al., 1991), and catechols (Nelson, 1988a) are all effective reductants of ferric lipoxygenase. A detailed study of the reactions of ferric lipoxygenase with a variety of substituted catechols having a range of reducing ability suggested that the reduction potential of the iron is in excess of 0.5 V (with respect to the normal hydrogen electrode) (Nelson, 1988a). This would make the iron site in ferric lipoxygenase among the most strongly oxidizing in nonheme iron enzymes; this fact needs to be considered in any discussion of the mechanism.

A fascinating feature of lipoxygenase, which is likely to be related to its activity, is the existence of a purple form of the enzyme. An absorption at about 350 nm, arising presumably from imidazole to iron charge transfer transitions (Zhang et al., 1991), gives ferric lipoxygenase a faintly yellow color. When treated with an excess of substrates (fatty acid and dioxygen) or product, solutions of the ferric enzyme turn purple as an additional transition at 585 nm appears (de Groot et al., 1975; Spaapen et al., 1979; Slappendel et al., 1983; Feiters et al., 1986). The structural origin of this transition is uncertain, but the high extinction coefficient ($>1000 \text{ M}^{-1}\text{cm}^{-1}$) suggests that the absorption has ligand-to-metal charge transfer character, and the relatively low energy is appropriate for a tyrosine phenoxide, cysteine thiolate, or, most tantalizingly, an alkylperoxide coordinated to $Fe^{3+}$ (Nishida and Akamatsu, 1991; Zang et al., 1993). Each of these possibilities requires some ligand exchange chemistry to occur at the iron in the presence of substrate or product. Kinetic studies of the rate of formation of the purple species indicate that it forms at a rate comparable to that of the dioxygenase reaction; furthermore, the primary deuterium kinetic isotope effect (see below) on the H-11 protons of linoleic acid is reported to be the same for both the formation of the purple color and the dioxygenase reaction (Egmond et al., 1977). These data suggest that the purple species is a mechanistic intermediate in the reaction and allow the speculation that this intermediate is a $Fe^{3+}$–fatty acid peroxide complex.

The iron is, of course, only one of the parts of the enzyme that may be participating in catalysis. Information about amino acid side chains in the active site that play a role in the oxygenation reaction is relatively scant. A methionine has been shown to be oxidized to a methionine sulfoxide during self-inactivation of a reticulocyte lipoxygenase (Rapoport et al., 1984), suggesting that this residue is proximate to the active site. Chemical modification studies have shown that both methionine and cysteine are likely to be present in or near the active site of soybean lipoxygenase (Spaapen et al., 1980; Grossman et al., 1984; Michaud-Soret and Chottard, 1992). Recently, site-directed mutagenesis of a methionine resulted in an enzyme with altered substrate specificity, suggesting a role for that residue in the positioning of the substrate in the active site (Sloane et al., 1991). Definitive information about the active site will have to await a more detailed exposition of the crystal structure.

## Mechanistic Studies of Lipoxygenases

*The anaerobic reaction.* The dioxygenase reaction is a bisubstrate reaction; that is, it requires both the fatty acid diene and dioxygen. Significant information frequently is learned about the mechanism of a bisubstrate enzyme by examining the partial reactions—the reactions of individual substrates with the enzyme. Treatment of ferric lipoxygenase with linoleic acid in the absence of dioxygen leads to reduction of the iron in a reaction that probably occurs rapidly enough to be a kinetically competent step in the oxygenase mechanism (De Groot et al., 1975). Substitution with deuterium at C-11 leads to a substantial kinetic isotope effect on the rate of reduction, comparable to that reported for turnover (see below) (Egmond et al., 1977). This result suggests very strongly that oxidation of the fatty acid substrate by $Fe^{3+}$, which we know to be a strongly oxidizing center, is involved in the dioxygenation reaction.

After the reduction of the iron, some of the oxidized fatty acid can be trapped by spin traps in solution. Analysis of the trapped species showed that the primary products were 9- and 13-adducts of linoleic acid, demonstrating that the radical(s) formed in the reaction had substantial spin density at carbons 9 and 13 (Garssen et al., 1972; de Groot et al., 1973). Removal of the equivalent of a hydrogen atom from C-11 would be expected to yield a pentadienyl radical with spin concentrated on C-9, C-11, and C-13, consistent with the spin-trapped products.

The coordination environment of the iron after the reduction by substrate is somewhat less clear. This is an important point because of the potential for $Fe^{2+}$ (but not $Fe^{3+}$) to bind and activate dioxygen.

The iron in native (as-isolated) ferrous lipoxygenase has not been shown to react with dioxygen (Petersson et al., 1985), although it binds the dioxygen analog nitric oxide tightly (Nelson, 1987). This is not surprising: the high $Fe^{3+}/Fe^{2+}$ reduction potential would preclude significant metal-to-ligand π-backbonding, effectively destabilizing the $Fe-O_2$ bond. Consequently, if the iron in lipoxygenase has a role in activating dioxygen, it may result from a difference between the reactivity of the native ferrous ion and that generated by reduction of the ferric ion by substrate during turnover. Magnetic susceptibility studies suggest a small difference between the native ferrous ion and that obtained by anaerobic reduction with substrate (Petersson et al., 1987). Mössbauer studies also indicate a difference between these two states (Funk et al., 1990), but in that experiment the substrate-reduced sample contained a small amount of methanol, which itself is known to perturb the iron environment (Slappendel et al., 1982a). EXAFS experiments detected no changes between the native and substrate-reduced ferrous states beyond those induced by methanol (Scarrow et al., 1994). At this time there are no convincing data to suggest that the ferrous site obtained via substrate reduction is substantially different from that in the native enzyme.

*Kinetic isotope effects.* A large primary kinetic isotope effect has been observed for the oxygenation reaction, demonstrating that the C—H bond is broken in the rate-determining step of the reaction (Egmond et al., 1973). For both the soybean and human platelet lipoxygenases, the primary isotope effect is seen on substitution of the pro (*S*) proton of the methylene of the diene to be oxygenated, but the corn germ enzyme evinces the isotope effect on substitution of the pro (*R*) proton (Egmond et al., 1972; Hamberg and Hamberg, 1980). Thus the hydrogen abstraction is stereospecific, but different lipoxygenases have different specificities. Very recently, it has been found that the magnitude of the primary isotope effect is actually >30, much larger than originally reported (Glickman et al., 1994; Hwang and Grissom, 1994). The mechanistic explanation for such a large, nonclassical isotope effect is unclear at this time.

A secondary kinetic isotope effect ($k_H/k_T = 1.19$) has been seen for the methylene hydrogen *not* removed during the oxygenation reaction (Brash et al., 1986). This is consistent with a rehybridization from $sp^3$ to $sp^2$ at that carbon during the rate-limiting hydrogen abstraction step and implies delocalization of one or both of the double bonds of the diene over that carbon. A secondary kinetic isotope effect ($k_H/k_T = 1.16$ for soybean lipoxygenase) was observed upon substitution of all of the vinylic protons of arachidonic acid, implying a change in the electronic configuration of one or both double bonds of the diene at or before the rate-determining step (Wiseman, 1989). An isotope effect of this magnitude is consistent with either the oxidation or deprotonation of the substrate to a delocalized radical or carbanion, respectively (Houk et al., 1992; Gajewski et al., 1993).

*Reactions with substrates.* "Normal" lipoxygenase substrates contain a bis-allylic methylene, the site of hydrogen abstraction. Under optimal conditions, the dominant reaction involves stereospecific hydrogen removal from this methylene, followed by regio- and stereospecific dioxygen addition; presumably the regiospecificity is determined by the position of the diene in the active site. Individual lipoxygenases utilize the distance from the carboxylate or the terminal methyl of the fatty acid to determine the site of oxygenation (Hamburg and Samuelsson, 1967; Holman et al., 1969; Veldink et al., 1972; Kuhn et al., 1985; Schewe et al., 1986; Hatanaka et al., 1989; Sloane et al., 1991). So, for example, soybean lipoxygenase at pH 9 removes the 11-pro (S) hydrogen from linoleic acid 2 and yields approximately 95% 13-(S)-hydroperoxide 3 with virtually no 13-(R) product (Hamburg, 1971; Nikolaev et al., 1990). The remainder of the product is approximately a 2:1 mix of 9-(S)- and 9-(R)-hydroperoxide, formed with removal of either the 11-pro (R) or pro (S) hydrogen nonspecifically (Egmond et al., 1972), implying a different mechanism for the synthesis of this minor product. (If it were merely that the 9-hydroperoxide is the result of the linoleic acid binding "backward" in the pocket, one would still expect both the hydrogen removal and dioxygen addition to be stereospecific.) With neutral substrates (i.e., with fatty acid esters or fatty alcohols), soybean lipoxygenase is much less regioselective and the 13-hydroperoxide formed is only about 70% S, while the 9-hydroperoxide is essentially racemic (Hatanaka et al., 1984, 1989). Thus an interaction between the fatty acid carboxylate anion and some group on the protein is important for determining the regio- and stereochemistry. At pH 7, the enzyme is also somewhat less regio- and stereoselective (Funk et al., 1987; Gardner, 1989). One can rationalize these results by hypothesizing a number of competing reactions:

1. A dominant reaction in which substrate binds in a preferred conformation, resulting in removal of the 11-pro (S) hydrogen (of linoleic acid) followed by addition of dioxygen stereospecifically to the activated intermediate to make the 13(S)-hydroperoxide.
2. A minor pathway in which the activated intermediate produced in (1)

adds dioxygen by a different mechanism to make the 9-hydroperoxide without control of the stereochemistry.
3. A second minor pathway in which the substrate binds backward, resulting in removal of the 11-pro (R) proton and synthesis of the 9(S)-hydroperoxide. This effect could be enhanced with neutral substrates and at lower pH.
4. Abortion of the enzyme reaction by dissociation of the activated intermediate from the enzyme into solution, where it could react with dioxygen, ultimately to yield all four isomeric hydroperoxides (9(R, S), 13(R, S)). This last reaction certainly occurs: dissociation of fatty acid radicals from the enzyme *is* observed in the anaerobic reaction (see above), and it has been estimated that a fatty acid radical dissociates from the enzyme once for every 200 dioxygenations under atmospheric dioxygen (Schilstra et al., 1993). One might predict that the intermediates formed from neutral substrates, or by substrates bound backward, would be more prone to dissociation, explaining the relative lack of stereospecificity seen at low pH or with fatty acid ester substrates.

The linoleic acid analogs (9E, 12Z)- and (9Z, 12E)-octadecadienoic acid are slow soybean lipoxygenase substrates (Funk et al., 1987). Both compounds yield a mixture of 9- and 13-hydroperoxide products, which for the (Z, E) isomer show little stereoselectivity in dioxygen addition; thus, the reactions of that isomer are probably not relevant to the dominant enzymatic reaction. For the (E, Z) isomer **31**, however, the major product is the 13-hydroperoxy-(9Z, 11E)-octadecadienoic acid (**32**), produced with 13(S):13(R) of 91:9 at pH 7. The stereoselectivity implies that this product is made by the dominant dioxygen addition pathway, and the isomerization of the 9 double bond from E to Z suggests lowering of the barrier to rotation of that double bond at some point in the mechanism, with sufficient thermodynamic pressure exerted by the enzyme to rotate the bond. [N.B.: the barriers to rotation of allyl and pentadienyl radicals are approximately 17 and 12 kcal/mol (Korth et al., 1981; MacInnes and Walton, 1985). Little isomerization of the intermediate pentadienyl radical is seen in the nonenzymatic production of fatty acid hydroperoxide from linoleic acid in solution (see above).] This result again supports the idea of C—H bond cleavage at C-11, followed by delocalization of the radical over (at least) the 9 double bond.

Another interesting substrate analog is one in which the "spectator" double bond, the one to which dioxygen is not added, is hydrated to form a ketone (**33**) (Wiseman and Nichols, 1988; Kuhn et al., 1991). The methylene hydrogen is still labile, although no longer bis-allylic, and these analogs are indeed substrates for lipoxygenases. In the one case in which the product was identified, the substrate was 12-keto-(9Z)-octadecenoic acid (Kuhn et al., 1991). The free acid is a very poor substrate, but the methyl ester **33** is oxygenated by lipoxygenase at about 10% of the rate of methyl linoleate. The product is methyl 9,12-diketo-(10E)-octadecenoate **35**, which is assumed to arise from decomposition of a 9-hydroperoxide initial product. This result suggests that a hydrogen at C-11 of the keto-ene is indeed removed by lipoxygenase, generating an activated intermediate. By analogy with linoleic acid, the free acid would normally be oxygenated regioselectively at C-13; however, that position is saturated in this substrate and no reaction occurs. Fatty acid esters, however, typically show lower regioselectivity (see above), and consequently the methyl ester yields the product oxygenated at C-9. Unfortunately, we cannot know the stereochemistry of dioxygen addition at C-9 because the instability of the initial hydroperoxide product precluded its isolation; consequently, we do not know whether the oxygenation occurred via the dominant mechanism (mechanism 1 above) or after release of an activated intermediate into solution.

Radical intermediates of enzyme reactions are often trapped and identified by the use of substrate analogs with appropriately positioned cyclopropyl groups (Ortiz de Montellano and Stearns, 1987). This type of experiment takes advantage of the propensity of cyclopropylcarbinyl radicals to decompose via ring-opening reactions (Griller and Ingold, 1980). Arachidonic acid analog **36** with a cyclopropyl group α to the preferred position for oxygenation is indeed a substrate for soybean lipoxygenase, but only under hyperbaric dioxygen (Corey and Nagata, 1987). The predominant products of the reaction are fatty acid hydroperoxides, with a small amount (14%) of the product resulting from ring opening of the cyclopropane. This result is open to two interpretations: (1) A species with radical character at C-13 is formed, but the cyclopropylcarbinyl radical ring-opening reaction does not compete effectively with dioxygen addition; (2) a species with radical character at C-13 is formed in a side reaction, one not on the dominant mechanistic pathway of the oxygenation reaction.

**36**

*Reactions with inhibitors.* A large number of compounds have been synthesized as lipoxygenase inhibitors because of the potential of such agents as anti-inflammatory drugs. We shall discuss those that react with lipoxygenase producing insight into the mechanism of the oxygenation reaction.

One class of inhibitors includes those able to reduce the active site iron from $Fe^{3+}$ to $Fe^{2+}$. Some of these have been mentioned above in the context of the high reduction potential of the iron, but for most, the relevance to the oxygenation mechanism is limited. Of special significance are catechols, which have been shown to reduce the iron via an inner sphere mechanism (formation of a bidentate complex before reduction), implying some lability of the iron coordination environment (Nelson, 1988a).

Linoleic acid analog **37** in which the 12 double bond is replaced by an episulfide is an inhibitor of soybean lipoxygenase and a reductant of the active-site $Fe^{3+}$ (Wright and Nelson, 1992). Interestingly, neither analog **38**, with a 9,10-episulfide, nor either of the saturated episulfides are reductants of the iron, although all are competitive inhibitors and thus probably have access to the active site. This result suggests a requirement for the 9-double bond, but not the 12-double bond, of linoleic acid for the iron reduction, a possibility supported by the observation that the mechanism of inhibition of lipoxygenase by 12-iodooleic acid **39** appears to involve reduction of the iron as well (Rotenberg et al., 1988). Thus, both double bonds of the substrate are not required to reduce the iron.

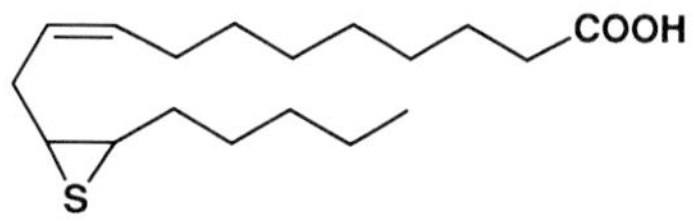

37

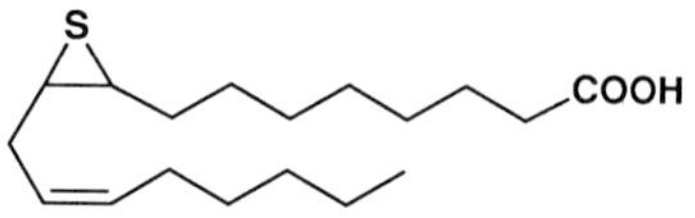

38

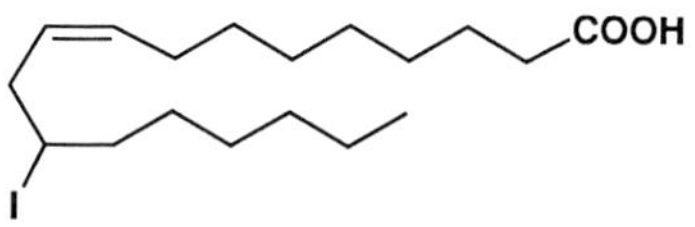

39

A second class of inhibitor acts as a substrate for lipoxygenase but generates a highly reactive species that chemically modifies the active site. Substrates in which the double bond to be oxygenated is instead a triple bond (e.g., 14,15-dehydroarachidonic acid **40**) irreversibly inactivate lipoxygenases in a reaction that requires dioxygen and results in oxidation of one methionine per protein to methionine sulfoxide (Corey et al., 1984; Kuhn et al., 1984). A similar result is obtained with hexanal phenylhydrazone **41** (Galey et al., 1988). In both cases, a highly reactive hydroperoxide is thought to be formed (a vinyl hydroperoxide in the former and an α-azo hydroperoxide in the latter) that then may oxidize a methionine to inactivate the enzyme. A less well-characterized example of this type of inhibitor is 13-thiaarachidonic acid **42**, in which the reactive methylene of the substrate is replaced by a sulfur (Corey et al., 1986). Inactivation of lipoxygenase by this compound results in covalent modification of the enzyme by the inhibitor, presumably the result of nucleophilic attack by an active-site base on the sulfonium ion product of oxidation of the inhibitor by the iron.

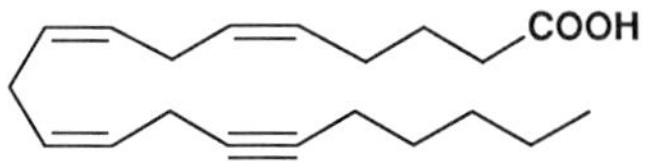

40

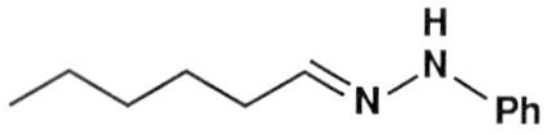

41

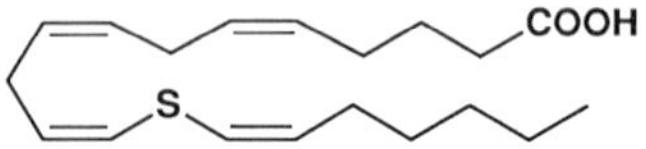

42

*Fatty acid radicals produced by lipoxygenase.* Fatty acid radicals have been observed as lipoxygenase products under various conditions. The experiments are of four main types: (1) use of spin traps to trap radicals produced from ferric lipoxygenase and substrates under anaerobic conditions; (2) direct observation of radicals produced under turnover conditions in solution; (3) direct observation of radicals in frozen samples of purple lipoxygenase; and (4) use of spin traps to trap the products of the lipoxygenase peroxidase reaction. The peroxidase reaction is the homolytic decomposition of the fatty acid hydroperoxides by reducing agents (most notably linoleic acid and other substrates) catalyzed by *ferrous* lipoxygenase (Chamulitrat and Mason, 1990; Chamulitrat et al., 1991). Experiment 1 above was discussed in the section on "The Anaerobic Reaction" and shows that pentadienyl radicals of linoleic acid are found in solution after anoxic oxidation of the substrate by ferric lipoxygenase, suggesting the possibility of one-electron oxidation of the diene as a partial reaction of the mechanism.

The second experiment is one in which a solution of lipoxygenase and substrates is passed through an EPR cell under turnover conditions. The signal observed was assigned to a peroxyl radical derivative of the fatty acid substrate (Chamulitrat and Mason, 1989). This was the first reported direct observation of a fatty acid peroxyl generated by lipoxygenase. However, it was not possible to tell if the peroxyl radical was the correct regioisomer to be the precursor of the dominant lipoxygenase product because the structural assignment was made by substitution of all of the vinylic protons of the substrate with deuterium.

Fatty acid radicals have also been observed in frozen solutions of purple lipoxygenase (Nelson and Cowling, 1990; Nelson et al., 1990). The hypothesis behind these experiments is that if the purple species is indeed an intermediate of the reaction with a significant lifetime, then purple solutions may contain equilibrium amounts of other intermediates as well. Both alkyl and peroxyl radicals have been observed, but the peroxyl radical has been shown to be the 9-peroxyl derivative of linoleic acid (Nelson et al., 1994). Consequently this peroxyl cannot be the precursor of the major lipoxygenase product but might be the precursor of the 9-hydroperoxide, the minor product. The alkyl radical has been shown to be an allyl radical, with radical character at C-9 and C-11 of linoleic acid, rather than the pentadienyl radical hypothesized as the intermediate of the dioxygenase reaction. One possibility is that this allyl radical is an intermediate of the dioxygenase reaction after dioxygen has added to the 12 double bond but before formation of the enzyme–product complex. This suggests a novel possibility—that the oxidation of linoleic acid by ferric lipoxygenase generates an allyl radical rather than a pentadienyl radical. We have alluded to this possibility (see above) in discussing the necessity of the 9 double bond, but not the 12 double bond, in linoleic acid analogs that reduce ferric lipoxygenase.

## HYPOTHESES FOR THE LIPOXYGENASE OXYGENASE REACTION

When considering a hypothetical mechanism for an enzyme, one must apply several tests. First, the reactions proposed must be chemically reasonable. Second, they must be kinetically competent; that is, they must occur at least as rapidly as the turnover of the enzyme. Third, they must be thermodynamically reasonable. This last requirement is the most difficult to apply because intuitions about thermodynamic reasonableness frequently fail in an enzyme active site, where conditions can vary dramatically from those found in aqueous solution.

Although there are several pieces of important data that provide information about the mechanism of lipoxygenase, we still cannot assemble one, and only one, picture that fits them all. Consequently we shall present multiple hypotheses and show where each fits or fails the data.

## Mechanisms Involving Singlet Dioxygen

As noted in the section on "Singlet Dioxygen," singlet dioxygen reacts readily with alkenes to yield allylic hydroperoxides. Indeed, singlet dioxygen has been reported to be produced by lipoxygenase, but the data show convincingly that the source is coupling of fatty acid peroxyl radicals leaking off the enzyme (Kanofsky and Axelrod, 1986). Were singlet dioxygen produced in the active site, its reaction with, for example, the 12 double bond of linoleic acid could explain the existence of primary and secondary kinetic isotope effects. However, it fails to explain the apparent delocalization of the 9 double bond during the reaction and fails to give a role for reduction of the iron by the substrate in the anoxic partial reaction, although the latter is conceivably only a coincidence. Finally, there is little precedent for the generation of singlet dioxygen by transition metals. On balance, this seems a very unlikely mechanism for the enzymatic reaction.

## Mechanisms Involving Nucleophilic $O_2$

Activation of the fatty acid diene could yield an electrophilic species (e.g., a carbonium ion or radical cation) competent to react with nucleophiles such as superoxide or peroxide. However, the source of reduced dioxygen is unclear; if the iron in the intermediate complex retains the high reduction potential of the native enzyme, it is unlikely to be able to reduce dioxygen to superoxide. Consequently, this mechanism also seems unlikely to be employed by lipoxygenase.

## Mechanisms Involving Iron–Carbon Complexes

As discussed in the section on "Carbanionic Routes," removal of the methylene proton between the double bonds of the substrate, with coordination of the resulting carbanion by the iron, would produce an organometallic intermediate. Insertion of dioxygen into the Fe—C bond to yield the ferric–peroxide complex and protonation completes the reaction.

This mechanism rationalizes most of the data in the literature. In particular, the existence of primary and secondary kinetic isotope effects is explained by abstraction of the methylene proton and concomitant rehybridization of C-11 (of linoleic acid) in the course of forming the alkyl iron complex. Insertion of dioxygen into the Fe—C bond imposes the stereospecificity of dioxygen addition, and the resulting ferric–peroxide complex could be the purple species.

Despite the attractive aspects of this mechanism, it does not provide a straightforward explanation for the high reduction potential of the iron. On the one hand, the high reduction potential would lead to inefficiency because the proposed intermediate $Fe^{3+}$-alkyl complex would readily dissociate to produce $Fe^{2+}$ and fatty acid radicals. The resulting competition between Fe—C homolysis and dioxygen insertion would siphon off active intermediates in the mechanism, especially at low dioxygen concentration. On the other hand, fatty acid radicals *are* seen in the anaerobic reaction between ferric lipoxygenase and substrate, and the high reduction potential may be a defense against a more active peroxidase activity. The iron does appear to have access to an open coordination site, as required by this mechanism.

The most telling evidence against the organometallic intermediate is the isomerization of the 9(E) analog of linoleic acid to the 9(Z) product. Rotation about the 9-double bond could occur in a discrete, delocalized carbanion intermediate, but the primary advantage in using the iron as a Lewis acid to promote substrate deprotonation comes in the *avoidance* of such a discrete intermediate, using the energy of the Fe—C bond as it forms to offset the energy required to break the C—H bond. Alternatively, the $\Delta^9$ isomerization could also be rationalized as a side reaction: if dioxygen insertion into the 9(E) organometallic intermediate were slow enough to allow transient Fe—C homolysis, then isomerization of the double bond could occur in the bound radical, followed by reformation of the Fe—C bond and dioxygen insertion.

## Mechanisms Involving Fatty Acid Radicals

These mechanisms have been the most popular with workers in the field since the early discovery of the anaerobic reduction of the active-site $Fe^{3+}$ by linoleic acid with release of fatty acid radicals into solution. The presumed intermediate is the most stable one-electron oxidation product of linoleic acid, the pentadienyl radical. The mechanism of this oxidation reaction is unclear. One might imagine the $Fe^{3+}$ oxidizing an amino acid (e.g., tyrosine or cysteine) to a radical that could abstract a hydrogen atom from the substrate. Alternatively, the $Fe^{3+}-OH^-$ unit could abstract the equivalent of a hydrogen atom, yielding $Fe^{2+}-OH_2$. The latter seems more likely on thermodynamic grounds.

In the section on "Autoxidation," we discussed the substantial precedent for the reaction of pentadienyl radicals with dioxygen to give peroxyl radicals; an intermediate peroxyl radical would be expected to reoxidize the $Fe^{2+}$ (Jovanovic et al., 1992), yielding the fatty acid hydroperoxide. Overall, this mechanism is consistent with the observed kinetic isotope effects, rationalizes the isomerization of the 9(E) analog of linoleic acid to the 9(Z) product during the reaction, and utilizes the unusually high reduction potential of the iron. If the oxidation of the iron by the peroxyl involves an inner-sphere electron transfer, this mechanism has a role for the purple complex as the penultimate enzyme–product complex, the $Fe^{3+}$–fatty acid peroxide complex.

A couple of points are not directly explained by this hypothesis. One is the regioselectivity and stereospecificity of the reaction. In solution, addition of dioxygen to linoleyl pentadienyl radicals ultimately yields all four 9(R, S)- and 13(R, S)-hydroperoxide products. Thus, the enzyme must prevent the approach of dioxygen to three of the four possible sites of attack in order to generate the 13(S)-hydroperoxide as the dominant product; such steric control of the reaction of dioxygen with substrates has been seen, for example, in the singlet dioxygen-ene reaction (see the section on "Singlet Dioxygen"). A further possibility is that interaction of dioxygen with the $Fe^{2+}$, even if weak, could direct the oxygenation to one end or face of the pentadienyl radical. The small amounts of other products generated could arise from release of the radical into solution for nonenzymatic oxygenation, binding of the substrate in a different orientation with respect to the active site, or a small amount of dioxygen getting past the steric interference of the protein.

Another result not directly explained is the relatively small amount of ring opening seen in the oxygenation of the cyclopropyl fatty acid. If radical character develops in the intermediate at the site of dioxygen attack, one would expect a substantial percentage of ring-opened products from that substrate, unless kinetically ring opening does not compete effectively with dioxygen addition. Examples of this exist, for example, in the reactions of cytochromes P-450 with cyclopropyl substrates (Ortiz de Montellano and Stearns, 1987). One can increase the rate of ring opening of cyclopropylcarbinyl radicals by attaching phenyl substituents to the cyclopropane ring (Castellino and Bruice, 1988a, 1988b); perhaps a phenyl-substituted cyclopropane fatty acid substrate for lipoxygenase would show an increase in the relative amount of ring opening resulting from more efficient competition of that reaction with oxygenation.

A final, less intuitive radical mechanism may be proposed from the available data. We have noted that substituted oleic acid derivatives are competent to reduce ferric lipoxygenase, implying that a monoene, rather than a diene, is sufficient to carry out this reaction. Together with the observation of allyl radicals in the purple samples, this result suggests that the intermediate in the reaction is not a pentadienyl radical but rather an allyl radical. Such a radical could arise from binding of the diene in the active site such that the two π-bonds cannot overlap, as the result of a rotation of the diene away from a planar structure. This mechanism is attractive because it imposes a difference between the two double bonds of the substrate and thus directly explains the regioselectivity of the reaction.

This hypothesis has a couple of difficulties. First, we require a mechanism for generation of an activated dioxygen species (e.g., singlet dioxygen or its electronic equivalent) to react with the alkene. This might occur via an interaction between the reduced iron (presumably $S = 2$) and triplet dioxygen; however, there is no precedent for the production of free singlet dioxygen from triplet dioxygen by $Fe^{2+}$. Second, the estimated difference in stabilization energy of simple allyl and pentadienyl radicals is between 7 and 10 kcal/mol, roughly 0.3 to 0.5 V (Green and Walton, 1984). Thus, on thermodynamic grounds, the allyl radical is a less attractive intermediate than the pentadienyl radical.

## CONCLUSIONS

It is clear that we do not yet have a definitive mechanism for the oxygenation reaction catalyzed by lipoxygenase. The data are most consistent with a radical mechanism, and the thermodynamically most reasonable radical intermediate is the pentadienyl radical generated by oxidation and deprotonation of the 1,4-diene unit of the substrate. The peroxyl radical formed by stereoselective addition of dioxygen would be reduced by the iron, in what is likely an inner sphere reaction, to form a ferric–alkyl peroxide complex from which the hydroperoxide product is released. Certain key experiments remain to be performed. It should be possible to design experiments with altered substrates or using kinetic isotope effects to show whether the activated fatty acid intermediate is an allyl or a pentadienyl radical. Detailed analysis of the crystal structure of soybean lipoxygenase will help to make clear the roles of amino acid side chains in binding the substrate and controlling the regio- and stereospecificity of the dioxygenation reaction.

Many enzymatic reactions have been shown to untilize radical intermediates. The closest parallel to lipoxygenase is prostaglandin H synthase (Smith and Marnett, 1991), which catalyzes the same dioxygenation of a fatty acid 1,4-diene as does lipoxygenase. Again, the reaction is thought to utilize addition of dioxygen to a pentadienyl radical. Although controversial, it may be that the radical is generated via hydrogen atom abstraction by an enzymatic tyrosyl radical (Tsai et al., 1992). The best-known examples of metallooxygenase enzymes that utilize radical intermediates are *mono*oxygenases: cytochrome P-450 (Ortiz de Montellano, 1986) and methane monooxygenase (Wilkins et al., 1992). The former contains a heme iron, the latter a binuclear iron cluster. Those enzymes are thought to utilize a high-valent Fe—O species formed by heterolytic O—O bond cleavage to abstract a hydrogen atom from the substrate to generate the radical. In these cases $[Fe=O]^{2+}$ is thought to add H· to become $[Fe-OH]^{2+}$, whereas we speculated above that in lipoxygenase the $[Fe-OH]^{2+}$ might abstract H· to become $[Fe-OH_2]^{2+}$. A similar hydrogen atom abstraction is envisioned for the thiazolidine ring cyclization reaction catalyzed by isopenicillin N synthase (Baldwin and Bradley, 1990), an enzyme with an iron in a very similar coordination environment to that in lipoxygenase (Randall et al., 1993). There are several other mononuclear nonheme iron enzymes (e.g., *p*-hydroxyphenylpyruvate dioxygenase) that use oxygen to hydroxylate organic substrates, presumably via P-450-type mechanisms. The chemical factors that determine the reactions catalyzed by a particular nonheme iron enzyme remain to be elucidated.

The best-characterized examples of enzymes that utilize radical intermediates are not dioxygenases. In the cases of the ribonucleotide reductases (Stubbe, 1990) and the vitamin $B_{12}$-dependent enzymes (Dowd, 1990), the substrate radical is apparently generated via hydrogen atom abstraction by an amino acid radical, probably a cysteinyl radical.

The iron-dependent ribonucleotide reductases appear to generate this amino acid radical with a tyrosyl radical (see Chapter 5), while the vitamin $B_{12}$-dependent enzymes utilize the adenosyl radical formed by homolytic Co—C bond cleavage of vitamin $B_{12}$. Two recent entries in this incomplete list of radical-generating enzymes are among the most interesting. In lysine aminomutase the radical chemistry may be initiated by an adenosyl radical formed from S-adenosylmethionine (Frey and Reed, 1993), while pyruvate formate-lyase apparently uses an enzymatic glycyl radical (Volker Wagner et al., 1992). Clearly, enzymes have developed a variety of ways to generate radical intermediates. Lipoxygenase, though intensively studied, is still among the least understood.

*Note Added in Proof*: Two recent papers present spectroscopic evidence that the metal ion in ferrous lipoxygenase exists in an equilibrium between 5- and 6-coordinate forms and that the equilibrium may be shifted toward the 6-coordinate state by addition of alcohols or, very interestingly, substrate (Pavlosky and Solomon, 1994; Scarrow et al., 1994). These results clarify the coordination number of the iron in this state. In another recent paper it was shown that aquoferrous units react very rapidly with alkyl radicals to form σ-organometallic complexes (van Eldik et al., 1994). This suggests the fascinating possibility that the active site $Fe^{3+}$—$OH^-$ in lipoxygenase might oxidize the substrate to an alkyl radical via hydrogen atom abstraction, and the resulting $Fe^{2+}$—$OH_2$ and alkyl radical combine to give the σ-organometallic complex predicted in the Corey mechanism (Corey and Nagata, 1987). Such a mechanism could explain most of the experimental results presented in the discussion of reactions with substrates.

## Acknowledgments

The authors would like to thank Professors Charles Grissom and Ken Houk for disclosing results in advance of publication.

## REFERENCES

ARASASINGHAM, R. D., BALCH, A. L., CORNMAN, C. R., and LATOS, G. L. (1989) Dioxygen Insertion into Iron(III)–Carbon Bonds. NMR Studies of the Formation and Reactivity of Alkylperoxo Complexes of Iron(III) Porphyrins. *J. Am. Chem. Soc.*, 111, 4357–4363.

ARASASINGHAM, R. D., BALCH, A. L., HAR, R. L., and LATOS, G. L. (1990) Reactions of Aryliron(III) Porphyrins with Dioxygen. Formation of

Aryloxyiron(III) and Aryliron(IV) Complexes. *J. Am. Chem. Soc.*, 112, 7566–7571.

BALDWIN, J. E., and BRADLEY, M. (1990) Isopenicillin N Synthase: Mechanistic Studies. *Chem. Rev.*, 90, 1079–1088.

BATT, D. G. (1992) 5-Lipoxygenase Inhibitors and Their Anti-inflammatory Activities. *Prog. Med. Chem.*, 29, 1–63,

BOYINGTON, J. C., GAFFNEY, B. J., and AMZEL, L. M. (1993) The Three-dimensional Structure of an Arachidonic Acid 15-Lipoxygenase. *Science (Washington, D.C.)*, 260, 1482–1486.

BRASH, A. R., INGRAM, C. D., and MAAS, R. L. (1986) A Secondary Isotope Effect in the Lipoxygenase Reaction. *Biochim. Biophys. Acta*, 875, 256–261.

CASTELLINO, A. J., and BRUICE, T. C. (1988a) Intermediates in the Epoxidation of Alkenes by Cytochrome P-450 Models. 2. Use of the *trans*-2, *trans*-3-Diphenylcyclopropyl Substituent in a Search of Radical Intermediates. *J. Am. Chem. Soc.*, 110, 7512–7519.

CASTELLINO, A. J., and BRUICE, T. C. (1988b) Radical Intermediates in the Epoxidation of Alkenes by Cytochrome P-450 Model Systems. The Design of a Hypersensitive Radical Probe. *J. Am. Chem. Soc.*, 110, 1313–1315.

CHAMULITRAT, W., HUGHES, M. F., ELING, T. E., and MASON, R. P. (1991) Superoxide and Peroxyl Radical Generation from the Reduction of Polyunsaturated Fatty Acid Hydroperoxides by Soybean Lipoxygenase. *Arch. Biochem. Biophys.*, 290, 153–159.

CHAMULITRAT, W., and MASON, R. P. (1989) Lipid Peroxyl Radical Intermediates in the Peroxidation of Polyunsaturated Fatty Acids by Lipoxygenase. Direct Electron Spin Resonance Investigations. *J. Biol. Chem.*, 264, 20968–20973.

CHAMULITRAT, W., and MASON, R. P. (1990) Alkyl Free Radicals from the Beta-scission of Fatty Acid Alkoxyl Radicals as Detected by Spin Trapping in a Lipoxygenase System. *Arch. Biochem. Biophys.*, 282, 65–69.

CHEESBROUGH, T. M., and AXELROD, B. (1983) Determination of the Spin State in Native and Activated Soybean Lipoxygenase 1 by Paramagnetic Susceptibility. *Biochemistry*, 22, 3837–3840.

CLAPP, C. H., BANERJEE, A., and ROTENBERG, S. A. (1985) Inhibition of Soybean Lipoxygenase 1 by N-alkylhydroxylamines. *Biochemistry*, 24, 1826–1830.

COREY, E. J. (1987) Enzymic Lipoxygenation of Arachidonic Acid: Mechanism, Inhibition, and Role in Eicosanoid Biosynthesis. *Pure Appl. Chem.*, 59, 269–278.

COREY, E. J., D'ALARCAO, M., and MATSUDA, S. P. T. (1986) A New Irreversible Inhibitor of Soybean Lipoxygenase: Relevance to Mechanism. *Tetrahedron Lett.*, 27, 3585–3588.

COREY, E. J., LANSBURY, P. T. J., CASHMAN, J. R., and KANTNER, S. S. (1984) Mechanism of the Irreversible Deactivation of Arachidonate 5-

Lipoxygenase by 5,6-Dehydroarachidonate. *J. Am. Chem. Soc.*, 106, 1501–1503.

COREY, E. J., and MEHROTRA, M. M. (1983) Stereochemistry of the Lipoxygenase-catalyzed Allylic Hydroperoxide → Oxiranylcarbinol Rearrangement. *Tetrahedron Lett.*, 24, 4921–4922.

COREY, E. J., and NAGATA, R. (1987) Evidence in Favor of an Organoiron-mediated Pathway for Lipoxygenation of Fatty Acids by Soybean Lipoxygenase. *J. Am. Chem. Soc.*, 109, 8107–8108.

COREY, E. J., NICOLAOU, K. C., SHIBASAKI, M., MACHIDA, Y., and SHINER, C. S. (1975) Superoxide Ion as a Synthetically Useful Oxygen Nucleophile. *Tetrahedron Lett.*, 16, 3183–3186.

COREY, E. J., and WALKER, J. C. (1987) Organoiron-mediated Oxygenation of Allylic Organotin Compounds. A Possible Chemical Model for Enzymatic Lipoxygenation. *J. Am. Chem. Soc.*, 109, 8108–8109.

DE GROOT, J. J. M. C., GARSSEN, G. J., VELDINK, G. A., VLIEGENTHART, J. F. G., and BOLDINGH, J. (1975) On the Interaction of Soybean Lipoxygenase-1 and 13-L-Hydroperoxylinoleic Acid, Involving Yellow- and Purple-coloured Enzyme Species. *FEBS Lett.*, 56, 50–54.

DE GROOT, J. J. M. C., GARSSEN, G. J., VLIEGENTHART, J. F. G., and BOLDINGH, J. (1973) The Detection of Linoleic Acid Radicals in the Anaerobic Reaction of Lipoxygenase. *Biochim. Biophys. Acta*, 326, 279–284.

DE GROOT, J. J. M. C., VELDINK, G. A., VLIEGENTHART, J. F. G., BOLDINGH, J., WEVER, R., and VAN GELDER, B. F. (1975) Demonstration by EPR Spectroscopy of the Functional Role of Iron in Soybean Lipoxygenase-1. *Biochim. Biophys. Acta*, 377, 71–79.

DEWAR, M. J. S., and THIEL, W. (1977) MINDO/3 Study of the Addition of Singlet Oxygen ($^1\Delta_g$) to 1,3-Butadiene. *J. Am. Chem. Soc.*, 99, 2338–2339.

DIXON, R. A. F., JONES, R. E., DIEHL, R. E., BENNETT, C. D., KARGMAN, S., and ROUZER, C. A. (1988) Cloning of the cDNA for Human 5-Lipoxygenase. *Proc. Natl. Acad. Sci. U.S.A.*, 85, 416–420.

DOWD, P. (1990) On the Mechanism of Action of Vitamin $B_{12}$, in *Selective Hydrocarbon Activation* (J. A. Davies, P. L. Watson, A. Greenberg, and J. F. Liebman, Eds.), VCH, New York, pp. 265–303.

DUNHAM, W. R., CARROLL, R. T., THOMPSON, J. F., SANDS, R. H., and FUNK, M. O., JR. (1990) The Initial Characterization of the Iron Environment in Lipoxygenase by Mössbauer Spectroscopy. *Eur. J. Biochem.*, 190, 611–617.

DUSSAULT, M. H., and Hayden, M. R. (1992) Auxiliary-directed Dioxygenation: Stereoselective Synthesis of a Diene Hydroperoxide. *Tetrahedron Lett.*, 33, 443–446.

DUSSAULT, P. (1995) Reactions of Hydroperoxides and Peroxides, in *Active Oxygen in Chemistry* (C. S. Foote, J. S. Valentine, A. Greenberg, and J. F. Liebman, Eds.), Chapman & Hall, New York, pp. 141–203.

EGMOND, M. R., FASELLA, P. M., VELDINK, G. A., VLIEGENTHART, J. F. G., and BOLDINGH, J. (1977) On the Mechanism of Action of Soybean Lipoxygenase. A Stopped-flow Kinetic study of the Formation and Conversion of Yellow and Purple Enzyme Species. *Eur. J. Biochem.*, 76, 469–479.

EGMOND, M. R., VELDINK, G. A., VLIEGENTHART, J. F. G., and BOLDINGH, J. (1973) C-11 H-abstraction from Linoleic Acid, the Rate-limiting Step in Lipoxygenase Catalysis. *Biochem. Biophys. Res. Commun.*, 54, 1178–1184.

EGMOND, M. R., VLIEGENTHART, J. F. C., and BOLDINGH, J. (1972) Stereospecificity of the Hydrogen Abstraction at Carbon Atom n-8 in the Oxygenation of Linoleic Acid by Lipoxygenases from Corn Germs and Soya Beans. *Biochem. Biophys. Res. Commun.*, 48, 1055–1060.

FEITERS, M. C., AASA, R., MALMSTRÖM, B. G., VELDINK, G. A., and VLIEGENTHART, J. F. G. (1986) Spectroscopic Studies on the Interactions Between Lipoxygenase-2 and Its Product Hydroperoxides. *Biochim. Biophys. Acta*, 873, 182–189.

FEITERS, M. C., BOELENS, H., VELDINK, G. A., VLIEGENTHART, J. F. G., NAVARATNAM, S., ALLEN, J. C., NOLTING, H.-F., and HERMES, C. (1990) X-ray Absorption Spectroscopic Studies on Iron in Soybean Lipoxygenase: A Model for Mammalian Lipoxygenases. *Rec. Trav. Chim. Pays-Bas*, 109, 133–146.

FOOTE, C. S., and CLENNAN, E. (1995) Reactions of Singlet Dioxygen, in *Active Oxygen in Chemistry* (C. S. Foote, J. S. Valentine, J. F. Liebman, and A. Greenberg, Eds.), Chapman & Hall, New York, pp. 105–140.

FREY, P. A., and REED, G. H. (1993) Lysine 2,3-aminomutase and the Mechanism of the Interconversion of Lysine and β-lysine. *Adv. Enzymol.*, 66, 1–39.

FUNK, C. D., FURCI, L., and FITZGERALD, G. A. (1990) Molecular Cloning, Primary Structure, and Expression of the Human Platelet/Erythroleukemia Cell 12-Lipoxygenase. *Proc. Natl. Acad. Sci. U.S.A.*, 87, 5638–5642.

FUNK, M. O. JR., ANDRE, J. C., and OTSUKI, T. (1987) Oxygenation of *trans* Polyunsaturated Fatty Acids by Lipoxygenase Reveals Steric Features of the Catalytic Mechanism. *Biochemistry*, 26, 6880–6884.

FUNK, M. O. JR., CARROLL, R. T., THOMPSON, J. F., SANDS, R. H., and DUNHAM, W. R. (1990) Role of Iron in Lipoxygenase Catalysis. *J. Am. Chem. Soc.*, 112, 5375–5376.

GAFFNEY, B. J., MAVROPHILIPOS, D. V., and DOCTOR, K. S. (1993) Access of Ligands to the Ferric Center in Lipoxygenase-1. *Biophys J.*, 64, 773–783.

GAJEWSKI, J. J., OLSON, L. P., and TUPPER, K. J. (1993) Hydrogen-deuterium Fractionation for Hydrogen-sp$^2$ Carbon Bonds in Olefins and Allyl Radicals. *J. Am. Chem. Soc.*, 115, 4548–4553.

GALEY, J. B., BOMBARD, S., CHOPARD, C., GIRERD, J. J., LEDERER, F., THANG, D. C., NAM, N. H., MANSUY, D., and CHOTTARD, J. C. (1988) Hexanal Phenylhydrazone is a Mechanism-based Inactivator of Soybean Lipoxygenase 1. *Biochemistry*, 27, 1058–1066.

GARDNER, H. W. (1989) Soybean Lipoxygenase-1 Enzymically Forms Both (9S)- and (13S)-Hydroperoxides from Linoleic Acid by a pH-Dependent Mechanism. *Biochim. Biophys. Acta*, 1001, 274–281.

GARDNER, H. W. (1991) Recent Investigations into the Lipoxygenase Pathway of Plants. *Biochim. Biophys. Acta*, 1084, 221–239.

GARSSEN, G. J., VELDINK, G. A., VLIEGENTHART, J. F. G., and BOLDINGH, J. (1976) The Formation of *threo*-11-Hydroxy-*trans*-12:13-epoxy-9-*cis*-octadecenoic Acid by Enzymic Isomerisation of 13-L-hydroperoxy-9-*cis*, 11-*trans*-Octadecadienoic Acid by Soybean Lipoxygenase-1. *Eur. J. Biochem.*, 62, 33–36.

GARSSEN, G. J., VLIEGENTHART, J. F. G., and BOLDINGH, J. (1972) The Origin and Structures of Dimeric Fatty Acids from the Anaerobic Reaction Between Soya-bean Lipoxygenase, Linoleic Acid and Its Hydroperoxide. *Biochem. J.*, 130, 435–442.

GLICKMAN, M. H., WISEMAN, J. S., and KLINMAN, J. P. (1994) Extremely Large Isotope Effects in the Soybean Lipoxygenase-linoleic Acid Reaction. *J. Am. Chem. Soc.*, 116, 793–794.

GOLLNICK, K., and KUHN, H. J. (1979) Ene-reactions with Singlet Oxygen, in *Singlet Oxygen* (H. H. Wasserman, and R. W. Murray, Eds.), Academic Press, New York, pp. 287–429.

GORMAN, A. A., HAMBLETT, I. LAMBERT, C., SPENCER, B., and STANDEN, M. C. (1988) Identification of Both Preequilibrium and Diffusion Limits for Reaction of Singlet Oxygen, $O_2$ ($^1\Delta_g$), with Both Physical and Chemical Quenchers: Variable-temperature, Time-resolved Infrared Luminescence Studies. *J. Am. Chem. Soc.*, 110, 8053–8059.

GRDINA, S. B., ORFANOPOULOS, M., and STEPHENSON, L. M. (1979) Stereochemical Dependence of Isotope Effects in the Singlet Oxygen–Olefin Reaction. *J. Am. Chem. Soc.*, 101, 3111–3112.

GREEN, I. G., and WALTON, J. C. (1984) Electron Delocalization and Stabilization in Heptatrienyl and Polyenyl Radicals. *J. Chem. Soc. Perkin Trans. 2*, 1253–1257.

GRILLER, D., and INGOLD, K. U. (1980) Free-radical Clocks. *Acc. Chem. Res.*, 13, 317–323.

GROSSMAN, S., KLEIN, B. P., COHEN, B., KING, D., and PINSKY, A. (1984) Methylmercuric Iodide Modification of Lipoxygenase-1. Effects on the Anaerobic Reaction and Pigment Bleaching. *Biochim. Biophys. Acta*, 793, 455–462.

HAMBURG, M. (1971) Steric Analysis of Hydroperoxides Formed by Lipoxygenase Oxygenation of Linoleic Acid. *Anal. Biochem.*, 43, 515–526.

HAMBERG, M., and HAMBERG, G. (1980) On the Mechanism of the Oxygenation of Arachidonic Acid by Human Platelet Lipoxygenase. *Biochem. Biophys. Res. Commun.*, 95, 1090–1097.

HAMBURG, M., and SAMUELSSON, B. (1967) On the Specificity of the Oxgenation of Unsaturated Fatty Acids Catalyzed by Soybean Lipoxygenase. *J. Biol. Chem.*, 242, 5329–5335.

HARDING, L. B., and GODDARD, W. A. I. (1980) The Mechanism of the Ene Reaction of Singlet Oxygen with Olefins. *J. Am. Chem. Soc.*, 102, 439–449.

HARPEL, M. R., and LIPSCOMB, J. D. (1990) Gentisate 1,2-Dioxygenase from *Pseudomonas*. Substrate Coordination to Active Site $Fe^{2+}$ and Mechanism of Turnover. *J. Biol. Chem.*, 265, 22187–22196.

HATANAKA, A., KAJIWARA, T., MATSUI, K., and YAMAGUCHI, M. (1989) Product Specificity in an Entire Series of ($\omega$-6Z,$\omega$-9)-C13~C20-Dienoic Acids and -Dienols for Soybean Lipoxygenase. *Z. Naturforsch.*, 44c, 64–70.

HATANAKA, A., KAJIWARA, T., SEKIYA, J., and ASANO, M. (1984) Product Specificity During Incubation of Methyl Linoeate with Soybean Lipoxygenase-1. *Z. Naturforsch.*, 39c, 171–173.

HIATT, R. (1971) Hydroperoxides, in *Organic Peroxides* (D. Swern, Ed.), Wiley-Interscience, New York, pp. 1–152.

HILDEBRAND, D. F., and GRAYBURN, W. S. (1991) Lipid Metabolites: Regulators of Plant Metabolism?, in *Plant Biochemical Regulators* (H. W. Gausman, Ed.), Marcel Dekker, New York, pp. 69–95.

HOLMAN, R. T., EGWIM, P. O., and CHRISTIE, W. W. (1969) Substrate Specificity of Soybean Lipoxidase. *J. Biol. Chem.*, 244, 1149–1151.

HOUK, K. N., GUSTAFSON, S. M. E., and BLACK, K. A. (1992) Theoretical Secondary Kinetic Isotope Effects and the Interpretation of Transition State Geometries. 1. The Cope Rearrangement. *J. Am. Chem. Soc.*, 114, 8565–8572.

HWANG, C. C., and GRISSOM, C. B. (1994) Unusually Large Deuterium Isotope Effect in Soybean Lipoxygenase is Not Caused by a Magnetic Isotope Effect. *J. Am. Chem. Soc.* 116, 795–796.

IWAHASHI, H., ALBRO, P. W., MCGOWN, S. R., TOMER, K. B., and MASON, R. P. (1991a) Isolation and Identification of Alpha-(4-pyridyl-1-oxide)-N-tert-butylnitrone Radical Adducts Formed by the Decomposition of the Hydroperoxides of Linoleic Acid, Linolenic Acid, and Arachidonic Acid by Soybean Lipoxygenase. *Arch. Biochem. Biophys.*, 285, 172–180.

IWAHASHI, H., PARKER, C. E., MASON, R. P., and TOMER, K. B. (1991b) Radical Adducts of Nitrosobenzene and 2-Methyl-2-nitrosopropane with 12, 13-Epoxylinoleic Acid Radical, 12,13-Epoxylinolenic Acid Radical and 14,15-Epoxyarachidonic Acid Radical. Identification by h.p.l.c.-e.p.r. and Liquid Chromatography-thermospray-m.s. *Biochem. J.*, 276, 447–453.

JOVANOVIC, S. V., JANKOVIC, I., and JOSIMOVIC, L. (1992) Electron-transfer Reactions of Alkyl Peroxy Radicals. *J. Am. Chem. Soc.*, 114, 9018–9021.

KANOFSKY, J. R., and AXELROD, B. (1986) Singlet Oxygen Production by Soybean Lipoxygenase Isozymes. *J. Biol. Chem.*, 261, 1099–1104.

KORTH, H.-G., TRILL, H., and SUSTMANN, R. (1981) [1-$^2$H]Allyl Radical: Barrier to Rotation and Allyl Delocalization Energy. *J. Am. Chem. Soc.*, 103, 4483–4489.

KUHN, H., EGGERT, L., ZABOLOTSKY, O. A., MYAGKOVA, G. I., and SCHEWE, T. (1991) Keto Fatty Acids Not Containing Doubly Allylic Methylenes Are Lipoxygenase Substrates. *Biochemistry*, 30, 10269-10273.

KUHN, H., HEYDECK, D., WIESNAER, R., and SCHEWE, T. (1985) The Positional Specificity of Wheat Lipoxygenase; the Carboxyl Group as a Signal for the Recognition of the Site of the Hydrogen Removal. *Biochim. Biophys. Acta*, 830, 25-29.

KUHN, H., HOLZHUTTER, H.-G., SCHEWE, T., HIEBSCH, C., and RAPOPORT, S. M. (1984) The Mechanism of Inactivation of Lipoxygenases by Acetylenic Fatty Acids. *Eur. J. Biochem.*, 139, 577-583.

LEE-RUFF, E. (1977) The Organic Chemistry of Superoxide. *Chem. Soc. Rev.*, 6, 195-214.

MACINNES, I., and WALTON, J. C. (1985) Rotational Barriers in Pentadienyl and Pent-2-en-4-ynyl Radicals. *J. Chem. Soc. Perkin Trans. 2*, 1073-1076.

MATSUDA, S., SUZUKI, H., YOSHIMOTO, T., YAMAMOTO, S., and MIYATAKE, A. (1991) Analysis of Non-heme Iron in Arachidonate 12-Lipoxygenase of Porcine Leukocytes. *Biochim. Biophys. Acta*, 1084, 202-204.

MCMILLAN, R. M., and WALKER, E. R. H. (1992) Designing Therapeutically Effective 5-Lipoxygenase Inhibitors. *Trends Pharmacol. Sci.*, 13, 323-330.

MICHAUD-SORET, I., and CHOTTARD, J. C. (1992) Investigation of Sulfur Containing Amino Acids at the Lipoxygenase Active Site Using a Platinum Complex. *Biochem. Biophys. Res. Commun.*, 182, 779-785.

MINOR, W., STECZKO, J., BOLIN, J. T., OTWINOWSKI, Z., and AXELROD, B. (1993) Crystallographic Determination of the Active-site Iron and Its Ligands in Soybean Lipoxygenase. *Biochemistry*, 32, 6320-6323.

MURRAY, R. W. (1979) Chemical Sources of Singlet Oxygen, in *Singlet Oxygen* (H. H. Wasserman and R. W. Murray, Eds.), Academic Press, New York, pp. 59-114.

NAVARATNAM, S., FEITERS, M. C., AL-HAKIM, M., ALLEN, J. C., VELDINK, G. A., and VLIEGENTHART, J. F. G. (1988) Iron Environment in Soybean Lipoxygenase-1. *Biochim. Biophys. Acta*, 956, 70-76.

NELSON, M. J. (1987) The Nitric Oxide Complex of Ferrous Soybean Lipoxygenase-1. *J. Biol. Chem.*, 262, 12137-12142.

NELSON, M. J. (1988a) Catecholate Complexes of Ferric Soybean Lipoxygenase-1. *Biochemistry*, 27, 4273-4278.

NELSON, M. J. (1988b) Evidence for Water Coordinated to the Active Site Iron in Soybean Lipoxygenase-1. *J. Am. Chem. Soc.*, 110, 2985-2986.

NELSON, M. J., BATT, D. G., THOMPSON, J. S., and WRIGHT, S. W. (1991) Reduction of the Active-site Iron by Potent Inhibitors of Lipoxygenases. *J. Biol. Chem.*, 266, 8225-8229.

NELSON, M. J., and COWLING, R. A. (1990) Observation of a Peroxyl Radical in Samples of "Purple" Lipoxygenase. *J. Am. Chem. Soc.*, 112, 2820-2821.

NELSON, M. J., SEITZ, S. P., and COWLING, R. A. (1990) Enzyme-bound

Pentadienyl and Peroxyl Radicals in Purple Lipoxygenase. *Biochemistry*, 29, 6897–6903.

NELSON, M. J., COWLING, R. A., and SEITZ, S. P. (1994) Structural Characterization of Alkyl and Peroxyl Radicals in Solutions of Purple Lipoxygenase. *Biochemistry*, 33, 4966–4973.

NGUYEN, T., FALGUEYRET, J.-P., ABRAMOVITZ, M., and RIENDEAU, D. (1991) Evaluation of the Role of Conserved His and Met Residues Among Lipoxygenases by Site-directed Mutagenesis of Recombinant Human 5-Lipoxygenase. *J. Biol. Chem.*, 266, 22057–22062.

NIKOLAEV, V., REDDANNA, P., WHELAN, J., HILDENBRANDT, G., and REDDY, C. C. (1990) Stereochemical Nature of the Products of Linoleic Acid Oxidation Catalyzed by Lipoxygenases from Potato and Soybean. *Biochem. Biophys. Res. Commun.*, 170, 491–496.

NISHIDA, Y., and AKAMATSU, T. (1991) Model Compounds for Purple Lipoxygenase. *Chem. Lett.*, 2013–2016.

ORFANOPOULOS, M., SMONOU, I., and FOOTE, C. S. (1990) Intermediates in the Ene Reactions of Singlet Oxygen and N-phenyl-1,2,4-triazoline-3,5-dione with Olefins. *J. Am. Chem. Soc.*, 112, 3607–3614.

ORTIZ DE MONTELLANO, P. R. (1986) *Cytochrome P-450: Structure, Mechanism, and Biochemistry*, Plenum Press, New York.

ORTIZ DE MONTELLANO, P. R., and STEARNS, R. A. (1987) Timing of the Radical Recombination Step in Cytochrome P-450 Catalysis with Ring-strained Probes. *J. Am. Chem. Soc.*, 109, 3415–3420.

ORVILLE, A. M., CHEN, V. J., KRIAUCIUNAS, A., HARPEL, M. R., FOX, B. G., MÜNCK, E., and LIPSCOMB, J. D. (1992) Thiolate Ligation of the Active Site $Fe^{2+}$ of Isopenicillin N Synthase Derives from Substrate Rather Than Endogenous Cysteine: Spectroscopic Studies of Site-specific Cys → Ser Mutated Enzymes. *Biochemistry*, 31, 4602–4612.

PAVLOSKY, M. A., and SOLOMON, E. I. (1994) Near-IR CD/MCD Spectral Elucidation of Two Forms of the Non-heme Active Site in Native Ferrous Soybean Lipoxygenase 1: Correlation to Crystal Structures and Reactivity. *J. Am. Chem. Soc.* 116, 11610–11611.

PERCIVAL, M. D. (1991) Human 5-Lipoxygenase Contains an Essential Iron. *J. Biol. Chem.*, 266, 10058–10061.

PETERSSON, L., SLAPPENDEL, S., FEITERS, M. C., and VLEIGENTHART, J. F. G. (1987) Magnetic Susceptibility Studies on Yellow and Anaerobically Substrate-treated Yellow Soybean Lipoxygenase-1. *Biochim. Biophys. Acta*, 913, 228–237.

PETERSSON, L., SLAPPENDEL, S., and VLIEGENTHART, J. F. G. (1985) The Magnetic Susceptibility of Native Soybean Lipoxygenase-1. Implications for the Symmetry of the Iron Environment and the Possible Coordination of Dioxygen to Fe(II). *Biochim. Biophys. Acta*, 828, 81–85.

PISTORIUS, E. K., AXELROD, B., and PALMER, G. (1976) Evidence for the

Participation of Iron in Lipoxygenase Reaction from Optical and Electron Spin Resonance Studies. *J. Biol. Chem.*, 251, 7144–7148.

PORTER, N. A. (1986) Mechanisms for the Autoxidation of Polyunsaturated Lipids. *Acc. Chem. Res.*, 19, 262–268.

PORTER, N. A., LEHMAN, L. S., WEBER, B. A., and SMITH, K. J. (1981) Unified Mechanism for Polyunsaturated Fatty Acid Autoxidation. Competition of Peroxy Radical Hydrogen Atom Abstraction, β-Scission, and Cyclization. *J. Am. Chem. Soc.*, 103, 6447–6455.

QUE, L., JR., LAUFFER, R. B., LYNCH, J. B., MURCH, B. P., and PYRZ, J. W. (1987) Elucidation of the Coordination Chemistry of the Enzyme–Substrate Complex of Catechol 1,2-Dioxygenase by NMR Spectroscopy. *J. Am. Chem. Soc.*, 109, 5381–5385.

RAMACHANDRAN, S., CARROLL, R. T., DUNHAM, W. R. and FUNK, M. O., JR., (1992) Limited Proteolysis and Active-site Labelling Studies of Soybean Lipoxygenase-1. *Biochemistry*, 31, 7700–7706.

RANDALL, C. R., ZANG, Y., TRUE, A. E., QUE, L., JR., CHARNOCK, J. M., GARNER, C. D., FUJISHIMA, Y., SCHOFIELD, C. J., and BALDWIN, J. E. (1993) X-ray Absorption Studies of the Ferrous Active Site of Isopenicillin N Synthase and Related Model Complexes. *Biochemistry*, 32, 6664–6673.

RAPOPORT, S., HÄRTEL, B., and HAUSDORF, G. (1984) Methionine Sulfoxide Formation: The Cause of Self-inactivation of Reticulocyte Lipoxygenase. *Eur. J. Biochem.*, 139, 573–576.

ROTENBERG, S. A., GRANDIZIO, A. M., SELZER, A. T., and CLAPP, C. H. (1988) Inactivation of Soybean Lipoxygenase 1 by 12-Iodo-*cis*-9-octadecenoic Acid. *Biochemistry*, 27, 8813–8818.

SAMUELSSON, B. E., DAHLÉN, S.-E., LINDGREN, J. A., ROUZER, C. A., and SERHAN, C. N. (1987) Leukotrienes and Lipoxins: Structures, Biosynthesis, and Biological Effects. *Science (Washington, D.C.)*, 237, 1171–1176.

SCARROW, R. C., TRIMITSIS, M. G., BUCK, C. P., GROVE, G. N., COWLING, R. A., and NELSON, M. J. (1994) X-ray Spectroscopy of the Iron Site in Soybean Lipoxygenase-1: Changes in Coordination Upon Oxidation or Addition of Methanol. *Biochemistry* 33, 15023–15035.

SCHEWE, T., and KÜHN, H. (1991) Do 15-Lipoxygenases Have a Common Biological Role? *Trends Biochem. Sci.*, 16, 369–373.

SCHEWE, T., KUHN, H., and RAPOPORT, S. M. (1986) Positional Specificity of Lipoxygenases and Their Suitability for Testing Potential Drugs. *Prost. Leukot. Med.*, 23, 155–160.

SCHILSTRA, M. J., VELDINK, G. A., VERHAGEN, J., and VLIEGENTHART, J. F. G. (1992) Effect of Lipid Hydroperoxide on Lipoxygenase Kinetics. *Biochemistry*, 31, 7692–7699.

SCHILSTRA, M. J., VELDINK, G. A., and VLIEGENTHART, J. F. G. (1993) Kinetic Analysis of the Induction Period in Lipoxygenase Catalysis. *Biochemistry*, 32, 7686–7691.

SHIBATA, D., STECZKO, J., DIXON J. E., ANDREWS, P. C., HERMODSON, M. and AXELROD, B. (1988) Primary Structure of Soybean Lipoxygenase L-2. *J. Biol. Chem.*, 263, 6816–6821.

SHIBATA, D., STECZKO, J., DIXON, J. E., HERMODSON, M., YAZDANPARAST, R., and AXELROD, B. (1987) Primary Structure of Soybean Lipoxygenase-1. *J. Biol. Chem.*, 262, 10080–10085.

SIEDOW, J. N. (1991) Plant Lipoxygenase: Structure and Function. *Annu. Rev. Plant Physiol. Plant Mol. Biol.*, 42, 145–188.

SIGAL, E., CRAIK, C. S., HIGHLAND, E., GRUNBERGER, D., COSTELLO, L. L., DIXON, R. A. F., and NADEL, J. A. (1988) Molecular Cloning and Primary Structure of Human 15-Lipoxygenase. *Biochem. Biophys. Res. Commun.*, 157, 457–464.

SLAPPENDEL, S., AASA, R., MALMSTRÖM, B. G., VERHAGEN, J., VELDINK, G. A., and VLIEGENTHART, J. F. G. (1982a) Factors Affecting the Line-shape of the EPR Signal of High-spin Fe(III) in Soybean Lipoxygenase. *Biochim. Biophys. Acta*, 708, 259–265.

SLAPPENDEL, S., MALMSTRÖM, B. G., PETERSSON, L., EHRENBERG, A., VELDINK, G. A., and VLIEGENTHART, J. F. G. (1982b) On the Spin and Valence State of the Iron in Native Soybean Lipoxygenase-1. *Biochem. Biophys. Res. Commun.*, 708, 673–677.

SLAPPENDEL, S., VELDINK, G. A., VLIEGENTHART, J. F. G., AASA, R., and MALMSTRÖM, B. G. (1983) A Quantitative Optical and EPR Study on the Interaction Between Soybean Lipoxygenase-1 and 13-L-hydroperoxylinoleic Acid. *Biochim. Biophys. Acta*, 747, 32–36.

SLOANE, D. L., LEUNG, R., CRAIK, C. S., and SIGAL, E. (1991) A Primary Determinant for Lipoxygenase Positional Specificity. *Nature*, 354, 149–152.

SMITH, W. L., and MARNETT, L. J. (1991) Prostaglandin Endoperoxide Synthase: Structure and Catalysis. *Biochim. Biophys. Acta*, 1083, 1–17.

SPAAPEN, L. J. M., VELDINK, G. A., LIEFKINS, T. J., VLIEGENTHART, J. F. G., and KAY, C. M. (1979) Circular Dichroism of Lipoxygenase-1 from Soybeans. *Biochim. Biophys. Acta*, 574, 301–311.

SPAAPEN, L. J. M., VERHAGEN, J., VELDINK, G. A., and VLIEGENTHART, J. F. G. (1980) The Effect of Modification of Sulfhydryl Groups in Soybean Lipoxygenase-1. *Biochim. Biophys. Acta*, 618, 153–162.

STECZKO, J., and AXELROD, B. (1992) Identification of the Iron-binding Histidine Residues in Soybean Lipoxygenase-1. *Biochem. Biophys. Res. Commun.*, 186, 686–689.

STECZKO, J., DONOHO, G. P., CLEMENS, J. C., DIXON, J. E., and AXELROD, B. (1992) Conserved Histidine Residues in Soybean Lipoxygenase: Functional Consequences of Their Replacement. *Biochemistry*, 31, 4053–4057.

STUBBE, J. (1990) Ribonucleotide Reductases: Amazing and Confusing. *J. Biol. Chem.*, 265, 5329–5332.

SWERN, D. (1979) Peroxides, in *Comprehensive Organic Chemistry* (J. F. Stoddart, Ed.), Pergamon Press, New York, pp. 909–940.

TSAI, A. L., PALMER, G., and KULMACZ, R. J. (1992) Prostaglandin H Synthase: Kinetics of Tyrosyl Radical Formation and of Cyclooxygenase Catalysis. *J. Biol. Chem.*, 267, 17753–17759.

VAN DER HEIJDT, L. M., FEITERS, M. C., NAVARATNAM, S., NOLTING, H.-F., HERMES, C., VELDINK, G. A., and VLIEGENTHART, J. F. G. (1992) X-ray Absorption Spectroscopy of Soybean Lipoxygenase-1. Influence of Lipid Hydroperoxide Activation and Lyophilization on the Structure of the Non-heme Iron Active Site. *Eur. J. Biochem.*, 207, 793–802.

VAN ELDIK, R., COHEN, H., and MEYERSTEIN, D. (1994) Ligand Interchange Controls Many Oxidations of Divalent First-Row Transition Metal Ions by Free Radicals. *Inorg. Chem.* 33, 1566–1568.

VELDINK, G. A., GARSSEN, G. J., VLIEGENTHART, J. F. G., and BOLDINGH, J. (1972) Positional Specificity of Corn Germ Lipoxygenase as a Function of pH. *Biochem. Biophys. Res. Commun.*, 47, 22–26.

VELDINK, G. A., and VLIEGENTHART, J. F. (1984) Lipoxygenases, Nonheme Iron-containing Enzymes. *Adv. Inorg. Biochem.*, 6, 139–161.

VOLKER WAGNER, A. F., FREY, M., NEUGEBAUER, F. A., SCHÄFER, W., and KNAPPE, J. (1992) The Free Radical in Pyruvate Formate-lyase is Located on Glycine-734. *Proc. Natl. Acad. Sci. U.S.A.*, 89, 996–1000.

WALLING, C. (1995) Liquid Phase Autoxidation, in *Active Oxygen in Chemistry* (C. S. Foote, J. S. Valentine, A. Greenberg, and J. F. Liebman, Eds.), Chapman & Hall, New York, pp. 24–65.

WANG, Z. X., KILLILEA, S. D., and SRIVASTAVA, D. K. (1993) Kinetic Evaluation of Substrate-dependent Origin of the Lag Phase in Soybean Lipoxygenase-1 Catalyzed Reactions. *Biochemistry*, 32, 1500–1509.

WASSERMAN, M. A., SMITH, E. F., III, UNDERWOOD, D. C., and BARNETTE, M. A. (1991) Pharmacology and Pathophysiology of 5-Lipoxygenase Products, in *Lipoxygenases and Their Products* (S. T. Crooke and A. Wong, Eds.), Academic Press, San Diego, pp. 1–50.

WHITTAKER, J. W., and SOLOMON, E. I. (1988) Spectroscopic Studies on Ferrous Non-heme Iron Active Sites: MCD of Mononuclear Sites in Superoxide Dismutase and Lipoxygenase. *J. Am. Chem. Soc.*, 110, 5329–5339.

WILKINS, P. C., DALTON, H., PODMORE, I. D., DEIGHTON, N., and SYMONS, M. C. R. (1992) Biological Methane Activation Requires the Intermediacy of Carbon-centered Radicals. *Eur. J. Biochem.*, 210, 67–72.

WISEMAN, J. S. (1989) Alpha-secondary Isotope Effects in the Lipoxygenase Reaction. *Biochemistry*, 28, 2106–2111.

WISEMAN, J. S., and NICHOLS, J. S. (1988) Ketones as Electrophilic Substrates of Lipoxygenase. *Biochem. Biophys. Res. Commun.*, 154, 544–549.

WISEMAN, J. S., SKOOG, M. T., and CLAPP, C. H. (1988) Activity of Soybean Lipoxygenase in the Absence of Lipid Hydroperoxide. *Biochemistry*, 27, 8810–8813.

WRIGHT, S. W., and NELSON, M. J. (1992) Episulfide Inhibitors of Lipoxygenase. *Bioorg. Med. Chem. Lett.*, 2, 1385–1390.

YAMAGUCHI, K., YABUSHITA, S., and FUENO, T. (1981a) Geometry Optimizations of the Dioxetane, Perepoxide and 1,4-Diradicals for the Ethylene Plus Molecular Oxygen System: Mechanism of Photooxygenation of Olefins. *Chem. Phys. Lett.*, 78, 572–577.

YAMAGUCHI, K., YABUSHITA, S., FUENO, T., and HOUK, K. N. (1981b) Mechanism of Photooxygenation Reactions. Computational Evidence Against the Diradical Mechanism of Singlet Oxygen Ene Reactions. *J. Am. Chem. Soc.*, 103, 5043–5046.

ZANG, Y., ELGREN, T. E., DONG, Y., and QUE, L., JR. (1993) A High-potential Ferrous Complex and Its Conversion to an Alkylperoxoiron(III) Intermediate. A Lipoxygenase Model. *J. Am. Chem. Soc.*, 115, 811–813.

ZHANG, Y., GEBHARD, M. S., and SOLOMON, E. I. (1991) Spectroscopic Studies of the Non-heme Ferric Active Site in Soybean Lipoxygenase: Magnetic Circular Dichroism as a Probe of Electronic and Geometric Structure. Ligand-field Origin of Zero-field Splitting. *J. Am. Chem. Soc.*, 113, 5162–5175.

ZHANG, Y. Y., RÅDMARK, O., and SAMUELSSON, B. (1992) Mutagenesis of Some Conserved Residues in Human 5-Lipoxygenase; Effects on Enzyme Activity. *Proc. Natl. Acad. Sci. U.S.A.*, 89, 485–489.

# 7
# The Biological Significance of Oxygen-Derived Species

BARRY HALLIWELL

### INTRODUCTION: OXYGEN TOXICITY

When living organisms first appeared on Earth, they did so under an atmosphere containing very little $O_2$, i.e., they were essentially anaerobes. Anaerobic microorganisms survive to this day, but their growth is inhibited and they can often be killed by exposure to 21% $O_2$, the current atmospheric level. As the $O_2$ content of the atmosphere rose (due to the evolution of organisms with photosythetic water-splitting capacity), many primitive organisms probably died. Present-day anaerobes are presumably the descendants of those primitive organisms that followed the evolutionary path of adapting to rising atmospheric $O_2$ levels by restricting themselves to environments that the $O_2$ did not penetrate. However, other organisms began the evolutionary process of developing *antioxidant defense systems* to protect against $O_2$ toxicity. In retrospect, this was a fruitful path to follow since organisms that tolerated the presence of $O_2$ could also evolve to use it for metabolic transformation (oxidases, oxygenases, etc.) and for efficient energy production by using electron transport chains with $O_2$ as the terminal electron acceptor, such as those present in mitochondria.

However, even present-day aerobes suffer demonstrable injurious effects if exposed to $O_2$ at concentrations greater than 21% (Balentine, 1982). Oxygen toxicity has been demonstrated in plants, animals, and microorganisms. For example, exposure of adult humans to pure $O_2$ at 1 atm pressure for as little as 6 hours causes chest soreness, cough, and sore throat in some subjects, and longer periods of exposure lead to damage to the alveoli of the lungs. The form of ocular damage in babies known as *retrolental fibroplasia* (formation of fibrous tissue behind the lens) increased abruptly in incidence in the early 1940s among babies born prematurely and often led to blindness. Not until 1954 was it realized that retrolental fibroplasia is associated with the use of high $O_2$ concentrations in incubators for premature babies. More careful control of $O_2$ concentrations (continuous transcutaneous $O_2$ monitoring, with supplementary $O_2$ given only where necessary) and administration of the lipid-soluble antioxidant α-tocopherol have decreased its severity, but the problem has not disappeared since many premature infants need high $O_2$ in order to survive at all (Ehrenkranz, 1989).

The damaging effects of elevated $O_2$ on aerobes vary considerably with the organism studied, age, physiological state, and diet, and different tissues are affected in different ways. Thus, cold-blooded animals such as turtles and crocodiles are relatively resistant to $O_2$ at low environmental temperatures but become more sensitive at higher temperatures. Young rats resist lung damage in an atmosphere of 100% $O_2$ far more effectively than do adult rats (Balentine, 1982).

The earliest explanation of $O_2$ toxicity was that $O_2$ is a direct inhibitor of enzymes, thereby interfering with metabolism. However, very few targets of direct damage by $O_2$ have been identified in aerobes. In 1954, Gerschman et al. stated that the damaging effects of $O_2$ could be attributed to the formation of oxygen radicals. This hypothesis was popularized and converted into the *superoxide theory* of $O_2$ toxicity following the discovery of superoxide dismutase (SOD) enzymes by McCord and Fridovich (reviewed by Fridovich, 1989). In its simplest form, this theory states that $O_2$ toxicity is due to excess formation of superoxide radical ($O_2^-$) and that the SOD enzymes are important antioxidant defenses.

## REACTIVE OXYGEN SPECIES *IN VIVO*

*Reactive oxygen species* (ROS) is a collective term used by biologists to include not only the oxygen-centered radicals ($O_2^-$, superoxide, and OH·, hydroxyl) but also some nonradical derivatives of $O_2$, such as hydrogen peroxide ($H_2O_2$), $^1O_2$ ($^1\Delta_g$), hypochlorous acid (HOCl)* and ozone ($O_3$). *Reactive* is, of course, a relative term: neither $O_2^-$ nor $H_2O_2$

---

*HOCl may be equally well regarded as an active chlorine species.

is particularly reactive in aqueous solution (Bielski and Cabelli, 1995), especially when compared with OH·.

All organisms suffer some exposure to OH· because it is generated by homolytic fission of O—H bonds in water driven by background ionizing radiation (von Sonntag, 1987). This radical is so reactive with all biological molecules that it is impossible to evolve a specific scavenger of it; almost everything in living organisms reacts with OH· with second-order rate constants of $10^9$–$10^{10}$ $M^{-1}s^{-1}$ (Anbar and Neta, 1967), i.e., collision of OH· with the molecules almost always results in reaction. Damage caused by OH·, once this radical has been formed, is probably unavoidable and is dealt with by repair processes (Table 7-1).

As a result of the pioneering work of many groups (reviewed by Chance et al., 1979; Fridovich, 1989; Halliwell and Gutteridge, 1989; Sies, 1985, 1991; Gutteridge and Halliwell, 1995), it has become well established that $O_2^-$ and $H_2O_2$ are produced in aerobes, although the precise amounts generated in mammals and the steady-state concentrations achieved are still uncertain. Generation of these species can occur by two types of reaction:

1. *Accidental generation.* This encompasses such mechanisms as "leakage" of electrons onto $O_2$ from mitochondrial electron transport chains, microsomal cytochromes P-450 and their electron-donating enzymes, and other systems (Fridovich, 1989; Imlay and Fridovich, 1991). For example, the mitochondrial electron transport chain is a gradient of redox potential, from the highly reducing NADH/NAD$^+$ couple to the oxidizing $O_2$. Thermodynamically, there is nothing to prevent constituents of the early part of the chain (nonheme iron proteins, flavoproteins, cytochromes b) from reducing $O_2$ directly. Fortunately, such reactions are kinetically restricted, so that most of the electrons entering mitochondrial electron transport arrive at cytochrome oxidase. It is often stated that "only" 1–3% of the electrons may leak to $O_2$ to form $O_2^-$. Since we breathe in such large amounts of $O_2$, however, a lot of radicals are generated (Scheme 1).

    Superoxide and $H_2O_2$ may also be generated by so-called autoxidation reactions in which compounds such as catecholamines, tetrahydrofolates, and reduced flavins react directly with $O_2$ to form $O_2^-$. The $O_2^-$ then oxidizes more of the compound and sets up a chain reaction. In fact, most so-called autoxidation reactions in biological systems usually depend on the presence of catalytic transition metal ions.

2. *Deliberate synthesis,* Examples include the production of $O_2^-$ and $H_2O_2$ by activated phagocytes (Babior and Woodman, 1990) and of $H_2O_2$ by several oxidase enzymes, such as glycolate oxidase and D-amino acid oxidase, and the oxidation of fatty acids in peroxisomes (Chance et al., 1979). Evidence is accumulating that $O_2^-$ is also produced *in vivo* by

several cell types other than phagocytes, including lymphocytes (Maly, 1990) and fibroblasts (Meier et al., 1990; Murrell et al., 1990). Superoxide produced by such cells has been proposed to be involved in intercellular signaling and growth regulation, but this has yet to be substantiated (Burdon and Rice-Evans, 1989). Vascular endothelial cells also generate $O_2^-$ (Arroyo et al., 1990; Babbs et al., 1991; Britigan et al., 1992). Generation of $O_2^-$ and $H_2O_2$ by phagocytes is known to play an important part in the killing of several bacterial strains. If formation of these species is defective (as happens in chronic granulomatous disease), a syndrome of persistent multiple infections by several bacterial strains, especially *Staphylococcus aureus*, results (Babior and Woodman, 1990). Reactive oxygen species generated by phagocytes may also influence the behavior of other cells at sites of inflammation (Meier et al., 1990; Murrell et al., 1990; Kubes et al., 1991).

An adult at rest utilizes 3.5 ml $O_2$/kg/min
or 352.8 l/day
or 14.7 moles/day.
If 1% makes $O_2^-$,
this is 0.147 moles/day
or 53.66 moles/year
or about 1.72 kg/year (of $O_2^-$).
During exertion, this can increase up to 20-fold.

**Scheme 7-1**

Some other metabolic roles for $H_2O_2$ have been proposed (e.g., Dupuy et al., 1991; Riley and Behrman, 1991; Shapiro, 1991). For example, $H_2O_2$ generated in the thyroid gland is used by a peroxidase enzyme to iodinate the thyroid hormones. $H_2O_2$ may up-regulate the expression of certain genes by leading to displacement of an inhibitory subunit from the cytoplasmic transcription factor NF-κB (Schreck et al., 1992). Displacement of the inhibitory subunit from this multiprotein complex causes it to migrate to the nucleus and activate expression of multiple genes. There is particular interest in this area at the moment because of proposals that reactive oxygen species can activate expression of human immunodeficiency virus I (HIV-I), the most common cause of acquired immunodeficiency syndrome (AIDS). When HIV-I infects its target cells, virally specified DNA becomes integrated into the host cell DNA. Activation of NF-κB can cause expression of this DNA, leading to viral multiplication in the cell (Schreck et al., 1992).

## ANTIOXIDANT DEFENSES

Several types of antioxidant defense occur *in vivo*. There are the low molecular mass scavengers of radicals, such as α-tocopherol, ascorbic

**TABLE 7-1** Repair of Oxidative Damage

| Substrate of Damage | Repair System | Representative Recent References |
|---|---|---|
| *DNA*: All components of DNA can be attacked by OH·, whereas singlet $O_2$ attacks guanine preferentially. $H_2O_2$ and $O_2^-$ do not attack DNA. | A wide range of enzymes exists that recognize abnormalities in DNA and remove them by excision, resynthesis and rejoining of the DNA strand. | Breimer (1991) |
| *Proteins*: Many ROS can oxidize -SH groups. OH· attacks many amino acid residues. Proteins often bind transition metal ions, making them a target of attack by site-specific OH· generation. | Oxidized methionine residues may be repaired by methionine sulfoxide reductase. Other damaged proteins may be recognized and preferentially destroyed by cellular proteases. | Stadtman and Oliver (1991), Brot and Weissbach (1991), Salo et al. (1988) |
| *Lipids*: Some ROS (excluding $O_2^-$ and $H_2O_2$) can initiate lipid peroxidation. | Chain-breaking antioxidants (especially α-tocopherol) remove chain-propagating peroxyl radicals. Phospholipid hydroperoxide glutathione peroxidase can remove peroxides from membranes, as can some phospholipases. Normal membrane turnover can release damage lipids. | Burton and Traber (1990), Maiorino et al. (1991), Sevanian et al. (1988) |

acid (vitamin C), uric acid and reduced glutathione, GSH (discussed by Halliwell and Gutteridge, 1989; Burton and Traber, 1990; Halliwell, 1990; Jain et al., 1992). Repair systems are also important (Table 7-1): products of free radical damage to the purine and pyrimidine bases of DNA can be measured in significant amounts in human urine, suggesting that oxidative DNA damage is constantly being repaired *in vivo* (Ames, 1989; Breimer, 1991). Hydroxyl radicals form a multiplicity of products from all four DNA bases; products include cytosine and thymine glycols, hydantoins, ring-opened purines, 2-hydroxyadenine, and 8-hydroxyguanine. By contrast, $H_2O_2$ and $O_2^-$ do not attack DNA bases, whereas $^1O_2$ appears to attack guanine specifically, generating 8-hydroxyguanine and ring-opened guanine (Dizdaroglu, 1991).

At least as important as these radical scavengers, and probably more so, are the antioxidant defense enzymes: superoxide dismutases, catalases, and glutathione peroxidases (Table 7-2). Experimental manipulations of antioxidant defense enzymes have shown that removal of most $O_2^-$ and $H_2O_2$ appears to be essential *in vivo*. The clearest data have come from experiments using modern molecular techniques (Fridovich, 1989; Touati, 1989; Amstad et al., 1991; Chang et al., 1991; White et al., 1991). White et al. (1991) showed that transgenic mice synthesizing human copper-zinc SOD were resistant to the toxic effects of elevated $O_2$ (Figure 7-1 illustrates the principles by which transgenic animals are produced.) Some forms of amyotrophic lateral sclerosis (Lou Gehrig's disease), a fatal degenerative disorder of motor neurons in the brain and spinal cord, are related to defects in the gene encoding CuZnSOD (discussed by McNamara and Fridovich, 1993), although exactly how is uncertain.

The antioxidant defense enzymes act as a *coordinated* system: hence, for example, too much SOD in relation to the activities of $H_2O_2$-

**TABLE 7-2**  Reactions Catalyzed by Antioxidant Defense Enzymes

| Enzyme | Reaction | Comment |
| --- | --- | --- |
| SOD | $2O_2^- + 2H^+ \rightarrow H_2O_2 + O_2$ | CuZnSOD largely in cytosol, Mn-SOD in mitochondria of animals. |
| Catalase | $2H_2O_2 \rightarrow 2H_2O + O_2$ | Located only in peroxisomes in most animal tissues; destroys $H_2O_2$ generated by peroxisomal oxidase enzymes. |
| Glutathione peroxidases | $2GSH + H_2O_2 \rightarrow GSSG + 2H_2O$ | Selenoproteins. Major route of $H_2O_2$ removal in human tissues. Glutathione reductase enzymes reduce GSSG to 2GSH, using NADPH as a source of reducing power. |

Down's syndrome (DS) is the most common human genetic disorder, occurring once in every 600–800 live births. Defects that may be suffered by DS individuals include short stature, malformation of the skin around the eyes, and mental retardation. Patients who survive beyond their thirties have an increased risk of developing Alzheimer's disease (this dementing condition normally affects over 10–15% of individuals of age 65 and over and perhaps 20% of those over the age of 80. It is marked by gradually developing loss of memory combined with growing confusion and disorientation). DS is usually caused by having three copies of chromosome 21 (trisomy 21), one of whose genes encodes CuZnSOD.

*Question*: Does the elevated level of CuZnSOD cause or contribute to DS?

*Approach to an Answer*: Produce transgenic mice that overexpress human CuZnSOD.
1. Isolate embryos from the reproductive tracts of female mice.
2. Microinject DNA carrying the gene for human CuZnSOD and the DNA sequences needed for its expression into the embryo pronucleus.
3. Transfer embryos into the reproductive tracts of "pseudopregnant" foster mothers.
4. Allow them to develop to term.
5. Select and breed the progeny expressing human CuZnSOD in tissues, thereby raising total SOD activity.
6. The resulting mice have elevated levels of CuZnSOD and
   a. are more resistant than controls to $O_2$ toxicity.
   b. are more resistant than controls to certain toxic agents.
   c. show abnormal neuromuscular junctions in the tongue.
   d. may show some of the other neurological defects characteristic of DS.

**FIGURE 7-1** Application of transgenic animal technology to study the biological effects of elevated SOD.

removing enzymes can be deleterious (Scott et al., 1989; Groner et al., 1990; Amstad et al., 1991). For example, mice transgenic for human CuZnSOD (Figure 7-1) are resistant to elevated $O_2$ and to certain toxins, but they show certain neuromuscular abnormalities resembling those found in patients with Down's syndrome (Groner et al., 1990). The gene encoding CuZnSOD is located on chromosome 21 in humans, and Down's syndrome is usually caused by trisomy of this gene, raising tissue CuZnSOD levels by about 50%. The limited data available at present are consistent with the view that the excess of CuZnSOD may contribute to at least some of the abnormalities in patients with Down's syndrome. (In all transgenic studies, one must be careful to check that overexpression of the transferred gene does not lead to secondary metabolic responses that could explain the events observed.)

Overall, antioxidant defenses seem to be approximately in balance with generation of oxygen-derived species *in vivo*, although some oxidative damage does occur continuously (Table 7-3). There appears to be no great reserve of antioxidant defenses in mammals. Indeed, if we had a large excess of antioxidants, we would be less dependent on repair systems (Table 7-1). However, antioxidant defenses can often

**TABLE 7-3** Evidence That Oxidative Stress Occurs *in vivo*

| Target of Damage | Evidence |
|---|---|
| DNA | Urinary excretion of DNA base damage products (Degan et al., 1991). Defects in DNA repair produce severe metabolic consequences (Breimer, 1991). |
| Protein | Attack of free radicals on proteins produces protein carbonyls. Low levels of these can be detected in animal tissues and body fluids (Stadtman and Oliver, 1991). |
| Lipid | Accumulation of "age pigments" in tissues. Lipid peroxidation in atherosclerotic lesions (Steinberg et al., 1989) (See also Chapter 10). Presence of end products of peroxidation in animal body fluids (for a recent review of the methodological problems in measuring such products, see Halliwell and Chirico, 1993). |

be induced in response to increased oxidative stress (Fridovich, 1989; Tartaglia et al., 1992). Perhaps one explanation of our inadequate antioxidant defenses is that some oxygen-derived species perform useful metabolic roles.

## TOXICITY OF SUPEROXIDE AND HYDROGEN PEROXIDE

Experimental data show clearly that removal of $O_2^-$ and $H_2O_2$ by antioxidant defense systems is essential for healthy aerobic life (Chance et al., 1979; Fridovich, 1989; Halliwell and Gutteridge, 1989; Touati, 1989). Why is this? In organic media, $O_2^-$ can be very reactive (Bielski and Cabelli, 1995), but in aqueous media, it is not, acting mainly as a moderate reducing agent. However, superoxide can react with some targets. In particular, $O_2^-$ can combine with nitric oxide and the rate constant is $>10^9 M^{-1} s^{-1}$ (Huie and Padmaja, 1993). Nitric oxide is also a free radical, although not a very reactive one.

$$O_2^- + NO \rightarrow \underset{\text{peroxynitrite}}{ONOO^-} \tag{1}$$

Nitric oxide is known to be produced *in vivo* by vascular endothelial cells and phagocytes. It performs useful physiological processes, such as regulation of vascular smooth muscle tone (hence, controlling blood pressure) and the killing of some parasites by macrophages (Liew et al., 1990; Moncada et al., 1991). Since NO· acts on smooth muscle cells in vessel walls to produce relaxation, $O_2^-$, by removing NO, can act as a vasoconstrictor, and this might have deleterious effects in some clinical situations (Laurindo et al., 1991; Nakazono et al., 1991). For example, excess vascular $O_2^-$ production could lead to hypertension.

There is considerable debate in the literature as to the biological significance of reaction (1), especially as vascular endothelial cells can produce not only NO but also $O_2^-$. Peroxynitrite could itself be damaging to cells (Radi et al., 1990). Another possibility is that peroxynitrite forms OH· or a species resembling OH· (Beckman et al., 1990; Hogg et al., 1992).

$$OONO^- + H^+ \rightarrow NO_2^. + OH^. \qquad (2)$$

Like NO, $O_2^-$ in aqueous solution is not very reactive. However, it has been claimed to be capable of inactivating several bacterial enzymes, such as *Escherichia coli* dihydroxyacid dehydratase, aconitase, or 6-phosphogluconate dehydratase (Fridovich, 1989; Gardner and Fridovich, 1991, 1992; Imlay and Fridovich, 1991). It has been suggested to attack iron–sulfur clusters at the enzyme active sites, although the precise chemistry of such reactions remains to be defined. Another unanswered question is whether such reactions of $O_2^-$ occur in animals. However, in isolated submitochondrial particles, $O_2^-$ has been claimed to inactivate the NADH dehydrogenase complex of the mitochondrial electron transport chain (Zhang et al., 1990).

The protonated form of $O_2^-$, perhydroxyl radical ($HO_2^.$), is much more reactive than $O_2^-$ *in vitro* (Bielski and Cabelli, 1995), but very little $O_2^-$ is protonated at pH 7.4 (the p$K_a$ is ~ 4.8). For example, $HO_2^.$ can initiate peroxidation of polyunsaturated fatty acids, which $O_2^-$ cannot (Bielski et al., 1983).

$$L—H + HO_2^. \rightarrow L^. + H_2O_2 \qquad (3)$$

$HO_2^.$ can also convert lipid hydroperoxides to peroxyl radicals (Aikens and Dix, 1991)

$$HO_2^. + L—OOH \rightarrow H_2O_2 + \underset{\text{peroxyl radical}}{L—OO^.} \qquad (4)$$

However, there is no *direct* evidence that $HO_2^.$ exerts damaging effects *in vivo*. $H_2O_2$ at micromolar levels also appears to be poorly reactive, but higher (>50 µM) levels of $H_2O_2$ can attack certain cellular targets.

For example, it can oxidize an essential -SH group on the glycolytic enzyme glyceraldehyde-3-phosphate dehydrogenase, blocking glycolysis and interfering with cell energy metabolism (Cochrane, 1991).

Beauchamp and Fridovich (1970) proposed that the toxicity of $O_2^-$ and $H_2O_2$ could involve their conversion into OH˙. Several mechanisms have been proposed to explain this. One involves the interaction of $O_2^-$ and NO (eqs. 1 and 2). An earlier proposal (Halliwell, 1978; McCord and Day, 1978) was the superoxide-driven Fenton reaction:

$$O_2^- + Fe^{III} \rightarrow Fe^{II} + O_2 \qquad (5)$$

$$Fe^{II} + H_2O_2 \rightarrow OH˙ + OH^- + Fe^{III} \qquad (6)$$

Despite repeated controversy in the literature, the author feels that the evidence for the formation of OH˙ in Fenton reactions is overwhelming (for reviews, see Halliwell and Gutteridge, 1990a, 1990b, 1992; Burkitt, 1993). This does not, of course, preclude the formation of reactive species additional to OH˙, such as oxo–iron complexes (ferryl, perferryl). However, none of these has yet been characterized chemically in Fenton systems.

Iron (and other transition metal ions) in chemical forms that can decompose $H_2O_2$ to OH˙ are in very short supply *in vivo*. Animals are very careful to ensure that as much iron and copper as possible is kept safely bound to transport or storage proteins. Indeed, this "sequestration" of metal ions is an important antioxidant defense mechanism (Halliwell and Gutteridge 1989, 1990b). Sequestration of metal ions deters the growth of bacteria in human blood plasma (Weinberg, 1990); it also ensures that plasma will not convert $O_2^-$ and $H_2O_2$ into OH˙. This may allow $O_2^-$ and $H_2O_2$ released into the extracellular environment (e.g., from endothelial cells, lymphocytes, and phagocytes) to perform useful metabolic roles rather than causing damage by forming OH˙. The presence of non-protein-bound iron or copper ions in plasma, as happens in certain metal overload diseases, produces severe damage to many body tissues.

In any case, any transition metal ions that do become available to catalyze free radical reactions *in vivo* will not exist in the free state for very long. Thus, if iron ions are liberated from storage sites, they must bind to a biological molecule or else eventually precipitate out of solution as ferric hydroxides, oxyhydroxides, and phosphates [body fluids and the intracellular environment operate at an alkaline pH (7.4) and are rich in phosphate and bicarbonate ions]. However, if transition metal ions bind to a target, that target can become a site of free radical damage. If bound metal ions react with $O_2^-$ and $H_2O_2$, damage to the ligand will be site specific. The $OH^{\cdot}$ generated will react rapidly with the target, whose concentration in the immediate vicinity of the $OH^{\cdot}$ will be very great. Site-specific $OH^{\cdot}$ damage does not usually resemble the damage caused by $OH^{\cdot}$ generated in free solution, e.g., by radiolysis of aqueous solutions (Borg and Schaich, 1984), which can attack any site on the target randomly. Site-specific damage involving $OH^{\cdot}$ has been demonstrated for glycoproteins (Cooper et al., 1985), nucleic acids (Halliwell and Aruoma, 1991), and proteins (Gutteridge and Wilkins, 1983; Marx and Chevion, 1986; Stadtman and Oliver, 1991).

Hence, a major determinant of the nature of the damage done by excess generation of reactive oxygen species *in vivo* may be the availability and location of metal ion catalysts of $OH^{\cdot}$ radical formation (Halliwell and Gutteridge, 1989, 1990a). If, for example, catalytic iron salts are bound to DNA in one cell type and to membrane lipids in another, then excessive formation of $H_2O_2$ and $O_2^-$ will, in the first case, fragment the DNA and in the second case could initiate lipid peroxidation. Evidence for $OH^{\cdot}$ formation in the nucleus of cells treated with $H_2O_2$ has been obtained by showing that all four DNA bases are modified in the way characteristic of $OH^{\cdot}$ attack (Dizdaroglu et al., 1991). If this $OH^{\cdot}$ is formed by metal ion-dependent reactions, then the catalytic metal ions must be bound to the DNA itself.

*E. coli* mutants lacking SOD activity are hypersensitive to damage by $H_2O_2$ (Touati, 1989), and extra SOD can protect liver cells against damage by $H_2O_2$, provided that it can enter the cell (Kyle et al., 1988). These data are consistent with a role of $O_2^-$ in facilitating damage by $H_2O_2$, and reactions (5) and (6) provide a possible explanation. However, many scientists are reluctant to believe that $O_2^-$ serves only as a reducing agent for $Fe^{III}$, since, in general, mammalian tissues are fairly reducing environments (containing ascorbate, GSH, NADH, and NADPH, for example). The arguments have been presented in detail (Halliwell and Gutteridge, 1989, 1990a; Sutton and Winterbourn, 1989), but the point is not yet settled. Superoxide can also participate in providing the iron needed for Fenton chemistry (see below).

## OXIDATIVE STRESS

Generation of reactive oxygen species and the activity of antioxidant defenses appear approximately balanced *in vivo*, although there seems to be ongoing oxidative damage. If they become seriously imbalanced in favor of the former, *oxidative stress* is said to result (Sies, 1985, 1991).

Most aerobes can tolerate mild oxidative stress; indeed, they often respond to it by increasing the synthesis of antioxidant defense enzymes. For example, if rats are gradually acclimatized to elevated $O_2$, they can tolerate pure $O_2$ for much longer periods than control rats, apparently due to the increased synthesis of antioxidant defense enzyme and of GSH in the lung (e.g., Frank et al., 1989; Iqbal et al., 1989). Another example is the complex response of *E. coli* to treatment with low concentrations of $H_2O_2$ (Figure 7-2; Storz and Tartaglia 1992).

However, severe oxidative stress can produce major interrelated derangements of cell metabolism, including DNA strand breakage (often an early event), rises in intercellular free $Ca^{2+}$, damage to membrane ion transporters and/or other specific proteins, and peroxidation of lipids (Orrenius et al., 1989; Cochrane, 1991; Halliwell and Aruoma, 1991; Stadtman and Oliver, 1991). Damage can be direct (e.g., if high levels of $H_2O_2$ directly oxidize thiol groups on glyceraldehyde-3-phosphate dehydrogenase or on membrane transport proteins, or if OH˙ is formed close to DNA and fragments it) and/or indirect. Oxidative stress damages the mechanisms which maintain intracellular $Ca^{2+}$ concentrations at submicromolar concentrations. The resulting rise in free $Ca^{2+}$ can activate proteases, which may attack the cytoskeleton of the cell and cause it to collapse, as well as nucleases which can fragment DNA (Orrenius et al., 1989). Protein kinases and thiol-containing proteins (such as transport proteins and thioredoxin) can respond to oxidative stress in ways that perturb cell metabolism. Thus, there is a fine balance between the regulatory properties of ROS (e.g., produced by phagocytes at sites of acute controlled inflammation) and

---

When *Escherichia coli* (a bacterium found in the human colon) is treated with low doses of $H_2O_2$, it adapts and becomes resistant to high levels of $H_2O_2$ that would normally kill it. One of the adaptation mechanisms is shown below. $H_2O_2$ also activates expression of other protective genes in *E. coli* by mechanisms not involving *oxy* R.

Treat with $H_2O_2$
↓
Oxidation of *oxy* R protein
↓
Oxidized protein switches on genes to increase the synthesis of at least eight proteins, including catalase and glutathione reductase.

Treat with $O_2^-$
↓
Superoxide forms $H_2O_2$ and activates the above $H_2O_2$ response system.
↓
Activates the *sox*R and *sox*S genes, leading to increased synthesis of at least nine proteins, including MnSOD and a DNA repair enzyme.

**FIGURE 7-2** Regulation of antioxidant defenses in a bacterium.

damaging properties, depending on the extent of ROS generation, the target subjected to oxidative stress, and the activity of antioxidant defenses. Another important factor to be considered is the presence or absence of transition metal ions (Halliwell and Gutteridge, 1990a, 1990b). Their presence will tend to worsen damage by converting fairly unreactive species into more reactive ones. Thus, $O_2^-$ and $H_2O_2$ can be converted into $OH^\cdot$. Transition metal ions can decompose lipid hydroperoxides into peroxyl and alkoxyl radicals, which can damage membrane proteins and propagate the chain reaction of lipid peroxidation.

The relative importance of damage to different molecular targets (DNA, proteins, lipids, carbohydrates) in causing cell injury or death resulting from severe oxidative stress depends on what degree of oxidative stress occurs, by what mechanism it is imposed, for how long, and the nature of the systems stressed. For example, considerable evidence indicates that lipid peroxidation occurs in human atherosclerotic lesions and makes a substantial contribution to development of atherosclerosis (Steinberg et al., 1989).

Several halogenated hydrocarbons (such as $CCl_4$ and bromobenzene) appear to exert some or all of their toxic effects by stimulating lipid peoxidation *in vivo* (Comporti, 1987). This stimulation is consequent on the metabolism of the compounds. For example, the cytochrome P-450 system of mammalian endoplasmic reticulum converts $CCl_4$ to trichloromethyl radical

$$CCl_4 + e^- \rightarrow CCl_3^\cdot + Cl^- \tag{7}$$

This carbon-centered radical reacts very rapidly with $O_2$:

$$CCl_3^\cdot + O_2 \rightarrow CCl_3O_2^\cdot \tag{8}$$

The resulting trichlomethylperoxyl radical efficiently abstracts hydrogen from polyunsaturated fatty acid side chains in membranes and initiates lipid peroxidation (Comporti, 1987).

However, for many other toxic agents, lipid peroxidation is not a major mechanism of primary cell injury by oxidative stress; damage to proteins and DNA is probably more important (Halliwell, 1987; Cochrane, 1991).

## OXIDATIVE STRESS AND TRANSITION METALS

Transition metal ions appear to be involved in at least part of the damage caused by oxidative stress. For example, Figure 7-3 sum-

## A. Fenton Chemistry

Oxidative stress → ˙OH generation upon DNA by reaction of $H_2O_2$ with transition metal ions already bound to DNA → Strand breakage / DNA base modification / Deoxyribose fragmentation

↓

Release of catalytic copper or iron ions within the cell → Binding of ions to DNA ↑

## B. Nuclease Activation

Oxidative stress → Inactivation of $Ca^{2+}$ binding by endoplasmic reticulum. Inhibition of plasma membrane $Ca^{2+}$ extrusion systems. Release of $Ca^{2+}$ from mitochondria.

↓

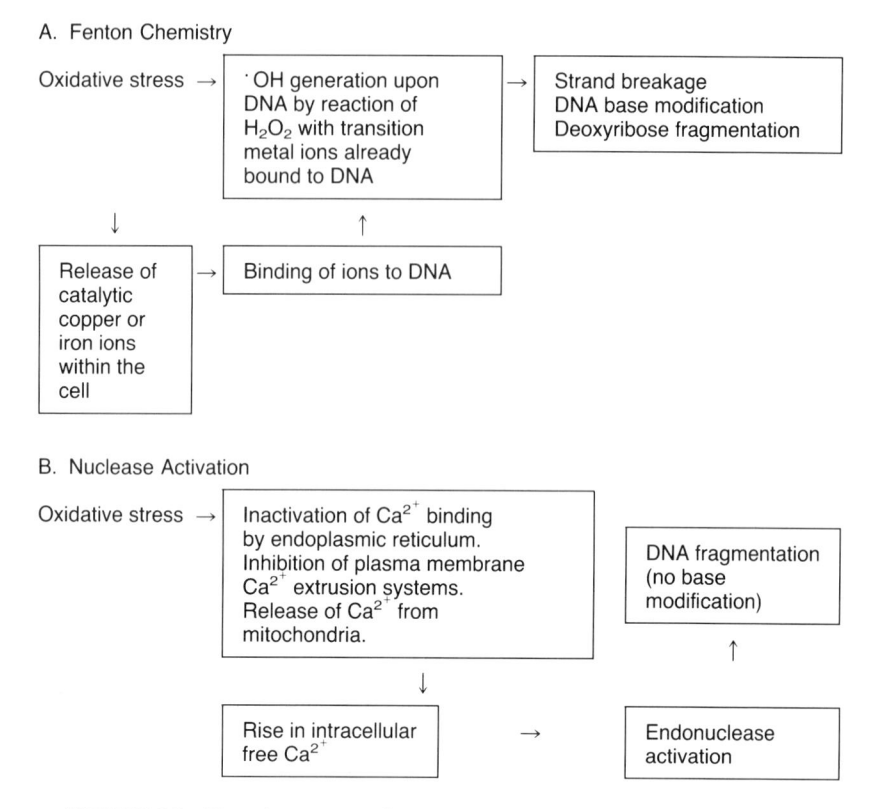

Rise in intracellular free $Ca^{2+}$ → Endonuclease activation → DNA fragmentation (no base modification)

**FIGURE 7-3** Hypotheses to explain DNA damage resulting from exposing cells to oxidative stress.

marizes the two mechanisms that have been put forward to explain why oxidative stress causes DNA damage. These mechanisms may be distinguished by examining the pattern of chemical changes in the DNA bases: nucleases only fragment the DNA backbone, whereas OH˙ attack on DNA produces a wide range of modified purines and pyrimidines (Dizdaroglu, 1991; Halliwell and Aruoma, 1991).

Halliwell (1987) has argued that, just as oxidative stress can cause rises in intracellular concentrations of free $Ca^{2+}$ ions by interfering with normal $Ca^{2+}$-sequestering mechanisms (Orrenius et al., 1989), oxidative stress may increase the iron ion concentration available within cells to catalyze free radical reactions. There is much evidence consistent with this view. To take one recent example, Ferrali et al. (1990) showed that iron ion release plays an important part in mediating the toxic effects of allyl alcohol in mice. Indeed, several iron ion chelating agents, the best-known example being desferrioxamine (Gutteridge et al., 1979), are able to inhibit at least some of the consequences of oxidative stress in cell, organ, and whole animal systems (reviewed by Halliwell, 1989).

Where could this released iron come from? One suggestion is that mitochondria might take up free $Ca^{2+}$, generated as a result of oxidative stress, and release $Fe^{II}$ in exchange (Merryfield and Lardy, 1982). Superoxide can reductively mobilize ferrous iron from the iron storage protein ferritin (Biemond et al., 1984), although the amount of superoxide-releasable iron is very small, and so ferritin-bound iron is much safer than an equivalent amount of free iron (Bolann and Ulvik, 1990). $H_2O_2$ can degrade hemoglobin, myoglobin, and other heme proteins to release iron from the heme ring (Gutteridge, 1986; Puppo and Halliwell, 1988a, 1988b).

## OXIDATIVE STRESS AND HUMAN DISEASE

Does oxidative damage play a role in human disease? Many of the biological consequences of excess radiation exposure may be due to OH·-dependent damage to proteins, DNA, and lipids (von Sonntag, 1987). Oxidative damage (resulting from exposure to elevated $O_2$ in incubators) may account for damage to the retina of the eye (retinopathy of prematurity) in premature babies. Indeed, several controlled clinical trials have documented the efficiency of α-tocopherol in minimizing the severity of the retinopathy, *consistent with* (but by no means proving) a role for lipid peroxidation (reviewed by Ehrenkranz, 1989). However, there are many papers in the biomedical literature suggesting a role for oxidative stress in other human diseases (over 100 at last count).

Tissue damage by disease, trauma, toxic agents, and other causes usually leads to formation of increased amounts of putative "injury mediators," such as prostaglandins, leukotrienes, interleukins, interferons, and tumor necrosis factors (TNFs). All of these have at various times been suggested to play important roles in different human dieases. Currently, for example, there is much interest in the roles played by TNFα, NO, and interleukins in adult respiratory distress syndrome and septic shock. ROS can be placed in the same category, i.e., *tissue damage will usually lead to increased ROS formation* and oxidative stress. Figure 7-4 summarizes some of the reasons for this reaction. Indeed, in *most* human diseases, oxidative stress is a *secondary* phenomenon, a consequence of the disease activity. That does not mean that it is unimportant! Its importance varies in different disease states

## 328  The Biological Significance of Oxygen-Derived Species

Heat
Trauma
Ultrasound
Infection
Radiation
Elevated $O_2$
Toxins
Exercise to excess
Ischemia
$\longrightarrow$ Tissue damage $\longrightarrow$

Increase in radical-generating enzymes (e.g., xanthine oxidase) and/or their substrates (e.g., hypoxanthine)

Activation of phagocytes

Activation of phospholipases, cyclooxygenases, and lipoxygenases

Dilution and destruction of antioxidants

Release of free metal ions from sequestered sites

Release of heme proteins (hemoglobin, myoglobin)

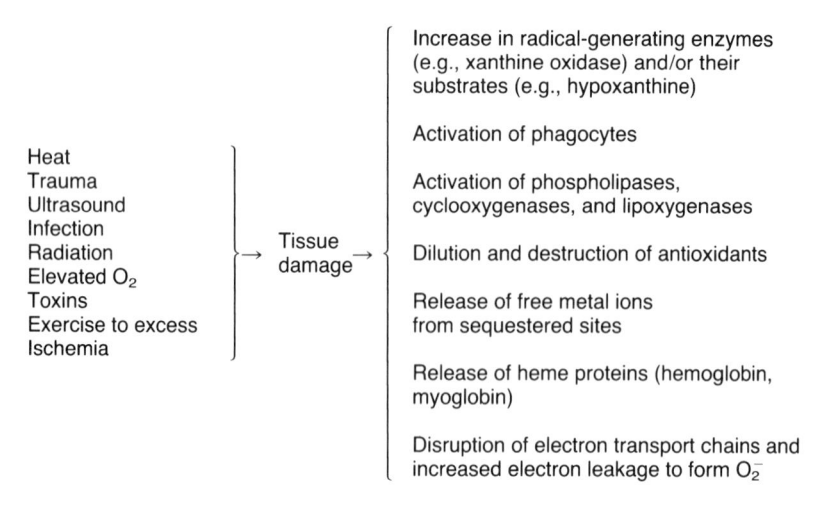

Disruption of electron transport chains and increased electron leakage to form $O_2^-$

**FIGURE 7-4** Some of the mechanisms by which cell injury leads to oxidative stress. Oxidative stress is an inevitable accompaniment of many human diseases. In some, it makes a significant contribution to worsening the disease pathology; in others, it may be an insignificant epiphenomenon of the disease process.

(reviewed by Halliwell et al., 1992). For example, oxidative damage to lipids in arterial walls seems to be a significant contributor to the development of atherosclerosis (Steinberg et al., 1989) (see Chapter 10). Degan et al. (1991) have suggested that free radical damage to DNA is a major contributor to the development of cancer.

## CONCLUDING COMMENTS

Free radicals and other nonradical, oxygen-derived species (such as $H_2O_2$) are part of normal human metabolism. Often they are useful, but in excess they can become harmful. They are probably involved in most human diseases but play important pathological roles in only some of them.

## REFERENCES

AIKENS, J., and DIX., T. A. (1991) Perhydroxyl Radical (HOO˙) Initiated Lipid Peroxidation. The Role of Fatty Acid Hydroperoxides. *J. Biol. Chem.*, 266, 15091–15098.

AMES, B. N. (1989) Endogenous Oxidative DNA Damage, Aging, and Cancer. *Free Radical Res. Commun.*, 7, 121–128.

AMSTAD, P., PESKIN, A., SHAH, G., MIRAULT, M. E., MORET, R., ZBINDEN, I., and CERUTTI, P. (1991) The Balance Between Cu,Zn-superoxide Dismutase and Catalase Affects the Sensitivity of Mouse Epidermal Cells to Oxidative Stress. *Biochemistry* 30, 9305–9313.

ANBAR, M., and NETA, P. (1967) A Compilation of Specific Bimolecular Rate Constants for the Reactions of Hydrated Electrons, Hydrogen Atoms and Hydroxyl Radicals with Inorganic and Organic Compounds in Aqueous Solutions. *Int. J. Appl. Radiat. Isotopes*, 18, 493–523.

ARROYO, C. M., CARMICHAEL, A. J., BOUSCAREL, B., LIANG, J. H., and WEGLICKI, W. B. (1990) Endothelial Cells as a Source of Oxygen Free Radicals. An ESR Study. *Free Radical Res. Commun.*, 9, 287–296.

BABBS, C. F., CREGOR, M. D., TUREK, J. J., and BADYLAK, S. F. (1991) Endothelial Superoxide Production in Buffer Perfused Rat Lungs, Demonstrated by a New Histochemical Technique. *Lab. Invest.*, 65, 484–496.

BABIOR, B. M., and WOODMAN, R. C. (1990) Chronic Granulomatous Disease. *Semin. Hematol.*, 27, 247–259.

BALENTINE, J. (1982) *Pathology of Oxygen Toxicity*, Academic Press, New York.

BEAUCHAMP, C., and FRIDOVICH, I. (1970) A Mechanism for the Production of Ethylene from Methional. The Generation of the Hydroxyl Radical by Xanthine Oxidase. *J. Biol. Chem.*, 243, 4641–4646.

BECKMAN, J. S., BECKMAN, T. W., CHEN, J., MARSHALL, P. A., and FREEMAN, B. A. (1990) Apparent Hydroxyl Radical Production by Peroxynitrite: Implications for Endothelial Injury from Nitric Oxide and Superoxide. *Proc. Natl. Acad. Sci. U.S.A.*, 87, 1620–1624.

BIELSKI, B. H. J. and CABELLI, D. E. (1995) Superoxide and Hydroxyl Radical Chemistry in Aqueous Solution, in *Active Oxygen in Chemistry* (C. S. Foote, J. S. Valentine, A. Greenberg, and J. F. Liebman, Eds.), Chapman & Hall, New York, pp. 66–104.

BIELSKI, B. H. J., ARUDI, R. L., and SUTHERLAND, M. W. (1983) A Study of the Reactivity of $HO_2/O_2^-$ with Unsaturated Fatty Acids. *J. Biol. Chem.*, 258, 4759–4761.

BIEMOND, P., VAN EIJK, H. G., SWAAK, A. J. G., and KOSTER, J. F. (1984) Iron Mobilization from Ferritin by Superoxide Derived from Stimulated Polymorphonuclear Leukocytes. Possibile Mechanism in Inflammation Diseases. *J. Clin. Invest.*, 73, 1576–1579.

BOLANN, B. J., and ULVIK, R. J. (1990) On the Limited Ability of Superoxide to Release Iron from Ferritin. *Eur. J. Biochem.*, 193, 899–904.

BORG, D. C., and SCHAICH, K. M. (1984) Cytotoxicity from Coupled Redox Cycling of Autoxidizing Xenobiotics and Metals; a Selective Critical Review and Commentary on Work in Progress. *Isr. J. Chem.*, 24, 38–53.

BREIMER, L. (1991) Repair of DNA Damage Induced by Reactive Oxygen Species. *Free Radical Res. Commun.*, 14, 159–171.

BRITIGAN, B. E., ROEDER, T. L., and SHASBY, D. M. (1992) Insight Into the Nature and Site of Oxygen-centered Free Radical Generation by Endothelial Cell Monolayers Using a Novel Spin Trapping Technique. *Blood*, 79, 699–707.

BROT, N., and WEISSBACH, H. (1991) Biochemistry of Methionine Sulfoxide Residues in Proteins. *Biofactors*, 3, 91–96.

BURDON, R. H., and RICE-EVANS, C. (1989) Free Radicals and the Regulation of Mammalian Cell Proliferation. *Free Radical Res. Commun.*, 6, 345–358.

BURKITT, M. J. (1993) ESR Spin Trapping Studies into the Nature of the Oxidizing Species in the Fenton Reaction: Pitfalls Associated with the Use of 5,5-Dimethyl-1-pyrroline-*N*-oxide in the Detection of Hydroxyl Radical. *Free Radical Res. Commun.*, 18, 43–57.

BURTON, G. W., and TRABER, M. G. (1990) Vitamin E: Antioxidant Activity Biokinetics and Bioavailability. *Annu. Rev. Nutr.*, 10, 357–382.

CHANCE, B., SIES, H., and BOVERIS, A. (1979) Hydroperoxide Metabolism in Mammalian Organs. *Physiol. Rev.*, 59, 527–605.

CHANG, E. C., CRAWFORD, B. F., HONG, Z., BILINSKI, T., and KOSMAN, D. J. (1991) Genetic and Biochemical Characterization of Cu,Zn-superoxide Dismutase Mutants in *Saccharomyces cerevisiae*. *J. Biol. Chem.*, 266, 4417–4424.

COCHRANE, C. G. (1991) Mechanisms of Oxidant Injury of Cells. *Mol. Aspects Med.*, 12, 137–147.

COMPORTI, M. (1987) Glutathione-depleting Agents and Lipid Peroxidation. *Chem. Phys. Lipids*, 45, 143–169.

COOPER, B., CREETH, J. M., and DONALD, A. S. R. (1985) Studies of the Limited Degradation of Mucus Glycoproteins. The Mechanism of the Peroxide Reaction. *Biochem. J.*, 228, 615–626.

DEGAN, P., SHIGENAGA, M. K., PARK, E. M., ALPAIN, P. E., and AMES, B. N. (1991) Immunoaffinity Isolation of Urinary 8-hydroxy-2′-deoxyguanosine and 8-hydroxyguanine and Quantitation of 8-hydroxy-2′-deoxyguanosine in DNA by Polyclonal Antibodies. *Carcinogenesis*, 12, 865–871.

DIZDAROGLU, M. (1991) Chemical Determination of Free Radical-induced Damage to DNA. *Free Rad. Biol. Med.*, 10, 225–242.

DIZDAROGLU, M., NACKERDIEN, Z., CHO, B. C., GAJEWSKI, E., and RAO, G. (1991) Chemical Nature of *in vivo* DNA Base Damage in Hydrogen Peroxide-treated Mammalian Cells. *Arch. Biochem. Biophys.*, 285, 388–390.

DUPPY, C., VIRION, A., OHAYON, R., KANIEWSKI, J., DEME, D., and POMMIER, J. (1991) Mechanism of Hydrogen Peroxide Formation Catalyzed by NADPH Oxidase in Thyroid Plasma Membrane. *J. Biol. Chem.*, 266, 3739–3743.

EHRENKRANZ, R. A. (1989) Vitamin E and Retinopathy of Prematurity: Still Controversial. *J. Pediat.*, 114, 801–803.

FERRALI. M., CICCOLI, L., SIGNORINI, C., and COMPORTI, M. (1990) Iron

Release and Erythrocyte Damage in Allyl Alcohol Intoxication in Mice. *Biochem. Pharmacol.*, 40, 1485–1490.

FRANK, L., IQBAL, J., HASS, M., and MASSARO, D. (1989) New "Rest Period" Protocol for Inducing Tolerance to High $O_2$ Exposure in Adult Rats. *Am. J. Physiol.*, 257, L226–L231.

FRIDOVICH, I. (1989) Superoxide Dismutases: An Adaption to a Paramagnetic Gas. *J. Biol. Chem.*, 264, 7761–7764.

GARDNER, P. R., and FRIDOVICH, I. (1991) Superoxide Sensitivity of the *Escherichia coli* 6-Phosphogluconate Dehydratase. *J. Biol. Chem.*, 266, 1478–1483.

GARDNER, P. R., and FRIDOVICH, I. (1992) Inactivation-reactivation of Aconitase in *Escherichia coli*. A Sensitive Measure of Superoxide Radical. *J. Biol. Chem.*, 267, 8757–8763.

GERSCHMAN, K., GILBERT, D. L., NYE, S. W., DWYER, P., and FENN, W. O. (1954) Oxygen Poisoning and X-irradiation: A Mechanism in Common. *Science*, 119, 623–626

GRONER, Y., ELROY-STEIN, O., AVRAHAM, K. B., YAROM, R., SCHICKLER, M., KNOBLER, H., and ROTMAN, G. (1990) Down Syndrome Clinical Symptoms Are Manifested in Transfected Cells and Transgenic Mice Overexpressing the Human CuZn-Superoxide Dismutase Gene. *J. Physiol. (Paris)*, 84, 53–77.

GUTTERIDGE, J. M. C. (1986) Iron Promoters of the Fenton Reaction and Lipid Peroxidation Can Be Released from Haemoglobin by Peroxides. *FEBS Lett.*, 201, 291–295.

GUTTERIDGE, J. M. C., and HALLIWELL, B. (1995) *Antioxidants in Nutrition, Health and Disease*, Oxford University Press, Oxford, UK.

GUTTERIDGE, J. M. C., RICHMOND, R., and HALLIWELL, B. (1979) Inhibition of the Iron-catalyzed Formation of Hydroxyl Radicals from Superoxide and of Lipid Peroxidation by Desferrioxamine. *Biochem. J.*, 184, 469–472.

GUTTERIDGE, J. M. C., and WILKINS, S. (1983) Copper Salt-dependent Hydroxyl Radical Formation. Damage to Proteins Acting as Antioxidants. *Biochim. Biophys. Acta* 759, 38–41.

HALLIWELL, B. (1978) Superoxide-dependent Formation of Hydroxyl Radicals in the Presence of Iron Chelates. *FEBS Lett.*, 92, 321–326.

HALLIWELL, B. (1987) Oxidants and Human Disease: Some New Concepts. *Faseb J.*, 1, 358–364.

HALLIWELL, B. (1989) Protection Against Tissue Damage *in vivo* by Desferrioxamine. What Is Its Mechanism of Action? *Free Rad. Biol. Med.*, 7, 645–651.

HALLIWELL, B. (1990) How to Characterize a Biological Antioxidant. *Free Rad. Res. Commun.*, 9, 1–32.

HALLIWELL, B., and ARUOMA, O. I. (1991) DNA Damage by Oxygen-derived

Species. Its Mechanism and Measurement in Mammalian Systems. *FEBS Lett.*, 281, 9–19.

HALLIWELL, B., and CHIRICO, S. (1993) Lipid Peroxidation: Its Mechanism, Measurement and Significance. *Am. J. Clin. Nutr.*, 57, 715S–724S.

HALLIWELL, B., and GUTTERIDGE, J. M. C. (1989) *Free Radicals in Biology and Medicine*, 2nd ed., Clarendon Press, Oxford.

HALLIWELL, B., and GUTTERIDGE, J. M. C. (1990a) Role of Free Radicals and Catalytic Metal Ions in Human Disease. *Meth. Enzymol.*, 186, 1–85.

HALLIWELL, B., and GUTTERIDGE, J. M. C. (1990b) The Antioxidants of Human Extracellular Fluids. *Arch. Biochem. Biophys.*, 200, 1–8.

HALLIWELL, B., and GUTTERIDGE, J. M. C. (1992) Biologically-relevant Metal Ion-dependent Hydroxyl Radical Generation. An Update *FEBS Lett.*, 300, 108–112.

HALLIWELL, B., GUTTERIDGE, J. M. C., and CROSS, C. E. (1992) Free Radicals, Antioxidants, and Human Disease: Where Are We Now? *J. Lab. Clin. Med.*, 119, 598–620.

HOGG, N., DARLEY-USMAR, V. M., WILSON, M. T., and MONCADA, S. (1992) Production of Hydroxyl Radicals from the Simultaneous Generation of Superoxide and Nitric Oxide. *Biochem. J.*, 281, 419–424.

HUIE, R. E., and PADMAJA, S. (1993) The Reaction of NO with Superoxide. *Free Radical Res. Commun.*, 18, 195–199.

IMLAY, J. A., and FRIDOVICH, I. (1991) Assays of Metabolic Superoxide Production in *Escherichia coli*. *J. Biol. Chem.*, 266, 6957–6965.

IQBAL, J., CLERCH, L. B., HASS, M. A., FRANK, L., and MASSARO, D. (1989) Endotoxin Increases Lung Cu, Zn Superoxide Dismutase mRNA: $O_2$ Raises Enzyme Synthesis. *Am. J. Physiol.*, 257, L61–L64.

JAIN, A., MARTENSSON, J., MEHTA, T., KRAUSS, A. N., AULD, P. A., and MEISTER, A. (1992) Ascorbic Acid Prevents Oxidative Stress in Glutathione-deficient Mice—Effects on Lung Type-2 Cell Lamellar Bodies, Lung Surfactant, and Skeletal Muscle. *Proc. Natl. Acad. Sci. U.S.A.*, 89, 5093–5097.

KUBES, P., GRISHAM, M. B., BARROWMAN, J. A., GAGINELLA, T., and GRANGER, D. N. (1991) Leukocyte-induced Vascular Protein Leakage in Cat Mesentery. *Am. J. Physiol.*, 261, H1872–H1879.

KYLE, M. E., NAKAE, D., SAKAIDA, I., MICCADEI, S., and FARBER, J. L. (1988) Endocytosis of Superoxide Dismutase is Required in Order for the Enzyme to Protect Hepatocytes from the Cytotoxicity of Hydrogen Peroxide. *J. Biol. Chem.*, 263, 3784–3789.

LAURINDO, F. R. M., DA LUZ, P. L., UINT, L., ROCHA, T. F., JAEGER, R. G., and LOPES, E. A. (1991) Evidence for Superoxide Radical-dependent Coronary Vasospasm After Angioplasty in Intact Dogs. *Circulation*, 83, 1705–1715.

LIEW, F. Y., MILLOTT, S., PARKINSON, C., PALMER, R. M., and MONCADA, S.

(1990) Macrophage Killing of *Leishmania* Parasite *in vivo* is Mediated by Nitric Oxide. *J. Immunol.*, 144, 4794–4797.

MAIORINO, M., CHU, F. F., URSINI, DAVIES, K. J., DOROSHOW, J. H., and ESWORTHY, R. S. (1991) Phospholipid Hydroperoxide Glutathione Peroxidase is the 18-kDa Selenoprotein Expressed in Human Tumor Cell Lines. *J. Biol. Chem.*, 266, 7728–7732.

MALY, F. E. (1990) The B-lymphocyte: A Newly-recognized Source of Reactive Oxygen Species with Immunoregulatory Potential. *Free Radical Res. Commun.*, 8, 143–148.

MARX, G., and CHEVION, M. (1986) Site-specific Modification of Albumin by Free Radicals. Reaction with Copper(II) and Ascorbate. *Biochem. J.*, 236, 397–400.

MCCORD, J. M., and DAY, E. D. (1978) Superoxide-dependent Production of Hydroxyl Radical Catalysed by an Iron–EDTA Complex. *FEBS Lett.*, 86, 139–142.

MCNAMARA, J. O., and FRIDOVICH, I. (1993) Did Radicals Strike Lou Gehrig? *Nature*, 362, 20–21.

MEIER, B., RADEKE, H., SELLE, S., RASPE, H. H., SIES, H., RESCH, K., and HABERMEHL, G. G. (1990) Human Fibroblasts Release Reactive Oxygen Species in Response to Treatment with Synovial Fluids from Patients Suffering from Arthritis. *Free Radical Res. Commun.*, 8, 149–160.

MERRYFIELD, M. L., and LARDY, H. A. (1982) $Ca^{2+}$-mediated Activation of Phosphoenolpyruvate Carboxykinase Occurs Via Release of $Fe^{2+}$ from Rat Liver Mitochondria. *J. Biol. Chem.*, 257, 3628–3635.

MONCADA, S., PALMER, R. M. J., and HIGGS, E. A. (1991) Nitric Oxide: Physiology, Pathophysiology, and Pharmacology. *Pharm. Rev.*, 43, 109–142.

MURRELL, G. A. C., FRANCIS, M. J. O., and BROMLEY, L. (1990) Modulation of Fibroblast Proliferation by Oxygen Free Radicals. *Biochem. J.*, 265, 659–665.

NAKAZONO, K., WATANABE, N., MATSUNO, K., SASAKI, J., SATO, T., and INOUE, M. (1991) Does Superoxide Underlie the Pathogenesis of Hypertension? *Proc. Natl. Acad. Sci. U.S.A.*, 88, 10045–10048.

ORRENIUS, S., MCCONKEY, D. J., BELLOMO, G., and NICOTERA, P. (1989) Role of $Ca^{2+}$ in Toxic Cell Killing. *Trends Pharm. Sci.*, 10, 281–285.

PUPPO, A., and HALLIWELL, B. (1988a) Formation of Hydroxyl Radicals from Hydrogen Peroxide in the Presence of Iron. Is Haemoglobin a Biological Fenton Catalyst? *Biochem. J.*, 249, 185–190.

PUPPO, A., and HALLIWELL, B. (1988b) Formation of Hydroxyl Radicals in Biological Systems. Does Myoglobin Stimulate Hydroxyl Radical Production from Hydrogen Peroxide? *Free Radical Res. Commun.*, 4, 415–422.

RADI, R., BECKMAN, J. S., BUSH, K. K. M., and FREEMAN, B. A. (1990)

Peroxynitrite Oxidation of Sulfhydryls. The Cytotoxic Potential of Superoxide and Nitric Oxide. *J. Biol. Chem.*, 266, 4244–4250.

RILEY, J. C. M., and BEHRMAN, H. R. (1991) Oxygen Radicals and Reactive Oxygen Species in Reproduction. *Proc. Soc. Exp. Biol. Med.*, 5, 781–791.

SALO, D. C., LIN, S. W., PACIFICI, R. E., and DAVIES, K. J. A. (1988) Superoxide Dismutase is Preferentially Degraded by a Proteolytic System from Red Blood Cells Following Oxidative Modification by Hydrogen Peroxide. *Free Rad. Biol. Med.*, 5, 335–339.

SCHRECK, R., ALBERMANN, K. A. J., and BAEUERLE, P. A. (1992) Nuclear Factor κB: An Oxidative Stress-responsive Transcription Factor of Eukaryotic Cells (a Review). *Free Rad. Res. Commun.*, 17, 221–237.

SCOTT, M. D., EATON, J. W., KUYPERS, F. A., CHIU, D. T., and LUBIN, B. H. (1989) Enhancement of Erythrocyte Superoxide Dismutase Activity: Effects on Cellular Oxidant Defense. *Blood*, 74, 2542–2549.

SEVANIAN, A., WRATTEN, M. L., MCLEOD, L. L., and KIM, E. (1988) Lipid Peroxidation and Phospholipid A2 Activity in Liposomes Composed of Unsaturated Phospholipids: A Structural Basis for Enzyme Activation. *Biochim. Biophys. Acta*, 961, 316–327.

SHAPIRO, B. M. (1991) The Control of Oxidative Stress at Fertilization. *Science*, 252, 533–536.

SIES, H. (Ed.) (1985) *Oxidative Stress*, Academic Press, New York.

SIES, H. (Ed.) (1991) *Oxidative Stress, Oxidants and Antioxidants*, Academic Press, New York.

STADTMAN, E. R., and OLIVER, C. N. (1991) Metal-catalyzed Oxidation of Proteins. Physiological Consequences. *J. Biol. Chem.*, 266, 2005–2008.

STEINBERG, D., PARTHASARATHY, S., CAREW, T. E., KHOO, J. C., and WITZTUM, J. L. (1989) Beyond Cholesterol. Modifications of Low-density Lipoprotein That Increase Atherogenicity. *N. Engl. J. Med.*, 320, 915–924.

STORZ, G., and TARTAGLIA, L. (1992) OxyR: A Regulator of Antioxidant Genes. *J. Nutr.*, 122, 627–630.

SUTTON, H. C., and WINTERBOURN, C. C. (1989) On the Participation of Higher Oxidation States of Iron and Copper in Fenton Reactions. *Free Radical Biol. Med.*, 6, 53–60.

TARTAGLIA, L. A., GIMENO, C. J., STORZ, G., and AMES, B. N. (1992) Multidegenerate DNA Recognition by the Oxy-R Transcriptional Regulator. *J. Biol. Chem.*, 262, 2038–2045.

TOUATI, D. (1989) The Molecular Genetics of Superoxide Dismutase in *E. coli*. An Approach to Understanding the Biological Role and Regulation of SODs in Relation to Other Elements of the Defence System Against Oxygen Toxicity. *Free Radical Res. Commun.*, 8, 1–9.

VON SONNTAG, C. (1987) *The Chemical Basis of Radiation Biology*, Taylor and Francis, London.

WEINBERG, E. D. (1990) Cellular Iron Metabolism in Health and Disease. *Drug Metab. Rev.*, 22, 531–579.

WHITE, C. W., AVRAHAM, K. B., SHANLEY, P. F., and GRONER, Y. (1991) Transgenic Mice with Expression of Elevated Levels of Copper-zinc Superoxide Dismutase in the Lungs are Resistant to Pulmonary Oxygen Toxicity. *J. Clin. Invest.*, 87, 2162–2168.

ZHANG, Y., MARCILLAT, O., GIULIVI, ERNSTER, L., and DAVIES, K. J. (1990) The Oxidative Inactivation of Mitochondrial Electron Transport Chain Components and ATPase. *J. Biol. Chem.*, 265, 16330–16336.

# 8
# Metal-Complex-Catalyzed Cleavage of Biopolymers

ROSEMARY A. MARUSAK AND CLAUDE F. MEARES

## INTRODUCTION

Oxygen radical species such as the superoxide ($O_2^-$), hydroxyl ($\cdot OH$), and hydroperoxyl ($\cdot OOH$) radicals are well known for their deleterious effects in living systems. Their buildup leads to oxidative stress, initiating lipid peroxidation, protein oxidation, DNA damage, and other cellular disorders (Fontecave and Pierre, 1991; Halliwell and Aruoma, 1991; Stadtman and Oliver, 1991) (see also Chapter 7). For example, site-specific generation of radicals plays a major role in carcinogenesis. The highly reactive $\cdot OH$ radicals act on DNA by abstracting H atoms from the deoxyribose sugar-phosphate backbone, causing strand scission and the release of bases and producing "alkali-labile" sites (strand breakage occurs on addition of base). Hydroxyl radicals can also add to DNA bases, forming radical adducts, which can undergo further chemical reactions. These changes to the DNA can lead to harmful mutations.

Similarly, site-specific radical damage to proteins (discussed in the section on "Protein Cleavage") is a major contributor to the onset of disease. Hydroxyl radicals react with protein side chains and the peptide backbone close to the metal binding site, resulting in biomolecules that are less active (in the case of enzymes), more sensitive to proteolytic degradation, and more thermolabile. Protein oxidation is thought to play a key role in the aging process, inflammatory diseases (e.g., rheumatoid arthritis), and neurological disorders.

Oxygen radical species can be generated in a variety of ways, such as through ionizing radiation or dioxygen activation by transition metal ions. The latter process is particularly important since trace metals are present in living systems. Healthy living systems, however, have natural defense mechanisms [e.g., superoxide dismutase (SOD) and catalase] for protection against these highly reactive and damaging species. When oxidative damage does occur, efficient repair mechanisms (e.g., DNA hydrolases and ligases) come into action.

Recently, researchers have begun to turn the chemistry of metal-ion-induced, site-specific oxygen radical generation in the presence of biopolymers to their advantage. In the past three decades, the degradative reactions associated with oxygen radicals have been used in the development of new reagents. In particular, this chemistry has been instrumental in the design of new tools for molecular biology and also in the development of chemotherapeutic agents. These techniques utilize metal ion *complexes*, in contrast to free metal ions in solution. The chelating ligands of the complex specifically alter the chemical and physical properties of metal ions, permitting the metal ion to be delivered to a desired location with controlled reactivity. This chapter will examine several different metal complexes that are used to carry out controlled dioxygen activation with subsequent biopolymer cleavage (i.e., artificial nucleases and proteases). Particular attention will be given to investigating the mechanism by which the complexes activate dioxygen, and the new applications of this chemistry in the biological field.

## DIOXYGEN ACTIVATION BY METAL COMPLEXES

The nature of the chelating ligand has an important influence on the physical and chemical properties of a metal ion. Physical properties are governed by the overall charge of the complex and the nature of the ligand framework. Chemical reactivity is reflected not only in the thermodynamic driving force for the overall oxidation-reduction process but also in kinetic reactivity, given by an experimentally determined rate constant. To help understand the roles of each reactant in the mechanism of dioxygen activation and biopolymer cleavage, pertinent kinetic and thermodynamic data for metal complexes will be referred to throughout the text. For information on physiological potentials and reactivities, see Burkitt and Gilbert (1990).

## General Mechanisms for Metal-Complex-Induced Dioxygen Activation in Aqueous Solution

Metal complexes typically activate molecular oxygen through a series of oxidation-reduction steps. A one-electron reduction of a neutral dioxygen molecule by a metal complex results in the formation of a superoxide radical, $O_2^-$ (eq. 1) (Fontecave and Pierre, 1991). The superoxide radical is often a reductant for metal ions in biological systems, mediating the production of more reactive species. Addition of a second electron to dioxygen, via $O_2^-$ dismutation (eq. 3) or reaction at a reduced metal center (eq. 4), reduces $O_2^-$ to the peroxide form, $O_2^{2-}$. It is often these metal-peroxide intermediates (eqs. 4 and 5) that are responsible for the production of highly reactive oxygen species.

$$LM^n + O_2 \rightarrow LM^{(n+1)} - O_2^- \qquad (1)$$

$$LM^{(n+1)} - O_2^- \rightarrow LM^{(n+1)} + O_2^- \qquad (2)$$

$$2O_2^- + 2H^+ \rightarrow H_2O_2 + O_2 \qquad (3)$$

$$LM^n + O_2^- \rightarrow LM^{(n+1)} - O_2^{2-} \qquad (4)$$

$$LM^n + H_2O_2 \rightarrow LM^n - O_2^{2-} + 2H^+ \qquad (5)$$

Commonly accepted mechanisms for metal complex-peroxide activation are summarized in eqs. (6–9), where $LM^n$ is the metal complex with metal oxidation state $n+$, $H_2O_2$ is the oxidant, and $R^{red}$ is a reductant (with oxidized form $R^{ox}$) that acts on the metal ion. Equations (6) and (7) depict a one-electron, metal-mediated redox process that results in homolytic cleavage of the O—O bond and formation of the highly reactive hydroxyl radical, ·OH. Reaction (6) is often called a *Fenton-like* reaction, the original Fenton reaction being $Fe^{2+} + H_2O_2$, yielding ·OH and $OH^-$ (Fenton, 1894). Hydroxyl radicals react rapidly with biopolymers, for example, abstracting H atoms from C—H bonds.

$$LM^n + HO-OH \rightarrow \cdot OH + OH^- + LM^{(n+1)} \qquad (6)$$

$$R^{red} + LM^{(n+1)} \rightarrow R^{ox} + LM^n \qquad (7)$$

An alternative mechanism for dioxygen activation involves heterolytic cleavage of the O—O bond. This is a consequence of two-electron reduction of the O—O bond by the metal center, producing a high oxidation state metal–oxo complex, $LM^{(n+2)}=O$. This concept was first introduced by Bray and Gorin (1932) as an alternative to the Fenton reaction. Spectroscopic evidence for this species has been obtained in porphyrin systems (Mandon et al., 1989; Bowry and Ingold, 1991). The electron-deficient $LM^{(n+2)}=O$ acts analogously to $\cdot OH$ in hydrogen atom abstraction reactions. This species can also act as an oxygen atom transfer reagent in its reactions with organic substrates, as demonstrated by metal porphyrin complexes (Meunier, 1992).

$$LM^n + HO-OH \rightarrow LM^{(n+2)}=O + H_2O \qquad (8)$$

$$LM^{(n+2)}=O + R^{red} + H_2O \rightarrow LM^{(n+1)}(OH) + OH^- + R^{ox} \qquad (9)$$

The reaction of an electronically excited metal complex with dioxygen can result in the excitation of $O_2$ to a highly reactive singlet-oxygen molecule, $^1O_2$ ($^1\Delta_g^+ O_2$). Singlet oxygen reacts with biological molecules (e.g., oxidation of DNA bases), inducing cleavage.

## Determining the Reaction Mechanism for Dioxygen Activation by Metal Complexes

The mechanism of dioxygen activation by metal complexes can be probed in several ways. The most direct evidence for a particular reaction pathway is obtained through the isolation of stable intermediates formed along the reaction path. This, however, is often difficult, and various methods of detecting transient intermediates are usually employed. Spectroscopic means for detection include electron paramagnetic resonance (EPR), nuclear magnetic resonance (NMR), Raman, and ultraviolet (UV)-visible spectroscopy. These tools have been employed to determine the nature of the metal complex, as well as to observe the reactive oxygen species. Oxygen radicals such as $\cdot OH$ often have short lifetimes and small steady-state concentrations, precluding direct detection by EPR methods. These species are first stabilized through reaction with another chemical species (called *a spin-trapping reagent*) that is itself stable in the radical form and can be subsequently analyzed by EPR. A commonly employed spin trap is 5,5-dimethyl-1-pyrroline-N-oxide (DMPO). The effects of radical scavengers on product formation are important in these systems, as scavengers are often used to distinguish between diffusible oxygen radicals and metal-bound electrophiles. It must be noted, however, that a lack of radical scavenger inhibition does not preclude radical production; radical species may still be formed in an environment that is inaccessible to the scavenger molecule. Typical radical scavengers are alcohols (e.g., methanol, *t*-butanol, mannitol, glycerol), sulfur compounds (e.g., organic thiols such as dithiothreitol), and natural biological molecules (e.g., superoxide dismutase). The determination of H atom abstraction selectivities, especially for alcohol substrates, helps to identify the reactive species (Winterbourn, 1987). When intermediates cannot be isolated or detected spectroscopically, the study of concentration effects on the reaction rate (determination of the rate law) may provide insight into the mechanism. Activated species can be distinguished by comparing relative rates of reaction with a common substrate. Finally, product analysis gives secondary evidence for reaction pathways.

## DNA CLEAVAGE

This section describes the interactions and cleavage reactions of various metal complexes with DNA and RNA substrates. Double-stranded DNA adopts many different helical conformation forms. Due to hydration within the minor groove, the right-handed helical B-form predominates in aqueous solution, and it is referred to most often throughout this text. Its structure is illustrated in Figure 8-1a (Cantor and Schimmel, 1980). The four nucleotides, adenine (A), thymine (T), cytosine (C), and guanine (G), comprise each strand of the double helix and are joined by phosphodiester linkages. Bases on opposite strands are paired by hydrogen bonding interactions, and the bases stack nearly perpendicular to the right-handed helical axis. The surface of the B-form contains opposed minor and major grooves, exposing the bases and providing favorable binding sites for proteins and small molecules such as metal complexes. The majority of complexes described in this chapter bind preferentially in the minor groove. On binding, metal complexes can cleave circular supercoiled DNA (form I) into form II (one strand nicked), and/or form III (linear) (Figure 8-1b). Form II occurs when one strand of the DNA is cleaved, still maintaining the circular conformation; in form III, both strands have been cleaved.

Redox active coordination complexes that produce strand breaks in DNA are commonly termed *chemical nucleases*. With the exception of the tris-phenanthroline complexes (described in the section on "Inert Chiral Metal *tris*-Phenanthroline Complexes"), the cleavage mechanism involves oxidation of the deoxyribose or ribose sugar. The mechanism consists of two steps, a metal complex–DNA binding reaction (Scheme 8-1a) followed by a DNA cleavage reaction (Scheme 8-1b). Since the sugar moiety is attacked, the cleavage reaction shows little base specificity, and any selectivity in the reaction is governed by the metal complex–DNA binding interaction. This step is described by the equilibrium constant for the binding reaction, $K_a$. The structure,

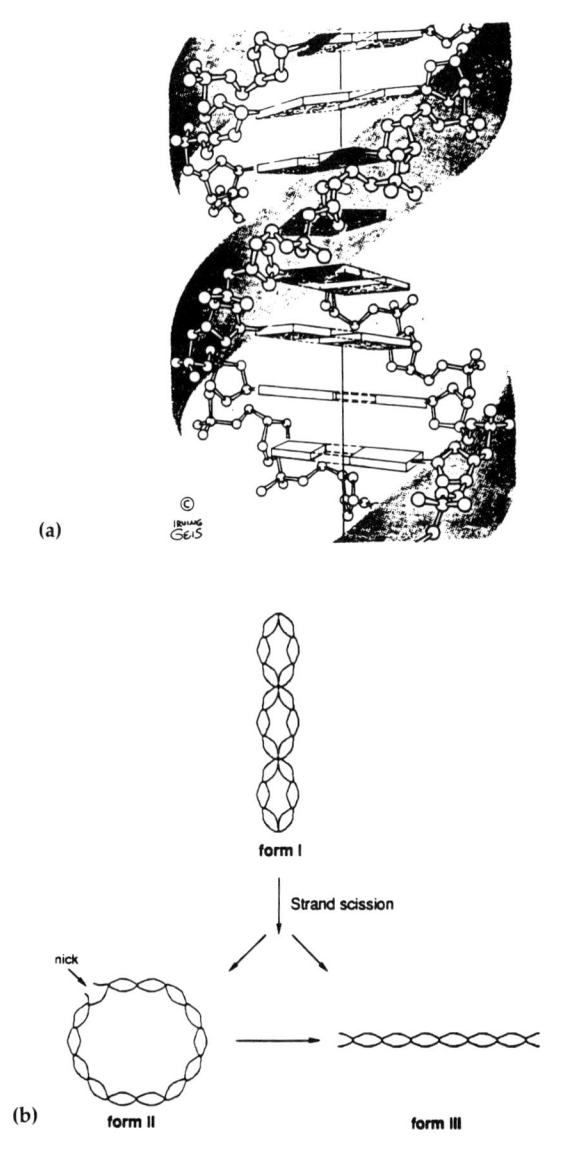

**FIGURE 8-1** (a) Structure of B DNA. Reproduced with permission from Cantor and Schimmel (1980). (b) Schematic of the different forms of DNA: closed circular, superhelical DNA (form I), nicked circular (form II), linear duplex (form III). Form II may arise from a single-strand break in form I, and form III may arise from a double-stranded cleavage of I or on an additional cut to form II.

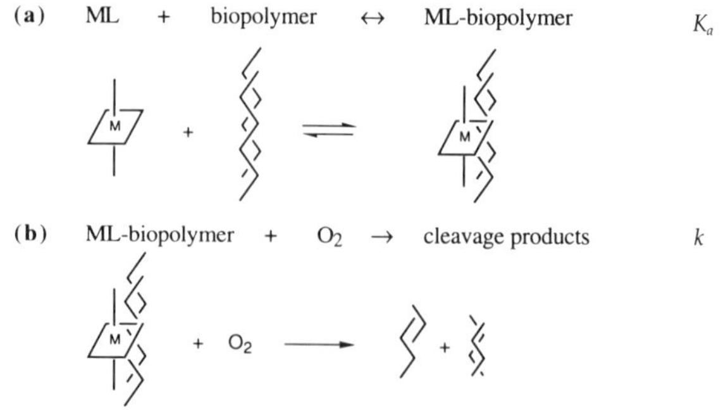

(a) ML + biopolymer ↔ ML-biopolymer        $K_a$

(b) ML-biopolymer + O₂ → cleavage products        $k$

Scheme 8-1

charge, and size of the metal complex can be important for the binding interaction. For example, complexes with overall positive charges interact favorably with the negatively charged phosphate backbone of DNA. Molecules possessing planar aromatic rings can be good intercalators (ligands that bind to DNA by insertion of the aromatic moiety between stacked base pairs). The size of the complex is also important for complementary binding to different substrate conformations. Substrate complementarity is important and can be sequence selective. When the binding step is specific, the metal complex system provides a sensitive means for probing subtle structural changes in the substrate. For structural determination, the information obtained using a particular metal chelate probe is restricted by how well the specificity of this reagent for its substrate is known. When both binding and cleavage are nonspecific, the system is suitable for analyzing protein– and drug–nucleic acid interactions. Binding sites can be mapped with high resolution (see the section on "Metal–EDTA Complexes").

The rate of the DNA cleavage reaction is described by the overall second-order rate constant, $k$, for dioxygen activation and product formation. One can visualize a variety of steps that need to be included to describe $k$. The rate constant encompasses the activation of $O_2$ to a reactive species, as well as the interaction of this species with the biopolymer. To carry out efficient cleavage, the reactive oxygen species must be generated close to the biopolymer; otherwise, it could be deactivated through either its natural decay or quenching by other reagents such as the buffer.

## Metallobleomycins

A metal complex that exemplifies dioxygen activation and its effect on DNA is the natural antibiotic bleomycin (BLM) (see Chapter 5).

BLM was first isolated from *Streptomyces verticillus* by Umezawa and coworkers (1966) and proved to be an effective tumor inhibitor. The drug is thought to kill cells by cleaving DNA in the nucleus.

*BLM structure and binding selectivity of metallobleomycins to DNA.* BLM is a glycopeptide comprised of an unusual peptide, a bithiazole, a terminal amine, and a disaccharide (Takita et al., 1978). Figure 8-2 partitions the BLM molecule into four regions, each of which plays an important role in the overall ternary DNA–metalloBLM–$O_2$ interaction and DNA chemistry (Otsuka et al., 1990). The metal-binding domain is where $O_2$ activation takes place. The sugar (mannosylgulose) moiety is thought to aid in metal-ion orientation and binding, as well as in the dioxygen activation process. The flexible linker is thought to be important for promoting a facile binding interaction and to be partly responsible for proper orientation of the metal center to the DNA backbone. The bithiazole region can bind in a sequence-dependent way to the minor groove of DNA. The positively charged terminal amine would have favorable interactions with negatively charged phosphate groups on the DNA backbone. These characteristics seem to be key to BLM's effectiveness as a DNA cutter.

BLM binds with moderate affinity ($K_{Blm-DNA} = 1.2 \times 10^5 M^{-1}$;

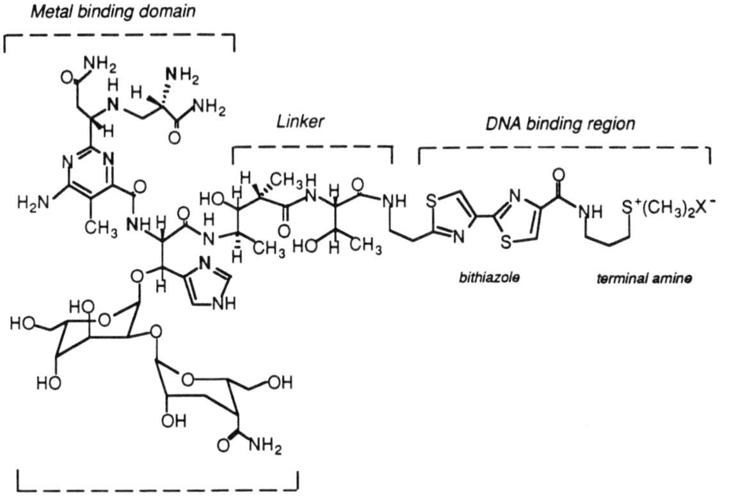

**FIGURE 8-2** Structure of bleomycin. Metal-coordinating nitrogens are labeled in bold face.

Chien et al., 1977; Kasai et al., 1978; Umezawa et al., 1984) to the minor groove of double-stranded B DNA, preferentially at guanine (G) bases. Single-stranded DNA is a poor substrate for the drug, and only those regions that have double-stranded character are cleaved (Kross et al., 1982; Carter et al., 1991). The binding of the bithiazole portion to the minor groove of double-stranded B DNA is shown in Figure 8-3 (Dickerson, 1986). The steric properties of 5' GpC and 5' GpT sequences favor a hydrogen-bonding interaction of the DNA guanine 2-amino group with the ring nitrogens or with the carboxamide oxygen adjacent to the bithiazole.

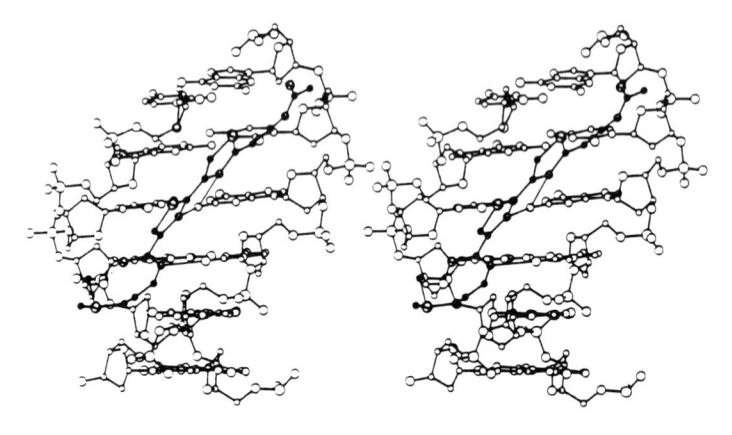

**FIGURE 8-3** Stereoview of the proposed binding interaction of the bithiazole moiety of BLM to the minor groove of B DNA. The helix is composed of alternating GC bases. Reproduced with permission from Dickerson (1986).

**FIGURE 8-4** (a) Sequencing gel electrophoresis of a φX174 DNA fragment following cleavage by FeBLM (lane 4). Reproduced with permission from Takeshita et al. (1978). Sequencing of DNA bands is performed as $^{32}$P end-labeled fragments of DNA migrate (in an electric field) down an acrylamide gel according to molecular weight (decreasing down the gel). The gel contains four standard markers alongside the cleavage reaction (first four lanes), consisting of fragments that end with the bases A, T, G, or C, which have been selectively produced from end-labeled DNA (not subjected to treatment with FeBLM) by chemical modification and cleavage (Maxam-Gilbert method). Each reaction cleavage site matches a unique base fragment in one of the marker lanes, allowing the base at which the FeBLM-induced cleavage occurs to be identified. Lanes 1–3 and 5 are control lanes in which BLM (1, 3, 5) or iron(II) has been omitted from the reaction mixture. (b) Analysis of the base release from the double-stranded restriction fragments of φX174 DNA shown in (a). The bases released and their relative intensities are marked with asterisks (*). Note that on the top (5' to 3') strand, pyrimidines located on the 3' side of G bases are preferentially cleaved.

DNA Cleavage 345

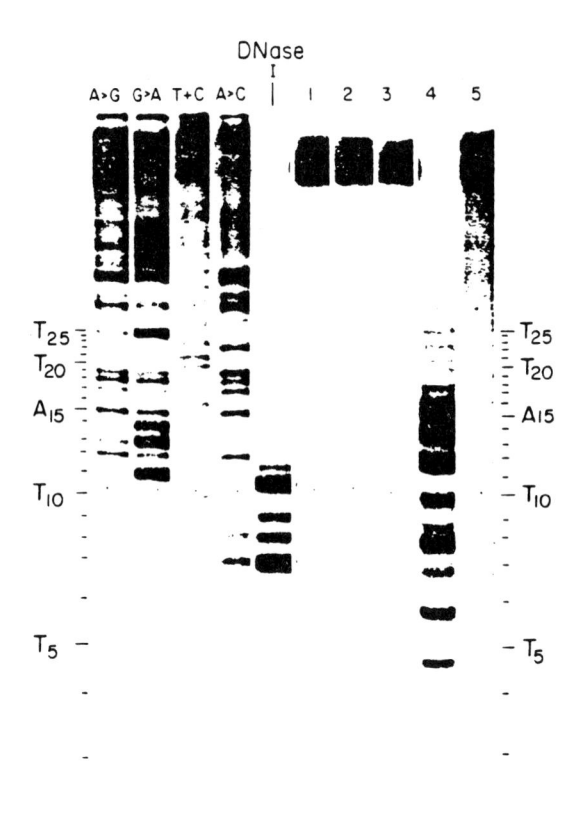

(a)

(b)
```
      (5')pCCCCTTACTTGAGGATAAATTATGTCTAAT-
           GGGGAATGAACTCCTATTTAATACAGATTA-
                    10         20         30

           ATTCAAACTGGCGCCGAGCGTATGCCGCAT···
           TAAGTTTGACCGCGGCTCGCATACGGCGTA···
                    40         50         60
```

*Specificity of DNA strand cleavage.* FeBLM cleaves supercoiled DNA (form I) into both the nicked circular (II) and linear duplex (III) forms (Love et al., 1981). In isolated double-stranded DNA, the cleavage occurs specifically at GpC(5'→3') and GpT(5'→3') (5'-G-pyrimidine-3') sequences (Takeshita et al., 1978; Umezawa et al., 1984; Kuwahara and Sugiura, 1988). To demonstrate this specificity, a sequencing gel of a double-stranded restriction fragment of φX174 DNA is shown in Figure 8-4 (Takeshita et al., 1978). Pyrimidines adjacent to the 3' side of the interaction are preferentially cleaved. BLM binding (Henichart et al., 1982; Kilkuskie et al., 1985) and orientation (Hamamichi et al., 1992), as dictated by the bithiazole moiety, as well as the metal-binding domain (Sugiyama et al., 1986; Carter et al., 1990, 1991), are responsible for the cleavage specificity; an illustration of FeBLM binding and orientation to double-stranded B DNA is presented in Figure 8-5.

*The mechanism of dioxygen activation and product formation in the iron BLM system*

*Mechanism of dioxygen activation.* Within a few years of BLM's discovery and isolation, the mechanism of its endonuclease-like action on DNA and RNA had been extensively probed (Nagai et al., 1969a; Shirakawa et al., 1971). Although BLM is isolated as the $Cu^{II}$ complex, Sausville and coworkers (1976, 1978a, 1978b) discovered that trace amounts of ferrous ion ($Fe^{2+}$) and molecular oxygen were likely to be responsible for DNA strand scission. The chemistry was shown to be enhanced in the presence of both oxidizing agents such as hydrogen peroxide (Nagai et al., 1969b) and reducing agents such as dithiothreitol, β-mercaptoethanol, and ascorbate (Muller et al., 1972; Lown and Sim, 1977). Reductants promote redox recycling of the iron center (continuous reduction of the oxidized metal center back to the ferrous state) making the process catalytic. Radical scavengers showed no significant effects on product formation, although radical species could be detected by spin-trapping experiments. Elucidation of the exact mechanism of dioxygen activation by FeBLM is ongoing, and some steps still remain controversial. Each step of the process is examined in detail below.

*Step 1. The binding of iron(II) to bleomycin.* The first step in dioxygen activation is the binding of $Fe^{II}$ to the BLM ligand. NMR, Mössbauer, and EPR data (Gupta et al., 1979; Sugiura, 1980; Albertini et al., 1982) all indicate that the initial BLM–$Fe^{II}$ complex is high spin, with absorption maxima at 370 and 476 nm, and is EPR silent. Although no crystal structure is available for the BLM–Fe complex, crystallographic

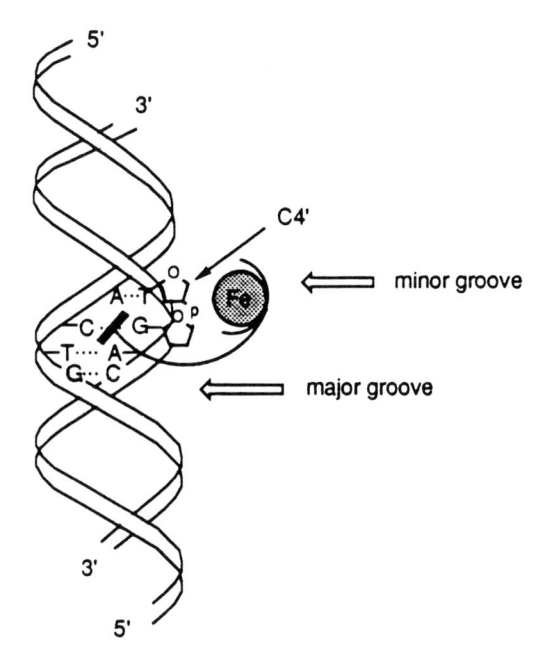

**FIGURE 8-5** A sketch of the FeBLM binding at a G base in the minor groove of B DNA. The metal center is in position to induce cleavage at the C4' site on the sugar of the pyrimidine (C) nucleotide adjacent to the 3' side of the metal complex binding interaction. The "p" represents the phosphodiester linkage on the sugar-phosphate backbone between the G and C bases on the 5' to 3' strand, hence the nomenclature GpC(5' → 3') (see text).

evidence from other metallobleomycin analogs (Iitaka et al., 1978) suggests a similar representation for the BLM–Fe$^{II}$ complex (Takita et al., 1978a; Kittaka et al., 1988). The crystal structure of PMA–Co$^{II}$ (Brown et al., 1989a; Tan et al., 1992), shown in Figure 8-6, is a model for the metal-binding region of BLM. A primary feature is the square planar metal center created by the four (5, 5, 5, and 6-membered) rings. The ligation is comprised of the α-amine, the secondary amine, the pyrimidine, the deprotonated amide, and the imidazole groups. These donor atoms provide a strong ligand field similar to those of porphyrins. Ligand binding has also been studied by NMR methods (Oppenheimer et al., 1979).

The binding of the axial NH$_2$ donor ligand plays an important role, controlling the spin state of the iron metal center and assisting in the O$_2$ binding and reduction processes. For desamido BLM, replacement of the axial NH$_2$ ligand by a carboxyl group or a water molecule

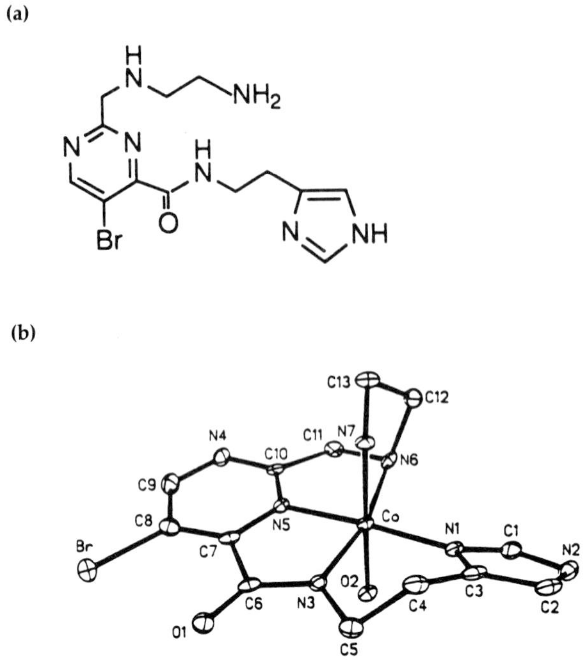

**(a)**

**(b)**

**FIGURE 8-6** Structure of the PMAH ligand (a) and thermal ellipsoid plot of the aqua-Co$^{II}$ complex, [Co(PMA)(H$_2$O)]$^{2+}$ (b). Reproduced with permission from Tan et al. (1992).

significantly decreases dioxygen activation. Because deamidation results in a higher p$K_a$ for the ligating α-NH$_2$ group, ligand binding and dioxygen activation become efficient only at pH values greater than the p$K_a$ (~9.3) for this complex (Sugiura, 1980a).

*Step 2. O$_2$ binding and reduction.* Under aerobic conditions, O$_2$ covalently binds to the sixth coordination site of the metal center ($k_f$ = 6 mM$^{-1}$s$^{-1}$ at 2 °C) in a reaction that is first order with respect to both O$_2$ and BLM–Fe$^{II}$ (Sausville et al., 1976; Burger et al., 1979, 1981; Kuwahara and Sugiura, 1988). The sugar moiety may act as a protecting pocket for dioxygen, which is thought to bind by end-on O coordination to the metal center (this is referred to as a *linear*, η$_1$ mode as opposed to a *side-on*, η$_2$ mode, where both O atoms coordinate to the metal center) (Kenani et al., 1988). The Fe$^{II}$, in its strong ligand field, readily donates an electron to the bound oxygen and is oxidized to a low-spin Fe$^{III}$ state (Oppenheimer et al., 1979; Sugiura, 1980a). The electronic structure immediately after $O_2$ binding is best described as the complexed superoxide anion species, BLM—$Fe^{III}$—$O_2^-$ (Burger et al., 1983). Due to the rapid rate of this process and the instability of the intermediate, evidence for this first reduction to superoxide, $O_2^-$, is inconclusive. The chemistry involved in this step is discussed in detail for BLM—$Co^{II}$ in the section on "Dioxygen Activation by Other Metallobleomycins."

The addition of a second electron to the BLM—$Fe^{III}$—$O_2^-$ complex yields a peroxide-bound species, BLM—$Fe^{III}$—OOH. Evidence for this species is more conclusive. Small rhombic splittings, $g_z = 2.254$, $g_y = 2.171$, and $g_x = 1.937$, observed in the EPR reflect the low-spin $Fe^{III}$ species (Sugiura, 1980a; Burger et al., 1981; Sugiura et al., 1982b). The absorption spectrum exhibits maxima at 365 and 384 nm (Burger et al., 1981). In the absence of added reductants (such as ascorbate—Burger et al., 1983; Natrajan et al., 1990a), the second reducing equivalent can be supplied by another BLM—$Fe^{II}$ complex. It has been proposed that the electron can be transferred through a BLM—Fe—O—O—Fe—BLM dimer, resulting in the products BLM—$Fe^{III}$—OOH and BLM—$Fe^{III}$—OH. Consistent with this model, the electron transfer reaction is inhibited by high concentrations of DNA (where all BLM–Fe is DNA bound) but is independent of the DNA concentration in the presence of added reductants.

In the absence of DNA, an overall 4 $e^-$ reduction of $O_2$ yields the thermodynamically stable, rhombic low-spin BLM–$Fe^{III}$OH species ($g = 2.45$, 2.18, 1.89) and $H_2O$ (Barr et al., 1990). Fe–BLM also undergoes rapid self-inactivation in the absence of DNA, yielding high-spin, rhombic, $Fe^{III}$ ($g = 4.3$)—see step 4 (Burger et al., 1981; Nakamura and Peisach, 1988; Buettner and Moseley, 1992). In the presence of DNA, however, an intimate interaction between the Fe–oxygen complex and DNA occurs, stabilizing the complex from rapid self-degradation (Burger et al., 1981). The bound dioxygen, in a relatively constrained conformation relative to the DNA, leads to a precisely oriented complex (Shields et al., 1982; Shields and McGlumphy, 1984; Chikira et al., 1989). The DNA—BLM—Fe—OOH ternary complex is now ready for the final stage of dioxygen activation, the severing of the O—O bond, creating a reactive electrophile close to the DNA backbone.

*Step 3. O—O bond scission.* As discussed in the section on "Dioxygen Activation by Metal Complexes," scission of the O—O bond can occur by either a heterolytic or a homolytic mechanism. Both pathways can result in high-valence Fe-oxo species which differ only in oxidation state; ·OH is produced in the homolytic mechanism. Elucidation of this process is needed to identify the species responsible for DNA cleavage. Attempts to clarify this issue have focused on investigating reactions with a variety of oxidants whose O—O bond cleavage mechanisms are well defined. The pathway for O—O bond scission depends on the p$K_a$ of the leaving group (Balasubramanian and Bruice, 1987) and the redox potential of the metal complex (Sheldon and Kochi, 1976). One limitation to this approach has been a lack of unambiguous proof for identical activated complexes and similar reactivity mechanisms in both the $H_2O_2$ and the other oxidant systems. To date, the nature of this cleavage is still controversial. For discussion, consider the mechanisms presented in Scheme 8-2.

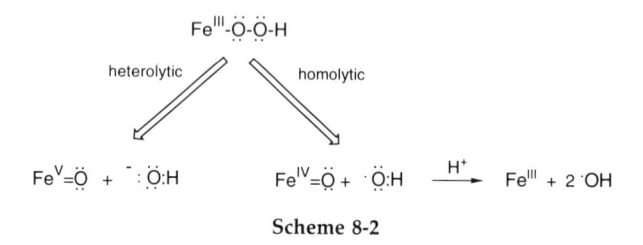

$$Fe^{III}\text{-}\ddot{O}\text{-}\ddot{O}\text{-H}$$

heterolytic / homolytic

$$Fe^V{=}\ddot{O} + {}^-{:}\ddot{O}{:}H \qquad Fe^{IV}{=}\ddot{O} + {}^{\cdot}\ddot{O}{:}H \xrightarrow{H^+} Fe^{III} + 2\,{}^\cdot OH$$

Scheme 8-2

Evidence for homolytic O—O scission has involved comparisons of oxygen atom transfer reactions with organic substrates using $H_2O_2$ and alkylhydroperoxides (Padbury and Sligar 1986; Heimbrook et al., 1987; Padbury et al., 1988). Subtle differences in the reactivities of the two types of oxidants toward organic substrates, however, suggest that the activated complex formed in each case may differ (Natrajan et al., 1990b). The alternative, and currently favored, mechanism for O—O bond scission is via the heterolytic pathway, in which the metal center donates two electrons to the O—O bond and ends up with a formal +5 charge; the activated complex is an iron-oxo intermediate, ($Fe^V{=}O$) BLM. Evidence for this pathway includes the lack of inhibition by radical scavengers and DNA cleavage by single-oxygen donors such as potassium monoperoxysulfate, $KHSO_5$, and sodium metaperiodate, $NaIO_4$ (Burger et al., 1985; Murugesan and Hecht, 1985; Pratviel et al., 1986). The latter argument assumes that the activated complexes formed by single-oxygen donors and $H_2O_2$ are similar.

Hydrogen-bonding properties inherent in the BLM—Fe—OOH precursor complex are thought to be important in the O—O bond cleavage process (Kenani et al., 1988; Shepherd et al., 1992). Recent molecular mechanics studies using the model ligand, PMA (Figure 8-

6a), suggest the importance of internal hydrogen bonding of the OOH group with the secondary NH of the BLM ligand (Wu et al., 1992). It is suggested that on O—O ligation and reduction, a rearrangement of the BLM ligand takes place in which the imidazole group replaces the NH$_2$ group at the axial site *trans* to the O—O bond, allowing an H-bonding interaction with the NH group to take place and enhancing heterolytic cleavage. The mechanism involves a *push-pull* (general acid/base catalysis) interaction dictated by a strong electron donor in the axial position (push), as well as an H-bonding interaction (pull) of the departing electronegative O atom, analogous to that at porphyrin cytochrome-P450 systems (Yamaguchi et al., 1992).

*Step 4. The fate of the Fe–BLM complex.* The final iron species after one complete cycle of DNA cleavage is BLM–Fe$^{III}$OH. Radical scavenger and inhibition studies indicate that the FeBLM intermediate partially decomposes, yielding ·OH and possibly O$_2^-$, the latter of which could lead to ·OH in the presence of BLM–Fe$^{III}$ and H$_2$O$_2$ (Rabow et al., 1990; Gajewski et al., 1991). The radicals produced lead to minor damage to DNA bases and the formation of 8-hydroxyguanine residues (Kohda et al., 1989; Gutteridge et al., 1990) but are not involved in DNA strand scission or in the formation of base propenals. Intact BLM–Fe$^{III}$ can be re-reduced to the Fe$^{II}$ form by external reductants (Sugiura et al., 1982a), making the cycle catalytic. DNA cleavage is terminated through self-inactivation of the BLM complex in which, for example, the bithiazole portion is altered (Nakamura and Peisach, 1988), or there is some degradation of the amino acid side chains of the metal-binding region (Owa et al., 1990).

*Mechanism of product formation.* Product analysis demonstrates (D'Andrea and Haseltine, 1978) that activated FeBLM cleaves double-stranded DNA preferentially at 5'-GC, 5'-GT and 5'-GA bases, with release of pyrimidine bases located to the 3' side of G residues (Takeshita et al., 1978; Hecht, 1986). The activated complex carries out a radical-like abstraction of the C4' hydrogen atom, resulting in a C4'-based radical intermediate, **I**, in Scheme 8-3. A large tritium isotope effect in the homolytic C4'–H bond cleavage ($k_H/k_T = 7$ to 11) indicates that this is the rate-limiting step (Stubbe and Kozarich, 1987; Kozarich et al., 1989).

Intermediate **I** degrades by two competitive pathways (Wu et al., 1983, 1985; Stubbe and Kozarich, 1987), one dependent on and one independent of O$_2$ (Giloni et al., 1981; Burger et al., 1982). An alternative *in vivo* mechanism has been presented by Ciriolo et al. (1989). In the oxygen-dependent pathway (path **a**), an extra molecule of O$_2$ is

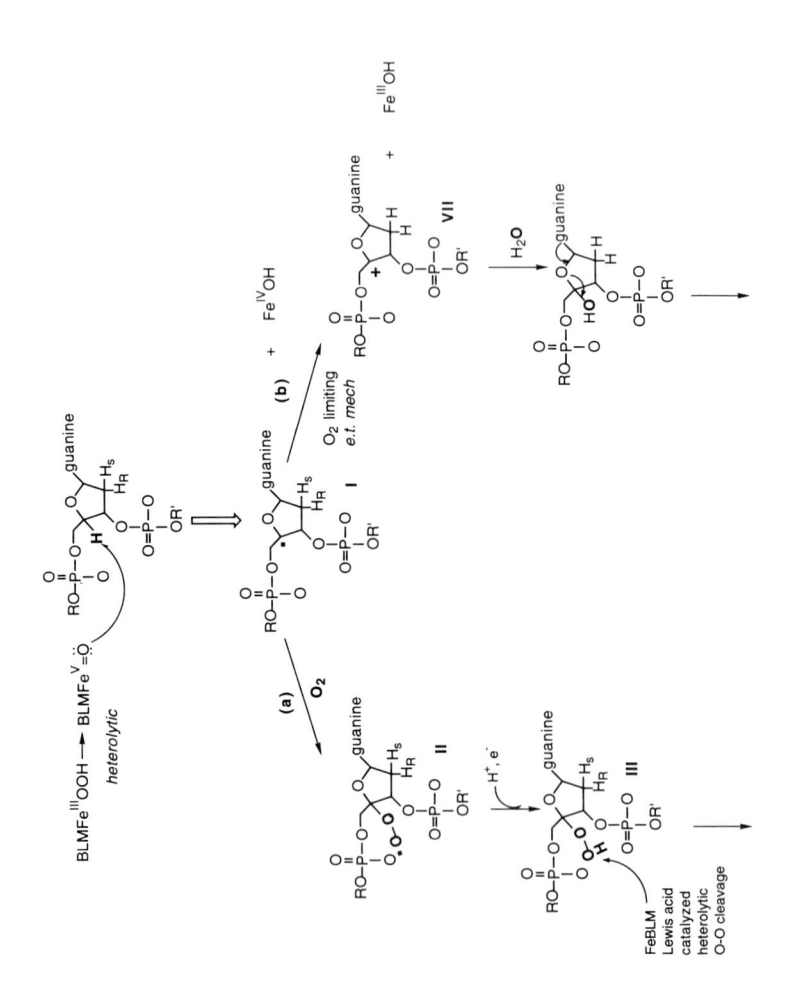

352  *Metal-Complex-Catalyzed Cleavage of Biopolymers*

DNA Cleavage 353

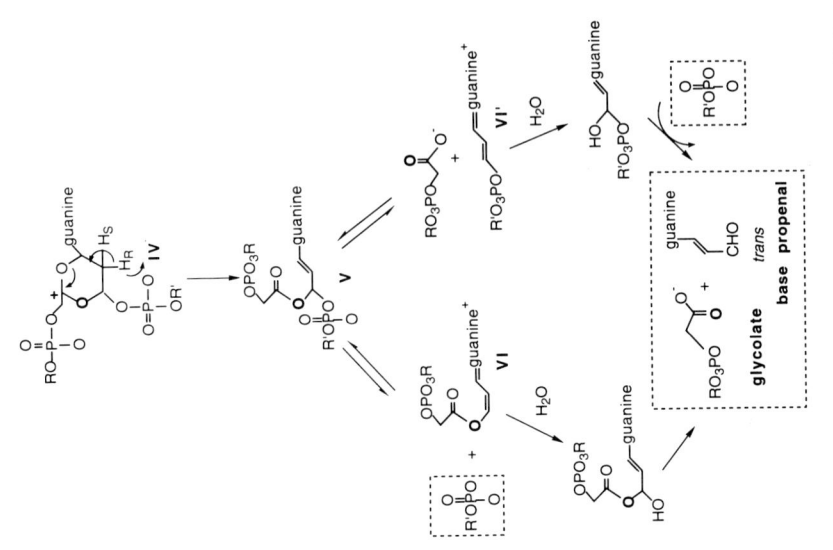

Scheme 8-3

required for stoichiometric strand scission, with the formation of base propenals derived from the deoxyribose sugar and the release of 5'-phosphate and 3'-phosphoglycolate modified ends. Murugesan et al. (1985) report that exclusively *trans* base propenals are produced. The mechanism for this process has been recently refined (Stubbe and Kozarich 1987; McGall et al., 1992). After the C4' H abstraction, $O_2$ reacts with **I** at the free radical site, producing the peroxyl radical **II**. A single electron donated to the peroxyl radical ·O—O— of **II** by an added reductant (e.g., RSH) and protonation yields the organic hydroperoxide **III**. The BLM–Fe$^{III}$ product could then act as a Lewis acid, catalyzing the heterolytic cleavage of the O—O bond and forming a 4' carbonium intermediate, **IV** (McGall et al., 1992; Kharasch et al., 1950). Based on extensive $^{18}O$ labeling studies (labels are in bold face), the mechanism for base propenal production has been proposed (McGall et al., 1992). When $^{18}O$ is introduced as molecular $O_2$ after formation of the reactive FeBLM intermediate, one oxygen in the glycolate moiety is derived from the $^{18}O_2$ and the other from the deoxyribose moiety; the aldehyde oxygen in the base propenal is from the solvent. Also, Ajmera et al. (1986) showed that when the 3'-phosphate group is initially labeled with $^{18}O$, all $^{18}O$ is detected in the free phosphate released in solution. All labeling studies are in accord with the following mechanism: The 2'-*pro* R proton (replacement of this proton yields the R configuration) is removed from intermediate **IV**, and heterolytic cleavage of the O4'—Cl' bond to give **V** follows. Elimination of the 3' phosphate group (to yield a 5'-phosphate product) or the 3'-phosphoglycolate moiety give either **VI** or **VI'**. Addition of $H_2O$ to either of these intermediates produces unstable products that decompose to the expected base propenal.

Under limiting $O_2$ conditions, path **b** predominates, resulting in the immediate release of free nucleic acid bases plus an intact but alkali-labile DNA backbone (due to an oxidatively modified sugar). After alkali treatment (pH ~12), 3'-phosphate and 5'-phosphate termini are released. The oxidized sugar, **VIII**, is unstable, but identification has been attempted (Sugiyama et al., 1985; Stubbe and Kozarich, 1987). One method involves reduction of the sugar intermediate with $NaBH_4$; after purification and enzymatic degradation, stable products are obtained. The resulting products have been separated and identified by HPLC (Rabow et al., 1986). By this method, **VIII** has been identified as the 4'-keto-1'-aldehyde and is shown to be produced in near-quantitative yield relative to base release (Rabow et al., 1990). (A minor pathway involves radical formation at the 1' position.) Using $^{18}O$-

labeled $O_2$ or $H_2O$, solvent has been shown to be the source of the keto oxygen in the oxidized sugar. A role for the final form of the activated BLM–Fe complex in this reaction has been proposed (Sugiyama et al., 1991). An electron transfer reaction between BLM–Fe and the C4' sugar radical generates a C4' carbonium ion intermediate, **VII**. A heterolytic mechanism for the cleavage of the original BLM—Fe—$OO^{2-}$ yields a final FeBLM species of +3 oxidation state. An alternative pathway that accounts for solvent $^{18}O$ incorporation into the oxidized sugar is a nonelectron transfer, oxygen rebound mechanism requiring lability of the iron-oxo oxygen for rapid exchange with the solvent (Stubbe et al., 1987).

*Dioxygen activation by other metallobleomycins.* BLM forms stable complexes with a number of metal ions (Sugiura et al., 1979a; Petering et al., 1990), showing stability constants (log K given in parentheses) in the order $Cu^{II}(18.1) > Fe^{III}(16) > Zn^{II}(9.7) > Co^{II} > Fe^{II}$ (<9.7). Surprisingly, the most reactive $Fe^{II}$ metal ion shows the least favorable binding. Indeed, early studies suggested inhibition of DNA strand scission in the presence of other metal ions such as $Cu^{II}$, $Zn^{II}$, $Co^{II}$, and $Mg^{II}$. More recently, it has been shown that certain metal ions simply require different conditions for the induction of DNA strand scission and that these requirements are dependent on the redox potentials of BLM complexes. The redox potential of the $Fe^{III/II}$ BLM couple is favorable for redox cycling ($E^{\circ\prime} = 0.129$ V; Melnyk et al., 1981), hence its facile dioxygen activation. The metals discussed below show decreased efficiency in DNA cleavage with respect to the iron complex system. This presumably reflects the differences in redox potentials, since the mechanisms of action as well as DNA binding are thought to be similar to those of BLM–Fe.

*Cobalt(III/II).* *Reaction of $O_2$ with $Co^{II}BLM$: a model for $O_2$ activation steps 1–3 in the BLM–Fe system.* $Co^{II}BLM$ shows no DNA cleavage activity in the presence of $O_2$ and reducing agents, and $Co^{II}$ ions are inhibitors of BLM–Fe chemistry (Sausville et al., 1978a). The lack of activity is due to the irreversible nature of the redox couple ($E^{\circ\prime} \leq -1.5$ V; Sugiura, 1980b). Although no DNA cleavage is observed in the BLM–$Co^{II}/O_2$ system, relatively slow kinetics and long-lived intermediates in the initial steps of dioxygen activation (eqs. 10–12) give valuable insight into mechanistic details which are difficult to assess in the BLM–Fe system.

The EPR spectrum of BLM–$Co^{II}$ depicts a low-spin 5-coordinate square pyramidal Co environment ($g_\perp = 2.272$ and $g_\parallel = 2.025$). The reaction of BLM–Co$^{II}$ with O$_2$ results in a monomeric dioxygen complex (Sugiura, 1980b), at which time the unpaired spin density is transferred from the metal center to the oxygen. The new $g$ values ($g_\perp$ = 2.006 and $g_\parallel$ = 2.080–2.103) reflect formation of the superoxide anion. Kinetic experiments using UV-visible, $^1$H NMR, and EPR spectroscopy have defined the reaction mechanism in the absence of DNA substrate (eqs. 10–12) (Xu et al., 1992a). The redox chemistry takes place via the paramagnetic binuclear complex, BLMCo$^{II}$–O$_2$–Co$^{II}$BLM, which decomposes to yield the two oxidized monomeric products. Importantly, the reduction of O$_2$ to O$_2^{2-}$ is a one-step, two-electron process mediated through the binuclear complex. A similar mechanism is expected for BLM–Fe$^{II}$ in the absence of added reductants.

$$\text{BLM—Co}^{II}\text{—H}_2\text{O} + \text{O}_2 \rightleftharpoons \text{BLM—Co}^{II}\text{—O}_2 \leftrightarrow \text{BLM—Co}^{III}\text{—O}_2^- \text{ (fast)} \quad (10)$$

$$2\text{BLM—Co—O}_2 \rightarrow \text{BLMCo}^{III}\text{—O}_2\text{—Co}^{III}\text{BLM} + \text{O}_2 \quad (11)$$

$$\text{BLMCo}^{III}\text{—O}_2\text{—Co}^{III}\text{BLM} + \text{H}_2\text{O} \rightarrow \text{BLM—Co}^{III}\text{—O}_2\text{H} + \text{Co}^{III}\text{BLM}^+\text{—H}_2\text{O} \quad (12)$$

In the presence of excess (10:1) DNA, the BLMCo–O$_2$ complex is stable (Xu et al. 1992b). EPR evidence suggests that this is best defined as the DNA · BLM–Co$^{III}$–O$_2^-$ complex. Like that for BLM–Fe, DNA binding followed by O$_2$ binding and reduction are dependent on the DNA concentration (Sugiura, 1980b). Slow dimerization does occur, but no dimer breakdown into the monomeric products and DNA · BLM–Co$^{III}$O$_2$H and DNA · BLMCo–H$_2$O is detected kinetically. (At low pH these end products are detected.) This evidence reflects the BLM–Fe system, which also shows inhibition of DNA cleavage by DNA itself. Unlike BLM–Fe, however, under oxidizing conditions neither DNA nor Co complex degradation takes place in the presence of DNA (Subramanian and Meares, 1986).

*Co$^{III}$BLM: DNA cleavage.* While the Co$^{II}$ complex is ineffective in DNA cleavage chemistry, BLM–Co$^{III}$ exhibits photochemical strand scission. Several different types of BLM–Co$^{III}$ complexes have been isolated and characterized. Individual signatures in the UV-visible spectrum are due to the nature of the sixth coordinating ligand (Chang et al., 1983). The aquo complex is brown, as are the formato and acetato complexes. A brown superoxide complex has also been proposed (Vos et al., 1980; Albertini and Garnier-Suillerot., 1982). The crystal structure of a model aquo Co$^{III}$ complex has been discussed in detail in the section on

"The Mechanism of Dioxygen Activation and Product Formation in the Iron BLM System" (Figure 8-6) (Tan et al., 1992). The green form is attributed to hydroperoxide ligation (Chang et al., 1983). The binding constants for these two complexes to doubled-stranded DNA are in the $10^7 M^{-1}$ range, considerably larger than for BLM–Fe$^{II}$ (Chang and Meares, 1984); the green form shows higher affinity than the brown form. The BLM–Co$^{III}$ orange form, a distinct complex, has all six coordination sites filled by the BLM ligand, and shows decreased binding to DNA and little or no strand scission chemistry.

Chang and Meares (1982) first defined the conditions necessary to carry out DNA cleavage with BLM–Co$^{III}$, i.e., activation by UV radiation (or by visible light in the presence of a photosensitizer ([Ru(bpy)$_3$]$^{2+}$) (Subramanian and Meares, 1986)). DNA cleavage (under aerobic or anaerobic conditions) is effected using either the green or brown form, the former being somewhat more efficient (Subramanian and Meares, 1986; Brown et al., 1989a). Photoreduction in these ligand systems occurs through stimulation of charge transfer or ligand transitions with near-UV light (330–450 nm), and action spectra indicate that the effective wavelengths for BLM–Co$^{III}$ are those which excite ligand-to-metal charge-transfer electronic transitions (Chang and Meares, 1984). This excitation most likely results in the reduction of Co$^{III}$ to Co$^{II}$ and an oxidized ligand (Adamson, 1979; Stubbe and Kozarich, 1987).

As observed with BLM–Fe$^{II}$ (Saito et al., 1989), strand breakage at the 3' side of guanine residues predominates. However, unlike BLM–Fe$^{II}$, DNA cleavage by BLM–Co$^{III}$ is independent of O$_2$ concentration (Chang and Meares, 1984). Furthermore, only free bases and alkali-labile lesions resulting in 3'-phosphoglycolate modified ends are produced; no base propenals are detected. BLM–Co$^{III}$ also differs from BLM–Fe$^{II}$ in its cleavage pattern of supercoiled DNA. Irradiation of BLM–Co$^{III}$ in the presence of supercoiled DNA completely converts the substrate to the nicked circular form (II), also producing alkali-labile sites. Unlike cleavage with BLM–Fe$^{II}$, no linear duplex form (III) is produced (see the section on "Specificity of DNA Strand Cleavage").

Two mechanisms have been proposed for the action of BLM–Co$^{III}$ on DNA. The first involves dioxygen activation by the photoreduced metal center analogous to that of the BLM–Fe system and assumes that the hydroperoxide complex is the active intermediate. The mechanism involves a C4'H abstraction by "activated BLM–Co$^{III}$" followed by hydroxylation at the C4' site (Saito et al., 1989). BLM–Co$^{III}$–OH$_2$ must convert to the hydroperoxide form prior to strand scission, accounting for the greater efficiency of the green form over the brown. The second mechanism suggests that DNA oxidation is carried out by ·OH radicals that are produced by ligand-based radicals initially formed upon photolysis (Tan et al., 1992).

*Copper(II/I).* Although it was known that BLMCu$^I$ in the presence of O$_2$ produced reactive oxygen species (Sugiura, 1979b), early work using the Cu$^{II}$ form showed no DNA cleavage activity in the presence of added reductants. It was thought that any reactivity exhibited by isolated BLM was due to reduction of the Cu$^{II}$ metal center to Cu$^I$ by thiol reducing agents (RSH) such as cysteine (CySH), glutathione (GSH), and dithiothreitol (DTT), followed by rapid substitution by trace amounts of Fe$^{II}$ (Takahashi et al., 1977).

Hecht's group was the first to describe experimental conditions needed for the activation of O$_2$, with subsequent cleavage of DNA by both Cu$^{II}$ and BLM–Cu$^I$ (Oppenheimer et al., 1981; Ehrenfeld et al., 1985, 1987). Important to their findings is the order of addition of the reagents (eq. 13). Either Cu$^{II}$ must first be reduced by the thiol to Cu$^I$ prior to incubation with the BLM ligand, or relatively long incubation times of the complex with the reductant are needed. Reductants such as ascorbate and NADH, used in the BLM–Fe system, are ineffective (Freedman et al., 1982) with BLM–Cu. Two equivalents of RSH are required (one for binding, the other for electron transfer) and the rates are relatively slow; $t_{1/2}$ values range from 2 to 80 minutes in the presence of 10-fold excess reductant.

$$Cu^{II} \xrightarrow{RSH} Cu^{I} \xrightarrow{BLM} BLM-Cu^{I} \qquad (13)$$

A large barrier to facile oxidation-reduction chemistry in copper systems (reflected in the negative E$^{°\prime}$ value ($-0.33$ to $-0.49$ V; Ishizu, et al., 1981; Melnyk et al., 1987) for the BLM–Cu$^{II/I}$ couple) is induced by the substantial differences in preferred coordination geometries for the two oxidation states. The preferred Cu$^{II}$ geometry is a distorted octahedron resembling that shown in Figure 8-6 for Co$^{II}$ (Brown et al., 1989a, 1989b). $^1$H and $^{13}$C NMR has been employed to elucidate the geometry of the Cu$^I$ complex (Ehrenfeld et al., 1985; Petering et al., 1990). Binding involves the β-aminoalaninamide, pyrimidinyl portions, and possibly the imidazole moiety. The N$^α$ of β-hydroxyhistidine is not a ligand. Despite differences in geometry, studies using the model ligand, PMAH (see the section on "The Mechanism of Dioxygen Activation and Product Formation in the Iron BLM System", Figure 8-

6a), suggest that rearrangement of the ligand coordination sphere for the Cu$^{II/I}$ couple could be relatively facile at neutral pH (the model may be less rigid than that of BLM), an important criterion for redox reactivity. Also contributing to the dioxygen activation barrier is competition for binding between the thiol-reducing agents and molecular O$_2$ (Petering et al., 1990).

High concentrations of BLM-Cu$^I$ carry out DNA cleavage in a reaction that is dependent on O$_2$. However, experiments using $^3$H DNA indicate that no propenal products are formed. Supercoiled DNA is cleaved by BLM-Cu$^I$ into the circular nicked and the linear duplex forms (Ehrenfeld et al., 1985). Hydroxyl radicals are thought to be the reactive species produced from BLM-Cu$^I$ and O$_2$.

*Other Metallobleomycins.* Vanadyl, V$^{IV}$O$^{2+}$ (Kuwahara et al., 1985; Butler and Carrano, 1991), manganese(III/II) (Burger et al., 1984; Ehrenfeld et al., 1984; Suzuki et al., 1984, 1985), and ruthenium(II) (Subramanian and Meares, 1985) BLMs have also been reported to cleave DNA.

## Metal-EDTA Complexes

The unique structural features and DNA cleavage chemistry of BLM-Fe have influenced the development of many new synthetic chelate/carrier molecules that can activate molecular oxygen and cleave biopolymers. The metal-EDTA chelates discussed below are employed as analytical tools in molecular biology, probing the structure and function of DNA and RNA molecules.

*The EDTA-Fe$^{III/II}$ system: Structure and mechanism of O$_2$ activation.* The crystal structure of EDTA-Fe$^{III}$ (Lind et al., 1964), (Figure 8-7), shows the heptacoordinate environment about the metal center. The H$_2$O moiety is relatively labile and is readily replaced by other ligands such as ascorbate and H$_2$O$_2$. The EDTA-Fe$^{III/II}$ couple has a potential (E$^{o\prime}$ = 0.117 V; Ogino et al., 1989) in the range for both facile reduction to the +2 state and oxidation back to the +3 state by many reagents. The reactions between EDTA-Fe$^{III/II}$, ascorbate, and O$_2$ or H$_2$O$_2$ are some of the most thoroughly studied metal complex systems in terms of mechanism. This chemistry was first investigated by Udenfriend et al. (1954) as a model for peroxidase chemistry; EDTA-Fe$^{III}$ is known to hydroxylate aromatic compounds in the presence of a reductant (ascorbate) and H$_2$O$_2$ or O$_2$ (Grinstead, 1960a, 1960b). An overview of

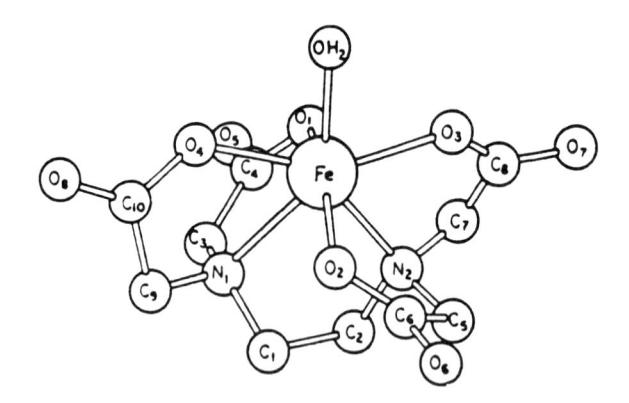

**FIGURE 8-7** Structure of [Fe(EDTA)(OH$_2$)]$^-$. Reproduced with permission from Lind et al. (1964).

the dioxygen activation mechanism for EDTA–Fe$^{III}$ in the presence of ascorbate is discussed below (eqs. 14–21).

*Reduction of EDTA–Fe$^{III}$ by ascorbate.* The first step involves the reduction of the ferric complex to ferrous by ascorbate (eq. 14), with an overall rate for reaction with the ionized ascorbate species (AscH$^-$) determined to be $18 \times 10^2 \, M^{-1} s^{-1}$ (pH 3; Kahn and Martell, 1967).

$$\text{EDTA–Fe}^{III} + \text{AscH}^- \rightarrow \text{EDTA–Fe}^{II} + \text{Asc}^{-\cdot} (+ \, H^+) \quad (14)$$

*O$_2$ Activation to the peroxide level.* The reduction of O$_2$ by EDTA–Fe$^{II}$ has been studied in detail by Zang and van Eldik (1990). The first step involves the association of O$_2$ with the metal chelate. A value for $K_{15}$ has been determined to be $2 \times 10^{-3} M^{-1}$ (Purmal et al., 1980). A rapid reduction ($k_{16} = 1.1 \times 10^5 \, s^{-1}$) of O$_2$ to O$_2^-$ then follows. Under conditions of excess O$_2$, competing reactions of O$_2^-$ with Fe$^{III}$ or Fe$^{II}$ can occur, the former (eq.17; $k_{17} = 2 \times 10^6 \, M^{-1} s^{-1}$) leading to oxidation of the O$_2^-$ to form O$_2$ and the latter (eq. 18) resulting in a second electron addition to O$_2$. Reaction (18) proceeds with a rate constant of $2 \times 10^6$ to $1 \times 10^7 \, M^{-1} s^{-1}$ to form the Fe$^{III}$ peroxide-bound intermediate. EPR evidence (Fujii et al., 1990) for this intermediate depicts the peroxide bound in an $\eta_1$ (see the section on "The Mechanism of Dioxygen Activation and Product Formation in the Iron BLM System," step 2) fashion at pH 7.6. In the presence of H$^+$, this quickly dissociates, yielding H$_2$O$_2$. H$_2$O$_2$ (and O$_2$) are also produced by the disproportionation of two O$_2^-$ in the presence of H$^+$.

$$\text{EDTA-Fe}^{II} + O_2 \rightleftharpoons \text{EDTA-Fe}^{III}(O_2^-) \quad (15)$$

$$\text{EDTA-Fe}^{III}(O_2^-) \rightarrow \text{EDTA-Fe}^{III} + O_2^- \quad (16)$$

$$\text{EDTA-Fe}^{III} + O_2^- \rightarrow \text{EDTA-Fe}^{II} + O_2 \quad (17)$$

$$\text{EDTA-Fe}^{II} + O_2^- \rightarrow \text{EDTA-Fe}^{III}(O_2^{2-}) \quad (18)$$

*Cleavage of the O—O bond.* The mechanism of the reaction of $H_2O_2$ with EDTA-Fe$^{II}$ (eq. 19), like that with BLM-Fe, is controversial (Rush et al., 1990). Evidence based on kinetics, radical scavenger studies, and product analysis has been presented for the reactive oxygen species being either the ·OH radical (Grinstead, 1960a; Walling, 1975; Bull et al., 1983) or an iron–oxo complex (Rush and Koppenol, 1986; Rahhal & Richter, 1988). Unfortunately, the instability of EDTA–Fe–O type intermediates precludes direct spectroscopic evidence for either the homolytic or heterolytic mechanism. Under certain conditions, $H_2O_2$ can also be a mild reductant (eq. 20) (Zang and van Eldik, 1990).

$$\text{EDTA-Fe}^{II} + H_2O_2 \rightarrow \text{EDTA-Fe}^{III} + \cdot OH + OH^- \quad (19a)$$

or

$$\text{EDTA-Fe}^{IV}O + H_2O \quad (19b)$$

$$\text{EDTA-Fe}^{III} + H_2O_2 \rightarrow \text{EDTA-Fe}^{II} + O_2^- + 2H^+ \quad (20)$$

Since ascorbate is a two-electron reductant, the radical anion produced after the first electron transfer, AscH·, can then go on to reduce another EDTA-Fe$^{III}$ or EDTA-Fe$^{IV}$O (eq. 21), yielding dehydroascorbate, Asc.

$$\text{EDTA-Fe}^{III} + \text{AscH}\cdot \rightarrow \text{EDTA-Fe}^{II} + \text{Asc} + H^+ \quad (21a)$$

or

$$\text{EDTA-Fe}^{IV}O \rightarrow \text{EDTA-Fe}^{III} + \text{Asc} + OH^- \quad (21b)$$

*Tethered EDTA complexes—DNA affinity reagents*

*Methidiumpropyl–EDTA–Fe$^{II}$ (MPE): A non-sequence-specific probe*

*Structure and substrate binding.* The first synthetic DNA cleavage reagent was designed to show little or no base sequence specificity in binding to and cleavage of DNA (Hertzberg and Dervan, 1982). This involved an EDTA chelate derivatized via a short hydrocarbon linker through one of its carboxylate arms to the DNA intercalator methidium. The iron complex of methidiumpropyl–EDTA (MPE–Fe) is shown below, **X**, MPE–Fe$^{II}$ (at concentrations as low as $10^{-8}$ M), in the presence of O$_2$ and DTT cleaves supercoiled circular DNA into nicked circular form II with efficiencies similar to that of BLM–Fe$^{II}$.

The DNA binding and cleavage specificities for the complex were determined by inhibition experiments involving DNA intercalators with well-defined binding specificities (Van Dyke et al., 1982). These included actinomycin D (intercalates with a preference for guanine-rich sequences), distamycin A and netropsin (minor groove binders with a preference for A + T-rich sequences), and daunomycin (intercalates with no base specificity). In the absence of a sequence-selective drug, cleavage occurs at all sites to create a uniform electrophoretic band pattern of DNA fragments. The fact that the site of maximum cleavage on the lower strand is shifted to its 3' side indicates that the complex intercalates in the minor groove of both A and B DNA (Schultz et al., 1982). The intercalation affinity for MPE–M$^{2+}$ complexes at a two base pair site is ~1.2–1.5 × 10$^5$ M$^{-1}$ (Hertzberg and Dervan, 1984). Intercalation of this complex into supercoiled DNA unwinds the double helix, with an unwinding angle of 11° per bound MPE–M$^{2+}$. MPE–Fe has found much use as a footprinting reagent (see below).

**X**

*Mechanism of O$_2$ activation.* In the presence of O$_2$, MPE–Fe$^{2+}$ (0.01–0.1 µM) carries out oxidative strand scission of DNA (10 µM in base pairs). The reaction is enhanced in the presence of a reductant (1–5 mM, with efficiency decreasing in the order of ascorbate > DTT > NADH). The optimum conditions have been determined to be pH 7.4, 22 °C, 3.5 hours, 5 mM NaCl with a turnover number (single-strand scission per MPE molecule) of ~1.1–1.4 (Hertzberg and Dervan, 1984). Both SOD and catalase inhibit the reaction, indicating an intermediate role for O$_2^-$ and H$_2$O$_2$, respectively. DNA cleavage occurs in the reaction of

MPE–Fe$^{III}$ with iodosobenzene (PhIO), suggesting that the formation of a high-valent Fe$^V$=O species may be possible. Based on the diffusible nature of the cleaving species [spanning over ca. four base pairs (Schultz et al., 1982)] and quenching effects of high buffer (HEPES, Tris, 0.1 M) and reductant (>1–5 mM DTT) concentrations, however, the reactive species is thought to be ·OH. Product analysis, discussed in the following section, also supports the argument for favoring ·OH over FeO.

*Product analysis and mechanism of DNA strand scission.* Product analysis using gel electrophoresis and high pressure liquid chromatography (HPLC) (Hertzberg and Dervan, 1984) reveals that all four bases are produced stoichiometrically with single-strand scission. A statistical release of the free bases affirms the low specificity of the reagent. In addition, and similarly to BLM–Fe, 5'-phosphorylated termini are released. Unlike the O$_2$-dependent reaction of BLM–Fe, free bases rather than base propenals are released, and relatively equal amounts of 3'-phosphoryl and 3'-phosphoglycolic acid groups are detected, which is similar to γ-radiolysis. Different activated intermediates are proposed for the BLM–Fe and MPE–Fe systems: Fe$^V$=O and ·OH, respectively.

## Applications

*DNA footprinting.* MPE–Fe$^{II}$ has been used to map (determine the size and location of) the binding sites of small DNA-binding molecules (Van Dyke and Dervan, 1983a, 1983b). In this *footprinting* technique, the chemical reagent maps the binding site of another molecule bound to the biopolymer. The chemical probe either binds nonspecifically to or simply comes into close proximity to DNA. Figure 8-8a describes chemical footprinting of a DNA-bound substrate by a metal complex probe such as MPE–Fe$^{II}$ (Dervan, 1986). In the presence of O$_2$, the metal complex produces reactive oxygen species that *nonspecifically* and oxidatively cleave the biopolymer backbone. A uniform ladder of nucleic acid fragments is observed on a high-resolution sequencing gel. In the presence of a bound substrate, the oxidizing species cleaves *around* the area covered by the substrate. Sequencing gel electrophoresis reveals a ladder of cleaved fragments except where the DNA strand was protected by the bound molecule, and the binding site is revealed. For example, MPE–Fe$^{II}$ has been used to probe the binding interaction of the anticancer drug chromomycin A$_2$ to defined segments (restriction fragments) of DNA. The footprint shows a minimum 3-bp recognition

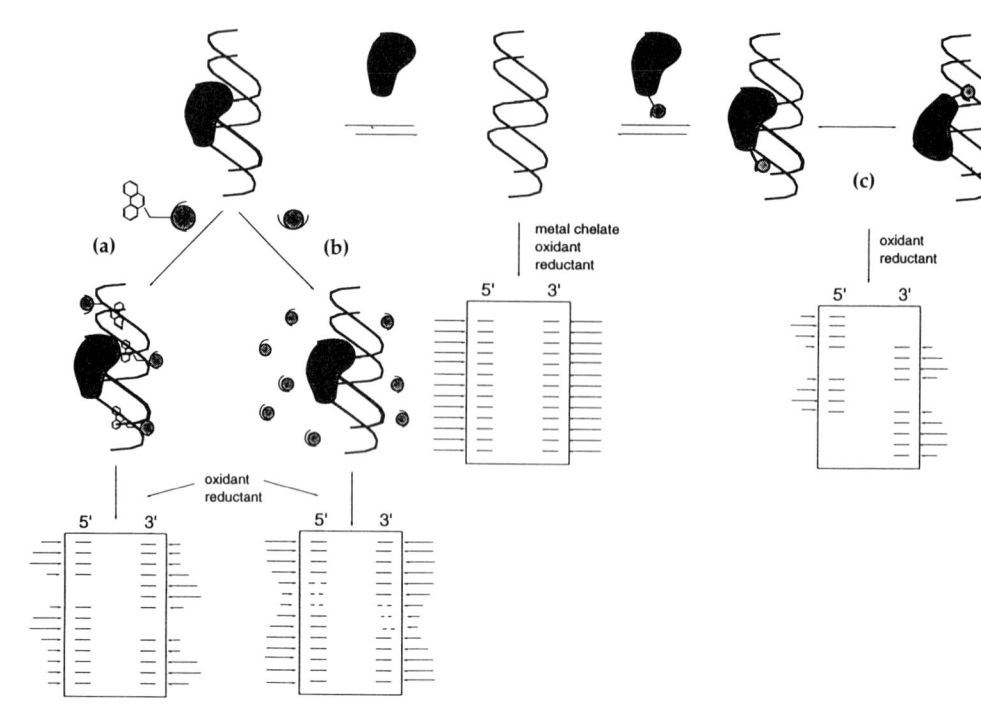

**FIGURE 8-8** Schematic of drug/protein–metal chelate–DNA interactions and resulting sequencing gels for (a) the DNA footprinting technique using DNA-bound metal chelates (e.g., MPE-Fe); (b) the DNA footprinting technique involving nonbinding metal chelates (e.g., EDTA-Fe); (c) the DNA affinity cleavage technique, in which the metal chelate is covalently attached to the drug–protein substrate (e.g., DE–Fe). The substrate is shown bound to the minor groove of B DNA. On addition of oxidant and reductant, metal complexes produce reactive oxygen species that cleave the biopolymer backbone. In the absence of a bound substrate, a uniform ladder of nucleic acid fragments is observed (center scheme). In the presence of substrate (a, b, and c), the species cleave the DNA *around* the area covered by the substrate and the binding site is mapped. The sequencing gels show lanes of both 5' and 3' end-labeled DNA fragments; the arrows indicate the relative intensity of cleavage. It can be seen that the 5' and 3' cleavage fragments are offset from and symmetrical to each other due to binding of the substrate in one groove of the helical DNA. Cleavage minima on each strand are approximately 3 bp apart, a pattern indicative of the substrate binding within the minor groove of B DNA. (Usually, a small organic substrate binds in the minor groove and proteins bind in the major groove.) Note that (1) more cleavage detail is obtained in (b) as opposed to (a) since the cleavage path for the diffusible species is not limited by metal complex binding; (2) in the affinity cleavage experiment (c), cleavage ladders are produced only at the site of bound substrate. In this technique, the attached chelate accurately determines the orientation and recognition site size for the substrate.

site with preferential binding at the 5'-AGC < CGG, GCC < AGC, GGG-3' regions.

*Structural probe.* MPE–Fe$^{II}$ has been used as a structural probe for RNA molecules (Vary and Vournakis, 1984; Kean et al., 1985; Tanner and Cech, 1985). The interaction with the well-characterized tRNA$^{phe}$ molecule reveals preferential binding of MPE–Fe$^{II}$ to double helical regions $K_a \sim 10^5$–$10^6\,M^{-1}$). Specific cleavage on opposite strands, and shifted to the 3' side, corresponds to a prominent intercalating site in the anticodon region (C27–G43 and C28–G42; C17–A66 and C49–G65). Other cleavage patterns reveal binding sites within the TψC loop and the D stem. The tRNA$^{phe}$ studies have allowed the mapping of helical domains in other RNA molecules such as 16 S ribosomal RNA (Kean et al., 1985) and IVS RNA (Tanner and Cech, 1985), as well as the probing of chromatin structure (Cartwright et al., 1983).

*Sequence specific probes—nucleic acid affinity cleavage.* In *affinity cleavage* systems, the EDTA–Fe$^{II}$ moiety is attached to the binding substrate. The general features of this method are presented in Figure 8-8c. Double-stranded cleavage by the attached chelate accurately determines the orientation and recognition site size for the substrate. If the base sequence specificity of the binding molecule is known, sequence mapping in a large nucleic acid is possible.

*Oligopeptides and proteins.* The first system developed involved attachment of EDTA to the N- and C-termini of the antibiotic distamycin (DE and ED, respectively) (Schultz et al., 1982; Taylor et al., 1984). Distamycin binds with sequence specificity in the minor groove of DNA, showing preference for A + T-rich sites and spanning at least 3 bp of AT composition (Kopka et al., 1985; Dervan, 1986). In the presence of O$_2$ and DTT, De–Fe$^{II}$ and ED–Fe$^{II}$ carry out single-strand cleavage of DNA on *both* sides of their recognition sites (two A + T-rich areas) (Taylor et al., 1984). This reflects two different orientations of binding for the drug. The preferred orientation for the drug is 5'-AATTT-3', with the N-terminus at the 3' end of the T rich segment.

Longer molecules (increasing N-methyl pyrrole carboxamide groups) were designed to see if increasing the number of base pair interactions would increase the size of the recognition site. A comparative study using oligopeptides containing three, four, five, and six N-methylpyrrolecarboxamide groups (containing four, five, six, and seven amide groups, respectively) show that these molecules bind Λ + T-rich regions in the minor groove containing 5, 6, 7, and 8 bp, respectively

(Youngquist and Dervan, 1985a). The substrates bind in two orientations and cleave both strands of linear DNA.

Oligopeptides can be coupled via linking groups that are compatible with the DNA conformation and groove, creating larger recognition moieties (Schultz and Dervan, 1983; Youngquist and Dervan, 1985b). For example, a trimer of tetra-$N$-methyl pyrrolecarboxamides linked together by β-alanine allows the affinity reagent to span one and one-half turns of the helix (Youngquist and Dervan, 1987). Depending on the nature of the linker, different binding specificities can be built into the affinity complex. For example, chiral recognition is introduced into these molecules with linkers comprised of enantiomers of tartaric acid (Griffin and Dervan, 1987a). Metalloregulation is incorporated by using a pseudo macrocyclic linker based on 18-crown-6 coordination (Griffin and Dervan, 1987b).

The covalent attachment of metal chelates to proteins provides a means for obtaining intricate details about protein–DNA contacts. To this end, Ebright et al. (1992) have covalently attached the FeEDTA chelate to the cro and catabolite gene activator proteins, and DNA–protein contact sites have been elucidated by affinity cleavage (other metal–protein DNA affinity reagents are discussed in the sections on "Cu$^{II/I}$(phen)$_2$" and "Metallo–Amino Acid Complexes."

*Oligonucleotides.* The attachment of FeEDTA chelates to the end of oligo- and oligodeoxynucleotides, followed by cleavage reactions, has enhanced targeting specificity for single-site binding to large segments of double- and single-stranded DNA (Chu and Orgel, 1985; Dreyer and Dervan, 1985; Moser and Dervan, 1987; Povsic and Dervan, 1989; Singleton and Dervan, 1992). For example, the oligonucleotide can bind with sequence complementarity to the major groove of double-stranded DNA, forming a triple helix and mapping its binding site. Triple helices form in natural DNAs at physiological pH and are thought to play a role in regulation of gene expression.

*Non-sequence-specific probes—simple EDTA coordination complexes*

*EDTA–Fe/substrate interaction.* The complexes examined thus far have shown some preferential binding to the biopolymer. While these reagents are sensitive probes for drug- and dye-binding sites, as well as for conformational changes in the biopolymer, some bias is introduced during their use as a general tool for elucidating small molecule–protein interactions. For example, MPE–Fe, due to its intercalating properties, shows a double strand preference. Tullius and Dombroski (1985) have introduced the first chemical nuclease with no specificity for individual bases, sequences, or conformations: EDTA–Fe$^{II}$. The metal-binding site is similar to that of Dervan's tethered complexes, but all of the carboxylates can serve as ligands to the metal center in EDTA-Fe$^{II}$ (Figure 8-7). In Dervan's complex, the carboxamide might serve as an Fe ligand yielding charges of 0 and −1 for the Fe$^{III}$ and Fe$^{II}$ complexes, respectively; the overall charges of the EDTA metal complexes are −1 for EDTA–Fe$^{III}$ and −2 for EDTA–Fe$^{II}$. This subtle change plays an important role. The negative charge on these complexes keeps the reagents from associating with the negatively charged DNA (or RNA) backbone and prevents probe binding; therefore, the probe will neither cause conformational changes in the nucleic acid nor compete with other molecules for binding sites. No preference for double- vs. single-stranded regions in RNA is observed (Celander and Cech, 1990). The EDTA–Fe$^{II}$ probe is referred to as a *solvent-based* reagent since cutting takes place at every solvent-accessible sugar residue. This makes EDTA–Fe$^{II}$ an ideal footprinting reagent. An illustration of DNA cleavage by this system is shown in Figure 8-8b.

*Dioxygen activation and biopolymer cleavage by EDTA-Fe.* The reductant used in this system is ascorbate; hence, the Tullius system is actually the Udenfriend system with addition of a biopolymer substrate (eqs. 15–22). The reactive species is generally considered to be ·OH (Tullius and Dombroski, 1985; Jezewska et al. 1989; Shafter et al., 1989). The ability to cleave small stretches of accessible DNA residues and strong cleavage inhibition by radical scavengers give evidence for ·OH. More definitive proof for ·OH is the similar cleavage profiles obtained with both gamma radiation and the EDTA–Fe$^{II}$ system.

*Applications; footprinting.* EDTA–Fe$^{II}$ is a powerful footprinting reagent (Tullius and Dombroski 1986; Tullius et al., 1987). Due to the small size and nonspecificity of both the metal complex and the ·OH species, high-resolution footprints of protein–DNA contacts are produced. As an example, the sequencing gel of products from the footprinting of bacteriophage λ repressor protein cI bound to the O$_R$1 operator DNA sequence is shown in Figure 8-9. In the absence of added protein, EDTA–Fe$^{II}$ cleaves at all sites (lanes 2 and 7). On addition of protein, symmetrical footprints on the 5' (lane 4) and 3' (lane 5) labeled strands reveal the binding of the protein to only one side of the helix on each strand. The 3-bp offset of the footprints defines substrate binding in two major grooves and bridging over the intervening minor groove. Such details have not been obtained with

**FIGURE 8-9** Hydroxyl radical footprinting gel of the λ repressor-O$_R$1 DNA binding interaction. Lanes 1–4 and 5–8 contain DNA labeled at the 5' end of the noncoding strand and at the 3' end of the coding strand. Lanes: 1, 8—untreated DNA; 2, 7—hydroxyl radical DNA cleavage in the absence of repressor protein; 3, 6—Maxam-Gilbert G-specific sequencing products; 4, 5—hydroxyl radical DNA cleavage in the presence of repressor protein. In the absence of added protein, a uniform ladder is produced (lanes 2 and 7). Symmetrical and 3 bp offset footprints on the 5' and 3' strands reveal a binding interaction on only one side of the helix in which the substrate binds in the major groove. See Figure 8-7b for a schematic representation of this technique. Reproduced with permission from Tullius et al. (1987).

any other reagent (comparisons with other chemical nucleases are found in Tullius et al., 1987). More recently, EDTA–Fe$^{II}$ has been used to produce high-resolution footprints of the DNA-binding sites of distamycin and actinomycin with 5S ribosomal RNA genes of *Xenopus* (Churchill et al., 1990). The interaction between tRNA and its binding site on the ribosomal protein also has been examined using this method (Huttenhofer and Noller, 1992).

*Conformational probe.* EDTA–Fe$^{II}$ has been used extensively in the study of DNA and RNA structure. For example, this reagent has been used to determine quantitatively the number of base pairs per helical turn, as well as perturbations in the helical structure in HSV-1 thymidine kinase DNA (Tullius and Dombroski, 1985). This technique has been recently extended to the study of more unusual DNA conformations (Chen et al., 1988; Churchill et al., 1988; Burkhoff and Tullius, 1987, 1988; Tullius, 1991).

Lathan and Cech (1989) have probed the three-dimensional structure of tRNA$^{phe}$. Because of the nonspecific nature of the probe, protected sites correspond only to regions with small, accessible surface areas for 1' and 4' hydrogens (the expected sites of H abstraction by EDTA−Fe$^{II}$-generated ·OH). Cleavage studies on the *Tetrahymena* ribozyme, for which the structure is unknown, have revealed that, like folded proteins, these RNA enzymes have both an inside and an outside.

## Metal Phenanthroline Complexes: Conformational Probes

*Cu$^{II/I}$(phen)$_2$*. Thirteen years after the discovery and isolation of BLM, Sigman and coworkers (1979) found the nuclease properties of copper(II/I) phenanthroline complexes, Cu$^{II/I}$(phen)$_2$, **XI**. Like the other metal complex probes (with the exception of FeEDTA), Cu$^{II/I}$(phen)$_2$ follows the general reaction of Scheme 8-1. Dioxygen activation by Cu$^{II/I}$(phen)$_2$ in the immediate vicinity of an oligonucleotide substrate causes rapid oxidative scission of the phosphodiester backbone.

**XI**

*Complex geometry and redox chemistry.* The 1,10-phenanthroline (phen) ligand is responsible for controlling the redox potential of the Cu$^{II/I}$ couple. For example, the formal midpoint potential for the unsubstituted complex is 0.174 V (James and Williams, 1961), while that for the 2,9 dimethyl substituted complex increases to 0.594 V. Such substitutions at the 2 and 9 positions create a high degree of steric hindrance in the planar Cu$^{II}$ complexes, whereas in the tetrahedral Cu$^{I}$ state no hindrance is expected (James and Williams, 1961). The difference in the redox potentials reflects this steric hindrance (rather than, e.g., basicity). The unsubstituted bis-phen complexes, $Cu^{II/I}(phen)_2$, are the reactive species in cleavage reactions (Sigman, 1990). The tetrahedral cuprous ($Cu^+$) form binds to the substrate, as well as activating molecular oxygen. If copper(II) is the starting reagent, a reductant must be provided (Marshall et al., 1981).

The phenanthroline ligand seems to be unique among other similar bidentate ligands (D'Aurora et al., 1978). For example, although bipyridine (bpy) has a structure similar to that of phenanthroline and shows similar $O_2$ reduction capabilities (the formal midpoint potential for the bis complex is 0.120 V), the bis–bipyridyl $Cu^I$ complex does not cleave DNA. Substitution at the 5 position has produced useful reagents (Sigman, 1986). The 5-phenyl derivative, for example, yields a complex with an enhanced reactivity toward single-stranded DNA and looped regions of RNA. The ability to derivatize at this position without hindering reactivity provides a means for attaching a linker to the ligand for site-specific targeting (Chen and Sigman, 1988; Sigman, 1990).

*Substrate binding interaction.* The preferred sites of interaction for the bis–phen complex are double-stranded regions of DNA. The interaction is sensitive to the secondary structure governed by the local sequence. Binding is enhanced by the overall positive charge of the complex, as well as by hydrophobic interactions provided by the phenanthroline ring system. The $Cu^I(phen)_2$ complex interacts reversibly with double-stranded DNA substrates or single-stranded DNAs preferentially where base paired regions (such as hairpin loops; Pope and Sigman, 1984) can be formed. The complex binds through an interaction that is sequence dependent. Different $K_a$ (Scheme 8-1) values describe different binding interactions (Yoon et al., 1990). High-affinity binding sites (large $K_a$ values) are found in the minor groove of B DNA, and it has been shown that all four bases may be a part of hypersensitive sites (Spassky and Sigman, 1985; Suggs and Wagner, 1986). Although the binding of $Cu^I(phen)_2^+$ to RNA is weak, strand scission may be induced if the complex is oriented in an appropriate conformation (Pope and Sigman, 1984; Chen and Sigman, 1988; Murakawa et al., 1989).

The $Cu(phen)_2^+$ complex binds noncovalently in the minor groove of double helical DNA, with a strong dependence on secondary (helical) structure (Pope and Sigman, 1984). Although evidence for intercalation has been reported (Marshall et al., 1981; Williams et al., 1988), a lack of apparent DNA distortion in footprinting analyses suggests that $Cu^I(phen)_2^+$ is not an intercalator (Kuwabara et al., 1986). Rate studies based on product formation indicate that A DNA is cleaved at one-third the rate of B DNA, resulting in only 10–20% the cleavage efficiency, and Z DNA shows no cleavage at all. The lesser cleavage observed for the latter two conformations is possibly due to weak binding interactions (Sigman, 1990) which result in unproductively oriented activated intermediates. In such cases, oxidation of the reductant and the phenanthroline ligand results.

*Mechanism of dioxygen activation.* The cupric complex, $Cu(phen)_2^{2+}$, activates $O_2$ in the presence of reducing agents such as 3-mercapto-propionic acid, NADH, or ascorbate (Graham et al., 1980). Though the (two-electron) reduction potentials for thiol-containing reducing agents ($E^{o\prime} \approx -0.25$ to $-0.35$ V; Petering et al., 1990) would appear to make them superior to ascorbate ($E^{o\prime} = -0.06$ V; Creutz, 1981; Petering et al., 1990), the latter is much more reactive (Chiou, 1984). The $Cu(phen)_2^{2+}$ reaction is inhibited by superoxide scavengers such as SOD and also by catalase, suggesting that both $O_2^-$ and $H_2O_2$ are active participants. The overall reduction of $O_2$ to $O_2^-$ by the $Cu(phen)_2^{2+/+}$ system can be described by equations analogous to (14)–(16) given for FeEDTA. As in the previous systems, $O_2^-$ is not responsible for biopolymer cleavage but is an intermediate to the second reduction product, $H_2O_2$. Superoxide can also act as a reductant for $Cu^{II}(phen)_2^{2+}$, regenerating the active $Cu^I(phen)_2^+$.

With $H_2O_2$ as the oxidant and in the presence of reductants, DNA cleavage by micromolar $Cu^I(phen)_2^+$ occurs in ~1 minute. The kinetics and mechanism of the reaction of $Cu^I(phen)_2^+$ with $H_2O_2$ have been investigated to elucidate the O—O bond cleavage (Johnson and Nazhat, 1987). Data from pulse radiolysis studies in the absence of $H_2O_2$ allow a kinetic comparison of ·OH to the reactive species produced in reaction (22). Data from competition rate studies involving scavengers are consistent with a reactive copper-oxygen intermediate that reacts with the radical scavengers more slowly than ·OH.

$$Cu^I(phen)_2^+ + H_2O_2 \rightarrow Cu^{II}(phen)_2^{2+} + \text{(reactive oxygen species)} \quad (22)$$

*The activated complex.* The nature of the reactive species is difficult to assess. $Cu^I$ prefers a coordination number of 4 in a tetrahedral geometry. On $H_2O_2$ ligation, the complex may lose the coordination of one or both of the ligating N donors from one phenanthroline ligand. The intermediate can be described as either $Cu^I(phen)_2(H_2O_2)$, $Cu^{II}(phen)(\cdot OH)(OH^-)$, or $Cu^{III}(phen)(OH^-)_2$. The latter two intermediates result from metal-mediated scission of the O—O bond.

A metal-bound hydroxyl radical complex, $Cu^{II}(phen)_2(\cdot OH)$ or $Cu^{II}(phen)(\cdot OH)(OH^-)$, best fits the kinetic data obtained for this reaction. DNA cleavage studies (Kuwabara et al., 1986) also support a metal-bound reactive oxygen species. A lack of radical scavenger inhibition in the presence of DNA substrate indicates that if $\cdot OH$ is produced, it is not diffusible. Also, comparison with $\cdot OH$-induced cleavage products (from cobalt-60 irradiation) indicates that the ratios of oxidation products are different for the two systems.

*DNA cleavage.* The reactive copper–oxygen complex attacks the deoxyribose from its bound position in the minor groove. Cleavage is nonspecific with respect to the base but is dependent on the primary sequence, perhaps because of its effect on the DNA conformation (Spassky and Sigman 1985; Suggs and Wagner, 1986; Williams et al., 1988). The nucleotide 5' to the site of phosphodiester bond scission is most important in governing the kinetics (Yoon et al., 1990). Cleavage analysis indicates that any of three residues on the 3' side of the binding site can be oxidized (Drew and Travers, 1984; Williams et al., 1988). While a diffusible $\cdot OH$ reactive species offers a plausible explanation, this 3' offset scission pattern is thought to reflect different orientations of the active complex within the strand.

DNA cleavage can be monitored by the release of acid-soluble products (Pope et al., 1982). Products and intermediates isolated and characterized include 5' and 3' phosphorylated termini, free bases, 5-methylenefuranone, and minor amounts of 3' phosphoglycolate. The predominant mechanism (80–90%) involves abstraction of the C1' H by the reactive intermediate, releasing free base and leading to the production of a metastable 3' intermediate (detected by gel electrophoresis) and a 5'-phosphorylated terminus. The transient 3' intermediate decays into a stable 3'-phosphorylated terminus and 5-methyl furanone. The structure of B DNA suggests that abstraction at C1' predominates if the complex is activated from within the minor groove (Kuwabara et al., 1986; Sigman, 1990; Yoon et al., 1990).

A minor pathway results from C4' H abstraction. This is in contrast to Fe–BLM chemistry, in which this pathway, resulting in the production of 3'-phosphoglycolate termini, predominates. Another difference is that no base propenal formation accompanies this abstraction. Instead, free base, 3'- and 5'-phosphorylated termini, and an unidentified 3-carbon fragment are produced. The production of 3'-phosphoglycolate varies with the base sequence, giving evidence for a conformation-dependent orientation of the $\cdot OH$ radical before attack on the sugar moiety.

Base modification in this system has been analyzed by gas chromatography/mass spectrometry with selected-ion monitoring (GC/MS-SIM) (Dizdaroglu et al., 1990). Quantitative results indicate that while base modification proved to be only a minor reaction in the Fe–BLM system, the amount of base modification in the Cu(phen)$_2^+$ system is much greater. Using ascorbate or mercaptoethanol as reductants, the primary products (out of 13 identified) are 8-hydroxyguanine and, to a lesser extent, 8-hydroxyadenine, thymine glycol, and cytosine glycol. These species are well-known ·OH products.

*Applications. Footprinting DNA–protein complexes.* The lack of base specificity in the cleavage reaction and the small size of the complex render this system useful for footprinting. An example is the cleavage of the dodecamer 5'-CGCGAATTCGCG-3' bound by the drug netropsin (Kuwabara et al., 1986). The footprint reveals cleavage inhibition at the C-9, T-8, T-7, and A-6 base pairs, directly corresponding to the binding region for netropsin to this strand (netropsin binds at T-8—T-7—A-6—A-5 within the minor groove) and is consistent with binding and attack of the activated copper complex from the minor groove. Footprinting analysis of the enzyme Eco RI bound to the oligonucleotide TCGCGAATTCGCG also agrees with X-ray data. In this system, a *lack* of protein protection from degradation by the Cu(phen)$_2^+$ complex is consistent with major-groove binding of the Eco RI protein to the DNA strand. Tullius et al. (1987) have compared the DNA footprints of the λ-repressor by Fe–MPE, Cu(phen)$_2^+$, and Fe(EDTA)$^{2-}$.

*Conformational probe.* The Cu(phen)$_2^+$ system is useful for obtaining conformational information about interactions for which no structural data are available. A recent review by Sigman (1990) provides an excellent overview of this work. Among a number of studies discussed is the use of the bis(5-phenylphenanthroline)Cu$^I$ complex to probe *E. coli* RNA polymerase-induced conformational changes in the lac UV-5 promoter (Thederahn et al., 1990) and the detection of intermediate complexes during transcription initiation. This unique application for the Cu(phen) complexes appears to be a consequence of favorable hydrophobic interactions between the complex and the protein, as well as with DNA.

RNA cleavage by Cu(phen)$_2^+$ has been examined using the tRNA$^{phe}$ substrate (Murakawa et al., 1989). In contrast to MPE–Fe$^{II}$ (see the section on "Tethered EDTA Complexes—DNA Affinity Reagents"), Cu(phen)$_2^+$ shows a preference for single-stranded regions: the juncture of the T and D loops (13–20), the anticodon loop (32–38), the D loop (57) and the T-stem (50) are most heavily cleaved. It is thought that the lack of a B-form minor groove leads this complex to bind in a different mode than is seen with DNA.

*Affinity cleavage.* The $Cu^I$ phenanthroline complex can be made site specific by covalently attaching a linker moiety to the 5-position (Sigman, 1990). The phen ligand has been linked to oligonucleotides for triple helix formation (Sun et al., 1988) and to DNA-binding proteins. Some proteins targeted include the *E. coli* trp repressor protein (Chen and Sigman, 1988), the *E. coli* catabolite activator protein (Ebright et al., 1990), and the λ phase cro protein (Bruice et al., 1991). In the last system, the phenanthroline ligand is attached via a 5-iodoacetyl-linkage to a cysteine that has been engineered into the protein at the C-terminal alanine site (A66C cro-Cys-OP, where OP is *ortho*-phenanthroline). This modification does not affect the helix-turn-helix binding of the protein to the major groove of the DNA. On binding of the protein conjugate to the DNA, the metal chelate is directed into the minor groove. In the presence of $Cu^{II}$, 3-MPA, and $H_2O_2$, the $Cu(phen)^+$ complex cleaves the DNA *within* the recognition site of the protein. Complete cleavage inhibition of this area is obtained when free $Cu(phen)_2^+$ is used as a footprinting reagent for this interaction.

*Miscellaneous metallophenanthroline systems.* Chiou (1984) has examined other metallophenthroline systems and has revealed that the reaction for DNA scission with various metals and metal complexes (presumably 1:2 complexes) follows the order $Cu(phen)_2^{2+}$, $Cu(phen)_2^+ > Cu^{2+}$, $Cu^+ > Mn(phen)_2^{2+} > Zn(phen)_2^{2+} > Cr(phen)_2^{3+} > Fe(phen)_2^{3+} > Co(phen)_2^{2+} > Fe^{3+}$, $Fe^{2+} > Fe(phen)_2^{2+} > Ni(phen)_2^{2+}$. It is suggested that this ordering reflects the redox potentials for these systems. Surprisingly, the redox inactive $Zn^{2+}$ system exhibits relatively high reactivity. This may be due to metal ion impurities, or a different mechanism, perhaps involving a ligand radical species, may account for its reactivity. It is notable that oxygen radical formation catalyzed by $Zn(bpy)_2^{2+}$ has also been reported (Sawyer, 1991).

*Inert chiral metal tris-phenanthroline complexes:* Δ, Λ-$[Ru(TMP)_3]^{2+}$. Metal tris phenanthroline complexes are sensitive probes of DNA conformation, and because they are chiral molecules, they are sensitive probes of stereochemistry in DNA-binding interactions (Barton, 1985). Three bidentate ligands, such as phenanthroline or bipyridine, can bind to a metal center, resulting in two different octahedral stereoisomers; the stereochemistry is defined by the helicity about the $C_3$ axis. Schematic drawings of the two enantiomeric tris-bidentate chelate forms, called the Λ and Δ, are presented below, **XII(a)** and **XII(b)**, respectively. Metals such as Co$^{III}$, Rh$^{III}$, and Ru$^{II}$ form tris bidentate chelate complexes that are inert to both substitution and racemization. These unique properties allow these complexes to bind stereospecifically to DNA. For example, the Λ-form of the Ru$^{II}$ and Co$^{III}$ *tris*-(4,7-diphenylphenanthroline) complexes (Ru(DIP)$_3^{2+}$ and Co(DIP)$_3^{3+}$) bind selectively to Z DNA. Bis(phenanthroline)(phenanthrenequinone diimine)-Rh$^{III}$ shows shape selectivity, targeting the major groove of DNA, and is a sensitive probe for conformation within the groove. The Λ-form of *tris* (3,4,7,8-tetramethylphenanthroline)Ru$^{II}$ complexes (Ru(TMP)$_3^{2+}$) is specific for A-form helices. After binding, these complexes can be photoactivated to produce lesions in the DNA strand at the site of interaction. Co$^{III}$, Rh$^{III}$, and Ru$^{II}$ *tris*-phenanthroline complexes cleave DNA and have been used to probe DNA structure. Only the Ru$^{II}$ complexes, however, are thought to carry out this chemistry by a mechanism that involves dioxygen activation, and the following discussion will focus on the Ru$^{II}$ chemistry.

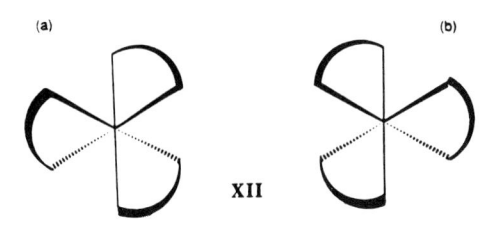

(a)   (b)

XII

*Substrate binding of Ru$^{II}$(phen)$_3^{2+}$ complexes*

*B-form helices.* Equilibrium binding, viscosity, $^1$H NMR, and spectroscopic studies have been employed to elucidate the binding of Ru(phen)$_3^{2+}$ complexes to DNA (Barton et al., 1986; Hiort et al., 1990; Rehmann and Barton, 1990; Eriksson et al., 1992; Satyanarayana et al., 1992). The binding affinity of Ru(phen)$_3^{2+}$ for DNA is relatively low; $K = 2 \times 10^3 \mathrm{M}^{-1}$ (0.1 M NaCl) and decreases with increasing ionic strength, indicating an electrostatic binding interaction. Early experiments indicated that Ru(phen)$_3^{2+}$ binds to double-stranded polynucleotides by two different binding modes; the Δ by intercalation in the major groove and the Λ by a surface-bound, solvent-accessible mode in the minor goove (Barton et al., 1986). Recent viscosity measurements, however, indicate that both isomers bind via an outside electrostatic interaction (Satyanarayana et al., 1992) that is not intercalation. Differences in viscosity changes for the two isomers suggest that the isomers bind with different orientations. Although the binding modes are not fully understood, $^1$H NMR studies have suggested that the $\Delta$ isomer binds preferentially in the major groove, while the $\Lambda$ isomer binds in the minor groove (Rehmann and Barton, 1990). Recent two-dimensional NMR studies have brought this viewpoint into question (Eriksson et al., 1992), indicating that both isomers bind to the minor groove. There is agreement that surface binding for Ru(phen)$_3^{2+}$ shows a preference for A-T-rich regions (Rehmann and Barton, 1990; Eriksson et al., 1992). This binding is entropy driven [hydrophobic displacement of bound water and bound (monovalent) cations] and is influenced by the charge of the complex (electrostatic attraction). Limited data are available on the details of this interaction (Rehmann and Barton, 1990; Eriksson et al., 1992).

When bulky substituents are placed on the periphery of the phenanthroline ligand, as in the Ru(DIP)$_3^{2+}$ complex, binding to B DNA is *stereospecific* for the $\Delta$ isomer (Barton, 1985). Z DNA, on the other hand, binds both $\Delta$ and $\Lambda$ forms. $\Delta$- and $\Lambda$-[Ru(DIP)$_3$]$^{2+}$ are sensitive probes for B- and Z-DNA.

*A-form helices.* Double-stranded RNA with its A-form conformation is a poor substrate for Ru(phen)$_3^{2+}$, but its slight preference for a surfacebound $\Lambda$-isomer interaction is nonetheless significant. The preferred interaction of the $\Lambda$-isomer for right-handed helices of A form DNA has been optimized using the tetramethylphenanthroline derivative, Ru(TMP)$_3^{2+}$. Ru(TMP)$_3^{2+}$ exhibits shape and symmetry recognition and binds with an enantiomeric preference for the $\Lambda$-isomer of 56–92% under low binding ratios. The complex interacts via electrostatic and hydrophobic interactions in the minor groove of the A-helix. Almost no binding to B DNA or Z DNA for this complex is observed under these conditions (Mei and Barton, 1986, 1988).

*Dioxygen activation and DNA cleavage by Ru(TMP)$_3^{2+}$.* Ru(TMP)$_3^{2+}$ cleaves DNA into acid-soluble products after irradiation of the metal to ligand charge transfer band with blue light (442 nm). Like binding, the cleavage is stereospecific, the $\Lambda$-form being the most efficient cleaver. Although the $\Lambda$-Ru$^{2+}$ complex binds both poly(rC)-poly(dG) and poly(rA)-poly(dT), only the former substrate is cleaved. This specificity for guanine residue cleavage is consistent with a dioxygen activation mechanism involving a diffusible singlet oxygen species; guanine residues are the most easily oxidized of the four bases (Mei and Barton, 1988), forming 8-hydroxy guanine and 2,6-diamino-4-hydroxy-5-

formamidopyrimidine (Halliwell and Aruoma, 1991). Enhanced cleavage intensities in $^2H_2O$, a solvent in which singlet oxygen is longer-lived, provides further evidence for this reactive species. Unfortunately, due to the specificity of the activated $O_2$ species, control experiments must be performed to subtract out this bias to obtain information about the metal complex as a conformational probe. This has been done by comparing the cleavage pattern of $Ru(TMP)_3^{2+}$ with that of $Ru(phen)_3^{2+}$. For pBR322 DNA, preferred cleavage at homopyrimidine sequences by $Ru(TMP)_3^{2+}$ is observed.

*RNA cleavage by $Ru(phen)_3^{2+}$ and $Ru(TMP)_3^{2+}$.* The tertiary structure of tRNA$^{phe}$ has recently been probed by $Ru(phen)_3^{2+}$ and $Ru(TMP)_3^{2+}$ (Chow and Barton, 1990). The Ru$^{II}$ complexes afford similar cleavage efficiency and shape recognition, as seen with double-stranded DNA. Alkaline hydrolysis reveals preferential attack at G residues, with no release of unmodified bases. This, along with the production of 5'- and 3'- or 2'-phosphate termini, is consistent with base removal by $^1O_2$. Although proceeding through a common mechanism, the two Ru$^{II}$ complexes show different cleavage selectivities, as expected for their different shape-dependent binding. Most G residues are cleaved by $Ru(phen)_3^{2+}$, with limited selectivity observed for the Δ conformation at low concentrations of $Ru(phen)_3^{2+}$; at high concentrations, no selectivity is observed. $Ru(TMP)_3^{2+}$, on the other hand, shows high selectivity for the Λ-isomer, and more residues are protected from cleavage due to the steric bulk of the $Ru(TMP)_3^{2+}$ complex.

## Miscellaneous Metal Complexes

*Metallo–amino acid complexes.* The $Cu^{2+}$ binding site of human serum albumin (HSA) is comprised of the tripeptide GlyGlyHis (GGH), in which the $Cu^{2+}$ ion is bound in a square planar environment (Camerman et al., 1976; Mack and Dervan, 1990). The $Cu^{2+}$GGH complex, in the presence of ascorbate, shows enhanced antitumor activity on Ehrlich Ascites and mouse sarcoma and neuroblastoma cells (Chiou, 1983). The activity is thought to be due to the production of reactive $O_2$ species generated at the metal center. At 0.5 mM concentrations of CuGGH and 1 mM of ascorbate (pH 7, 27 °C, 30 min), a relatively mild and specific cleavage of DNA occurs. The complex shows Fenton-like reactivity, and a lack of complete inhibition by radical scavengers suggests that the final oxidizing species is localized (Chiou, 1983).

Dervan and coworkers have taken advantage of the DNA cleavage properties of GGH by attaching this chelate to the NH$_2$ terminus of a DNA-binding domain (residues 139–190) of the Hin recombinase protein (GGH(Hin 139–190) (Mack et al., 1988). This protein binds with specificity to DNA, and in the presence of $H_2O_2$ and ascorbate, preferential cleavage at one of the four Hin binding sites is observed. At the micromolar concentrations used in this study, the metal-binding domain unattached to the protein shows no cleavage. The observed 3'- and 5'-phosphate product termini and 2-bp cleavage range indicate oxidative cleavage by a localized species. (For comparison with attached $Cu(phen)_2^+$ see Chen and Sigman, 1987; and with attached $Fe(EDTA)^-$, see Sluka et al., 1987.)

*Metal chelates with built-in reducing agents.* Imidazolyl and aminoalkyl derivatives of a bisresorcinol chelating agent, in the presence of $Cu^{2+}$ and $O_2$ (30°C, pH 7.4), cleave supercoiled ɸX174 RFI plasmid DNA into the open and linear forms (Motomura et al., 1992). Binding of the N donors on the backbone moieties to the $Cu^{2+}$ ion is a requirement for the chemistry. The activation of dioxygen to an oxy-radical species is believed to take place at the metal center after reduction of $Cu^{2+}$ to $Cu^+$ by the ligand itself; there is no need for an external reductant. This internal metal center reduction is evidenced by a slow decrease in the d-d transition absorption band at 740 nm ($Cu^{2+} \rightarrow Cu^+$) which is concomitant with oxidation of the bisresorcinol ligand. This band is restored on addition of $O_2$, and in the presence of nucleic acid substrate, strand cleavage occurs.

## PROTEIN CLEAVAGE

Protein-bound metal ions can activate dioxygen, yielding highly reactive oxygen species (e.g., ·OH). Analogous to DNA chemistry, these species abstract H atoms from the polypeptide substrate, resulting in carbonyl derivatives and a peptide backbone that is more susceptible to degradation (Scheme 8-4) (Bateman et al., 1985; other oxidation pathways are given in Stadtman and Berlett, 1991b). Perhaps because they are more exposed, the sugar-phosphate backbones of DNA and RNA are easier to cleave than the polypeptide backbone. Metal-catalyzed oxidation of proteins, however, can lead to degradation and inactivation (Jayco and Garrison, 1958; Chiou, 1983; Levine, 1983; Kim and Stadtman, 1985; Tabor and Richardson, 1987; Stadtman and Oliver, 1991). Proteolytic cleavage by metal *chelates* (free in solution) can be carried out, but high concentrations (~0.5 mM) of the metal complex are required. This has been shown for the Cu(GGH) chelate (Chiou, 1983); protein scission is thought to be partly responsible for the antitumor activity of

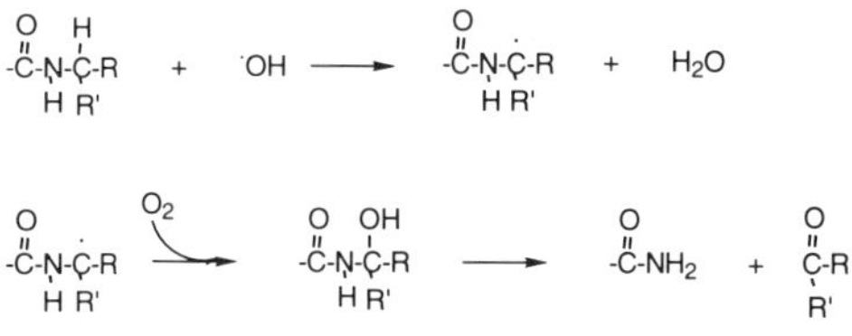

**Scheme 8-4**

the complex. Indeed, chelators such as EDTA are often used to remove trace metals from protein surfaces and serve as quenchers of oxidative degradation (Yamazaki and Piette, 1990).

Only recently have researchers discovered that by tethering a metal chelate to a carrier molecule or a protein surface, specific (and sometimes efficient) cleavage of the polypeptide backbone can be induced. As with nucleic acids, the chelate is used to target the redox-active metal to its substrate. The application to proteins, when fully developed, will aid in sequencing, structural mapping, and regulatory and active binding site determinations and will complement other high-resolution protein analytical techniques such as X-ray crystallography and NMR. It also has the potential for the development of site-specific chemotherapeutic agents. This section will describe the uses of metal chelates as artificial *proteases*.

## Hydrolytic Cleavage by Covalently Attached Metal Chelates

Unlike nucleic acids, which can be sequenced by their electrophoretic mobility, proteins must be sequenced by other methods, such as Edman degradation. Oxidative cleavage of a peptide backbone is usually accompanied by chemical modification of amino acid side chains and produces nonsequenceable, modified N-termini. This prevents determination of the exact site of cleavage by Edman degradation and makes data analysis difficult. To use protein cleavage by metal chelates effectively as a means for probing protein structure and function, systems that *hydrolytically* cleave the protein backbone to yield new, readily sequenceable, unmodified N- and C-termini are desirable.

*BSA–FeEDTA system.* The first system examined was that of the 66-kDa protein Bovine Serum Albumin (BSA) (Rana and Meares, 1990). As shown schematically in Figure 8-10, an iron EDTA chelate is covalently attached to the single free cysteine (Cys 34) on the molecule. In the presence of $H_2O_2$ and ascorbate at pH 7, the metal chelate efficiently

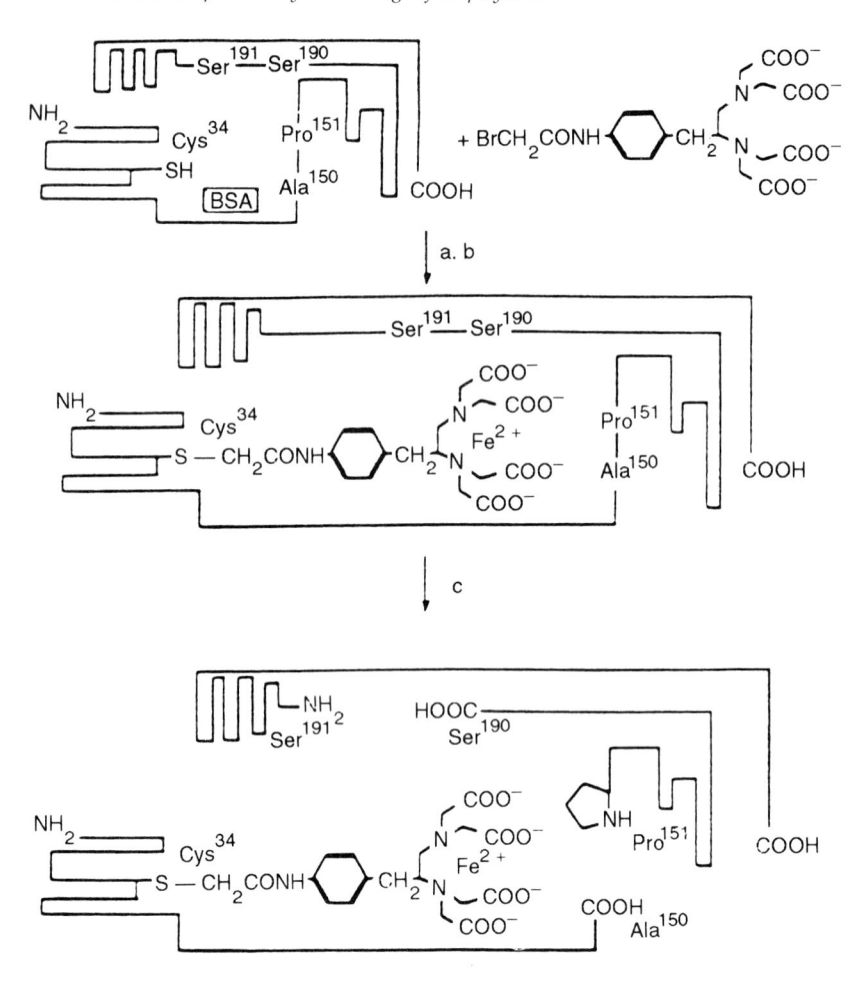

**FIGURE 8-10** Schematic of chelate conjugation to and hydrolytic cleavage of BSA: (a) and (b) chelate conjugation and ferrous ion loading; (c) addition of $H_2O_2$ and ascorbate, resulting in strand scission. Reproduced with permission from Rana and Meares (1990).

cleaves the peptide backbone at room temperature in 10 seconds at two specific sites, producing three new fragments. N-terminal and C-terminal sequencing of the new fragments (MW ≈ 45, 17, and 5 kDa) revealed that cleavage occurred between residues Ala(150)-Pro(151) and Ser(190)-Ser(191) on *each* protein molecule. Two scission events per protein indicate that the reaction is catalytic. Unfortunately, the crystal structure of BSA is not available, so the steric requirements and selectivity of the cleavage reaction cannot be studied easily. Although evidence for ·OH formation is apparent in the system (some degradation of the amino acid side chains occurs), the amino acids around the site of cleavage, for the most part, are unaltered. Also, the cleavage reaction is not inhibited by radical scavengers. These two results indicate a mechanism different from the reaction with ·OH.

*HCAI–Fe(EDTA) system.* To address the questions of steric influence, selectivity, and mechanism on the protein cleavage reaction, Rana and Meares (1991a, 1991b) investigated the chemistry using a protein of known crystal structure, human carbonic anhydrase I (HCAI). In this case, the Fe(EDTA) chelate was attached to the unique Cys 212. In the presence of $H_2O_2$ and ascorbate, rapid and highly efficient cleavage occurred at a single site. No detectable degradation of neighboring amino acids occurred. The site of cleavage was identified to be between Leu(189) and Asp(190). Interestingly, although the 145–146-peptide bond is closer to Cys 212, it was not cleaved, indicating a possible role of peptide bond/iron chelate orientation in this process. These and the previous results indicate that the cleavage is nonspecific toward individual amino acids and is dependent on peptide-chelate proximity.

*The mechanism.* To investigate the mechanism of this reaction, $^{18}O$-labeled $O_2$ and $H_2O_2$ were used in the HCAI system to follow the path of the activated species. High-resolution mass spectrometry indicated that $^{18}O$ is transferred with ~90% efficiency to the new carboxyl terminus of Leu 189. Also, titration studies with both $H_2O_2$ and ascorbate indicated that the reaction stoichiometry is 1:1:1 in cleaved peptide bond, oxidant and reductant. In contrast to the result on the electrophilic, oxidative species involved in practically all the systems mentioned above, these data can be explained in terms of a mechanism involving a *nucleophilic* peroxide intermediate. The suggested mechanism is shown in Scheme 8-5. The bound peroxide attacks the carbonyl carbon of the peptide bond that is in the most favorable orientation. The sulfur of Cys 212 lies above the plane of the Leu 189/Asp 190 peptide bond, 5.3 Å away from the C=O carbon of Leu 189. In the case of Gly 145, although closer to Cys 212, its C=O oxygen points toward Cys 212, making nucleophilic attack unfavorable. It is proposed that the O—O bond undergoes *heterolytic* cleavage to yield the leaving $OH^-$ as part of the new C-terminus and an electron-deficient oxygen bound to a high-oxidation Fe metal center ($Fe=O^{2+}$)(D). The net result is hydrolysis of one peptide bond. This chemistry is also in agreement with the observed stoichiometry. Finally, reduction of the predicted

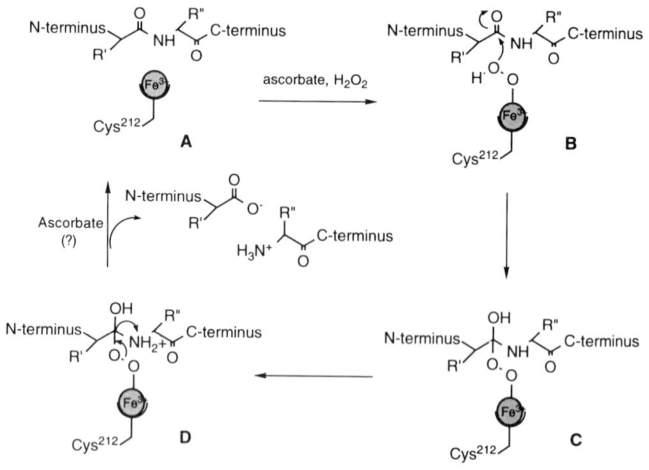

**Scheme 8-5**

(modified with permission from Rana and Meares, 1991)

high-oxidation Fe species back to $Fe^{III}$ (required for another catalytic cycle) could be supplied by the second reducing equivalent of ascorbate. It should be noted that superoxide radical can also be a nucleophile and is known to cleave esters (Sawyer, 1991). Despite the occurrence of protein cleavage in the presence of radical scavengers, attack by a localized $O_2^-$ cannot be ruled out. Several redox active metals have been investigated under various conditions (for the BSA system: $V=O^{2+}$, $Mn^{2+}$, $Co^{2+}$, $Ni^{2+}$, $Cu^{2+}$); only the $Fe^{3+/2+}$ couple has provided facile and specific cleavage (Marusak and Meares, unpublished results), reflecting the redox potentials of the various metal–EDTA chelates.

*Staphylococcal nuclease–Fe(EDTA) system.* One feature that covalently bound metal chelates offer is that cleavage occurs near the site of chelate attachment. Ermacora et al. (1992) have taken advantage of this to investigate nonnative states of the protein staphylococcal nuclease. The chelate, (EDTA-2-aminoethyl) 2-pyridyl disulfide (EPD), was synthesized and covalently attached to a Cys 28 side chain which was genetically engineered into the nuclease (K28C-EDTA-Fe). On addition of ascorbate and $O_2$ (or $H_2O_2$) and incubation at 4 °C and pH 7.2 for 15 minutes, intramolecular cleavage of the modified protein in its native state yielded small amounts of several new products. The new bands were 16 kDa (two bands), 14 kDa (one band), and 6–8 kDa (several bands) in molecular weight. It was concluded that hydrolytic cleavage had occurred at Lys 71-Ile 72, Lys 78-Gly 79, Gln 80-Arg 81, and Lys 84-Tyr 85. In the presence of the SDS denaturant, the 16-kDA and 6–8-kDa fragments were replaced by several cleavage fragments of 13–15 kDa. These were apparently cleaved at sites near the attached chelates in the primary sequence of the protein.

### Oxidative Cleavage by Noncovalently Localized Metal Chelates: Affinity Probes

Affinity cleavage probes have been prepared by covalently attaching a metal chelate to a reagent which specifically binds to a region of the protein (hormone-receptor and allosteric enzyme-effector, for example). The affinity moiety brings the metal chelate in close proximity to the polypeptide backbone, so cleavage occurs at a targeted site under relatively mild conditions. The metal complex chemistry discussed below is similar to that described in the section on "Metal–EDTA Complexes" for DNA. Addition of a reductant and an oxygen source to the chelate system produces highly reactive oxygen species which induce *oxidative* cleavage of the polypeptide backbone.

*Calmodulin–trifluoperazine system.* Schepartz and Cuenoud (1990) tethered an EDTA moiety to the calmodulin antagonist trifluoperazine (TFP), aiming to map the binding site for this molecule. Iron–trifluoperazine–EDTA (TFE–Fe), has been shown to bind (with a $Ca^{2+}$ dependence) to the active site of calmodulin, and on addition of DTT and $O_2$ (15 minutes, 7 °C), a small percentage of the calmodulin is cleaved, forming six distinct fragments. If the binding of TFE–Fe is inhibited (through preincubation with TFP or omission of $Ca^{2+}$ from the system), cleavage occurs, indicating the need for chelate-protein proximity. The reaction is also inhibited by catalase, supporting a mechanism involving the reduction of $O_2$ to the peroxide state. Diffusible radicals such as ·OH are thought to be the major reactive species. Unfortunately, due to the oxidative nature of the cleavage, the fragments (molecular weights of 11.3, 10.2, 9.1, 7.9, 6.4, and 5.3 kDa) have modified amino termini and are not sequenceable by Edman degradation. The cleavage and binding sites were not identified.

In attempting to increase the efficiency of calmodulin oxidative cleavage, Cuenoud et al. (1992) prepared the $Ni^{II}GGH$ analog (TFP-NiGGH), which has been shown to cleave DNA oxidatively (Mack and Dervan, 1990) via a nondiffusible reative oxygen species. Like TFE–Fe, the TFP–NiGGH complex binds in the active site of calmodulin, and in the presence of monoperoxyphthalate, it oxidatively cleaves the protein at a unique site. Amino acid analysis and molecular weight estimation tentatively identify this site as the second EF hand of the protein.

*Streptavidin–biotin system.* Hoyer and coworkers (1990) studied a similar affinity cleavage reaction involving streptavidin and biotin. Biotin–EDTA derivatives of different linker lengths have been shown to bind to streptavidin at the same site and with the same affinity ($K_a \approx 10^{15} M^{-1}$) as the parent complex. Both $Cu^{2+}$ and $Fe^{3+}$ chelates of short linker lengths are effective in carrying out dioxygen-activated cleavage of the streptavidin core (14 kDa) after incubation for 5 minutes in the presence of mercaptoethanol at 90 °C. One cleavage fragment (MW 7 kDa) was observed.

## Acknowledgment

Supported by research grants from the National Institutes of Health: Public Health Service, National Research Service Award 1 F32 CA09024-02,03 (RAM) and Research Grants of CFM. We thank Douglas P. Greiner and Dr. Angelo Gunasekera for helpful discussions and a careful reading of the manuscript.

## REFERENCES

ADAMSON, A. W. (1979) Whither Inorganic Photochemistry? A Parochial View. *Pure Appl. Chem.*, 51, 313–329.

AJMERA, S., WU, J. C., WORTH, L., RABOW, L. E., STUBBE, J., and KOZARICH, J. W. (1986) DNA Degradation by Bleomycin: Evidence for 2'R-proton Abstraction and for C—O Bond Cleavage Accompanying Base Propenal Formation. *Biochemistry*, 25, 6586–6592.

ALBERTINI, J-P., and GARNIER-SUILLEROT, A. (1982) Cobalt–Bleomycin–Deoxyribonucleic Acid System. Evidence of a Deoxyribonucleic Acid bound Superoxo and μ-Peroxo Cobalt-Bleomycin. *Biochemistry*, 21, 6777–6782.

ALBERTINI, J-P., GARNIER-SUILLEROT, A., and TOSI, L. (1982) Iron–Bleomycin · DNA · System—Evidence of a Long-lived Bleomycin · Iron · Oxygen Intermediate. *Biochem. Biophys. Res. Commun.*, 104, 557–563.

BALASUBRAMANIAN, P. N., and BRUICE, T. C. (1987) Oxygen Transfer Involving Non-heme Iron: The Influence of Leaving Group Ability on the Rate Constant for Oxygen Transfer to (EDTA)Fe(III) from Peroxycarboxylic Acids and Hydroperoxides. *Proc. Natl. Acad. Sci. U.S.A.*, 84, 1734–1738.

BARR, J. R., VAN ATTA, R. B., NATRAJAN, A., HECHT, S. M., VAN DER MAREL,

G. A., and VAN BOOM, J. H. (1990) Iron(II)-bleomycin-mediated Reduction of $O_2$ to Water: An $^{17}O$ Nuclear Magnetic Resonance Study. *J. Am. Chem. Soc.*, 112, 4058–4060.

BARTON, J. K. (1985) Simple Coordination Complexes: Drugs and Probes for DNA Structure. *Comments Inorg. Chem.*, 3, 321–348.

BARTON, J. K., GOLDBERG. J. M., KUMAR, C. V., and TURRO, N. J. (1986) Binding Modes and Base Specificity of *tris*-(Phenanthroline)ruthenium(II) Enantiomers with Nucleic Acids: Tuning the Stereoselectivity. *J. Am. Chem. Soc.*, 108, 2081–2088.

BATEMAN, R. C., YOUNGBLOOD, W. W., BUSBY, W. H., and KIZER, J. S. (1985) Nonenzymatic Peptide α-Amidation. Implications for a Novel Enzyme Mechanism. *J. Biol. Chem.*, 260, 9088–9091.

BOWRY, V. W., and INGOLD, K. U. (1991) A Radical Clock Investigation of Microsomal Cytochrome P-450 Hydroxylation of Hydrocarbons. Rate of Oxygen Rebound. *J. Am. Chem. Soc.*, 113, 5699–5707.

BRAY, W. G., and GORIN, M. (1932) Ferryl Ion, a Compound of Tetravalent Iron. *J. Am. Chem. Soc.*, 54, 2124–2126.

BROWN, S. J., HUDSON, S. E., and MASCHARAK, P. K. (1989a) Light-induced Nicking of DNA by a Synthetic Analogue of Cobalt(III)-bleomycin. *J. Am. Chem. Soc.*, 111, 6446–6448.

BROWN, S. J., HUDSON, S. E., STEPHAN, D. W., and MASCHARAK, P. K. (1989b) Syntheses, Structures, and Spectral Properties of a Synthetic Analogue of Copper(II)-bleomycin and an Intermediate in the Process of Its Formation. *Inorg. Chem.*, 28, 468–477.

BRUICE, T. W., WISE, J. G., ROSSER, D. S. E., and SIGMAN, D. S. (1991) Conversion of λ Phage Cro into an Operator-specific Nuclease. *J. Am. Chem. Soc.*, 113, 5446–5447.

BUETTNER, G. R., and MOSELEY, P. L. (1992) Ascorbate Both Activates and Inactivates Bleomycin by Free Radical Generation. *Biochemistry*, 31, 9784–9788.

BULL, C., MCCLUNE, G. J., and FEE, J. A. (1983) The Mechanism of Fe-EDTA Catalyzed Superoxide Dismutation. *J. Am. Chem. Soc.*, 105, 5290–5300.

BURGER, R. M., BLANCHARD, J. S., HORWITZ, S. B., and PEISACH, J. (1985) The Redox State of Activated Bleomycin. *J. Biol. Chem.*, 260, 15406–15409.

BURGER, R. M., FREEDMAN, J. H., HORWITZ, S. B., and PEISACH, J. (1984) DNA Degradation by Manganese(II)-bleomycin Plus Peroxide. *Inorg. Chem.*, 23, 2215–2217.

BURGER, R. M., HORWITZ, S. B., PEISACH, J., and WITTENBERG, J. B. (1979) Oxygenated Iron Bleomycin—a Short Lived Intermediate in the Reaction of Ferrous Bleomycin with $O_2$. *J. Biol. Chem.*, 254, 12299–12302.

BURGER, R. M. KENT, T. A., HORWITZ, S. B., MÜNCK, E., and PEISACH, J. (1983) Mössbauer Study of Iron Bleomycin and Its Activation Intermediates. *J. Biol. Chem.*, 258, 1559–1564.

BURGER, R. M., PEISACH, J., and HORWITZ, S. B. (1981) Activated Bleomycin—a Transient Complex of Drug, Iron, and Oxygen That Degrades DNA. *J. Biol. Chem.*, 256, 11636–11644.

BURGER, R. M., PEISACH, J., and HORWITZ, S. B. (1982) Stoichiometry of DNA Strand Scission and Aldehyde Formation by Bleomycin. *J. Biol. Chem.* 257, 8612–8614.

BURKHOFF, A. M., and TULLIUS, T. D. (1987) The Unusual Conformation Adopted by the Adenine Tracts in Kinetoplast DNA. *Cell*, 48, 935–943.

BURKHOFF, A. M., and TULLIUS, T. D. (1988) Structural Details of an Adenine Tract that Does Not Cause DNA to Bend. *Nature*, 331, 455–457.

BURKITT, M. J., and GILBERT B. C. (1990) Model Studies of the Iron-catalysed Haber-Weiss Cycle and the Ascorbate-driven Fenton Reaction. *Free Radical Res. Commun.*, 10, 265–280.

BUTLER, A., and CARRANO, C. J. (1991) Coordination Chemistry of Vanadium in Biological Systems. *Coord. Chem. Rev.*, 109, 61–105.

CAMERMAN, N., CAMERMAN, A., and SARKAR, B. (1976) Molecular Design to Mimic the Copper(II) Transport Site of Human Albumin. The Crystal and Molecular Structure of Copper(II)-glycylglycyl-L-histidine-N-methylamide Monoaquo Complex. *Can. J. Chem.*, 54, 1309–1316.

CANTOR, C. R., and SCHIMMEL, P. R. (1980) in *Biophysical Chemistry—Part I: The Conformation of Biological Macromolecules*, W. H. Freeman, New York, p. 180.

CARTER, B. J., MURTY, V. S., REDDY, K. S., WANG, S-N., and HECHT, S. M. (1990) A Role for the Metal Binding Domain in Determining the DNA Sequence Selectivity of Fe-bleomycin. *J. Biol. Chem.*, 265, 4193–4196.

CARTER, B. J., REDDY, K. S., and HECHT, S. M. (1991) Polynucleotide Recognition and Strand Scission by Fe-bleomycin. *Tetrahedron*, 47, 2463–2474.

CARTWRIGHT, I. L., HERTZBERG, R. P., DERVAN, P. B., and ELGIN, S. C. (1983) Cleavage of Chromatin with Methidiumpropyl-EDTA · iron(II). *Proc. Natl. Acad. Sci. U.S.A.*, 80, 3213–3217.

CELANDER, D. W., and CECH, T. R. (1990) Iron(II)-ethylenediaminetetraacetic Acid Catalyzed Cleavage of RNA and DNA Oligonucleotides: Similar Reactivity Toward Single- and Double-stranded Forms. *Biochemistry*, 29, 1355–1361.

CHANG, C.-H., DALLAS, J. L., and MEARES, C. F. (1983) Identification of a Key Structural Feature of Cobalt(III)-bleomycins: An Exogenous Ligand (e.g., Hydroperoxide) Bound to Cobalt. *Biochem. Biophys. Res. Commun.*, 110, 959–966.

CHANG, C.-H., and MEARES, C. F. (1982) Light-induced Nicking of Deoxyribonucleic Acid by Cobalt(III) Bleomycins. *Biochemistry*, 24, 6332–6334.

CHANG, C.-H., and MEARES, C.F. (1984) Cobalt-bleomycins and Deoxyribonucleic Acid: Sequence-dependent Interactions, Action Spectrum for Nicking, and Indifference to Oxygen. *Biochemistry*, 23, 2268–2274.

CHEN, C.-H. B., and SIGMAN, D. S. (1987) Chemical Conversion of a DNA-binding Protein Into a Site-specific Nuclease. *Science*, 237, 1197–1201.

CHEN, C.-H. B., and SIGMAN, D. S. (1988) Sequence-specific Scission of RNA by 1,10-Phenanthroline-copper Linked to Deoxyoligonucleotides. *J. Am. Chem. Soc.* 110, 6570–6572.

CHEN, J.-H., CHURCHILL, M. E. A., TULLIUS, T.D., KALLENBACH, N. R., and SEEMAN, N. C. (1988) Construction and Analysis of Monomobile DNA Junctions. *Biochemistry*, 27, 6032–6038.

CHIEN, M., GROLLMAN, A. P., and HORWITZ, S. B. (1977) Bleomycin–DNA Interactions: Fluorescence and Proton Magnetic Resonance Studies. *Biochemistry*, 16, 3641–3646.

CHIKIRA, M., ANTHOLINE, W. E., and PETERING, D. H. (1989) Orientation of Dioxygen Bound to Cobalt(II) Bleomycin-DNA Fibers. *J. Biol. Chem.*, 264, 21478–21480.

CHIOU, S.-H. (1983) DNA- and Protein-scission Activities of Ascorbate in the Presence of Copper Ion and a Copper–Peptide Complex. *J. Biochem.* 94, 1259–1267.

CHIOU, S.-H. (1984) DNA-scission Activities of Ascorbate in the Presence of Metal Chelates. *J. Biochem.*, 96, 1307–1310.

CHOW, C. S., and BARTON, J. K. (1990) Shape-selective Cleavage of tRNA$^{Phe}$ by Transition-metal Complexes. *J. Am. Chem. Soc.*, 112, 2839–2841.

CHU, B. C. F., and ORGEL, L. E. (1985) Nonenzymatic Sequence-specific Cleavage of Single-stranded DNA. *Proc. Natl. Acad. Sci. U.S.A.*, 82, 963–967.

CHURCHILL, M. E. A., HAYES, J. J., TULLIUS, T. D. (1990) Detection of Drug Binding to DNA by Hydroxyl Radical Footprinting. Relationship of Distamycin Binding Sites to DNA Structure and Positioned Nucleosomes on 5S RNA Genes of Xenopus. *Biochemistry*, 29, 6043–6050.

CHURCHILL, M. E. A., TULLIUS, T. D., KALLENBACH, N. R., and SEEMAN, N. C. (1988) A Holliday Recombination Intermediate Is Twofold Symmetric. *Proc. Natl. Acad. Sci. U.S.A.*, 85, 4653–4656.

CIRIOLO, M. R., PEISACH, J., and MAGLIOZZO, R. S. (1989) A Comparative Study of the Interactions of Bleomycin with Nuclei and Purified DNA. *J. Biol. Chem.*, 264, 1443–1449.

CREUTZ, C. (1981) The Complexities of Ascorbate as a Reducing Agent (Corrected for pH and H$^+$ Transfer). *Inorg. Chem.*, 20, 4449–4452.

CUENOUD, B., TARASOW, T. M., and SCHEPARTZ, A. (1992) A New Strategy for Directed Protein Cleavage. *Tetrahedron Lett.*, 33, 895–898.

D'ANDREA, A. D., and HASELTINE, W. A. (1978) Sequence Specific Cleavage of DNA by the Antitumor Antibiotics Neocarzinostatin and Bleomycin. *Proc. Natl. Acad. Sci. U.S.A.* 75, 3608–3612.

D'AURORA, V., STERN, A. M., and SIGMAN, D. S. (1978) 1,10-Phenanthroline-Cuprous Ion Complex, a Potent Inhibitor of DNA and RNA Polymerases. *Biochem. Biophys. Res. Commun.*, 80, 1025–1032.

DERVAN, P. B. (1986) Design of Sequence-specific DNA-binding Molecules. *Science*, 232, 464–471.

DICKERSON, R. E. (1986) DNA–Drug Binding and Control of Genetic Information, in *Mechanisms of DNA Damage and Repair* (M. G. Simic, L. Grossman, and A. C. Upton, Eds.), Plenum Press, New York, pp. 245–255.

DIZDAROGLU, M., ARUOMA, O. I., and HALLIWELL, B. (1990) Modification of Bases in DNA by Copper Ion-1,10-phenanthroline Complexes. *Biochemistry*, 29, 8447–8451.

DREW, H. R., and TRAVERS, A. A. (1984) DNA Structural Variations in the *E. coli tyrT* Promoter. *Cell*, 37, 491–502.

DREYER, G. B., and DERVAN, P. B. (1985) Sequence-specific Cleavage of Single-stranded DNA: Oligodeoxynucleotide-EDTA · Fe(II). *Proc. Natl. Acad. Sci. U.S.A.*, 82, 968–972.

EBRIGHT, Y. W., CHEN, Y., PENDERGRAST, P. S., and EBRIGHT, R. H. (1992) Incorporation of an EDTA–Metal Complex at a Rationally Selected Site Within a Protein: Application to EDTA-iron DNA Affinity Cleaving with Catabolite Gene Activator Protein (CAP) and Cro. *Biochemistry*, 31, 10664–10670.

EBRIGHT, R. H., EBRIGHT, Y. W., PENDERGRAST, P. S., and GUNASEKERA, A. (1990) Conversion of a Helix-turn-helix Motif Sequence-specific DNA Binding Protein into a Site-specific DNA Cleavage Agent. *Proc. Natl. Acad. Sci. U.S.A.*, 87, 2882–2886.

EHRENFELD, G. M., MURUGESAN, N., and HECHT, S. M. (1984) Activation of Oxygen and Mediation of DNA Degradation by Manganese-bleomycin. *Inorg. Chem.*, 23, 1498–1500.

EHRENFELD, G. M., RODRIGUEZ, L. O., and HECHT, S. M. (1985) Copper(I)–Bleomycin: Structurally Unique Complex That Mediates Oxidative DNA Strand Scission. *Biochemistry*, 24, 81–92.

EHRENFELD, G. M., SHIPLEY, J. B., HEIMBROOK, D. C., SUGIYAMA, H., LONG E. C., VAN BOOM, J. H., VAN DER MAREL, G. A., OPPENHEIMER, N. J., and HECHT, S. M. (1987) Copper-dependent Cleavage of DNA by Bleomycin. *Biochemistry*, 26, 931–942.

ERIKSSON, M., LEIJON, M., HIORT, C., NORDEN, B., and GRASLUND, A. (1992) Minor Groove Binding of [Ru(phen)$_3$]$^{2+}$ to [d(CGCGATCGCG)]$_2$ Evidenced by Two-dimensional Nuclear Magnetic Resonance Spectroscopy. *J. Am. Chem. Soc.*, 114, 4933–4934.

ERMACORA, M. R., DELFINO, J. M., CUENOUD, B., SCHEPARTZ, A., and FOX, R. O. (1992) Conformation-dependent Cleavage of Staphylococcal Nuclease with a Disulfide-linked Iron Chelate. *Proc. Natl. Acad. Sci. U.S.A.*, 89, 6383–6387.

FENTON, H. J. H. (1894) Oxidation of Tartaric Acid in the Presence of Iron. *J. Chem. Soc.*, 65, 899–910.

FONTECAVE, M., and PIERRE, J. L. (1991) Activation et Toxicité de l'Oxygêne Principés des Therapeutiques Antioxydantes. *Bull. Soc. Chim. Fr.*, 128, 505–520.

FREEDMAN, J. H., HORWITZ, S. B., and PEISACH, J. (1982) Reduction of Copper(II)-bleomycin: A Model for *in vivo* Drug Activity. *Biochemistry*, 21, 2203–2210.

FUJII, S., OHYA-NISHIGUCHI, H., and HIROTA, N. (1990) EPR Evidence of Intermediate Peroxo Complexes Formed in a SOD Model System. *Inorg. Chim. Acta*, 175, 27–30.

GAJEWSKI, E., ARUOMA, O. I., DIZDAROGLU, M., and HALLIWELL, B. (1991) Bleomycin-dependent Damage to the Bases in DNA Is a Minor Side Reaction. *Biochemistry*, 30, 2444–2448.

GILONI, L., TAKESHITA, M., JOHNSON, F., IDEN, C., and GROLLMAN, A. P. (1981) Bleomycin-induced Strand-scission of DNA. *J. Biol. Chem.*, 256, 8608–8615.

GRIFFIN, J. H., and DERVAN, P. B. (1987a) Sequence-specific Chiral Recognition of Right-handed Double-helical DNA by (2S,3S)- and (2R,3R)-dihydroxybis(netropsin)succinamide. *J. Am. Chem. Soc.*, 108, 5008–5009.

GRIFFIN, J. H., and DERVAN, P. B. (1987b) Metalloregulation in the Sequence Specific Binding of Synthetic Molecules to DNA. *J. Am. Chem. Soc.*, 109, 6840–6842.

GRAHAM, D. R., MARSHALL, L. E., REICH, K. A., and SIGMAN, D. S. (1980) Cleavage of DNA by Coordination Complexes. Superoxide Formation in the Oxidation of 1,10-Phenanthroline–Cuprous Complexes by Oxygen–Relevance to DNA-cleavage Reaction. *J. Am. Chem. Soc.*, 102, 5419–5421.

GRINSTEAD, R. (1960a) The Oxidation of Ascorbic Acid by Hydrogen Peroxide. Catalysis by Ethylenediaminetetraacetato-iron(III). *J. Am. Chem. Soc.*, 82, 3464–3471.

GRINSTEAD, R. (1960b) Oxidation of Salicylate by the Model Peroxidase Catalyst Iron-ethylenediaminetetraacetato-iron(III) Acid. *J. Am. Chem. Soc.*, 3472–3476.

GUPTA, R. K., FERETTI, J. A., and CASPARY, W. J. (1979) Location of Iron in the $Fe^{2+}$–Bleomycin Complex as Observed by $^{13}C$ NMR Spectroscopy. *Biochem. Biophys. Res. Commun.*, 89, 534–541.

GUTTERIDGE, J. M. C., WEST, M., ENEFF, K., and FLOYD, R. A. (1990) Bleomyciniron Damage to DNA with Formation of 8-Hydroxydeoxyguanosine and Base Propenals. Indications That Xanthine Oxidase Generates Superoxide from DNA Degradation Products. *Free Radical Res. Commun.*, 10, 159–165.

HALLIWELL, B., and ARUOMA, O. I. (1991) DNA Damage by Oxygen-derived Species. Its Mechanism and Measurement in Mammalian Systems. *FEBS Lett*, 281, 9–19.

HAMAMICHI, N., NATRAJAN, A., and HECHT, S. M. (1992) On the Role of Individual Bleomycin Thiazoles in Oxygen Activation and DNA Cleavage. *J. Am. Chem. Soc.*, 114, 6278–6291.

HECHT, S. M. (1986) The Chemistry of Activated Bleomycin. *Acc. Chem. Res.*, 19, 383–391.

HEIMBROOK, D. C., CARR, M. A., MENTZER, M. A., LONG, E. C., and HECHT, S. M. (1987) Mechanism of Oxygenation of *cis*-Stilbene by Iron Bleomycin. *Inorg. Chem.*, 26, 3835–3836.

HENICHART, J.-P., HOUSSIN, R., BERNIER, J.-L., and CATTEAU, J.-P. (1982) Synthetic Model of a Bleomycin Metal Complex. *J. Chem. Soc. Chem. Commun.*, 1295–1297.

HERTZBERG, R. P., and DERVAN, P. B. (1982) Cleavage of Double Helical DNA by (Methidiumpropyl-EDTA)iron(II). *J. Am. Chem. Soc.*, 104, 313–315.

HERTZBERG, R. P., and DERVAN, P. B. (1984) Cleavage of DNA with Methidiumpropyl-EDTA-iron(II): Reaction Conditions and Product Analyses. *Biochemistry*, 23, 3934–3945.

HIORT, C., NORDEN, B., and RODGER, A. (1990) Enantiopreferential DNA Binding of [Ru$^{II}$(1,10-phenanthroline)$_3$]$^{2+}$ Studied with Linear and Circular Dichroism. *J. Am. Chem. Soc.*, 112, 1971–1982.

HOYER, D., CHO, H., and SCHULTZ, P. G. (1990) A New Strategy for Selective Protein Cleavage. *J. Am. Chem. Soc.*, 112, 3249–3250.

HUTTENHOFER, A., and NOLLER, H. F. (1992) Hydroxyl Radical Cleavage of tRNA in the Ribosomal P Site. *Proc. Natl. Acad. Sci., U.S.A.*, 89, 7851–7855.

IITAKA, Y., NAKAMURA, H., NAKATANI, T., MURAOKA, Y., FUJII, A., TAKITA, T., and UMEZAWA, H. (1978) Chemistry of Bleomycin XX. The X-ray Structure Determination of P-3A Cu(II)-complex, a Biosynthetic Intermediate of Bleomycin. *J. Antibiot.*, 31, 1070–1072.

ISHIZU, K., MURATA, S., MIYOSHI, K., SUGUIRA, Y., TAKITA, T., and UMEZAWA, H. (1981) Electrochemical and ESR Studies on Cu(II) Complexes of Bleomycin and Its Related Compounds. *J. Antibiot.*, 34, 994–1000.

JAMES, B. R., and WILLIAMS, R. J. P. (1961) The Oxidation-reduction Potentials of Some Copper Complexes. *J. Chem. Soc.*, 2007–2019.

JAYCO, M. E., and GARRISON, W. W. (1958) Formation of >C=O Bonds in the Radiation-induced Oxidation of Protein in Aqueous Systems. *Nature (London)*, 181, 413–414.

JEZEWSKA, M. J., BUJALOWSKI, W., and LOHMAN, T. M. (1989) Iron(II)-ethylenediaminetetraacetic Acid Catalyzed Cleavage of DNA Is Highly Specific for Duplex DNA. *Biochemistry*, 28, 6161–6164.

JOHNSON, A. G. R., and NAZHAT, N. B. (1987) Kinetics and Mechanism of the Reaction of the *bis*(1,10-Phenanthroline)copper(I) Ion with Hydrogen Peroxide in Aqueous Solution. *J. Am. Chem. Soc.*, 109, 1990–1994.

KAHN, M. M. T., and MARTELL, A. E. (1967) Metal Ion and Metal Chelate

Catalyzed Oxidation of Ascorbic Acid by Molecular Oxygen. II. Cupric and Ferric Chelate Catalyzed Oxidation. *J. Am. Chem. Soc.*, 89, 7104–7111.

KASAI, H., NAGANAWA, H., TAKITA, T., and UMEZAWA, H. (1978) Chemistry of Bleomycin. XXII. Interaction of Bleomycin with Nucleic Acids, Preferential Binding to Guanine Base and Electrostatic Effect of the Terminal Amine. *J. Antibiot.*, 31, 1316–1320.

KEAN, J. M., WHITE, S. A., and DRAPER, J. M. (1985) Detection of High-affinity Intercalator Sites in a Ribosomal RNA Fragment by the Affinity Cleavage Intercalator Methidiumpropyl-EDTA-iron(II). *Biochemistry*, 24, 5062–5070.

KENANI, A., BAILLY, C., HELBECQUE, N., CATTEAU, J-P., HOUSSIN, R., BERNIER, J-L., and HENICHART, J-P. (1988) The Role of the Gulose-mannose Part of Bleomycin in Activation of Iron–Molecular Oxygen Complexes. *Biochem. J.*, 253, 497–504.

KHARASCH, M. S., FONO, A., and NUDENBERG, W. (1950) The Chemistry of Hydroperoxides. I. The Acid Catalyzed Decomposition of α,α-Dimethylbenzyl (α-Cumyl) Hydroperoxide. *J. Org. Chem.*, 15, 748–752.

KILKUSKIE, R. E., SUGUIRA, H., YELLIN, B., MURUGESAN, N., and HECHT, S. M. (1985) Oxygen Transfer by Bleomycin Analogues Dysfunctional in DNA Cleavage. *J. Am. Chem. Soc.*, 107, 260–261.

KIM, K., RHEE, S. G., and STADTMAN, E. R. (1985) Nonenzymatic Cleavage of Proteins by Reactive Oxygen Species Generated by Dithiothreitol and Iron. *J. Biol Chem.*, 260, 15394–15397.

KITTAKA, A., SUGANO, Y., OTSUKA, M., and OHNO, M. (1988) Man-designed Bleomycins. Synthesis of Dioxygen Activating Molecules and a DNA Cleaving Molecule Based on Bleomycin–Fe(II)–O$_2$ Complex. *Tetrahedron*, 44, 2821–2833.

KOHDA, K., KASAI, H., OGAWA, T., SUZUKI, T., and KAWAZOE, Y. (1989) Deoxyribonucleic Acid (DNA) Damage Induced by Bleomycin-Fe(II) *in vitro*: Formation of 8-Hydroxyguanine Residues in DNA. *Chem. Pharm. Bull.*, 37, 1028–1030.

KOPKA, M. L., YOON, C., GOODSELL, D., PJURA, P., and DICKERSON, R. E. (1985) The Molecular Origin of DNA-Drug Specificity in Netropsin and Distamycin. *Proc. Natl. Acad. Sci. U.S.A.*, 82, 1376–1380.

KOZARICH, J. W., WORTH, L., FRANK, B. L., CHRISTNER, D. F., VANDERWALL, D. E., and STUBBE, J. (1989) Sequence-specific Isotope Effects on the Cleavage of DNA by Bleomycin. *Science*, 245, 1396–1898.

KROSS, J., HENNER, D., HECHT, S. M., and HASELTINE, W. A. (1982) Specificity of Deoxyribonucleic Acid Cleavage by Bleomycin, Phleomycin, and Tallysomycin. *Biochemistry*, 21, 4310–4318.

KUWABARA, M., YOON, C., GOYNE, T., THEDERAHN, T., and SIGMAN, D. S. (1986) Nuclease Activity of 1,10-Phenanthroline-copper Ion: Reaction with CGCGAATTCGCG and Its Complexes with Netropsin and *Eco*RI. *Biochemistry*, 25, 7401–7408.

KUWAHARA, J., and SUGUIRA, Y. (1988) Sequence-specific Recognition and Cleavage of DNA by Metallobleomycin: Minor Groove Binding and Possible Interaction Mode. *Proc. Natl. Acad. Sci. U.S.A.*, 85, 2459–2463.

KUWAHARA, J., SUZUKI, T., and SUGUIRA, Y. (1985) Effective DNA Cleavage by Bleomycin–Vanadium(IV) Complex Plus Hydrogen Peroxide. *Biochem. Biophys. Res. Commun.*, 129, 368–374.

LATHAM, J. A., and CECH, T. R. (1989) Defining the Inside and Outside of a Catalytic RNA Molecule. *Science*, 245, 276–282.

LEVINE, R. L. (1983) Oxidative Modification of Glutamine Synthetase. I. Inactivation Is Due to Loss of One Histidine Residue. *J. Biol Chem.*, 258, 11823–11827.

LIND, M. D., HOARD, J. L., HAMOR, M. J., HAMOR, T. A., and HOARD, J. L. (1964) Stereochemistry of Ethylenediaminetetraacetato Complexes. II. The Structure of Crystalline Rb[Fe(OH$_2$)Y] · H$_2$O. III. The Structure of Crystalline Li[Fe(OH$_2$)Y] · 2H$_2$O. *Inorg. Chem.*, 3, 34–43.

LOVE, J. D., LIARAKOS, C. D., and MOSES, R. E. (1981) Nonspecific Cleavage of ϕX174 RFI Deoxyribonucleaic Acid by Bleomycin. *Biochemistry*, 20, 5331–5336.

LOWN, J. W., and SIM, S. (1977) The Mechanism of the Bleomycin-induced Cleavage of DNA. *Biochem. Biophys. Res. Commun.*, 77, 1150–1157.

MACK, D. P., and DERVAN, P. B. (1990) Nickel-mediated Sequence-specific Oxidative Cleavage of DNA by a Designed Metalloprotein. *J. Am. Chem. Soc.*, 112, 4604–4606.

MACK, D. P., IVERSON, B. L., and DERVAN, P. B. (1988) Design and Chemical Synthesis of a Sequence-specific DNA-cleaving Protein. *J. Am. Chem. Soc.*, 110, 7572–7574.

MANDON, D., WEISS, R., FRANKE, M., BILL, E., and, TRAUTWEIN, A. X. (1989) Oxoiron Porphyrin Species with High-valent Iron: Formation by Solvent-dependent Protonation of a Peroxoiron(III) Porphyrinate Derivative. *Angew. Chem., Int. Ed. Engl.*, 28, 1709–1711.

MARSHALL, L. E., GRAHAM, D. R., REICH, K. A., and SIGMAN, D. S. (1981) Cleavage of Deoxyribonucleic Acid by the 1,10-Phenanthroline–Cuprous Complex. Hydrogen Peroxide Requirement and Primary and Secondary Structure Specificity. *Biochemistry*, 20, 244–250.

MCGALL, G. H., RABOW, L. E., ASHLEY, G. W., WU, S. H., KOZARICH, J. W., and STUBBE, J. (1992) New Insight into the Mechanism of Base Propenal Formation During Bleomycin-mediated DNA Degradation. *J. Am. Chem. Soc.*, 114, 4958–4967.

MEI, H-Y., and BARTON, J. K. (1986) Chiral Probe for A-form Helices of DNA and RNA: *tris*(Tetramethylphenanthroline)ruthenium(II). *J. Am. Chem. Soc.*, 108, 7414–7416.

MEI, H-Y., and BARTON, J. K. (1988) *Tris*(Tetramethylphenanthroline)ruthenium(II): A Chiral Probe That Cleaves A-DNA Conformations. *Proc. Natl. Acad. Sci. U.S.A.*, 85, 1339–1343.

MELNYK, D. L., HORWITZ, S. B., and PEISACH, J. (1981) Redox Potential of Iron-bleomycin. *Biochemistry*, 20, 5327–5331.

MELNYK, D. L., HORWITZ, S. B., and PEISACH, J. (1987) The Oxidation-reduction Potential of Copper-bleomycin. *Inorg. Chim. Acta*, 138, 75–78.

MEUNIER, B. (1992) Metalloporphyrins as Versatile Catalysts for Oxidation Reactions and Oxidative DNA Cleavage. *Chem. Rev.*, 92, 1411–1456.

MOSER, H. E., and DERVAN, P. B. (1987) Sequence-specific Cleavage of Double Helical DNA by Triple Helix Formation. *Science*, 238, 645–650.

MOTOMURA, T., ARAKI, K., KOBAYASHI, K., TOI, H., and AOYAMA, Y. (1992) Artificial Metallonuclease. $Cu^{II}$-promoted DNA Strand Scission as Effected by Bisresorcinol Derivatives Having a Metal-ion Binding Site. *Chem. Lett.*, 963–966.

MÜLLER, W. E. G., YAMAZAKI, Z-I., BRETER, H-J., and ZAHN, R. K. (1972) Action of Bleomycin on DNA and RNA. *Eur. J. Biochem.*, 31, 518–525.

MURAKAWA, G. J., CHEN, C-B., KUWABARA, M. D., NIERLICH, D., and SIGMAN, D. S. (1989) Scission of RNA by the Chemical Nuclease of 1,10-Phenanthroline-copper Ion: Preference for Single-stranded Loops. *Nucl. Acids. Res.*, 17, 5361–5375.

MURUGESAN, N., and HECHT, S. M. (1985) Bleomycin as an Oxene Transferase. Catalytic Oxygen Transfer to Olefins. *J. Am. Chem. Soc.*, 107, 493–500.

MURUGESAN, N., XU, C., EHRENFELD, G. M., SUGIYAMA, H., KILKUSKIE, R. E., RODRIGUEZ, L. D., CHANG, L.-H., and HECHT, S. M. (1985) Analysis of Products Formed During Bleomycin-mediated DNA Degradation. *Biochemistry*, 24, 5735–5744.

NAGAI, K., SUZUKI, H., TANAKA, N., and UMEZAWA, H. (1969b) Decrease of Melting Temperature and Single Strand Scission of DNA by Bleomycin in the Presence of Hydrogen Peroxide. *J. Antibiot. Ser. A*, 22, 624–628.

NAGAI, K., YAMAKI, H., SUZUKI, H., TANAKA, N., and UMEZAWA, H. (1969a) The Combined Effects of Bleomycin and Sulfhydryl Compounds on the Thermal Denaturation of DNA. *Biochem. Biophys. Acta*, 179, 165–171.

NAKAMURA, M., and PEISACH, J. (1988) Self-inactivation of Fe(II)-bleomycin. *J. Antibiot.*, 41, 638–647.

NATRAJAN, A., HECHT, S. M., VAN DER MAREL, G. A., and VAN BOOM, J. H. (1990a) A Study of $O_2$-versus $H_2O_2$-Supported Activation of Fe·bleomycin. *J. Am. Chem. Soc.*, 112, 3997–4002.

NATRAJAN, A., HECHT, S. M., VAN DER MAREL, G. A., and VAN BOOM, J. H. (1990b) Activation of Fe(III)·bleomycin by 10-Hydroperoxy-8,12-octa-decadienoic Acid. *J. Am. Chem. Soc.*, 112, 4532–4538.

OGINO, H., TAKAKO, N., and OGINO, K. (1989) Redox Potentials and Related Thermodynamic Parameters of (Diamino Polycarboxylato)metal(III/II) Redox Couples. *Inorg. Chem.*, 28, 3656–3659.

OPPENHEIMER, N. J., CHANG, C., RODRIGUEZ, L. D., and HECHT, S. M. (1981) Copper(I)·Bleomycin—a Structurally Unique Oxidation-reduction Active Complex. *J. Biol. Chem.*, 256, 1514–1517.

OPPENHEIMER, N. J., RODRIGUEZ, L. O., and HECHT, S. M. (1979) Structural Studies of "Active Complex" of Bleomycin: Assignment of Ligands to the Ferrous Ion in a Ferrous–Bleomycin–Carbon Monoxide Complex. *Proc. Natl. Acad. Sci. U.S.A.*, 76, 5616–5620.

OTSUKA, M., MASUDA, T., HAUPT, A., OHNO, M., SHIRAKI, T., SUGUIRA, Y., and MAEDA, K. (1990) Man-designed Bleomycin with Altered Sequence Specificity in DNA Cleavage. *J. Am. Chem. Soc.*, 112, 838–845.

OWA, T., SUGIYAMA, T., OTSUKA, M., and OHNO, M. (1990) A Model Study on the Mechanism of the Autooxidation of Bleomycin. *Tetr. Lett.*, 31, 6063–6066.

PADBURY, G., and SLIGAR, S. G. (1986) Oxygen Activation by Bleomycin: Heteroatom Dealkylation. *Fed. Proc., Fed. Am. Soc. Exp. Biol.*, 45, 1520.

PADBURY, G., SLIGAR, S. G., LABEQUE, R., and MARNETT, L. J. (1988) Ferric Bleomycin Catalyzed Reduction of 10-Hydroperoxy-8,12-octadecadienoic Acid: Evidence for Homolytic O—O Bond Scission. *Biochemistry*, 27, 7846–7852.

PETERING, D. H., BYRNES, R. W., and ANTHOLINE, W. E. (1990) The Role of Redox-active Metals in the Mechanism of Action of Bleomycin. *Chem.–Biol. Interact.*, 73, 133–182.

POPE, L. M., REICH, K. A., GRAHAM, D. R., and SIGMAN, D. S. (1982) Products of DNA Cleavage by the 1,10-Phenanthroline Copper Complex. Inhibitors of *Escherichia coli* DNA Polymerase I. *J. Biol. Chem.*, 257, 12121–12128.

POPE, L. M., and SIGMAN, D. S. (1984) Secondary Structure Specificity of the Nuclease Activity of the 1,10-Phenanthroline–Copper Complex. *Proc. Natl. Acad. Sci. U.S.A.*, 81, 3–7.

POVSIC, T. J., and DERVAN, P. B. (1989) Triple Helix Formation by Oligonucleotides on DNA Extended to the Physiological pH Range. *J. Am. Chem. Soc.*, 111, 3059–3061.

PRATVIEL, G., BERNADOU, J., and MEUNIER, B. (1986) DNA Breaks Generated by the Bleomycin–Iron III Complex in the Presence of KHSO$_5$, a Single Oxygen Donor. *Biochem. Biophys. Res. Commun.*, 136, 1013–1020.

PURMAL, A. P., SKURLATOV, YU. I., and TRAVIN, S. O. (1980) Formation of an Intermediate Oxygen Complex in the Autooxidation of Iron(II)-ethylenediaminetetraacetate. *Bull. Acad. Sci. USSR Div. Chem. Sci. (Engl. Transl.)*, 29, 315–320.

RABOW, L. E., STUBBE, J., and KOZARICH, J. W. (1990) Identification and Quantification of the Lesion Accompanying Base Release in Bleomycin-mediated DNA Degradation. *J. Am. Chem. Soc.*, 112, 3196–3203.

RABOW, L., STUBBE, J., KOZARICH, J. W., and GERLT, J. A. (1986) Identification of the Alkaline-labile Product Accompanying Cytosine Release During Bleomycin-mediated Degradation of d(CGCGCG). *J. Am. Chem. Soc.*, 108, 7130–7131.

RAHHAL, S., and RICHTER, H. W. (1988) Reduction of Hydrogen Peroxide

by the Ferrous Iron Chelate of Diethylenetriamine-*N,N,N',N',N",N"*-pentaacetate. *J. Am. Chem. Soc.*, 110, 3126–3133 (see ref. within).
RANA, T. R., and MEARES, C. F. (1990) Specific Cleavage of a Protein by an Attached Iron Chelate. *J. Am. Chem. Soc.*, 112, 2457–2458.
RANA, T. R., and MEARES, C. F. (1991a) Iron Chelate Mediated Proteolysis: Protein Structure Dependence. *J. Am. Chem. Soc.*, 113, 1859–1861.
RANA, T. R., and MEARES, C. F. (1991b) Transfer of Oxygen from an Artificial Protease to Peptide Carbon During Proteolysis. *Proc. Natl. Acad. Sci. U.S.A.*, 88, 10578–10582.
REHMANN, J. P., and BARTON, J. K. (1990) $^1$H NMR Studies of *tris*-(Phenanthroline) Metal Complexes Bound to Oligonucleotides: Characterization of Binding Modes. *Biochemistry*, 29, 1701–1709.
RUSH, J. D., and KOPPENOL, W. H. (1986) Oxidizing Intermediates in the Reaction of Ferrous EDTA with Hydrogen Peroxide. *J. Biol. Chem.*, 261, 6730–6733.
RUSH, J. D., MASKOS, Z., and KOPPENOL, W. H. (1990) Distinction Between Hydroxyl Radical and Ferryl Species. *Meth. Enzymol.*, 186, 148–156.
SAITO, I., MORII, T., SUGIYAMA, H., MATSUURA, T., MEARES, C. F., and HECHT, S. M. (1989) Photoinduced DNA Strand Scission by Cobalt Bleomycin Green Complex. *J. Am. Chem. Soc.*, 111, 2307–2308.
SATYANARAYANA, S., DABROWIAK, J. C., and CHAIRES, J. B. (1992) Neither Delta- nor Lambda-*tris*(phenanthroline)ruthenium(II) Binds to DNA by Classical Intercalation. *Biochemistry*, 31, 9319–9324.
SAUSVILLE, E. A., PEISACH, J., and HORWITZ, S. B. (1976) A Role for Ferrous Ion and Oxygen in the Degradation of DNA by Bleomycin. *Biochem. Biophys. Res. Commun.*, 73, 814–822.
SAUSVILLE, E. A., PEISACH, J., and HORWITZ, S. B. (1978a) Effect of Chelating Agents and Metal Ions on the Degradation of DNA by Bleomycin. *Biochemistry*, 17, 2740–2746.
SAUSVILLE, E. A., STEIN, R. A., PEISACH, J., and HORWITZ, S. B. (1978b) Properties and Products of the Degradation of DNA by Bleomycin and Iron(II). *Biochemistry*, 17, 2746–2754.
SAWYER, D. T. (1991) *Oxygen Chemistry—International Series of Monographs on Chemistry 26*, Oxford University Press, New York, pp. 44, 172.
SCHEPARTZ, A., and CUENOUD, B. (1990) Site-specific Cleavage of the Protein Calmodulin Using a Trifluorperazine-based Affinity Reagent. *J. Am. Chem. Soc.*, 112, 3247–3249.
SCHULTZ, P. G., and DERVAN, P. B. (1983) Sequence-specific Double-strand Cleavage of DNA by *bis*(EDTA-distamycin·Fe$^{II}$) and EDTA-*bis*(distamycin·Fe$^{II}$). *J. Am. Chem. Soc.*, 105, 7748–7750.
SCHULTZ, P. G., TAYLOR, J. S., and DERVAN, P. B. (1982) Design and Synthesis of a Sequence-specific DNA Cleaving Molecule (Distamycin-EDTA)iron(II). *J. Am. Chem. Soc.*, 104, 6861–6863.

SHAFER, G. E., PRICE, M. A., and TULLIUS, T. D. (1989) Use of the Hydroxyl Radical and Gel Electrophoresis to Study DNA Structure. *Electrophoresis*, 10, 397–404.

SHELDON, R. A., and KOCHI, J. K. (1976) Metal-catalyzed Oxidations of Organic Compounds in the Liquid Phase: A Mechanistic Approach, in *Advances in Catalysis* (D. D. Eley, H. Pines, and P. B. Weisz, eds.), Academic Press, New York, 25, pp. 272–413.

SHEPHERD, R. E., LOMIS, T. J., and KOEPSEL, R. R. (1992) A Ferryl(V) Pathway in DNA Cleavage Induced by $Fe^{II}$(haph) with $O_2$ or $H_2O_2$. *J. Chem. Soc. Chem. Commun.*, 222–224.

SHIELDS, H., and MCGLUMPHY, C. (1984) The Orientation of the Ligands in Iron(III)-bleomycin Intercalated with DNA. *Biochim. Biophys. Acta*, 800, 277–281.

SHIELDS, H., MCGLUMPHY, C, and HAMRICK, P. J. (1982) The Conformation and Orientation of Copper(II) Bleomycin Intercalated with DNA. *Biochim. Biophys. Acta*, 697, 113–120.

SHIRAKAWA, I., AZEGAMI, M., ISHII, S.-I., and UMEZAWA, H. (1971) Reaction of Bleomycin with DNA Strand Scission of DNA in the Absence of Sulfhydryl or Peroxide Compounds. *J. Antibiot. Ser. A*, 24, 761–766.

SIGMAN, D. S. (1986) Nuclease Activity of 1,10-Phenanthroline-copper Ion. *Acc. Chem. Res.*, 19, 180–186.

SIGMAN, D. S. (1990) Chemical Nucleases. *Biochemistry*, 29, 9097–9105.

SIGMAN, D. S., GRAHAM, D. R., D'AURORA, V., and STERN, A. M. (1979) Oxygen-dependent Cleavage of DNA of the 1,10-Phenanthroline·Cuprous Complex—Inhibition of *Escherichia coli* DNA Polymerase I. *J. Biol. Chem.*, 254, 12269–12272.

SINGLETON, S. F., and DERVAN, P. B. (1992) Thermodynamics of Oligodeoxyribonucleotide-directed Triple Helix Formation: An Analysis Using Quantitative Affinity Cleavage Titration. *J. Am. Chem. Soc.*, 114, 6957–6965.

SLUKA, J. P., HORVATH, S. J., BRUIST, M. F., SIMON, M. I., and DERVAN, P. B. (1987) Synthesis of a Sequence-specific DNA-cleaving Peptide. *Science*, 238, 1129–1132.

SPASSKY, A., and SIGMAN, D. S. (1985) Nuclease Activity of 1,10-Phenanthroline-copper Ion. Conformational Analysis and Footprinting of the *Lac* Operon. *Biochemistry*, 24, 8050–8056.

STADTMAN, E. R., and BERLETT, B. S. (1991b) Fenton Chemistry. Amino Acid Oxidation. *J. Biol. Chem.* 266, 17201–17211.

STADTMAN, E. R., and OLIVER, C. N. (1991) Metal-catalyzed Oxidation of Proteins. Physiological Consequences. *J. Biol. Chem.*, 266, 2005–2008.

STUBBE, J., and KOZARICH, J. W. (1987) Mechanisms of Bleomycin-induced DNA Degradation. *Chem. Rev.*, 87, 1007–1136.

SUBRAMANIAN, R., and MEARES, C. F. (1985) Photo-induced Nicking of

Deoxyribonucleic Acid by Ruthenium(II)-bleomycin in the Presence of Air. *Biochem. Biophys. Res. Commun.*, 133, 1145–1151.

SUBRAMANIAN, R., and MEARES, C. F. (1986) Photosensitization of Cobalt Bleomycin. *J. Am. Chem. Soc.*, 108, 6427–6429.

SUGGS, J. W., and WAGNER, R. W. (1986) Nuclease Recognition of an Alternating Structure in a d(AT)$_{14}$ Plasmid Insert. *Nucleic Acids Res.*, 14, 3703–3716.

SUGIYAMA, H., KILKUSKIE, R. E., CHANG, L-H., MA, L-T., and HECHT, S. M. (1986) DNA Strand Scission by Bleomycin: Catalytic Cleavage and Strand Selectivity. *J. Am. Chem. Soc.*, 108, 3852–3854.

SUGIYAMA, H., SERA, T., DANNOUE, Y., MARUMOTO, R., and SAITO, I. (1991) Bleomycin-mediated Degradation of Aristeromycin-containing DNA. Novel Dehydrogenation Activity of Iron(II)-bleomycin. *J. Am. Chem. Soc.*, 113, 2290–2295.

SUGIYAMA H., XU, C., MURUGESAN, N., and HECHT, S. M. (1985) Structure of the Alkali-labile Product Formed During Iron(II)-bleomycin-mediated DNA Strand Scission. *J. Am. Chem. Soc.*, 107, 4104–4105.

SUGUIRA, Y. (1979b) The Production of Hydroxyl Radical from Copper(I) Complex Systems of Bleomycin and Tallysomycin: Comparison with Copper(II) and Iron(II) Systems. *Biochem. Biophys. Res. Commun.*, 90, 375–383.

SUGIURA, Y. (1980a) Bleomycin–Iron Complexes. Electron Spin Resonance Study, Ligand Effect, and Implication for Action Mechanism. *J. Am. Chem. Soc.*, 102, 5208–5215.

SUGIURA, Y. (1980b) Monomeric Cobalt(II)-oxygen Adducts of Bleomycin Antibiotics in Aqueous Solution. A New Ligand Type of Oxygen Binding and Effect of Axial Lewis Base. *J. Am. Chem. Soc.*, 102, 5216–5221.

SUGIURA, Y., ISHIZU, K., and MIYOSHI, K. (1979a) Studies of Metallobleomycins by Electronic Spectroscopy, Electron Spin Resonance Spectroscopy, and Potentiometric Titration. *J. Antibiot.*, 32, 453–461.

SUGIURA, Y., SUZUKI, T., KUWAHARA, J., and TANAKA, H. (1982a) On the Mechanism of Hydrogen Peroxide-, Superoxide-, and Ultraviolet Light-induced DNA Cleavages of Inactive Bleomycin–Iron(III) Complex. *Biochem. Biophys. Res. Commun.*, 105, 1511–1518.

SUGIURA, Y., SUZUKI, T., and TANAKA, H. (1982b) Some Properties, Oxygen Activation, and DNA Cleavage of Iron Complexes of Bleomycin and Its Synthetic Analog, in *Oxygenases and Oxygen Metabolism* (M. NOZAKI), Ed., Academic Press, New York, pp. 511–519.

SUN, J. S., FRANÇOIS, J-C., LAVERY, R., SAISON-BEHMOARAS, T., MONTENAY-GARESTIER, T., NGUYEN, T. T., and HÉLÈNE, C. (1988) Sequence-targeted Cleavage of Nucleic Acids by Oligo-alpha-thymidylate-phenanthroline Conjugates: Parallel and Antiparallel Double Helices Are Formed with DNA and RNA, Respectively. *Biochemistry*, 27, 6039–6045.

SUZUKI, T., KUWAHARA, J., GOTO, M., and SUGUIRA, Y. (1985) Nucleotide Sequence Cleavages of Manganese-bleomycin Induced by Reductant, Hydrogen Peroxide and Ultraviolet Light. Comparison with Iron- and Cobalt-bleomycins. *Biochim. Biophys. Acta*, 824, 330–335.

SUZUKI, T., KUWAHARA, J., and SUGUIRA, Y. (1984) DNA Cleavages of Bleomycin–Transition Metal Complexes Induced by Reductant, Hydrogen Peroxide, and Ultraviolet Light: Characteristics and Biological Implication. *Nucleic Acids Res. Symp Ser. No. 15*, 161–164.

TABOR, S., and RICHARDSON, C. C. (1987) Selective Oxidation of the Exonuclease Domain of Bacteriophage T7 DNA Polymerase. *J. Biol Chem.*, 262, 15330–15333.

TAKAHASHI, K., YOSHIOKA, O., MATSUDA, A., and UMEZAWA, H. (1977) Intracellular Reduction of the Cupric Ion of Bleomycin Copper Complex and Transfer of the Cuprous Ion to a Cellular Protein. *J. Antibiot.*, 30, 861–869.

TAKESHITA, M., GROLLMAN, A. P., OHTSUBO, E., and OHTSUBO, H. (1978) Interaction of Bleomycin with DNA. *Proc. Natl. Acad. Sci. U.S.A.*, 75, 5983–5987.

TAKITA, T., MURAOKA, Y., NAKATANI, T., FUJII, A., IITAKA, Y., and UMEZAWA, H. (1978a) Chemistry of Bleomycin. XXI Metal-complex of Bleomycin and its Implication for the Mechanism of Bleomycin Action. *J. Antibiot.*, 31, 1073–1077.

TAKITA, T., MURAOKA, Y., NAKATANI, T., FUJII, A., UMEZAWA, Y., NAGANAWA, H., and UMEZAWA, H. (1978b) Chemistry of Bleomycin. XIX. Revised Structures of Bleomycin and Phleomycin. *J. Antibiot.*, 31, 801–804.

TAN, J. D., HUDSON, S. E., BROWN, S. J., OLMSTEAD, M. M., and MASCHARAK, P. K. (1992) Syntheses, Structures, and Reactivities of Synthetic Analogues of the Three Forms of Co(III)-bleomycin: Proposed Mode of Light-induced DNA Damage by the Co(III) Chelate of the Drug. *J. Am. Chem. Soc.*, 114, 3841–3853.

TANNER, K., and CECH, T. R. (1985) Self-catalyzed Cyclization of the Intervening Sequence RNA of *Tetrahymena*: Inhibition by Methidiumpropyl·EDTA and Localization of the Major Dye Binding Sites. *Nucl. Acids Res.*, 13, 7759–7779.

TAYLOR, J. S., SCHULTZ, P. G., and DERVAN, P. B. (1984) DNA Affinity Cleaving. Sequence Specific Cleavage of DNA by Distamycin-EDTA·Fe(II) and EDTA-distamycin·Fe(II). *Tetrahedron*, 40, 457–465.

THEDERAHN, T., SPASSKY, A., KUWABARA, M. D., and SIGMAN, D. S. (1990) Chemical Nuclease Activity of 5-Phenyl-1,10-phenanthroline-copper Ion Detects Intermediates in Transcription Initiation by *E. coli* RNA Polymerase. *Biochem. Biophys. Res. Commun.*, 168, 756–762.

TULLIUS, T. D. (1991) DNA Footprinting with the Hydroxyl Radical. *Free Radical Res. Commun.*, 12–13, 521–529.

TULLIUS, T. D., and DOMBROSKI, B. A. (1985) Iron(II) EDTA Used to Measure the Helical Twist Along Any DNA Molecule. *Science*, 230, 679–681.

TULLIUS, T. D., and DOMBROSKI, B. A. (1986) Hydroxyl Radical "Footprinting": High-resolution Information About DNA–Protein Contacts and Application to λ Repressor and Cro Protein. *Proc. Natl. Acad. Sci. U.S.A.*, 83, 5469–5473.

TULLIUS, T. D., DOMBROSKI, B. A., CHURCHILL, M. E. A., and KAM, L. (1987) Hydroxyl Radical Footprinting; A High-resolution Method for Mapping Protein–DNA Contacts. *Meth. Enzymol.*, 155, 537–559.

UDENFRIEND, S., CLARK, C. T., AXELROD, J., and BRODIE, J. J. (1954) Ascorbic Acid in Aromatic Hydroxylation. I. A Model System for Aromatic Hydroxylation. *J. Biol. Chem.*, 208, 731–739.

UMEZAWA, H., MAEDA, K., TAKEUCHI, T., and OKAMI, Y. (1966) New Antibiotics, Bleomycin A and B. *J. Antibiot.*, Ser. A., 19, 200–209.

UMEZAWA, H., TAKITA, T., SUGUIRA, Y., OTSUKA, M., KOBAYASHI, S., and OHNO, M. (1984) DNA–Bleomycin Interaction. Nucleotide Sequence-specific Binding and Cleavage of DNA by Bleomycin. *Tetrahedron*, 40, 501–509.

VAN DYKE, M. W., and DERVAN, P. B. (1983a) Chromomycin, Mithramycin, and Olivomycin Binding Sites on Heterogeneous Deoxyribonucleic Acid. Footprinting with (Methidiumpropyl-EDTA)iron(II). *Biochemistry*, 22, 2373–2377.

VAN DYKE, M. W., and DERVAN, P. B. (1983b) Methidiumpropyl-EDTA·Fe(II) and DNase I Footprinting Report Different Small Molecule Binding Site Sizes on DNA. *Nucl. Acids Res.*, 11, 5555–5567.

VAN DYKE, M. W., HERTZBERG, R. P., and DERVAN, P. B. (1982) Map of Distamycin, Netropsin, and Actinomycin Binding Sites on Heterogeneous DNA: DNA Cleavage Inhibition Patterns with Methidiumpropyl-EDTA·Fe(II). *Proc. Natl. Acad. Sci. U.S.A.*, 79, 5470–5474.

VARY, C. P. H., and VOURNAKIS, J. N. (1984) RNA Structure Analysis Using Methidiumpropyl-EDTA·Fe(II): A Base-pair-specific RNA Structure Probe. *Proc. Natl. Acad. Sci. U.S.A.*, 81, 6978–6982.

VOS, C. M., WESTERA, G., and ZANTEN, B. (1980) Different Forms of the Cobalt–Bleomycin A$_2$ Complex. *J. Inorg. Biochem.*, 12, 45–55.

WALLING, C. (1975) Fenton's Reagent Revisited. *Accs. Chem. Res.*, 8, 125–131.

WILLIAMS, L. D., THIVIERGE, J., and GOLDBERG, I. H. (1988) Specific Binding of o-Phenanthroline at a DNA Structural Lesion. *Nucleic Acids Res.*, 16, 11607–11615.

WINTERBOURN, C. C. (1987) The Ability of Scavengers to Distinguish Hydroxyl Radical Production in the Iron-catalyzed Haber-Weiss Reaction: Comparison of Four Assays for Hydroxyl Radical. *Free Radical Biol. Med.*, 3, 33–39.

Wu, J. C., and Kozarick, J. W. (1985) Mechanism of Bleomycin: Evidence for a Rate-determining 4'-Hydrogen Abstraction from Poly (dA-dU) Associated with the Formation of Both Free Base and Base Propenal. *Biochemistry*, 24, 7562–7568.

Wu, J. C., Kozarich, J. W., and Stubbe, J. (1983) The Mechanism of Free Base Formation From DNA by Bleomycin. *J. Biol. Chem.*, 258, 4694–4697.

Wu, Y., Houk, K. N., Valentine, J. S., and Nam, W. (1992) Is Intramolecular Hydrogen-bonding Important for Bleomycin Reactivity? A Molecular Mechanics Study. *Inorg. Chem.*, 31, 720–722.

Xu, R. S., Antholine, W. E., and Petering, D. H. (1992a) Reaction of Co(II)bleomycin with Dioxygen. *J. Biol. Chem.*, 267, 944–949.

Xu, R. S., Antholine, W. E., and Petering, D. H. (1992b) Reaction of DNA-bound Co(II)bleomycin with Dioxygen. *J. Biol. Chem.*, 267, 950–955.

Yamaguchi, K., Watanabe, Y., and Morishima, I. (1992) Push Effect on the Heterolytic O—O Bond Cleavage of Peroxoiron(III) Porphyrin Adducts. *Inorg. Chem.*, 31, 156–157.

Yamazaki, I., and Piette, L. H. (1990) ESR Spin-trapping Studies on the Reaction of $Fe^{2+}$ Ions with $H_2O_2$-reactive Species in Oxygen Toxicity in Biology. *J. Biol. Chem.*, 265, 13589–13594.

Yoon, C., Kuwabara, M. D., Spassky, A., and Sigman, D. S. (1990) Sequence Specificity of the Deoxyribonuclease Activity of 1,10-Phenanthroline-copper Ion. *Biochemistry*, 29, 2116–2121.

Youngquist, R. S., and Dervan, P. B. (1985a) Sequence-specific Recognition of B-DNA by Oligo(N-methylpyrrolecarboxamide)s. *Proc. Natl. Acad. Sci. U.S.A.*, 82, 2565–2569.

Youngquist, R. S., and Dervan, P. B. (1985b) Sequence-specific Recognition of B DNA by bis(EDTA-distamycin)fumaramide. *J. Am. Chem. Soc.*, 107, 5528–5529.

Youngquist, R. S., and Dervan, P. B. (1987) A Synthetic Peptide Binds 16 Base Pairs of A,T Double Helical DNA. *J. Am. Chem. Soc.*, 109, 7564–7566.

Zang, V., and van Eldik, R. (1990) Kinetics and Mechanism of the Autoxidation of Iron(II) Induced Through Chelation by Ethylenediaminetetraacetate and Related Ligands. *Inorg. Chem.*, 29, 1705–1711.

# 9

# Exploration of Selected Pathways for Metabolic Oxidative Ring Opening of Benzene Based on Estimates of Molecular Energetics

ARTHUR GREENBERG

## INTRODUCTION

### General Aspects and Limitations of Thermochemical Estimations

Relatively simple methods of thermochemical estimation have been demonstrated to be powerful tools for the prediction of favorable mechanistic pathways in organic chemistry. The purpose of this chapter is to demonstrate their use in estimating the molecular energetics of metabolic reactions of oxygen with organic substrates. In employing this approach, we demonstrate that it is possible to evaluate the likelihood of postulated pathways through analysis of the possible structures and associated energies of intermediates, transition states, and stable products.

*Brief introduction to group-increment and other methods.* The Benson group-increment approach (Benson, 1976) and other estimation methods (Stull et al., 1969; Cox and Pilcher, 1970; Pedley et al., 1986; Liebman and Van Vechten, 1987; Lias et al., 1988; Liebman and Greenberg, 1989) are first introduced very briefly for readers unfamiliar with these approaches toward evaluating molecular properties as well as potential reaction mechanisms. The thermochemical data employed are all for the gas phase at 298 K—the frame of reference for studying isolated molecules not involved in intermolecular, solvent, or crystalline-state associations. The standard gas-phase enthalpy of formation of ethane [$\Delta H_f^\circ$(g) (CH$_3$—CH$_3$)] is −20.0 kcal/mol (unless otherwise noted, all $\Delta H_f^\circ$(g) values are taken from Pedley et al., 1986, where data in kilojoules per mole are divided by 4.184 kJ/kcal). This value could be apportioned as −10.0 kcal per methyl group attached to tetravalent carbon, [C(H)$_3$(C)]. The $\Delta H_f^\circ$(g) value for propane is −25.0 kcal/mol; thus, the additional C(H)$_2$(C)$_2$ increment is −5.0 kcal/mol. Virtually the same C(H)$_2$(C)$_2$ increment appears in transforming propane (−25.0) into n-butane (−30.0), n-pentane (−35.1) and n-hexane (−39.9). Since there is a large data set of $\Delta H_f$ values for hydrocarbons, Benson was able to derive a series of "unstrained" group increments through regression analysis (Benson, 1976). Thus, the group increments for C(H)$_3$(C) (−10.20 kcal/mol) and C(H)$_2$(C)$_2$ (−4.93 kcal/mol) may be summed to yield $\Delta H_f^\circ$(g) (C$_3$H$_8$) = −25.33 kcal/mol. Cyclohexane ($\Delta H_f^\circ$(g) = −29.5 kcal/mol) is well calculated by the sum of group increments [6 × C(H)$_2$(C)$_2$ (6 × −4.93 = −29.6 kcal/mol)], consistent with the belief that this cyclic compound is virtually strain-free. In contrast, the experimental $\Delta H_f^\circ$(g) value of cyclopropane (12.7 kcal/mol) is 27.5 kcal/mol higher than the estimated value (−14.8 kcal/mol), and the discrepancy is attributed to strain energy due to both ring strain and steric repulsion from the six eclipsed hydrogens. This strain energy is termed a *ring correction factor* (Benson, 1976).

The Benson approach assumes no non-next-nearest neighbor effects in the group increments, and this assumption restricts them to a large but manageable number. Thus, the C(H)$_2$(C)$_2$ increment used is the same for n-pentane, 1-chloropropane, 1,3-dichloropropane, and 1,3-propanediol. Corrections due to strain (see above) and steric interactions (e.g., *gauche* or *cis* versus *trans* repulsions) are added to the computed sum of group increments. This approach works extremely well except for the case where unanticipated long-range interactions are present. We illustrate this issue through comparison of the calculated and experimental $\Delta H_f^\circ$(g) values for 1,3-butanediol. The presumed (gas-phase) structure and the comparison between calculated and experimental values are depicted is Scheme 9-1. The rather large (9.6 kcal/mol) disparity between the experimental and calculated $\Delta H_f^\circ$(g) values indicates that the actual isolated molecule is considerably more stable than predicted. The source of this discrepancy is almost certainly the long-range hydrogen bond depicted in the scheme. This example illustrates both the insights derived from the approach and the need for chemical intuition in anticipating such corrections.

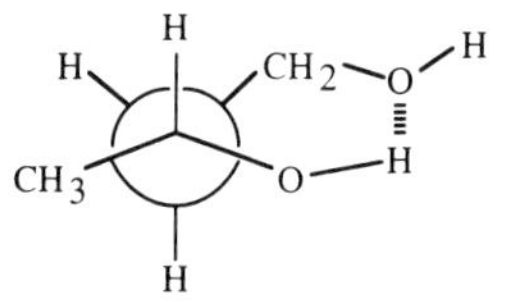

Calculated

$2 \times O(H)(C) = 2(-37.9) = -75.8$
$C(H)(O)(C)_2 = -7.2$
$C(H)_2(C)_2 = -4.93$
$C(H)_2(O)(C) = -8.1$
$C(H)_3(C) = -10.20$
1 *gauche* O---C = +0.5

*Experimental* $\Delta H_f^\circ(g) = -115.3$ kcal/mol
*Calculated* $\Delta H_f(g) = -105.7$
Discrepancy = $-115.3 - (-105.7) = -9.6$ kcal/mol (due to H-bonding)

**Scheme 9-1**

It is important to realize that, in addition to the simplicity of its application, one great strength of the Benson group-increment approach is that, in principle, it relies on a large database having many more experimental points than independent parameters, and application of regression analysis is therefore relatively free of investigator bias. Hydrocarbons best exemplify this situation. However, many other classes of compounds and structural types have much more limited data sets and are more subject to errors due to discrepancies in the data set, as well as to the investigator's choice of data.

Another method which we will employ for estimations is termed *macroincrementation* (Liebman and Van Vechten, 1987). In this approach, entire molecules or "chunks" of molecules are added and subtracted to yield an estimated value. This approach is illustrated by the calculation of $\Delta H_f^\circ(g)$ for the unknown allene oxide from the known value of methylenecyclopropane, using the difference between ethylene oxide and cyclopropane (eq. 1) or methoxyethylene and 1-butene (eq. 2) to estimate the effect of converting $CH_2$ to O (Liebman and Greenberg, 1974) (the data below the molecules are $\Delta H_f^\circ(g)$ values; Pedley et al., 1986), and the underlined value is estimated (Liebman and Greenberg, 1974). The two approaches to replacing $CH_2$ by O are judged to be equally valid, and the $\Delta H_f^\circ(g)$ value is taken as +21.5 kcal/mol, the average of the two calculated values. The advantages of this approach are (1) overt statement of the reference molecules and (2) the opportunity to employ informed intuition. The disadvantages are (1) a potentially arbitrary choice among different approaches and (2) the risk of employing misguided intuition.

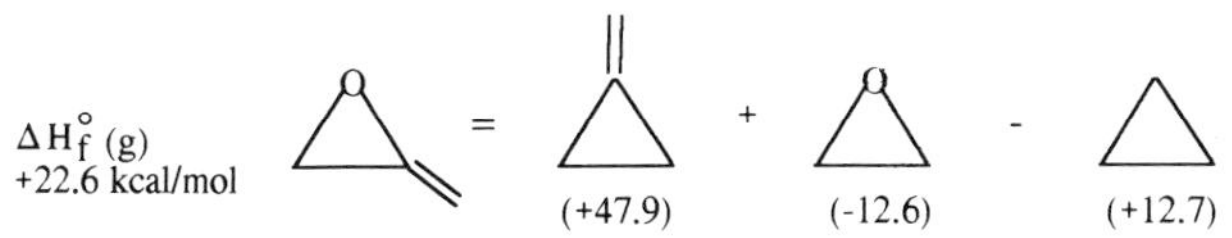

$\Delta H_f^\circ$ (g)
+22.6 kcal/mol

(+47.9)   (-12.6)   (+12.7)    (1)

$\Delta H_f^\circ(g)$
+20.2 kcal/mol

[cyclopropane] = [methylenecyclopropane] + CH₃OCH=CH₂ - CH₃CH₂CH=CH₂ (2)
(+47.9)                (−27.7)                  (0.0)

There is no reason for a determined researcher to be constrained by one approach or the other or, for that matter, by the absence of critical $\Delta H_f^\circ(g)$ or group-increment data. Indeed, in this regard, I have adopted the *Weltanschauung* (world view) of my long-time friend Professor Joel Liebman, who has labeled his approach *ab omnia* ("from everything")— a play on the theoretical calculational approach termed *ab initio* ("from the beginning"). This approach is better illustrated than described by rules. Below, the *ab omnia* approach is applied to the estimation of $\Delta H_f^\circ(g)$ for the molecule 2H-pyran (1) and a selected series of its derivatives (2–5) which will be utilized later in this chapter.

1    2    3    4    5

There are no thermochemical data for 2H-pyran. Thus, a variety of approaches may be compared, and these are summarized here. The first approach is straightforward application of the Benson group-increment scheme according to eq. (3) ("$C_d$" refers to doubly bonded carbon and "Ring Corr." stands for ring correction or strain). It is important to note that while there are group increments for alkoxy olefins, none exists for a 1-alkoxy-1,3-diene, nor is it clear what resonance interactions may be present in the cyclic structure.

*First approach*
$\Delta H_f^\circ(g)(1) = O(C)(C_d) + C_d(O)(H) + 2[C_d(C_d)(H)] + C_d(H)(C)$
−1.4 kcal/mol = (−30.5) + (8.6) + 2(6.8) + (8.6)  (3)
     + C(H)₂(O)(C_d) + Ring Corr.(1,3-cyclohexadiene)
        (−6.5)    +    (4.8)

A second approach employs macroincrementation, as described in eq. (4). Alkoxyalkene stabilization ("Stab." in eq. 4, −3.4 kcal/mol) is equated to the difference between the enthalpies (Pedley et al., 1986) of hydrogenation of ethyl vinyl ether (−26.6 kcal/mol) and 1-pentene (−30.0 kcal/mol).

## Second approach

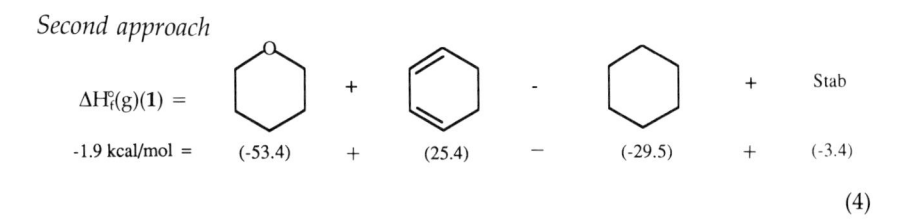

$\Delta H_f^{\circ}(g)(1)$ = [tetrahydropyran] + [benzene] − [cyclohexene] + Stab

−1.9 kcal/mol = (−53.4) + (25.4) − (−29.5) + (−3.4)   (4)

A third approach involves the macroincrementation scheme shown in eq. (5). Here the alkoxyalkene stabilization is explicit in one of the model compounds.

## Third approach

$\Delta H_f^{\circ}(g)(1)$ = [3,4-dihydro-2H-pyran] + [benzene] − [cyclohexene]   (5)

−3.3 kcal/mol = (−29.9) + (25.4) − (−1.2)

A fourth approach is to employ a macroincrementation estimate for $\Delta H_f^{\circ}(g)$ of 4H-pyran [−2 kcal/mol (Lias et al., 1988)], along with the calculated (6–31 G*//3–21G) total energy difference (Butt et al., 1990) between this molecule and its isomer 2H-pyran (**1**). [One must note at this point that the total energies obtained by *ab initio* molecular orbital calculations refer to hypothetical vibrationless, gas-phase molecules at 0 K. A frequency calculation allows thermal correction to 298 K. Although the published total energy difference between 2H- and 4H-pyran did not make this thermal correction, the similarity in structures argues that they are likely to be very small (≪1 kcal/mol).] This estimate is shown in eq. (6), where $\Delta E_T$ refers to the calculated *ab initio* total energy difference. Although enthalpies and energies are not strictly comparable, the differences in these parameters for isomeric substances, or for chemical equations in which *all* species are assumed to be in the gas phase, should be negligible. It is noteworthy that the Benson group-increment scheme yields $\Delta H_f^{\circ}(g)$ (4H-pyran) = −2.4 kcal/mol, in excellent agreement with the macroincrementation value of −2 kcal/mol cited above.

## Fourth approach

$\Delta H_f^{\circ}(g)(1)$ = [4H-pyran] + $\Delta E_T$ [[2H-pyran] − [4H-pyran]]   (6)

+3.3 kcal/mol = (−2) + (5.3)

A fifth approach employs the calculated *ab initio* total energy difference (6–31G*//3–21G) favoring 2H-pyran over E-2,4-pentadienal (the total energy difference = 5.2 kcal/mol; this value = 7.1 kcal/mol when thermally corrected). When this difference is combined with the Benson group increment $\Delta H_f^\circ(g)$ value for E-2,4-pentadienal (−4.3 kcal/mol), the value is 2.8 kcal/mol (eq. 7, where $\Delta E_T$ is the *ab initio* total energy difference).

*Fifth approach*

$$\Delta H_f^\circ(1) = E-CH_2=CH-CH=CH-CHO + \Delta E_T(1\text{-}E-CH_2=CH-CH=CH-CHO) \quad (7)$$
$$+2.8\,\text{kcal/mol} = \quad\quad (-4.3) \quad\quad\quad\quad\quad\quad (7.1)$$

The simple average of the five estimated values (−1.45, −1.9, −3.3, +3.3, +2.8) yields $\Delta H_f^\circ(g)(1)$ = −0.1 kcal/mol. The estimation for **2** is straightforward (eq. 8), and similarly $\Delta H_f^\circ(3)$ is estimated to be −90.1 kcal/mol.

$$\Delta H_f^\circ(g)(2) = \mathbf{1} + (CH_3)_2CH-CHO - (CH_3)_2CH_2 \quad (8)$$
$$-26.7\,\text{kcal/mol} = (-0.1) + \quad (-51.6) \quad - \quad (-25.0)$$

The issue of estimating $\Delta H_f^\circ(g)$ for **4** and **5** is complicated by the lack of any thermochemical data for 1,1-dialkoxyalkenes. However, comparison of the data in eq. (9) based on published 6–31G*//6–31G* total energies (not thermally corrected) (Gallinella and Cadioli, 1991; Leibovitch et al., 1991) yields a calculated energy difference of −2.2 kcal/mol, which is assigned to extra stabilization ($E_{stab}$) in the 1,1-isomer relative to two separated 1-alkoxyalkenes. If one equates these energy differences to enthalpy differences and uses published $\Delta H_f^\circ$ values for methylvinyl ether and ethylene, then the $\Delta H_f^\circ(g)$ can be derived for 1,1-dimethoxyethylene (eq. 10). Furthermore, the increment

$$E_T(CH_2=C(OCH_3)_2 = 2E_T(CH_2=CHOCH_3) - E_T(CH_2=CH_2) + E_{Stab};$$
$$E_{Stab} = -2.2 \quad (9)$$

$$\Delta H_f^\circ(CH_2=C(OCH_3)_2) = 2CH_2=CHOCH_3 - CH_2=CH_2 + E_{stab}$$
$$-66.1\,\text{kcal/mol} \quad = \quad 2(-25.7) \quad - \quad (12.5) \quad + (-2.2) \quad (10)$$

of −40.4 kcal/mol [which "transforms" methoxyethene ($\Delta H_f^\circ$ = −25.7 kcal/mol) to 1,1-dimethoxyethene (−66.1 kcal/mol)] is used to transform **1** to **5** (−40.5 kcal/mol). In order to estimate the value for **4**, one should note the $\Delta H_f^\circ$(g) value (Lias et al., 1988) for transforming CH$_2$=CH—OCH$_3$ to CH$_2$=CH—OH (−6 kcal/mol) and CH$_2$=C(CH$_3$)OCH$_3$ to CH$_2$=C(CH$_3$)OH (−7 kcal/mol). Thus, application of the average (−6.5 kcal/mol) to **5** yields a value of $\Delta H_f^\circ$(g)(**4**) = −47.0 kcal/mol. Scheme 9-2 shows preferred $\Delta H_f^\circ$(g) values for **1–5**; these data are also listed in Table 9-1. Values for **6–8** are readily obtained through inspection of the values for **1–5**, ignoring unanticipated steric effects. For example, the $\Delta H_f^\circ$(g) value for **6** is derived from **4** plus **2** minus **1**. The $\Delta H_f^\circ$(g) value for **9** is obtained by adding +5.0 kcal/mol to the value for **3** [$\Delta H_f^\circ$(CH$_3$CO$_2$CH$_3$) vs. $\Delta H_f^\circ$(CH$_3$CO$_2$H)] (Pedley et al., 1986).

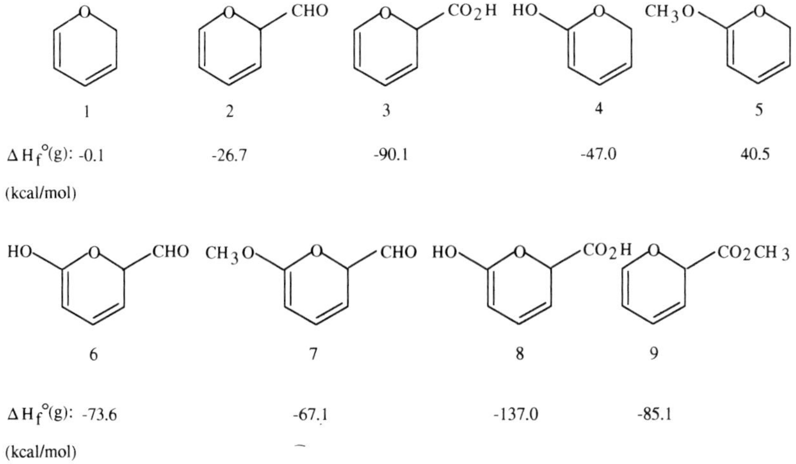

|  | 1 | 2 | 3 | 4 | 5 |
|---|---|---|---|---|---|
| $\Delta H_f^\circ$(g) (kcal/mol) | -0.1 | -26.7 | -90.1 | -47.0 | 40.5 |

|  | 6 | 7 | 8 | 9 |
|---|---|---|---|---|
| $\Delta H_f^\circ$(g) (kcal/mol) | -73.6 | -67.1 | -137.0 | -85.1 |

Scheme 9-2

Entropy effects may also play a role in these considerations, particularly where ring closures are involved. There are also group increments for total internal entropies ($S_{int}$ at 298 K) (Benson, 1976). These must be corrected by entropy changes due to symmetry ($S$ = −RlnN, where N is the molecule's symmetry number (e.g., N = 1 for a molecule with no symmetry or only a plane of symmetry, and N = 3 for a molecule with a threefold axis)), as well as other external factors. The effect of ring closure can be demonstrated by comparing the $S_{int}^\circ$ for n-hexane (98.5 cal/mol K or eu) with that of cyclohexane (74.9 eu) (Benson, 1976). In closure, five C—C bonds, previously freely rotating, have lost this degree of freedom. (The new C—C bond in cyclohexane did not

**TABLE 9-1** Estimated Gas Phase Standard Enthalpies of Formation [$\Delta H_f^\circ(g)$] in kcal/mol (1 kcal/mol = 4.184 kJ/mol)

| Compound | $\Delta H_f^\circ(g)$ |
| --- | --- |
| 2H-pyran (1) | $-0.1^a$ |
| 2H-pyran-2-carboxaldehyde (2) | $-26.7^a$ |
| 2H-pyran-2-carboxylic acid (3) | $-90.1^a$ |
| 6-Hydroxy-2H-pyran (4) | $-47.0^a$ |
| 6-Methoxy-2H-pyran (5) | $-40.5^a$ |
| 6-Hydroxy-2H-pyran-2-carboxaldehyde (6) | $-73.6^a$ |
| 6-Methoxy-2H-pyran-2-carboxaldehyde (7) | $-67.1^a$ |
| 6-Hydroxy-2H-pyran-2-carboxylic acid (8) | $-137.0^a$ |
| Methyl-2H-pyran-2-carboxylate (9) | $-85.1^a$ |
| Z,Z-2,4-Hexadienedioic acid (Z,Z-Muconic acid) (16a)* | $-154.6^b$ |
| Z,E-Muconic acid (16b) | $-155.6^b$ |
| E,E-Muconic acid (16c) | $-156.6^b$ |
| Z,Z-2,4-Hexadienedial (Z,Z-Muconaldehyde) (25a)* | $-32.6^b$ |
| Z,E-Muconaldehyde (25b) | $-33.6^b$ |
| E,E-Muconaldehyde (25c) | $-34.6^b$ |
| 2,8-Dioxabicyclo [5.1.0]octa-3,5-diene (26) | $-12.2^c$ |
| 7,8-Dioxabicyclo[4.2.0]octa-2,4-dine (32) | $+49.3^d$ |
| 7,8-Dioxabicyclo[2.2.2]octa-2,5-diene (36) | $+34.2^e$ |
| trans-Benzene-1,2:4,5-dioxide (39) | $+8.0^f$ |
| 4,8-Dioxabicyclo[5.1.0]octa-2,5-diene (40) | $+3.8^g$ |
| trans-3,4-Dimethyl-1,2-dioxetane | $-4.5^h$ |
| Bicyclo[5.1.0]octa-2,4-diene (42] | $+46.7^i$ |
| Z,Z-Octa-1,3,5,7-tetraene (43)* | $+55.1^j$ |

*The various Z,Z-dienes discussed in this chapter have extra steric repulsions which are relieved by significant twisting about the central C—C bond. In particular, the e,Z,z,Z,z-isomers initially formed from the various bicyclo[5.1.0]octane frameworks are about 9–10 kcal/mol higher in energy than the most stable stereoisomers (e,E,e,E,e) and, therefore, about 7–8 kcal/mol less stable than calculated here for Z,Z-isomers. Any meaningful ramifications of this feature are discussed in text.

$^a$ See this paper.

$^b$ Use of Benson group increments (Benson, 1976).

$^c$ −12.2 kcal/mol = bicyclo[5.1.0]octane (−4.0) + oxirane (−12.6) − cyclopropane (12.7) + hexahydropyran (−53.4) − cyclohexane (−29.5) + 1,3-cycloheptadiene (22.5) − cycloheptane (−28.2) + 1,3-dioxane (−81.8) − 1,4-dioxane (−75.5) + VINYLETHER STAB (−3.4). *Note*: VINYLETHER STAB calculated by comparing the heats of hydrogenation of ethylvinyl ether and 1-pentene. *Second estimate*: −19.3 kcal/mol based on the value for 8-oxabicyclo[5.1.0]octane (−36.4) (Wang and Kachurina, 1988), thus eliminating oxirane and cyclopropane from the above analysis. Although it is known that [(CH$_3$)$_2$CH]$_2$O is about 3 kcal/mol more stable than (CH$_3$CH$_2$CH$_2$)$_2$O, the second value is almost 7 kcal/mol and may be too low. Thus, we use the earlier number since the values are well established. See text for comparison with MP2/6–31G* data.

exist in n-hexane.) Thus, the entropy lost per bond is about (98.5 − 74.9)/5 or 4.7 eu. In this chapter, we frequently refer to ring closures typified by that depicted in Scheme 9-3. Since three of the acyclic connections are double bonds that lack rotational freedom, only two rotational degrees of freedom are lost on ring closure corresponding to an entropy decrease of 2 × 4.7 = 9.4 eu. Since $T\Delta S = -\Delta H$, this amounts to an *increase* in free energy of ~3 kcal/mol at 298 K accompanying this ring closure.

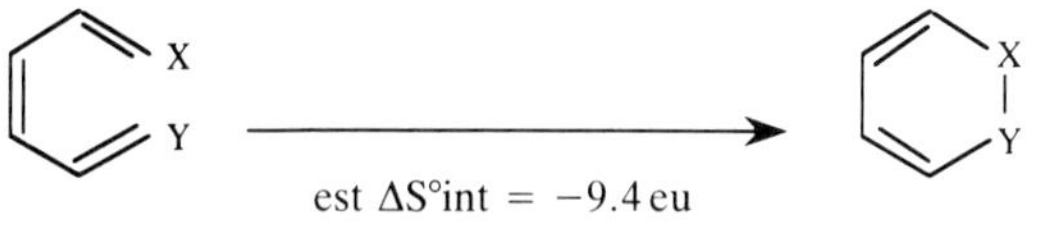

est $\Delta S°$int = −9.4 eu

Scheme 9-3

## Thermochemistry of 1,4-benzyne and its biochemical consequences

The coupling of the metabolic decomposition of organic nutrients to $CO_2$ and $H_2O$ with energy storage in the form of adenosine triphosphate (ATP) is fundamental to biological processes. Knowledge of the energy content of such nutrients and of associated metabolic intermediates

---

**TABLE 9-1** *Notes continued*

[d] 49.3 kcal/mol = bicyclo[4.2.0]octane (−6.3) + 1,3-cyclohexadiene (25.4) − cyclohexane (−29.5) + diethyl peroxide (−46.1) − n-hexane (−39.9) + *cis*-barrier in $H_2O_2$ [6.9 kcal/mol (Liebman and Van Vechten, 1987)].

[e] +34.2 kcal/mol = 2[bicyclo[2.2.2]oct-2-ene(4.9)] − bicyclo[2.2.2]octane (−23.7) + diethyl peroxide (−46.1) − n-hexane (−39.9) + *syn*-barrier (+6.9 kcal/mol).

[f] +8.0 kcal/mol = 2[bicyclo[4.1.0]heptane (0.4)] + cyclohexene (−1.2) −2[cyclohexane (−29.5)] + 2[oxirane(−12.6) − cyclopropane(12.7)].

[g] +3.8 kcal/mol = bicyclo[5.1.0]octane (−4.0) + oxirane (−12.6) − cyclopropane (12.7) + hexahydropyran (−53.4) − cyclohexane (−29.5) + divinyl ether (−3.3) − diethyl ether (−60.3).

[h] −4.6 kcal/mol = *trans*-dimethylcyclobutane (−5.3) + diethyl peroxide (−46.1) − n-hexane (−39.9) + *syn*-barrier (+6.9).

[i] +46.7 kcal/mol = bicyclo[5.1.0]octane (−4.0) + 1,3-cycloheptadiene (22.5) − cycloheptane (−28.2).

[j] +55.1 kcal/mol = Z,Z-2,4-hexadiene (+12.5) + 2[1,3-butadiene (+26.3)] − 2[propene (4.8)]. The $\Delta H_f°(g)$ value for Z,Z-2,4-hexadiene is obtained from Fang, W. and Rogers, D. W. (1992) Enthalpy of Hydrogenation of the Hexadienes and *cis*- and *trans*-1,3,5-Hexatriene. *J. Org. Chem.*, 57, 2294.

permits estimates to be made of the energy-storing capacities of individual reaction steps. In applying these methods, it is important to remember that many biological interconversions are catalyzed by enzymes whose influences on the energetics of individual reaction intermediates are neglected in this approach. In addition, group-increment schemes do not correct for the effects of solvation, which are so important in biological systems. Nevertheless, some biochemical mechanisms lend themselves to simple studies based on molecular energetics, at least as starting points. The mechanistic issue illustrated in this section relates to the hypothesized mode of action of a series of modern anticancer agents.

An elegant application of the thermochemical estimation approach by Jones and Bergman (1972) over 20 years ago made a strong case for the intermediacy of 1,4-benzyne (or 1,4-benzenediyl, **10**) in a molecular rearrangement. The scrambling of deuterium in labeled Z-1,5-hexadiyne-3-ene (**11**), shown in Scheme 9-4, was found to be complete in 5 minutes at 200 °C. In contrast, the E-isomer did not scramble deuterium. Other experiments by Jones and Bergman (1972) indicated the intermediacy of a species having free-radical character. Employing Benson group increments (Benson, 1976), Jones and Bergman calculated the $\Delta H_f^\circ(g)$ value for 1,5-hexadiyne-3-ene (**11**) and, employing the C—H dissociation energy of benzene along with appropriate heats of formation, also calculated $\Delta H_f^\circ(g)$ for 1,4-benzenediyl (**10**). Comparison of these values with the experimental activation energy indicated that **10** sits in an energy well of *at least* 18 kcal/mol (Jones and Bergman, 1972) and is therefore a metastable intermediate in this reaction. It is important to note that the estimate for **10** was a worst-case scenario involving a (presumably singlet) 1,4-benzenediyl having no stabilizing interaction between the radical centers at positions 1 and 4. If such stabilization does in fact exist, then the case for the discrete intermediacy of **10** is even stronger. Only recently has an experimental study yielded a value for $\Delta H_f^\circ(g)$ of 1,4-benzyne (Wenthold et al., 1991). The experimental number, 128 ± 3 kcal/mol, is, as expected, even lower than the estimated value (140 kcal/mol).

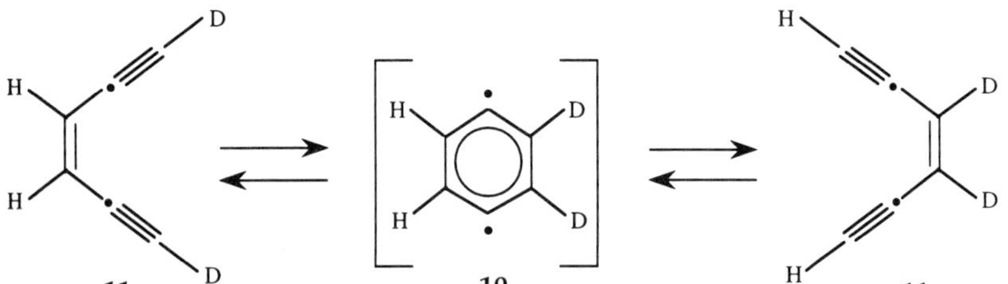

**Scheme 9-4**

This study wonderfully exemplifies the utility of enthalpy estimation, as well as the indeterminate nature of pure chemical research. Although **10** could have been perceived by mission-oriented funding agencies as an arcane bit of chemical exotica, it has strong current relevance in the study of highly potent antitumor agents such as calicheamicin (**12**) (Lee et al., 1987a, 1987b; Halcomb et al., 1992; Nicolaou et al., 1992a, 1992b) which are thought to form reactive 1,4-benzenediyls capable of attacking DNA in cancer cells. This case study illustrates some of the strengths and weaknesses of such estimation approaches. Thus, 1,5-hexadiyne-3-ene is probably well estimated since parameters for related species are quite accurate (Benson, 1976). The nature of the intermediate is somewhat indeterminate; therefore, assumptions grounded in experiment were made about its structure. The uncertainties in such a case are increased relative to conventional estimates, but in the present case they did not weaken the conclusion because the value estimated for 1,4-benzenediyl could reasonably be regarded as a worst-case scenario. Any additional stabilization would strengthen the argument for stability of **10**. Finally, no attempt was made to calculate the energy of the transition state leading to **10** solely by group increments. However, combination of the estimated $\Delta H_f^°(g)$ value for **11** with the experimental $E_a$ for its rearrangement provided an estimate for this value which led to the conclusion that **10** sits in an energy well at least 18 kcal/mol deep.

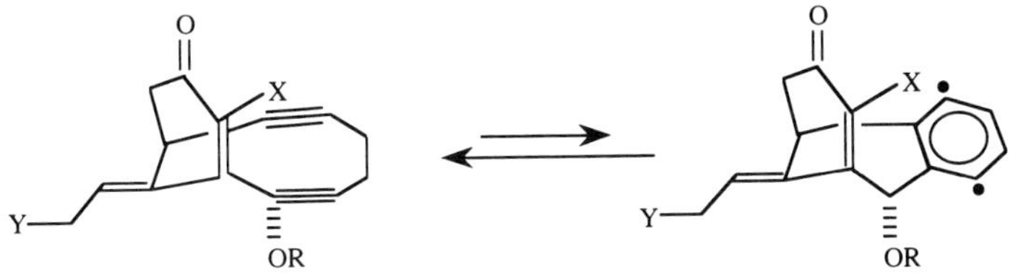

**Scheme 9-5**

### Ring-Opening Metabolism of Benzene

Benzene metabolism is a fascinating topic with relevance to human health, environmental contamination, and biotechnology. This high-octane component of gasoline has long been known to be a human carcinogen (Williams, 1959; NIEHS, 1989; Goldstein and Witz, 1991; Snyder et al., 1993). The nature of its metabolic fate and the associated cytotoxicity of its metabolites are active areas of investigation. Once benzene has contaminated the environment, it may decompose chemically as well as biologically. It is known that microorganisms adapt to

412  *Exploration of Selected Pathways*

contaminated waters and develop the ability to metabolize contaminants. Thus, bacterial "fishing expeditions" have taken investigators to lagoons contaminated with benzene in search of microorganisms capable of degrading it. In this fashion, strains of *Pseudomonas* and *Arthrobacter* bacteria capable of metabolizing benzene and various derivatives, including chlorinated species, have been cultured and isolated. Although the major eukaryotic metabolites of benzene are phenol (23) and catechol (15), some percentage is ring opened to generate E,E-muconic acid (16c). In contrast, prokaryotes metabolize benzene (and its derivatives) to Z,Z-muconic acid (16a) (see Scheme 9-6) (Gibson, 1968; Gibson and Champan, 1971; Gibson et al., 1982; Cerniglia, 1984;

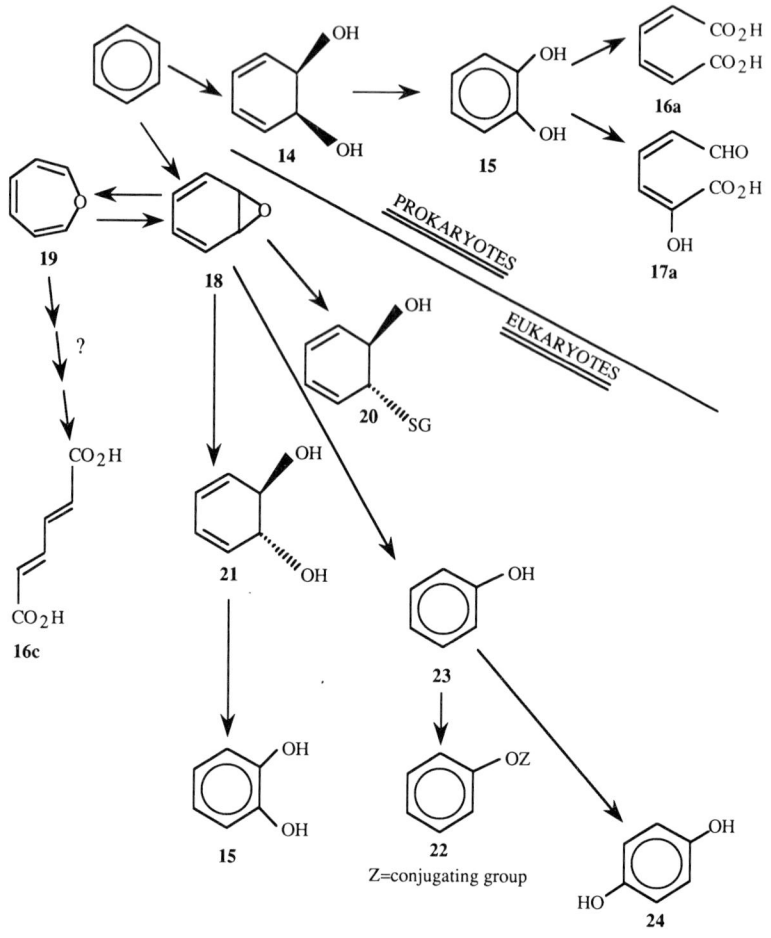

**Scheme 9-6**

Gibson and Subramanian, 1984; Dagley, 1986; Young et al., 1987; Harayama and Timmis, 1989; Haggblom, 1992), a chemical of significant potential industrial importance as a precursor of adipic acid. There are, of course, major differences in the metabolic pathways of eukaryotic and prokaryotic species. The former employ monooxygenases in the metabolism of aromatics, while the latter primarily employ dioxygenases, although monooxygenases are also known for bacteria (Whited and Gibson, 1991). There is, of course, an overwhelming strategic difference between prokaryotic and eukaryotic metabolism. Prokaryotes such as bacteria use hydrocarbons for energy and as a source of carbon. Eukaryotes, in sharp contrast, survive by eliminating these xenobiotics through their oxidation and ultimately conjugation with water-solubilizing species (e.g., sulfate, glucuronic acid, or glutathione).

## INVESTIGATION OF PATHWAYS TO MUCONALDEHYDE

Elucidation of human metabolic pathways for benzene and their relationships to toxicity remains a fundamental area of research activity. In eukaryotic species, most benzene is metabolized to phenol and catechol, with the former considered to be responsible for most of the cytotoxicity (Williams, 1959; NIEHS, 1989; Goldstein and Witz, 1991; Snyder et al., 1993). Although muconic acid has long been recognized as a metabolite of benzene, identification of muconaldehyde [presumably the Z,Z-isomer (**25a**) initially], by Goldstein, Witz, and coworkers, following incubation of mouse liver microsomes with benzene and spectroscopic analysis of a derivative (Goldstein et al., 1982; Latriano et al., 1986a), was important since muconaldehyde is a logical precursor of the diacid (Nakajima et al., 1959; Tomida and Nakajima, 1960; Goldstein et al., 1982; Latriano et al., 1986a, 1986b), exhibits bone marrow toxicity (myelotoxicity) similar to that of benzene, and is both cytotoxic and genotoxic to mammalian cells (Goldstein et al., 1982; Latriano et al., 1986a, 1986b; Kirley et al., 1989; Goon et al., 1992).

It is known that $\alpha,\beta$-unsaturated aldehydes react rapidly with sulfhydryl groups and nucleic acids and are associated with many types of toxicity (Schauenstein et al., 1977). $E,E$-Muconaldehyde itself reacts extremely rapidly with glutathione, presumably via a Michael reaction, in what appears to be a detoxification pathway (Kline et al., 1993). However, the reversibly formed initial adduct with glutathione may be stable enough to escape the liver and produce toxic effects at other organs and tissues (Bleasedale et al., 1993). The Z,Z-isomer was found to react with primary amines to generate pyrroles. Of particular interest is the observation of adducts with nucleosides suggestive of promutagenic activity (Bleasedale et al., 1993).

E,E-Muconaldehyde is known to be metabolized to the aldehyde-acid E,E-6-oxo-2,4-hexadienoic acid (OHC—CH=CH—CH=CH—COOH) in the presence of aldehyde dehydrogenase and NAD$^+$ (Kirley et al., 1989; Goon et al., 1992) and to the aldehyde-alcohol (OHC—CH=CH—CH=CH—CH$_2$OH) in the presence of NADH-fortified alcohol dehydrogenase (Goon et al., 1992). The interaction of oxidative and reductive routes produces the corresponding hydroxy-acid metabolite (HOOC—CH=CH—CH=CH—CH$_2$OH) (Goon et al., 1992). In contrast to the very short lifetime of muconaldehyde *in vivo*, the aldehyde-acid and the aldehyde-alcohol are much less reactive with glutathione and are thought to be stable enough to clear the liver (Kline et al., 1993). A series of mechanisms has been postulated for the formation of E,E-muconaldehyde (**25a**) (Goldstein et al., 1982; Latriano et al., 1986a), and some of these will be explored in this chapter.

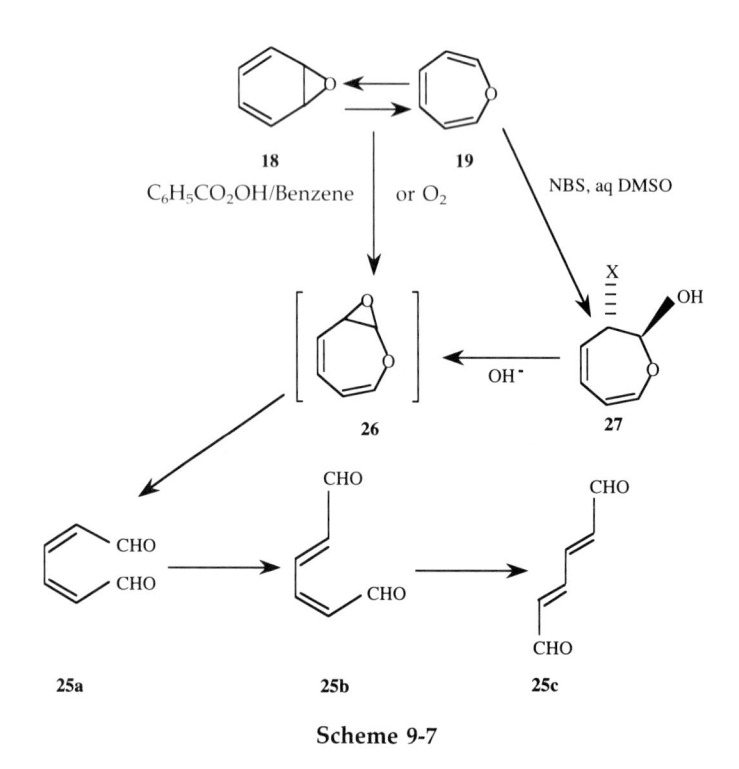

**Scheme 9-7**

## Model Reactions Which Imply the Existence of 2,3-Epoxy-oxepin

Davies and Whitham (1977) postulated the intermediacy of 2,3-epoxy-oxepin (**26**, 2,8-dioxabicyclo[5.1.0]octa-3,5-diene) formed by reaction of benzene oxide/oxepin (**18/19**) (1) with peroxybenzoic acid in benzene, (2) with N-bromosuccinimide in aqueous DMSO, and (3) by a slower reaction in the presence of air and absence of light (Scheme 9-7). They postulated that it is oxepin (**19**), rather than the valence tautomer benzene oxide (**18**), that yields muconaldehyde. The basis for this conclusion is the observation that the analogous reaction of 2,7-dimethyloxepin (**28**) with perbenzoic acid yields 3,5-octadiene-2,7-dione (eq. 11). 2,7-Dimethyloxepin is known to be significantly more stable than the corresponding benzene oxide isomer (Vogel and Gunther, 1967). In contrast, the bridged benzene oxide **30** is much more stable than its oxepin tautomer (which would violate Bredt's rule—prohibition of bridgehead double bonds in small bicyclic molecules) and forms products that leave the six-membered ring intact (eq. 12) (Davies and Whitham, 1977).

$$28 \xrightarrow{C_6H_5CO_3OH} 29 \qquad (11)$$

$$30 \xrightarrow{C_6H_5CO_3OH} 31 \qquad (12)$$

A more recent study by Mello et al. (1990) provides additional compelling *chemical* evidence for the pathway described in Scheme 9-7. These researchers noted that while benzene is not oxidized by dimethyldioxirane, it is oxidized by methyl (trifluoromethyl)-dioxirane, a stronger epoxidizing reagent. The resulting compound is muconaldehyde (eq. 13). As epoxidizing agents, the dioxirane family of reagents are known to deliver one oxygen at a time. It is also worthwhile to note that higher polycyclic aromatic hydrocarbons (PAH) do not give ring opening with this reagent but yield stable oxides or hydroxylation (Mello et al., 1990). We return to this issue later.

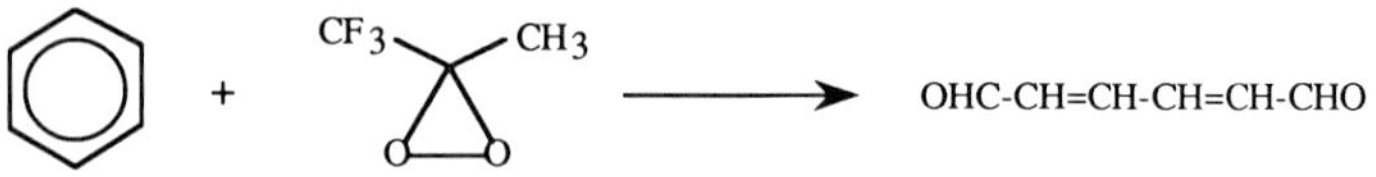

**416** *Exploration of Selected Pathways*

$$\text{C}_6\text{H}_6 + \underset{\underset{\text{O}-\text{O}}{\triangle}}{\overset{\text{CF}_3\diagdown\hspace{-3pt}\diagup\text{CH}_3}{}} \longrightarrow \text{OHC-CH=CH-CH=CH-CHO} \quad (13)$$

Combination of the earlier-postulated pathways with those derived from the work of Davies and Whitham (1977), as well as related observations on muconaldehyde Z-E isomerizations (Golding et al., 1988), yields the enhanced pathway chart depicted in Scheme 9-8, which includes the possibility of enzymatic ring opening of the dihydrodiol to muconaldehyde (Yang, 1993). The postulated pathways in Scheme 9-8 invoke at least three different active oxygen species (hydroxyl radical, monooxygenase, and singlet oxygen). The prevalent pathway should

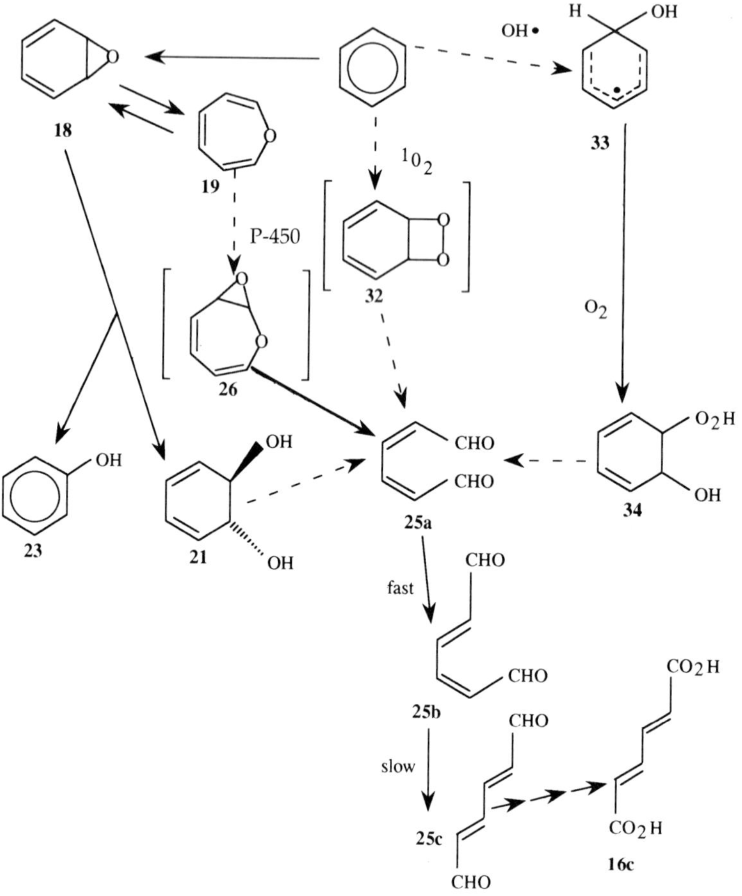

**Scheme 9-8**

depend, in part, on where (membrane, cytosol, etc.) the biological reactions take place and what oxygen species are actually present at the locus of reaction.

The principles of energetic estimations described above may be used to assess the viability of the 2,3-epoxy-oxepin (**26**) pathway depicted in Scheme 9-8. In the process, a number of interesting insights emerge which suggest further research directions. The fact that eukaryotes produce *E,E*-muconic acid (**16c**), while prokaryotes produce *Z,Z*-muconic acid (**16a**), is intriguing. The isomerization to *E,E*-muconic acid is likely to start with *Z* to *E* isomerization of *Z,Z*-muconaldehyde which has been explored in part using calculation methods (Bock et al., 1994). We will start, however, by briefly examining the viability of another postulated intermediate, the dioxetane **32**, in order to show the use of this approach in narrowing the field of mechanistic choices.

## Elimination of the 1,2-Dioxetane Intermediate 32

The suggestion (Goldstein et al., 1982; Latriano et al., 1986a) that singlet oxygen may react with benzene to form dioxetane **32** lacks precedent. While a published claim for the formation of muconaldehyde from photolysis of oxygen-saturated liquid benzene suggested the possible generation of **32** (Wei et al., 1967), no follow-up studies have appeared. Indeed, no direct evidence for **32** or for the existence of 1,2-dioxetanes derived from substituted benzenes has survived scrutiny. For example, while a 1,2-dioxetane structure had been postulated for reaction of singlet oxygen with polystyrene, it is not supported by the available chemical evidence (Kaplan and Trozzolo, 1979). In general, only electron-rich benzenes and PAH react with singlet oxygen, and these yield endoperoxides rather than 1,2-dioxetanes (Foote and Clennan, 1995).

In the context of the present discussion, it is worthwhile to make a prediction about the energetics of the hypothetical reaction of benzene with singlet oxygen to yield **32** (eq. 14).

(hypothetical rxn)

benzene + $^1O_2$ ⟶ **32**    $\Delta H_r^\circ = +7.6$ kcal/mol     (14)

(19.7)      (22.0)      32 (49.3)

The standard gas-phase enthalpies of formation employed in this and subsequent reactions are obtained from published compendia (Pedley et al., 1986; Liebman and Greenberg, 1989), and estimated values are collected in Table 9-1 unless otherwise noted.

Combination of the standard gas-phase enthalpy of formation, $\Delta H_f^o(g)$, for singlet oxygen [$^1\Delta_g$, 22.0 kcal/mol (Herzberg, 1950)] with the value for benzene and the estimated value for **32** (Table 9-1; estimated values used in equations such as 14 are underlined) indicates that the reaction is endothermic by 7.6 kcal/mol. Although estimation uncertainties do not rule out the possibility of very slight exothermicity, the strongly negative entropic contribution also makes this reaction highly unlikely.

It is worth noting noting that 1,2-dioxetanes such as **35** (eq. 15) have also been cited as potential intermediates in bacterial metabolism of benzene (Gibson, 1968). However, compelling arguments based on estimates of molecular energetics similar to those employed in this

$$\text{15} \xrightarrow{O_2, \text{ dioxygenase}} \text{35} \quad \text{(hypothetical rxn)} \tag{15}$$

chapter have ruled out such species (Hamilton, 1969, 1973). For comparison, we note that the corresponding hypothetical reaction of E-2-butene to form *trans*-3,4-dimethyl-1,2-dioxetane (eq. 16) is exothermic by 24 kcal/mol [an ene reaction occurs in reality (Foote and Clennan, 1995)]. Although Diels-Alder addition to benzene to form endoperoxide **36** also does not occur, the reaction is worthy of calculation for com-

$$\underset{(-2.7)}{\diagup\!\!\!\diagup} + \underset{(22.0)}{^1O_2} \xrightarrow{\text{(hypothetical rxn)}} \underset{(-4.5)}{\square} \quad \Delta H_r = -24 \text{ kcal/mol} \tag{16}$$

parison since endoperoxide formation does occur for electron-rich substituted benzenes and PAH (Foote and Clennan, 1995). Scheme 9-9 indicates the relative energetics of 1,2- versus 1,4-addition to benzene, as well as the relative energies of the potential products Z,Z-muconaldehyde (**25a**) and hydroquinone (**24**). Not surprisingly, the less strained 1,4-addition product is more stable. Reaction of hexamethylbenzene with singlet oxygen yields the endoperoxide (Van den Heuvel et al., 1980), and similar endoperoxides have been isolated from other aromatics (Foote and Clennan, 1995).

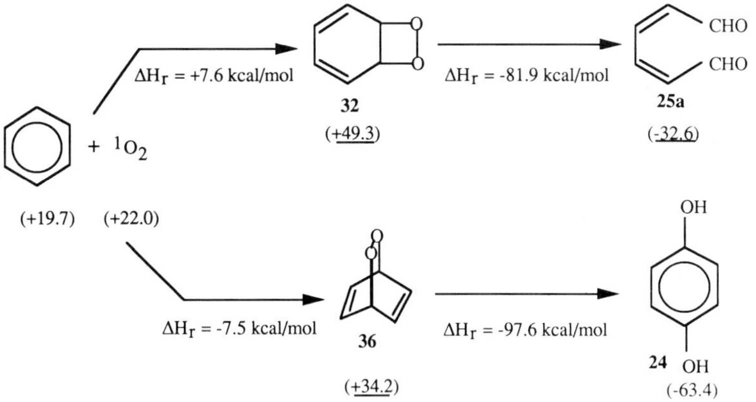

Scheme 9-9

The energetics depicted in Scheme 9-9 predict that ring opening of dioxetane **32** to Z,Z-muconaldehyde is highly exothermic (−81.9 kcal/mol). Furthermore, since typical activation energies for this type of reaction are about 25 kcal/mol (Bartlett and Landis, 1979; Baumstark and Wilson, 1981), the activated complex may be about 100 kcal/mol higher in energy than ground-state muconaldehyde—certainly enough energy to produce the excited singlet and triplet states. Earlier, Hamilton had made a similar point concerning the possibility of generating electronically excited muconic acid from **35** and noted that no such luminescence had ever been reported (Hamilton, 1973).

### 2,3-Epoxy-oxepin (26) Intermediate Derived from Oxepin (19)

As described above, the investigations of Davies and Whitham (1977) and Mello et al. (1990) strongly imply the formation of 2,3-epoxyoxepin (**26**) and its rapid ring opening to Z,Z-muconaldehyde (**25a**) (see Scheme 9-7). An attractive feature of this mechanistic pathway is that both oxidation steps could be catalyzed by the same monooxygenase.

*Energy relationship between benzene oxide (**18**) and oxepin (**19**).* In order to place the energetic arguments in proper perspective, it is worthwhile to examine first the benzene oxide/oxepin equilibrium (**18/19**, Schemes 9-7 and 9-8). Although the less polar oxepin is favored in the nonpolar solvents $CF_3Br$/pentane ($\Delta G° = -1.3$ kcal/mol) and isooctane (Vogel and Gunther, 1967) (and presumably in the gas phase), the enthalpy of benzene oxide (more rigid and lower in entropy) is lower than that of the oxepin by 1.7 kcal/mol (Vogel and Gunther, 1967). Thus, the more positive entropy of the flexible oxepin ring relative to that of the more rigid bicyclic benzene oxide is an important factor. In 85:15 water–methanol, the equilibrium favors the more polar benzene oxide (Vogel and Gunther, 1967). Clearly, the experimental energy differences are small. *Ab initio* molecular orbital calculations favor the oxepin *energetically* at the 6–31G* level, in contrast to the experimental findings (Cremer et al., 1984; Schulman et al., 1984; Bock et al., 1990; George et al., submitted). Only when electron correlation is added (e.g., MP2/6–31G*/6–31G* and MP3/6–31G*/6–31G*) do *ab initio* calculations predict the correct direction and magnitude of the total energy term (Cremer et al., 1984; Schulman et al., 1984; Bock et al., 1990), which one can equate with the experimental difference in enthalpy between the two isomers.

The benzene oxide/oxepin equilibrium is known to present difficulties for thermochemical group-incrementation (or bond-additivity) estimation approaches (Altenbach and Vogel, 1972). Reasonable estimates for the enthalpy of formation of benzene oxide can be obtained using model eqs. (17–19). The enthalpy of formation for norcaradiene (**38**) in eqs. (17) and (18) is obtained by adding the published value for tropylidene (**37**) the hydrocarbon analog of oxepin [43.2 kcal/mol (Pedley et al., 1986)]) to the experimental enthalpy difference (6.2 kcal/mol) (Anet, 1984) between tropylidene (**37**), and norcaradiene (**38**), the hydrocarbon analog of benzene oxide.

(17)

+ 19.1 kcal/mol    (49.5)    (−30.0)    (0.4)

(18)

+ 24.2 kcal/mol    (49.5)    (−12.6)    (12.7)

(19)

+ 24.9 kcal/mol    (−30.0)    (25.4)    (−29.5)

        37                    38

The value for 7-oxabicyclo[4.1.0]heptane (−30.0 kcal/mol) used in eqs. (17) and (19) is from published data (Wang and Kachurina, 1988). A simple average of the data calculated using eqs. (17–19) yields $\Delta H_f^\circ(g)$ = 22.7 ± 2.4 kcal/mol. Application of the Benson group-increment approach yields a value of 23.2 kcal/mol (Greenberg, 1993). If one then employs the experimental 1.7 kcal/mol increment and adds it to the value for benzene oxide, a value of 24.4 ± 2.4 kcal/mol is predicted for oxepin (see eq. 20).

$$\text{18} \rightleftharpoons \text{19} \tag{20}$$

   (22.7 ± 2.4 kcal/mol)      (24.4 ± 2.4 kcal/mol)

It is unfortunate but interesting that bond-additivity estimations do rather poorly for oxepin. The source of the difficulty is the absence of suitable model data. Thus, one might imagine two extreme models. The first (eq. 21) assumes that oxepin has the full resonance energy of divinyl ether despite its tub shape, which is expected to reduce resonance stabilization markedly. The second model (eq. 22) assumes no resonance stabilization at all—another unlikely scenario. In spite of this assumption, even eq. (22) predicts a value some 5 kcal/mol lower

$$\text{oxepin} = \text{cycloheptene-O} + \text{divinyl ether} − \text{1-butene} \tag{21}$$

 +14.7 kcal/mol (calc)    (43.2)       (−3.3)       (25.2)

$$\text{oxepin} = \text{cycloheptene-O} + \text{cycloheptyl ether} − \text{cycloheptane} \tag{22}$$

 +19.3 kcal/mol (calc)    (43.2)       (−53.4)      (−29.5)

than that which we are assuming (24.4 ± 2.4 kcal/mol). What is the origin of this sizable disparity? We posit that the ~7 kcal/mol [average using eqs. (21) and (22) and $\Delta H_f^o(g)$ **(19)** = 24.4 kcal/mol] disparity may reflect some destabilization in oxepin, which, if planar, would have eight pi electrons and would be antiaromatic. Presumably, the nonplanarity of oxepin reduces antiaromaticity relative to the planar transition state for oxepin ring inversion. However, it is worthwhile to note that benzene oxide has also been postulated to have some aromatic stabilization (Rao et al., 1993).

*Structure of 2,3-epoxy-oxepin and relative energetics.* In Scheme 9-10, estimated enthalpies of formation of selected products of monooxygenation of the oxepin/benzene oxide pair (**18/19**) are depicted relative to the enthalpies of **18** and **19**, as well as to the enthalpy of formation of benzene. We note that no differentiation of conformational energies has been made for **26** and **40** and that **39** is assumed to be *trans*. Obviously, the molecules at three different levels of oxygenation ($C_6H_6$, $C_6H_6O$, $C_6H_6O_2$) cannot be compared unless the equation is balanced (e.g., with the half-cell potential of P-450–Fe$^V$O). Nonetheless, the enthalpy differences between isomers at a given oxidation level, as well as the relative changes on oxidation, are of considerable interest.

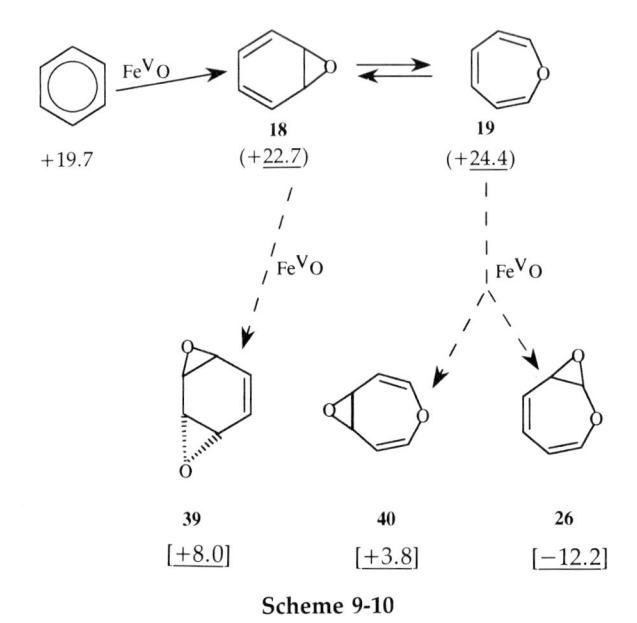

**Scheme 9-10**

Scheme 9-10 implies that the first monooxygenation of benzene is energetically more demanding than subsequent monooxygenation. At the second oxygenation level, generation of 2,3-epoxy-oxepin (**26**) is certainly energetically favored. Moreover, while the oxide-oxepin equilibrium favors benzene oxide in water (Vogel and Gunther, 1967), oxepin might well predominate in the hydrophobic cavity of an enzyme, another factor favoring formation of **26**. A related point is the rapidity of the rearrangement of the initially formed benzene oxide to oxepin [$E_a$ = 9.1 kcal/mol (Vogel and Gunther, 1967)] and its competition with enzyme-catalyzed hydration, rearrangement [the NIH shift (Daly et al., 1972; Jerina and Daly, 1974) spontaneous or otherwise] to phenol, and nucleophilic attack by biological molecules such as glutathione (Daly et al., 1972; Jerina and Daly, 1974). It is also interesting to speculate concerning the role that the microsomal membrane might play in kinetic stabilization of benzene oxide–oxepin in addition to possible enhancement of the oxepin concentration in the equilibrium. Thus, study of the equilibrium and chemical stability of benzene oxide–oxepin in micelles or vesicles is of interest.

Clearly, synthesis of 2,3-epoxy-oxepin (**26**) is a worthwhile goal, although the work of Davies and Whitham (1977) suggests that it is fragile. One may, however, imagine the use of more gentle reaction conditions such as those associated with the Murray reagent or some modification, as per Mello et al. (1990), but at low temperature (see eq. 23). The data in Scheme 10-10 indicate that 2,3-epoxide **26** is more likely to be isolated than the 4,5-epoxide **39**,

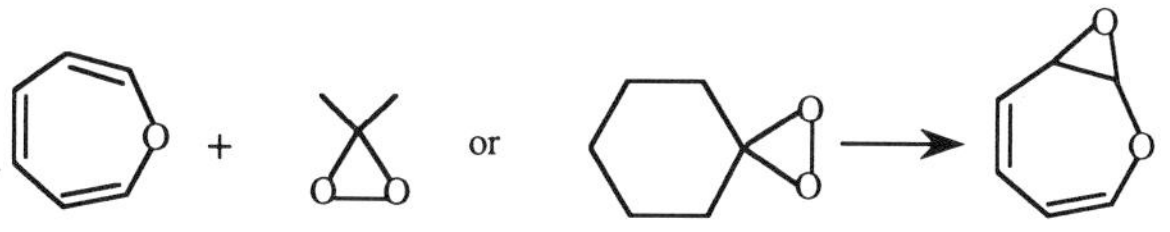

$$\text{(23)}$$

**26**

*Transition state for ring opening of 2,3-epoxy-oxepin.* Bond-additivity calculations are of very limited use in estimating the energetics of transition states unless (1) the structure of the transition state is well described and (2) the features of the transition state lend themselves to parametrization. Recently, some *ab initio* molecular orbital calculations at the MP2/6–31G* level have been performed, with the goal of assessing the stability of 2,3-oxepin oxide, as well as that of the transition state (**41**) for ring opening of 2,3-epoxyoxepin to muconaldehyde. The result (see eq. 24) is an energy of activation of 17.7 kcal/mol starting from the *cisoid* conformer and proceeding via the 1.2 kcal/mol less stable *transoid* conformer (Greenberg et al., 1993). This surprising result may explain why **26** has not been observed under previous experimental conditions. It also implies that **26** might be observable at lower temperatures.

$$\underset{26}{\text{[epoxy-cycloheptadiene]}} \xrightarrow{E_a = 17.7 \text{ kcal/mol}} \left[ \underset{41}{\text{[intermediate]}} \right]^{\ddagger} \longrightarrow \underset{25a}{\text{CHO-diene-CHO}} \quad (24)$$

In striking contrast to the thermal instability of 2,3-epoxy-oxepin, the hydrocarbon analog **42** does not open thermally to Z,Z-1,3,5,7-octatetraene (eq. 25) (Grimme and Doering, 1973; Bock et al., 1995). The bond additivity results suggest that ring opening to Z,Z-1,3,5,7-octatetraene **43** is endothermic by 8.4 kcal/mol. The presumably initially generated $e,Z,z,Z,e$-conformer **43** is calculated at the MP2/6–31G*//6–31G* level (with thermal corrections) to be 19.2 kcal/mol less stable

$$\underset{42\ (\underline{46.7})}{\text{[bicyclic]}} \longrightarrow \underset{43\ (\underline{55.1})}{\text{[octatetraene]}} \quad (25)$$

than **42** [$e$ and $z$ refer to the conformations about $C_2$-$C_3$, $C_4$-$C_5$ and $C_6$-$C_7$ single bonds in **43** and correspond to convention *s-trans* and *s-cis* nomenclature but are more convenient (see Greenberg et al., 1993, and earlier studies referred to in that paper)]. The difference in the ring-opening behavior between 2,3-epoxy-oxepin (**26**) and hydrocarbon **43** is in the generation of the very stable carbonyl groups, which cause rearrangement to muconaldehyde to be markedly exothermic.

## APPLICATION TO POTENTIAL RING OPENING OF NAPHTHALENE AND PAH

Our concluding discussion considers the potential for PAH to open the ring in monooxygenase-mediated systems. Naphthalene and higher PAH are known to be ring opened by bacteria (Gibson and Subramanian, 1984). If 2,3-epoxy-oxepin is a mandatory intermediate en route to ring opening, then its accessibility through the oxepin will determine whether ring opening occurs. An interesting *ab initio* molecular orbital study by Bock et al. (1991) predicted the relative stabilities of epoxides and oxepins derived from naphthalene. This study supported the earlier predictions of National Institutes of Health scientists who rationalized PAH metabolism in terms of the stabilities of derived epoxides (Daly et al., 1972). We have used their conclusions in order to construct Scheme 9-11 (Greenberg et al., 1993, 1994). In this scheme the calculated (6–31G//6–31G) total energies of the isomers (kcal/mol) (Bock et al., 1991) are placed in parentheses under them. Here it is apparent that **46**, the 1,2-epoxide, will be formed in preference to the two less stable epoxides, **44** and **48**, but that its rearrangement to the much more energetic oxepin **47** is unlikely. Thus, ring-opened dialdehydes will not be formed (Greenberg et al., 1993, 1994). This is consistent with the results of the study of Mello et al. (1990), who ring-opened benzene but not PAH with dioxiranes. There is some evidence, albeit indirect, that consecutive monooxygenase metabolic reactions of naphthalene produce diepoxide **50** (Stillwell et al., 1978). It is noteworthy that (gas-phase) reaction of naphthalene with hydroxyl radical produces some ring-opened products derived from naphthalene (Bunce and Zhu, 1994). Thus, careful exploration reveals a lack of ring-opened naphthalene metabolites in eukaryotes, and this result would suggest that the OH· route is relatively unimportant.

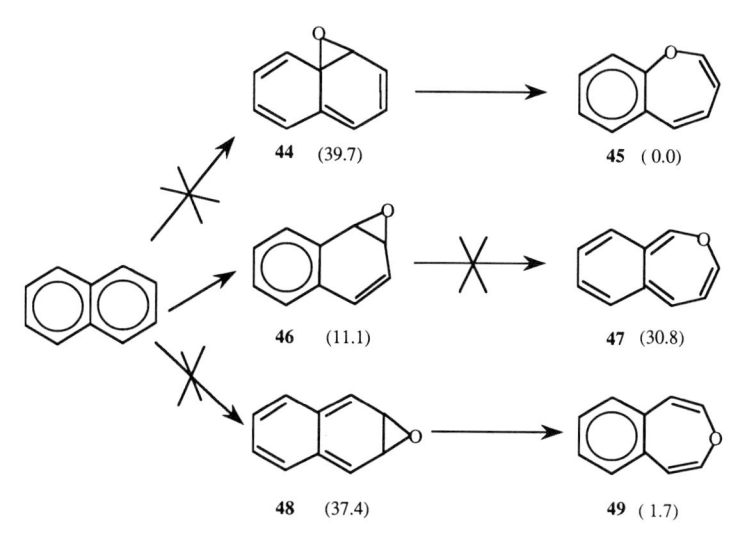

**Scheme 9-11**

50

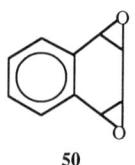

Among the higher PAH, it appears that the PAH epoxides normally derived from metabolism are generally more stable than the corresponding oxepins (Harvey, 1991). Thus, we would not expect these PAH to ring-open via monooxygenases (Greenberg et al., 1994). The PAH oxepin isomers that may be accessible would correspond to those having epoxide isomers which (1) are created in significant quantities and (2) racemize relatively rapidly through ring inversion of the energetically accessible oxepin (Greenberg et al., 1994). Such ring-opened PAH, however improbable, would provide additional pathways for reaction with DNA and proteins. We also note that, while muconaldehyde is effectively divalent and can, in principle, cross-link DNA (Moss, 1993), the majority of the potential dialdehydes derived from PAH would lack this ability since one of the potential Michael addition units will be tied up in the aromatic system.

## Acknowledgments

We are pleased to acknowledge discussions with Drs. Gisela Witz, Stan Kline, Joel F. Liebman, Charles W. Bock, Philip George, Jenny P. Glusker, Robert Snyder, C. S. Yang, and Bernard D. Goldstein.

## REFERENCES

ALTENBACH, H. J., and VOGEL, E. (1972) Valence Tautomerism of 1,4-Dioxicin with syn-Benzene Dioxide. *Angew. Chem. Int. Ed. Engl.*, 11, 937.

ANET, F. (1984) See citation in Schulman et al. (1984).

BARTLETT, P. D., and LANDIS, M. E. (1979) The 1,2-Dioxetanes, in *Singlet Oxygen* (H. H. Wasserman, and R. W. Murray, Eds.), Academic Press, New York, p. 252.

BAUMSTARK, A. L., and WILSON, C. F. (1981) The Thermolysis of 3,4-Dialkyl-1,2-dioxetanes: Effect of Cyclic Substituents. *Tetrahedron Lett.*, 22, 4363.

BENSON, S. W. (1976) *Thermochemical Kinetics*, 2nd ed., John Wiley & Sons, New York.

BLEASDALE, C., Golding, B. T., KENNEDY, G., MACGREGOR, J. O., and WATSON, W. P. (1993) Reactions of Muconaldehyde Isomers with Nucleophiles Including Tri-*O*-acetylguanosine: Formation of 1,2-Disubstituted Pyrroles from Reaction of the (Z,Z)-Isomer with Primary Amines. *Chem. Res. Toxicol.*, 6, 407.

BOCK, C. W., GEORGE, P., STEZOWSKI, J. J., and GLUSKER, J. P. (1990) Theoretical Studies of the Benzene Oxide-Oxepin Valence Tautomerism. *Struct Chem.*, 1, 33.

BOCK, C. W., GEORGE, P., and GLUSKER, J. P. (1991) A Computational Molecular Orbital Study of the Oxide and Oxepin Valence Tautomers of Naphthalene. *J. Molec. Struct. (Theochem.)*, 234, 227.

BOCK, C. W., GEORGE, P., GLUSKER, J. P., and GREENBERG, A. (1994) An *ab initio* Computational Molecular Orbital Study of the Conformers of Muconaldehyde, and the Possible Role of 2-Formyl-2H-Pyran in Bringing about the Conversion of a Z,Z-Muconaldehyde Structure into an E,Z-Muconaldehyde Structure. *Chem. Res. Toxicol.*, 7, 534–543.

BOCK, C. W., GREENBERG, A., GALLAGHER, J. P., GEORGE, P., and GLUSKER, J. P. (1995) Comparison of Structures, Energetics and Selected Reaction Pathways for 2,3-Epoxy-oxepin and its Hydrocarbon and Mono-oxa Analogues. *J. Org. Chem.*, in press.

BOHM, S., and KUTHAN, J. (1990) Quantum Chemical Prediction of 2H-Pyran Vibration Spectrum. *Collect. Czech. Chem. Commun.*, 55, 10.

BUNCE, N., and ZHU, J. (in press) Reaction Products from Photochemical Reactions of Naphthalene in Air. *Polycycl. Arom. Cmpds.*, 5, 123.

BUTT, G., MARRIOTT, S., and TOPSOM, R. D. (1990) Stability of Substituted Pyrilium Salts and the Corresponding 2-H and 4-H Pyrans. *Theochem. J. Molec. Struc.*, 64, 261.

CERNIGLIA, C. E. (1984) Microbial Transformations of Aromatic Hydrocarbons, in *Petroleum Microbiology* (R. M. Atlas, Ed.), Macmillan, New York.

COX, J. D., and PILCHER, G. (1970) *Thermochemistry of Organic and Organometallic Compounds*, Academic Press, London.

CREMER, D., DICK, B., and CHRISTEU, D. (1984) Theoretical Determination of Molecular Structure and Conformation. Part 12. Puckering of 1,3,5-Cycloheptatriene, 1H-Azepine, Oxepine and Their Norcaradiene-like Valence Tautomers. *Theochem. J. Molec. Struc.*, 116, 277.

DAGLEY, S. (1986) Biochemistry of Aromatic Hydrocarbon Degradation in Pseudomonads, in *The Bacteria: A Treatise on Structure and Function*, Vol. X, *The Biology of Pseudomonas* (J. R. Sokatch, Ed.), Academic Press, Orlando, FL, pp. 527–555.

DALY, J. W., JERINA, D. M., and WITKOP, B. (1972) Arene Oxides and the NIH Shift: The Metabolism, Toxicity and Carcinogenicity of Aromatic Compounds. *Experientia*, 28, 1129.

DAVIES, S. G., and WHITHAM, G. H. (1977) Benzene Oxide-oxepin. Oxidation to Muconaldehyde. *J. Chem. Soc. Perkin Trans.*, 1, 1346.

FOOTE, C. S., and CLENNAN, E. L. (1995) Properties and Reactions of Singlet Dioxygen, in *Reactive Oxygen in Chemistry* (C. S. Foote, J. S. Valentine, A. Greenberg, and J. F. Liebman, Eds.), Chapman and Hall, New York, Chap. 5.

GALLINELLA, E., and CADIOLI, B. (1991) Theoretical and Experimental Study of the Non-*s-cis* Form of Unsaturated Ethers. Molecular Structure and Vibrational Assignment of the High-energy Isomer of Methylvinylether. *J. Molec. Struct.*, 249, 343.

GIBSON, D. T. (1968) Microbial Degradation of Aromatic Compounds. *Science*, 161, 1093.

GIBSON, D. T., and CHAPMAN, P. J. (1971) The Microbial Oxidation of Aromatic Hydrocarbons, in *CRC Critical Reviews in Microbiology*, CRC Press, Boca Raton, FL, pp. 199–222.

GIBSON, D. T., and SUBRAMANIAN, V. (1984) Microbial Degradation of Aromatic Hydrocarbons, in *Microbial Degradation of Organic Compounds* (D. T. Gibson, Ed.), Marcel Dekker, New York, pp. 181–252.

GIBSON, D. T., YEH, W. K., LIU, T. N., and SUBRAMANIAN, V. (1982) Toluene Dioxygenase: A Multicomponent Enzyme System from *Pseudomonas putida*, in *Oxygenases and Oxygen Metabolism* (N. Nozaki, S. Yamamoto, Y. Ishimura, M. J. Coon, L. Ernster, and R. W. Estabrook, Eds.), Academic Press, New York, pp. 51–62.

GOLDING, B. T., KENNEDY, G., and WATSON, W. P. (1988) Simple Syntheses of Isomers of Muconaldehyde and 2-Methylmuconaldehyde. *Tetrahedron Lett.*, 29, 5991.

GOLDSTEIN, B. D., and WITZ, G. (1991) Benzene, in *Critical Reviews of Environmental Toxicants—Human Exposure and Their Health Effects* (M. Lippmann, Ed.), Van Nostrand Reinhold, New York, Chap. 3.

GOLDSTEIN, B. D., WITZ, G., JAVID, J., AMORUSO, M. A., ROSSMAN, T., and WOLDER, B. (1982) Muconaldehyde, a Potential Toxic Intermediate of Benzene Metabolism, in *Biological Reactive Intermediates* (R. Snyder, D. V. Parke, J. J. Kocsis, D. J. Jallow, C. G. Gibson, and C. M. Witmer, Eds.), Plenum Press, New York, 2, Part A, pp. 331–339.

GOON, D., CHENG, X., RUTH, J. A., PETERSEN, D. R., and ROSS, D. (1992) Metabolism of *trans,trans*-Muconaldehyde by Aldehyde and Alcohol Dehydrogenase: Identification of a Novel Metabolite. *Toxicol. Appl. Pharmacol.*, 114, 147.

GREENBERG, A. (1993) We have employed Benson group increments (Benson, 1976), with some very minor additions where data were lacking. To calculate $H_f(g)$ for benzene oxide (**18**) (in kcal/mol), the following increments were employed: O-(C)$_2$ ($-23.7$); C$_d$-(C)(H) ($+8.6$); C$_d$-(C$_d$)(H) ($+6.8$); C-O(C)(C$_d$)(H) ($-5.3$). The last value was estimated through comparison of C-(O)(C$_d$)(H)$_2$ ($-6.9$), C-(O)(C)(H)$_2$ ($-8.6$), and C-(O)(C)$_2$(H) ($-6.9$). While it is tempting to simply sum the ring strain corrections for bicyclo-[4.1.0]heptane (28.9) and 1, 3-cyclohexadiene (1.2), there is no *a priori* justification for this assumption. Instead, we compared the experimental $H_f(g)$ for bicyclo[4.1.0]hexa-2,4-diene (norcaradiene) (+49.5 kcal/mol; see text) with the value calculated using Benson group increments, without correction for ring strain: C-(C)$_2$(H)$_2$ (−4.9); C-(C)$_2$(C$_d$)(H) (−1.5); C$_d$-(C)(H) (+35.9); C$_d$-(C$_d$)(H) (+6.8). The sum of the group increments is +22.8 kcal/mol. The effective strain correction (49.5 − 22.7 = 26.8 kcal/mol) is then transferred to benzene oxide, implicitly assuming that any difference in strain between cyclopropane and ethylene oxide is very small relative to the errors in estimation. Thus, $H_f(\mathbf{18}) = -23.7 + 2(-5.3) + 2(8.6) + 2(6.8) + 26.8 = +23.3$ kcal/mol. The assumptions involved in our use of the group increment scheme are not completely independent of eqs. (17) and (18) since these equations use the experimental $H_f(g)$ for norcaradiene used in generating the 26.8 kcal/mol ring strain correction factor. It is interesting to remark on the result that the ring correction factor for norcaradiene (26.8) is smaller than the sum of the correction factors (33.7) and even smaller than the factor for bicyclo[4.1.0]heptane (28.9) alone. Perhaps this 2–7 kcal/mol discrepancy is due to some aromatic stabilization in norcaradiene wherein the cyclopropane ring acts as a "mitigated ethylene." Indeed, Rao et al. (1993) make a case for aromatic stabilization of benzene oxide which would, of course, reduce the apparent antiaromaticity of oxepin.

GREENBERG, A., BOCK, C., GEORGE, P., and GLUSKER, J. P. (1993) Energetics of the Metabolic Production of *E,E*-Muconaldehyde from Benzene Via the Intermediates 2,3-Epoxy-Oxepin, *Z,Z*- and *E,Z*-Muconaldehyde: Ab Initio Molecular Orbital Calculations. *Chem. Res. Toxicol.*, 6, 701.

GREENBERG, A., BOCK, C., GEORGE, P., and GLUSKER, J. P. (1994) Mechanism of Metabolic Ring Opening of Benzene and its Relation to Mammalian PAH Metabolism. *Polycycl. Arom. Cmpds.*, 7, 123.

GRIMME, W., DOERING, W. V. E. (1973) Eine Degenierte Butadienylcyclopropan-Umlagerung in Bicyclo-[5.1.0]octa-2,4-dien. *Chem. Ber.*, 106, 1765.

HAGGBLOM, M. (1992) Microbial Breakdown of Halogenated Aromatic Pesticides and Related Compounds. *FEMS Microbiol. Rev.*, 103, 29.

HALCOMB, R. L., BOYER, S. H., and DANISHEFSKY, S. J. (1992) Synthesis of the Calicheamicin Aryltetrasaccharide Domain Bearing a Reducing Terminus: Coupling of Fully Synthetic Aglycone and Carbohydrate Domains by the Schmidt Reaction. *Angew. Chem. Int. Ed. Engl.*, 31, 338.

HAMILTON, G. A. (1969) Mechanism of Two- and Four-electron Oxidations Catalyzed by Some Metalloenzymes, in *Advances in Enzymology* (F. F. Nord, Ed.), John Wiley and Sons, New York, 32, pp. 55–96.

HAMILTON, G. A. (1973) On the Oxygenated Intermediate in Enzymatic Oxygenation (discussion section), in *Oxidases and Related Redox Systems* (T. E. King, H. S. Mason, and M. Morrison, Eds.), University Park Press, Baltimore, 1, pp. 135–138.

HARAYAMA, S., and TIMMIS, K. N. (1989) Catabolism of Aromatic Hydrocarbons, in *Genetics of Bacteria Diversity* (D. A. Hopwood, and K. F. Chater, Eds.), Academic Press, London, pp. 151–174.

HARVEY, R. G. (1991) *Polycyclic Aromatic Hydrocarbons. Chemistry and Carcinogenicity.* Cambridge University Press, Cambridge, pp. 276–280.

HERZBERG, G. (1950) *Molecular Spectra and Molecular Structure, Spectra of Diatomic Molecules*, 2nd ed., Van Nostrand, New York, I, p. 560.

JERINA, D. M., and DALY, J. W. (1974) Arene Oxides: A New Aspect of Drug Metabolism. *Science*, 185, 573.

JONES, R. R., and BERGMAN, R. G. (1972) *p*-Benzyne Generation as an Intermediate in a Thermal Isomerization Reaction and Trapping Evidence for the 1,4-Benzenediyl Structure. *J. Am. Chem. Soc.*, 94, 660.

KAPLAN, M. L., and TROZZOLO, A. M. (1979) Role of Singlet Oxygen in the Degradation of Polymers, in *Singlet Oxygen* (H. H. Wasserman and R. W. Murray, Eds.), Academic Press, New York, p. 584.

KIRLEY, T. A., GOLDSTEIN, B. D., MANIARA, W. M., and WITZ, G. (1989) Metabolism of *trans,trans*-Muconaldehyde, a Microsomal Hematotoxic Metabolite of Benzene, by Purified Yeast Aldehyde Dehydrogenase and a Mouse Liver Soluble Fraction. *Toxicol. Appl. Pharmacol.*, 100, 360.

KLINE, S. A., XIANG, Q., GOLDSTEIN, B. D., and WITZ, G. (1993) Reaction of (*E,E*)-Muconaldehyde and Its Aldehydic Metabolites (*E,E*)-6-Oxohexadienoic Acid and (*E,E*)-6-Hydroxy-2,4-dienal with Glutathione. *Chem. Res. Toxicol.*, 6, 578.

LATRIANO, L., GOLDSTEIN, B. D., and WITZ, G. (1986a) Formation of Muconaldehyde, an Open-ring Metabolite of Benzene, in Mouse Liver Microsomes: An Additional Pathway for Toxic Metabolites. *Proc. Natl. Acad. Sci. U.S.A.*, 83, 8356.

LATRIANO, L., ZACCARIA, A., GOLDSTEIN, B. D., and WITZ, G. (1986b) Muconaldehyde Formation from Free $^{14}$C-Benzene in a Hydroxyl-radical Generating System. *J. Free Radical Biol. Med.*, 1, 363.

LEE, M. D., DUNNE, T. S., CHANG, C. C., ELLESTAD, G. A., SIEGEL, M. M., MORTON, G. O., MCGAHREN, W. J., and BORDERS, D. B. (1987b) Calichemicins, a Novel Family of Antitumor Antibiotics. 2. Chemistry and Structure of Calichemicin$_1$. *J. Am. Chem. Soc.*, 109, 3466.

LEE, M. D., DUNNE, T. S., SIEGEL, M. M., CHANG, C. C., MORTON, G. O., and BORDERS, D. B. (1987a) Calichemicins, a Novel Family of Antitumor Antibiotics. 1. Chemistry and Partial Structure of Calichemicin$_1$. *J. Am. Chem. Soc.*, 109, 3464.

LEIBOVITCH, M., KRESGE, A. J., PETERSON, M. R., and CSIZMADIA, I. G. (1991) Ab initio Investigation of the Structure and Reactivity of Vinyl Ether. *J. Molec. Struct. (Theochem.)*, 230, 349.

LIAS, S. G., BARTMESS, J. E., LIEBMAN, J. F., HOLMES, J. L., LEVIN, R. D., and MALLARD, W. G. (1988) Gas-Phase Ion and Neutral Thermochemistry. *J. Phys. Chem. Ref. Data*, 17, Supplement 1.

LIEBMAN, J. F., and GREENBERG, A. (1974) Estimation by Bond-additivity Schemes of the Relative Thermodynamic Stabilities of Three-membered Ring Systems and Their Open Dipolar Forms. *J. Org. Chem.*, 39, 123.

LIEBMAN, J. F., and GREENBERG, A. (1989) Survey of the Heats of Formation of Three-membered-ring Species. *Chem. Rev.*, 89, 1225.

LIEBMAN, J. F., and VAN VECHTEN, D. (1987) Universality: The Differences and Equivalences of Heats of Formation, Strain Energy and Resonance Energy, in *Molecular Structure and Energetics*, vol. 2, *Physical Measurements*, (J. F. Liebman, and A. Greenberg, Eds.), VCH, New York, pp. 315–374.

MELLO, R., CIMINALE, F., FIORENTINO, M., FUSCO, C., PRENCIPE, T., and CURCI, R. (1990) Oxidation by Methyl(trifluoromethyl)dioxirane. 4. Oxyfunctionalizations of Aromatic Hydrocarbons. *Tetrahedron Lett.*, 31, 6097.

MOSS, R. A. (1993) The author acknowledges the suggestion of Professor Robert A. Moss concerning the DNA cross-linking potential of muconaldehyde.

NAKAJIMA, M., TOMIDA, I., and TAKEI, S. (1959) The Chemistry of 3,5-Cyclohexadiene-1,2-diol. IV. Preparation of *cis*-3,5-Cyclohexadiene-1,2-diol, the Stereoisomeric Muconic Dialdehydes and Conduritol F. *Chem. Ber.*, 92, 163.

NICOLAOU, K. C., SCHREINER, E. P., IWABUCHI, Y., and SUZUKI, T. (1992b) Total Synthesis of Calichemaicin–Dynemicin Hybrid Molecules. *Angew. Chem. Int. Ed. Engl.* 31, 340.

NICOLAOU, K. G., SORENSEN, E. J., DISCORDIA, R., HWANG, C. K., MINTO, R. E., BHARUCHA, K. N., and BERGMAN, R. G. (1992a) Ten-membered Ring Enediynes with Remarkable Chemical and Biological Profiles. *Angew. Chem., Int. Ed. Engl.*, 31, 1044.

NIEHS (1989) Benzene Metabolism, Toxicity and Carcinogenesis, *Environ. Health Perspect.* 1989, 82.

PEDLEY, J. B., NAYLOR, R. D., and KIRBY, S. P. (1986) *Thermochemical Data of Organic Compounds*, 2nd ed., Chapman and Hall, London.

RAO, S. N., MORE O'FARRELL, R. A., KELLY, S. C., BOYD, D. R., and AGARWAL, R. (1993) Acid-catalyzed Aromatization of Arene Oxides and Arene Hydrates: Are Arene Oxides Homoaromatic? *J. Am. Chem. Soc.*, 115, 5458.

SCHAUENSTEIN, E., ESTERBAUER, H., and ZOLLNER, H. (1977) *Aldehydes in Biological Systems*, Methuen, New York, pp. 25–88.

SCHULMAN, J. M., DISCH, R. L., and SABIO, M. L. (1984) Energetics of Valence Isomerization in Seven-membered Rings. Cycloheptatriene-norcaradiene and Related Rearrangements. *J. Am. Chem. Soc.*, 106, 7696.

SNYDER, R., WITZ, G., and GOLDSTEIN, B. D. (1993) The Toxicology of Benzene. *Environ. Health Perspect.*, 100, 293.

STILLWELL, W. G., BOUWSMA, O. J., THENOT, J. P., HORNING, M. G., GRIFFIN,

G. W., ISHIKAWA, K., and TAKAKU, M. (1978) Methylthio Metabolites of Naphthalene Excreted by the Rat. *Res. Commun. Chem. Pathol. Pharmacol.*, 20, 509.

STULL, D. S., WESTRUM, E. F., JR., and SINKE, G. C. (1969) *The Chemical Thermodynamics of Organic Compounds*, John Wiley and Sons, New York.

TOMIDA, I., and NAKAJIMA, M. (1960) The Chemistry of 3,5-Cyclohexadiene-1,2-diol. VI. Metabolism of the Glycols and Muconic Dialdehyde. *Z. Physiol. Chem.*, 318, 171.

VAN DEN HEUVEL, C. J. M., HOFLAND, A., STEINBERG, H., and DE BOER, T. J. (1980) The Photo-oxidation of Hexamethylbenzene and by Singlet Oxygen. *Recl. Trav. Chim. Pays-Bas.*, 99, 275.

VOGEL, E., and GUNTHER, H. (1967) Benzene Oxide-Oxepin Valence Tautomerism. *Angew. Chem. Int. Ed. Engl.*, 6, 385.

WANG, C. H., and KACHURINA, N. S. (1988) Thermochemical Properties of Oxirane Derivatives. *Russ. J. Phys Chem.* (Engl. Trans.), 61, 622.

WEI, K., MANI, J. C., and PITTS, J. N., JR. (1967) The Formation of Polyenic Dialdehydes in the Photooxidation of Pure Liquid Benzene. *J. Am. Chem. Soc.*, 89, 4225.

WENTHOLD, P. G., PAULINO, J. A., and SQUIRES, R. R. (1991) The Absolute Heats of Formation of *o*-, *m*-, and *p*-Benzyne. *J. Am. Chem. Soc.*, 113, 7414.

WHITED, G. M., and GIBSON, D. T. (1991) Toluene-4-Monooxygenase, a Three-component Enzyme System That Catalyzes the Oxidation of Toluene to *p*-Cresol in *Pseudomonas mendocina* KR1. *J. Bacteriol.*, 173, 3010.

WILLIAMS, R. T. (1959) *Detoxification Mechanisms: The Metabolism of Detoxification of Drugs, Toxic Substances and Other Organic Compounds*, 2nd ed., John Wiley and Sons, New York, pp. 188–194.

YANG, C. S. (1993) The oxidation of the dihydrodiol was suggested as a possibility by Professor C. S. Yang, by analogy to the oxidation oberved for glycerol [Clejan, L. A., and Cederbaum, A. I. (1992) Role of Cytochrome P450 in the Oxidation of Glycerol by Reconstituted Systems and Microsomes. *FASEB J.*, 6, 765].

YOUNG, L. (1987) in *Toxic Chemicals, Health and the Environment* (L. B. Lave, and A. C. Upton, Eds.), Johns Hopkins University Press, Baltimore, p. 222.

# 10
# The Role of Oxidized Lipids in Cardiovascular Disease

JUDITH A. BERLINER AND ANDREW D. WATSON

### INTRODUCTION

Lipid oxidation products play an important role in normal biology and in a number of disease processes (Sevanian and Hochstein, 1985; Marx, 1987). Oxidation products have been shown to be abnormal in many of the major chronic diseases, including rheumatoid arthritis, cancer, Alzheimer's disease, and diabetes. In the last 5 years, the role of oxidized lipids in cardiovascular disease has been the subject of extensive studies (Steinberg et al., 1989; Witztum and Steinberg, 1991). Alterations in the synthetic rates of naturally occurring oxidized lipids and effects of lipid oxidation products that form as the result of non-physiological free-radical-generating processes have been shown to play a key role in blood vessel diseases such as atherosclerosis and thrombosis. Atherosclerosis leads to thickening of the wall of the blood vessel, reducing blood flow. Thrombosis is the formation of a blood clot which can completely stop blood flow in an already narrowed vessel. Ischemia (interruption of blood flow) following the development of atherosclerosis and thrombosis is the predominant cause of heart attacks, strokes (blood clots shutting off brain vessels), and ischemic syndromes (chest pains and irregular heartbeat) which are responsible for significant human morbidity and mortality. In this chapter, we will focus on the role of oxidized lipids in atherosclerosis and thrombosis.

**434**   *The Role of Oxidized Lipids in Cardiovascular Disease*

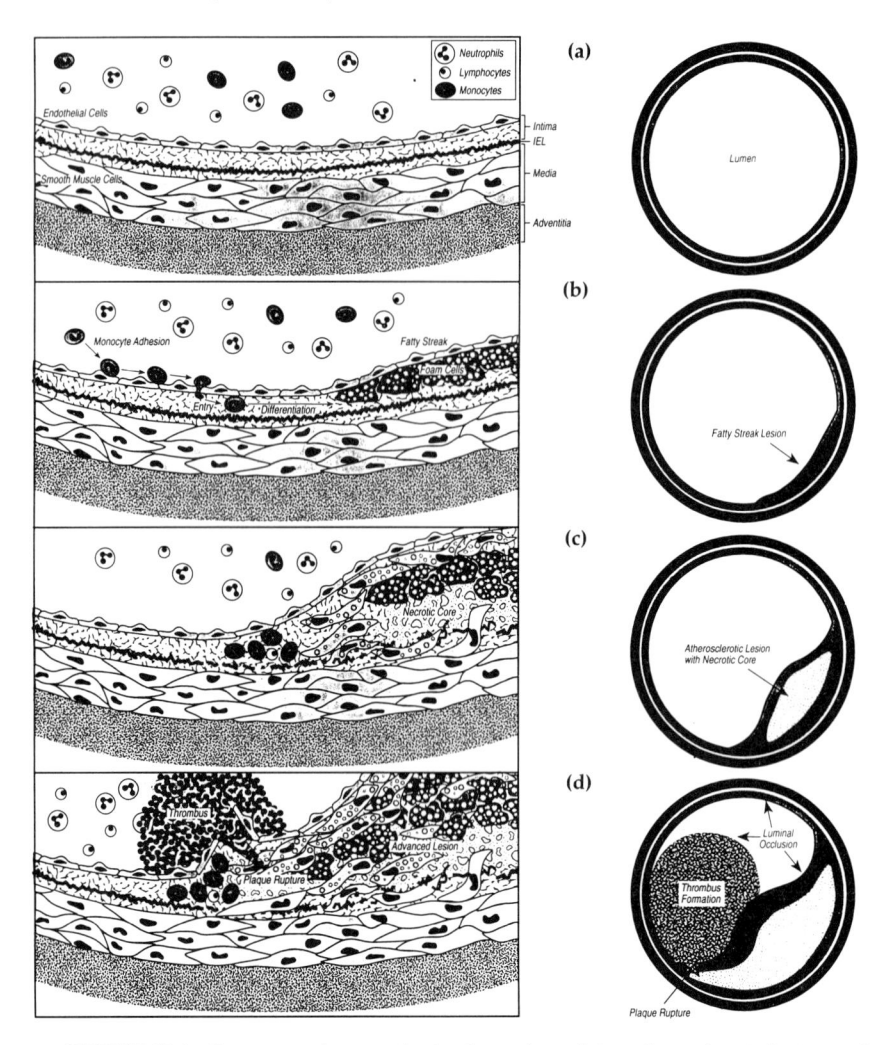

**FIGURE 10-1** Sequence of events in the formation of the atherosclerotic lesion and the associated thrombus. This schematic diagram has been derived from information aquired by a number of investigators based on the sequence of atherosclerotic and thrombotic events in the aorta and coronary arteries of experimental animals and from observations of human vessels. On the right, a gross cross-sectional view of the vessel is shown to demonstrate the amount of occlusion at the various stages. On the left, a more detailed cross-sectional view of the vessel shows the individual cells of the vessel wall. (a) The normal vessel wall is composed of the intima, which is lined with endothelial cells and contains a loose underlying matrix which in some vessels has scattered, undifferentiated cells. The intima is separated from the media by the internal elastic lamina (IEL), an elastin-rich membrane which gives the vessel resilience. The media contain several layers of smooth muscle cells. Outside of the media is the adventitia, a

## ATHEROSCLEROSIS AND THROMBOSIS

Cardiovascular disease is the result of deleterious interactions among lipoproteins from the blood, arterial endothelium, and cells in the blood, such as platelets and monocytes (Ross, 1986). Figure 10-1 shows the sequence of events in atherosclerosis. Many of these events also occur in other diseases, such as rheumatoid arthritis, gout, asthma, and bacterial infections and are collectively referred to as *inflammatory processes* (Belotsky et al., 1990; Lanchbury and Ptzalis, 1993; Sperber, 1993; Terkeltaub, 1993). The normal vessel wall is shown diagrammatically in Figure 10-1a. Atherosclerosis in cholesterol-fed animals is initiated by migration of monocytes from the blood into the vessel wall, where they become macrophages (differentiated monocytes which are highly phagocytic) and take up altered lipids present in the vessel wall (Zhang et al., 1993) (Figure 10-1b). These lipid-loaded macrophages, sometimes referred to as *foam cells* due to the appearance of multiple lipid droplets in the cytoplasm, produce peptide factors (cytokines) that stimulate the proliferation of smooth muscle cells to form the atherosclerotic plaque which thickens the vessel wall (Figure 10-1c). As cells in the center of the plaque die, a necrotic core is formed which may rupture due to the mechanical forces exerted on the vessel wall by flowing blood. Exposure of the necrotic core, as well as subendothelial matrices such as collagen, fibronectin, and von Willebrand factor, activates thrombus formation (Figure 10-1d). During any of these stages, vessels may exhibit spontaneous vasospasm (spasmodic contraction of the vessels) due to lowered levels of vasorelaxant or to increased levels of vasoconstrictor (Verbeuren et al., 1986; Lernam et al., 1991).

◄─────────────────────────────────────────────

loose aggregate of collagen and elastin fibers containing fibroblasts, variable numbers of adipocytes (fat storage cells), and capillaries (small blood vessels). The cells in the flowing blood, including lymphocytes, monocytes, and neutrophils, do not bind to the normal vessel. (b) The fatty streak lesion. Monocytes bind to the endothelium, migrate into the intima, and take up lipid, forming foam cells. (c) The atherosclerotic plaque. The IEL has broken down, and smooth muscle cells from the media migrate into the intima and cover the foam cells. Some of these foam cells die, forming a necrotic core (dead cell debris). Smooth muscle cells are still present in the media, and the adventitia appears as in the normal vessel. (d) A ruptured plaque on which a thrombus has formed. The plaque, which contains a fluid core due to the dead cells, has ruptured under the force of blood flow, causing the endothelium to tear. Platelets and clotting proteins in the blood are exposed to the intimal matrix, which causes a thrombus (clot) to form on the vessel wall.

The term *hemostasis* refers to the dynamic regulation of blood clotting factors so that there is neither a propensity to bleed nor a tendency to develop clots. Local or systemic perturbations that inhibit anticoagulant factors or stimulate coagulant factors may induce thrombosis (Walker, 1984). Under normal physiological conditions, circulating blood platelets do not adhere to the vessel wall or to each other because of the inert, negatively charged surface of endothelial cells and the limited exposure of platelet ligands and activators (Rosenberg, 1987). Activated platelets release mediators that recruit circulating platelets to aggregate to those adhered to the vessel wall. Activation of the extrinsic pathway of blood coagulation via tissue factor exposure on endothelial cells, and activation of the intrinsic pathway of blood coagulation via activation of factor XII by injured endothelial cells or collagen, initiates a cascade of events that result in the formation of fibrin (a fibrous protein that forms large aggregates). Platelet aggregation and fibrin deposition culminate in the formation of a thrombus at the site of injury which arrests the loss of blood and can cause obstruction of blood flow through vessels supplying vital organs. It has been suggested that thrombotic events may occur frequently but are normally resolved without incident due to the efficient regulation of thrombogenesis and thrombolysis (enzymatic breakdown of blood clots) (Astrup, 1956). The occurrence of a clinically significant cardiovascular event is probably related to the extent of the injury, as well as to the thrombotic propensity of the individual (Marcus, 1984; Kinsella, 1987).

## ENZYMATICALLY PRODUCED LIPID MEDIATORS OF VASCULAR FUNCTION

Lipids play a key role in the regulation of inflammatory responses and hemostasis (Moncada and Vane, 1979). Eicosanoids have recently been defined as a series of biologically active, oxygenated derivatives of arachidonic acid (AA) (Figure 10-2, structure A) and other polyunsaturated fatty acid (PUFA) precursors (Thiemermann et al., 1993). AA is a 20-carbon PUFA with a series of four double bonds beginning six carbons from the methyl end and is therefore abbreviated 20:4, $\omega$-6. The relative amount of eicosanoids produced during an inflammatory response is greatly influenced by the availability of AA, which is largely governed by the quantity of dietary $\omega$-6 PUFA consumed (Kinsella, 1987). AA is synthesized in the liver from linoleic acid (18:2, $\omega$-6) and transported in the circulation to various cells, where it is

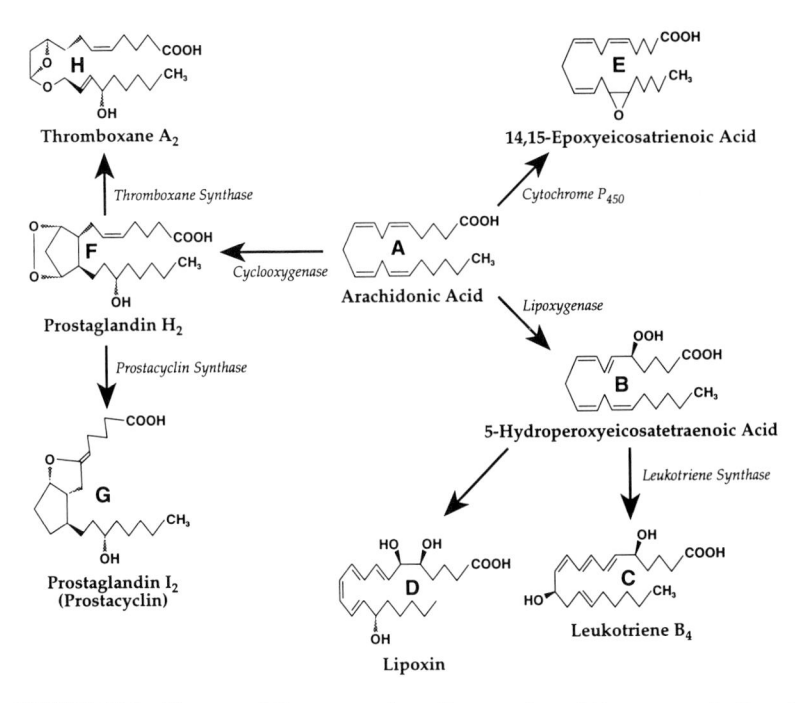

**FIGURE 10-2** Diagram of the enzymatic pathways of arachidonate metabolism that form lipids which influence atherosclerosis and thrombosis. Various structures referred to in the text are lettered.

esterified to phospholipids as a means of controlling the level of free AA in tissues and blood (Kinsella, 1987).

Lipoxygenase is an enzyme that catalyzes the addition of molecular oxygen to AA to form hydroperoxyeicosatetraenoic acid (HPETE) (Figure 10-2, structure B), which can be further metabolized by selective synthases to leukotrienes (LT) (Figure 10-2, structure C) and lipoxins (Figure 10-2, structure D). Leukotriene $B_4$ ($LTB_4$) is chemotactic (a substance which attracts) for leukocytes and increases vascular permeability (leakage of blood components out of the vessel) (Smith et al., 1980; Dahlén et al., 1981). Lipoxins are trihydroxytetraene derivatives of AA which can function as extracellular and intracellular signaling molecules involved in the regulation of vascular tone, inflammation, and immune responses (Serhan and Samuelsson, 1988; Serhan, 1991). Epoxyeicosatrienoic acids (EETs) can be produced from AA in some cell types in a process mediated by a particular cytochrome P-450 pathway (Figure 10-2, structure E). Cyclooxygenase, like lipoxygenase, is an

438   The Role of Oxidized Lipids in Cardiovascular Disease

enzyme that catalyzes the addition of molecular oxygen to AA to form the endoperoxide prostaglandin $H_2$ (PGH$_2$) (Figure 10-2, structure F) (Prichard, et al., 1990). Prostacyclin synthase, located in endothelial cells, and thromboxane synthase, located in platelets, convert PGH$_2$ to prostacyclin (PGI$_2$) (Figure 10-2, structure G) and thromboxane (TxA$_2$) (Figure 10-2, structure H), respectively. PGI$_2$ is a potent vasodilator that inhibits platelet aggregation, whereas TxA$_2$ is a vasoconstrictor that induces platelet aggregation (Hamberg et al., 1975; Moncada, 1982). Endothelial cells exposed to thrombin (which is formed by the coagulation cascade at the site of vessel damage) produce and release PGI$_2$ and nitric oxide, also known as *endothelium-derived relaxing factor*, which are thought to protect areas adjacent to sites of thrombus formation (Moncada et al., 1987; Palmer et al., 1987).

Platelet activating factor (1-O-alkyl-2-acetyl-3-phosphocholine, PAF) (Figure 10-3, structure A) is a potent bioactive phospholipid that induces platelet aggregation and secretion, neutrophil activation, vasodilatation, increased vascular permeability, and direct effects on the heart (McManus et al., 1981; O'Flaherty et al., 1981; Humphrey et

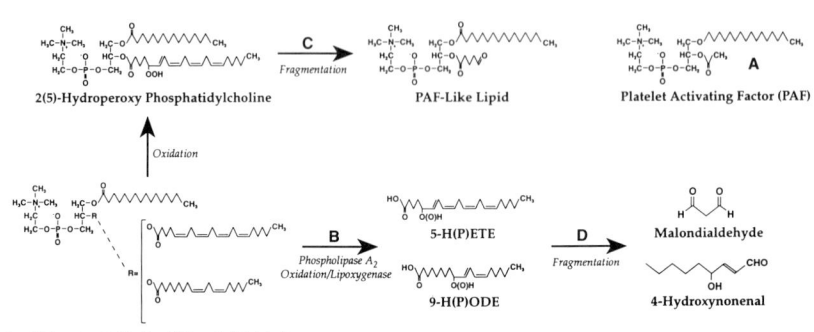

**FIGURE 10-3** Pathways of phospholipid derivatization lipoxygenase, phospholipase A$_2$, and nonenzymatic oxidation, forming products involved in atherosclerosis and thrombosis. Structure A is PAF, a vasoactive phospholipid shown for comparison. Pathways B and D show the fatty acid oxidation products formed from arachidonate and linoleic acid after phospholipase A$_2$ releases them from the phospholipid. Pathway B represents the initial oxidation products. 5-H(P)ETE represents the oxidation of the double bond at the 5-carbon of arachidonic acid to either a hydroxy or a hydroperoxy derivative. 9-H(P)ODE represents the oxidation of the double bond at the 9-carbon of linoleic acid to a hydroxy or a hydroperoxy derivative. Pathway D shows the results of the further oxidation and scission of these products. Pathway C represents the structure of the phospholipid after fragmentation of arachidonic acid esterified to the *sn*-2 position, forming PAF-like lipids.

al., 1982; Kramp et al., 1984). PAF is synthesized by stimulated inflammatory cells but does not exist preformed in unstimulated cells (Pinckard et al., 1982; Ludwig et al., 1984). The metabolism of eicosanoids and PAF in the blood plays a vital role in maintaining vascular tone and function.

## EXTRACELLULAR OXIDIZED LIPIDS

Insoluble lipids are transported in the blood tightly associated with apoproteins and are called *lipoproteins*. All lipoproteins consist of a core of neutral lipids (triglycerides and cholesteryl esters) surrounded by a shell of phospholipids, unesterified cholesterol, and protein. Lipoproteins are designated on the basis of buoyant density and type of associated apoproteins. Very low density lipoproteins (VLDL) are rich in triglycerides and contain apoproteins B and E. The low density lipoproteins (LDL) are rich in cholesterol and contain predominantly apoprotein B, whereas high density lipoproteins (HDL) are rich in protein relative to the other classes of lipoproteins and contain mostly apoproteins AI and AII. The lipids in these lipoproteins are protected from oxidation in the blood by a number of antioxidants, including vitamin E, β-carotene, vitamin C, ubiquinol, and metal-binding proteins such as ferritin (iron) and ceruloplasmin (copper) (Esterbauer et al., 1992). Lipophilic antioxidants, such as vitamin E and β-carotene, actually associate with the lipoprotein particle. Water-soluble antioxidants, such as vitamin C, are not intrinsic components of lipoproteins but appear to have a role in the redox cycling of lipoprotein-bound antioxidants. Retention of lipids, mainly LDL, in the vessel wall may lead to focal depletion of antioxidants. Oxidative stress induced by exposure to cigarette smoke, infections, or excess lipid uptake may initiate lipid oxidation (Halliwell et al., 1992). The cells and pathways involved in forming extracellular oxidized lipids are clear in some settings and unclear in others. Vascular wall cells have been shown to be capable of oxidizing lipids and lipoproteins. Superoxide radicals produced by activated leukocytes with high levels of NADPH oxidase are important in some areas of inflammation (Stein-brecher, 1988; Cathcart et al., 1989). Lipoxygenase and cyclooxygenase are also implicated in forming oxidized lipids in the cell membrane which transfer from the cell to the extracellular lipid, where they propagate oxidation (Steinbrecher et al., 1984). Free metal ions may also be involved in cell-mediated oxidation based on the observation that metal chelators block oxidation (Heinecke et al., 1984). It has been suggested that metal-ion-dependent oxidation requires cellular metabolism of cysteine to recycle iron and copper to a reduced state (Heinecke et al., 1987; Parthasarathy, 1987). Myeloperoxidase, an enzyme released from neutrophils and monocytes, is also capable of oxidizing lipids (Heinecke et al., 1993).

A number of active oxidized lipids are found in the extracellular space. Some, such as prostaglandins, fatty acid hydroperoxides, and EETs (Figure 10-2, structure E), are formed by the cyclooxygenase, lipoxygenase, and by P-450 pathways, respectively. The levels of these enzymatically synthesized lipids may be abnormal in diseased vessels due to changes in their regulatory pathways (Hajjar and Pomerantz, 1992). In the presence of reactive oxygen species, fatty acids can be oxidized independently of enzymatic pathways to produce hydroxy or hydroperoxy derivatives of arachidonic and linoleic acids (see Dussault, 1995 and Figure 10-3, pathway B). Oxidative scission of fatty acids esterified to phospholipids can form molecules which have biological activities similar to those of PAF (Figure 10-3, pathway C) (Smiley et al., 1991). Reactive aldehydes such as 4-hydroxynonenal and malondialdehyde form from fatty acid fragmentation. These products can form Schiff bases with lysine, arginine, and cysteine residues on proteins, which may result in protein crosslinking (Figure 10-3, pathway D). Cytotoxic oxidation products of cholesterol and cholesteryl esters have also been found to be present at high levels in the vessel wall at various stages of atherosclerosis (Figure 10-4) and are also believed to be carcinogenic (Pettersen et al., 1991; Sevanian et al., 1991).

## OXIDIZED LIPIDS IN CARDIOVASCULAR DISEASE

Endothelial cells, smooth muscle cells, monocytes, and macrophages are capable of oxidizing LDL *in vitro* (Heinecke et al., 1984, 1987, 1993; Steinbrecher et al., 1984; Parthasarathy, 1987; Steinbrecher, 1988; Cathcart et al., 1989), and there is evidence that oxidized LDL is present in atherosclerotic lesions and may contribute to formation of the plaque at the earliest stages (Haberland et al., 1988; Ylä-Herttuala et al., 1989; Rosenfeld et al., 1990). Oxidized phospholipids appear to stimulate the entry of monocytes into the vessel wall, forming the fatty streak (Berliner et al., 1990). EETs also stimulate monocyte binding to endothelial cells (Pritchard et al., 1990). Oxidative metabolites of AA such as 12-HETE and leukotriene $B_4$ may also act as chemotactic factors for monocytes (Smith et al., 1980; Grossi et al., 1989). Apolipoproteins

Cholesterol

Cholest-3,5-diene-7-one

Cholestan-5α,6α-epoxy-3β-ol

Cholestan-5β,6β-epoxy-3β-ol

Cholest-5-en-3β-ol-7-one

Cholest-5-en-3β,7α-diol

Cholestan-3β,5α-6β-triol

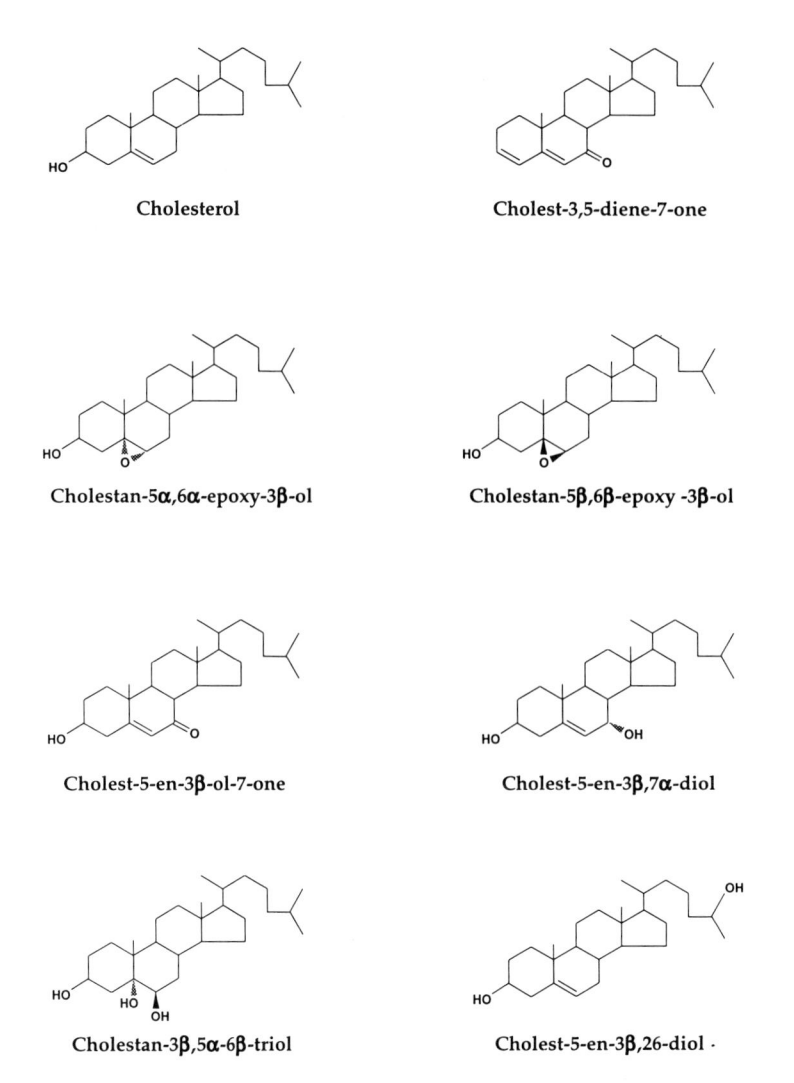

Cholest-5-en-3β,26-diol ·

**FIGURE 10-4** Cholesterol oxidation products formed as a result of enzymatic or nonenzymatic oxidation. The major oxidation products are cholestan-5β,6β-epoxy-3β-ol and cholestan-5α,6α-epoxy-3β-ol.

of LDL and VLDL that have been modified by forming Schiff bases with malondialdehyde or 4-hydroxynonenal are recognized by receptors that cause accumulation of lipids in macrophages forming foam cells (Esterbauer et al., 1992). Eicosanoids have been shown to play an important role in regulating esterification of cholesterol, which in turn affects foam cell formation (Hajjar and Pomerantz, 1992). Oxidized phospholipids can also stimulate macrophages to produce cytokines (low molecular weight bioactive peptides), which in turn stimulate the proliferation of smooth muscle cells to form the definitive atherosclerotic plaque (Ku et al., 1992). Studies suggest that lipid hydroperoxides can decrease the biosynthesis of $PGI_2$ (an antiaggregatory agent) in endothelial cells (Warso and Lands, 1983; Lands, 1985). $PGI_2$ inhibits smooth muscle cell proliferation and regulates vasoconstriction and relaxation (Moncada, 1982).

The role of oxidized lipids in promoting platelet activation, and ultimately thrombosis, is less well understood. Some associations between lipid peroxidation and thrombotic tendency have been established. Diabetic patients, women taking oral contraceptives, and cigarette smokers, who are clearly under oxidative stress, have an increased thrombotic tendency (Hawkins, 1972; Saba and Mason, 1975; Davis et al., 1978; Ciavatti and Renaud, 1991; Jennings et al., 1991). Aging is correlated with a decrease in plasma antioxidants and an increase in plasma lipid peroxides (Jorgensen et al., 1980). Compared to the young, older people have more rapid blood clotting, a reduced platelet life span, enhanced platelet reactivity, and increased platelet sensitivity to various aggregating agents, all of which indicate a thrombotic propensity (Abrahamsen, 1968; Johnson et al., 1975; Couch and Hassanein, 1976; Jorgensen et al., 1980; Zahavi et al., 1980; Sie et al., 1981). Age-related platelet hypersensitivity seems to be linked to the antioxidant status and "peroxide tone" of the individual (Varicel et al., 1988).

A number of oxidative products may impact the coagulation system. Decreases in $PGI_2$ can affect the plasma coagulation system to induce a thrombogenic, hypercoagulable state. This finding is supported by *in vivo* studies illustrating that $PGI_2$ formation is impaired in severely atherosclerotic lesions that contain high levels of lipid hydroperoxides (Szczeklik and Grglewski, 1991). In addition, *in vitro* coagulation studies have shown that oxidized derivatives of AA produced *in vitro* can increase thrombin formation and decrease thrombin degradation (Dahlén et al., 1981). PAF-like lipids may act similarly to PAF to stimulate platelet aggregation. The possibility of simultaneous production of PAF-like lipids and eicosanoids during oxidation of phospholipids is of particular interest, considering the synergy that exists between PAF and eicosaniods in evoking inflammatory responses and platelet activation (O'Flaherty, 1985; Altman and Scazziota, 1986; Rossi and O'Flaherty, 1991).

Several investigators have shown that LDL, particularly oxidized

LDL (Ox-LDL), activates platelet responsiveness. Unoxidized LDL at high concentrations (1.5–3.0 mg protein per milliliter) has been shown to cause platelet aggregation *in vitro* (Andrews et al., 1987) and to enhance platelet aggregation by acting synergistically with other platelet activators (Surya et al., 1992). $Cu^{2+}$-oxidized LDL induces platelet aggregation and the release of AA and $TxA_2$ concomitant with a decrease in platelet membrane fluidity (Ardlie et al., 1989). Aviram and colleagues have confirmed that platelet aggregation induced by Ox-LDL is directly related to changes in platelet membrane fluidity (Aviram, 1989). According to a recent study by Naseem et al. (1993), unoxidized LDL at 0.25–1.0 mg protein per milliliter does not cause platelet aggregation. LDL oxidized with $Cu^{2+}$ or by prolonged exposure to air, however, enhances platelet responsiveness to aggregatory stimuli at 0.5 mg protein per milliliter and induces platelet aggregation in the absence of other agonists at 1.0 mg protein per milliliter. Mildly oxidized LDL may also activate the coagulation cascade by inducing tissue factor expression on the endothelial cell surface (Drake et al.,

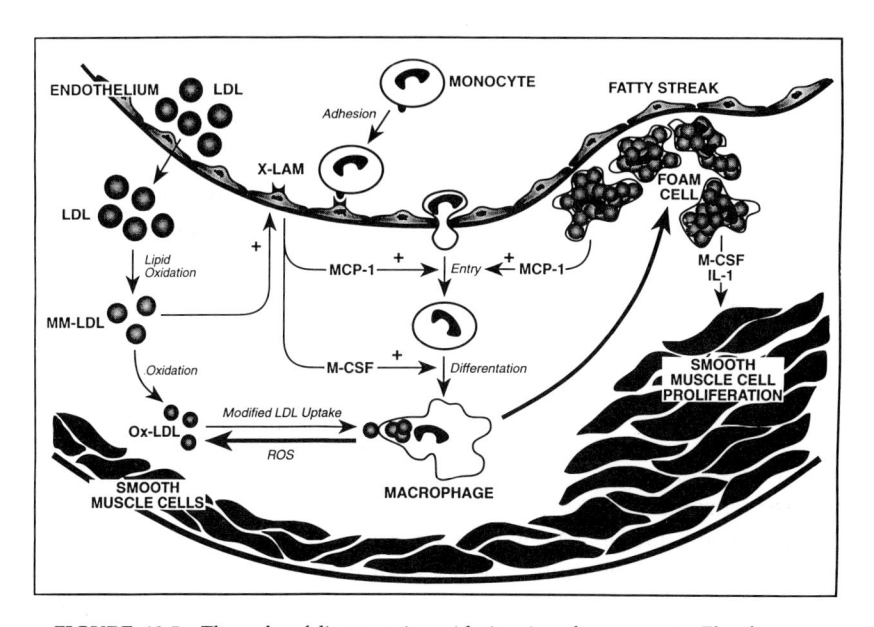

**FIGURE 10-5** The role of lipoprotein oxidation in atherogenesis. The figure combines the findings from studies by a number of groups to show the potential roles of oxidized lipoproteins in the formation of the fatty streak and fibrous plaque. IL-1, interleukin-1; MCP-1, monocyte chemotactic protein-1; M-CSF, macrophage colony-stimulating factor; MM-LDL, minimally oxidized LDL; Ox-LDL, highly oxidized LDL; ROS, reactive oxygen species: X-LAM, an unidentified monocyte adhesion molecule.

1991). Finally, cholesterol oxidation products may cause the death of cells in the center of the plaque, which then initiates thrombus formation.

From these studies, it is clear that oxidized lipids contribute to all stages in the formation of the atherosclerotic plaque and subsequent thrombus formation (Figure 10-5). Many different lipids produced by enzymatic and nonenzymatic oxidation play a role. These lipid-like cytokines activate cell signaling pathways to produce inflammatory and thrombotic responses.

## REFERENCES

ABRAHAMSEN, A. F. (1968) Platelet Survival Studies in Man with Special Reference to Thrombosis and Atherosclerosis. *Scand. J. Haematol.*, Suppl. 3, 1–53.

ALTMAN, R., and SCAZZIOTA, A. (1986) Synergistic Actions of PAF-acether and Sodium Arachidonate in Human Platelet Aggregation. 2. Unexpected Results After Asprin Intake. *Thromb. Res.*, 43, 113–120.

ANDREWS, H. E., AITAKEN, J. W., HASSALL, D. G., SKINNER, V. O., and BRUCKDORFER, K. R. (1987) Intracellular Mechanisms in the Activation of Human Platelets by Low Density Lipoproteins. *Biochem. J.*, 242, 559–564.

ARDLIE, N. G., SELLEY, M. L., and SIMMS, L. A. (1989) Platelet Activation by Oxidatively Modified Low Density Lipoproteins. *Atherosclerosis*, 76, 117–124.

ASTRUP, T. (1956) Analytical Review. Fibrinolysis in the Organism. *Blood*, 11, 781–806.

AVIRAM, M. (1989) Modified Forms of Low Density Lipoprotein Affect Platelet Aggregation *in vitro*. *Thromb. Res.*, 53, 561–567.

BELOTSKY, S. M., GUZU, E. V., KARLOV, V. A., DIKOVSKAYA, E. S., FILJOKOVA, O. B., and SNASTINA, T. I. (1990) Wound Tissue Respiratory Burst and Local Microbial Inflammation. *Inflammation*, 14, 663–668.

BERLINER, J. A., TERRITO, M. C., SEVANIAN, A., RAMIN, S., KIM, J. A., BARNSHAD, B., ESTERSON, M., and FOGELMAN, A. M. (1990) Minimally Modified Low Density Lipoprotein Stimulates Monocyte Endothelial Interactions. *J. Clin. Invest.*, 85, 1260–1266.

CATHCART, M. K., MCNALLY, A. K., MOREL, D. W., and CHISOLM, G. M. (1989) Superoxide Anion Participation in Human Monocyte-mediated Oxidation of Low Density Lipoprotein and Conversion of Low Density Lipoprotein to a Cytotoxin. *J. Immunol.*, 142, 1963–1969.

CIAVATTI, M., and RENAUD, S. (1991) Oxidative Status and Oral Contraceptive. Its Relevance to Platelet Abnormalities and Cardiovascular Risk. *Free Radical Biol. Med.*, 10, 325–338.

COUCH, J. R., and HASSANEIN, R. S. (1976) Platelet Aggregation, Stroke, and Transient Ischemic Attack in Middle-aged and Elderly People. *Neurology*, 26, 888–895.

DAHLÉN, S. E., BJÖRK, J., HEDQVIST, P., ARFORS, D. E., HAMMARSTRÖM, S., LINDGREN, J. A., and SAMMUELSSON, B. (1981) Leukotrienes Promote Plasma Leakage and Leukocyte Adhesion in Postcapillary Venules: *in vivo* Effects with Relevance to the Acute Inflammatory Response. *Proc. Natl. Acad. Sci. U.S.A.*, 78, 3887–3891.

DAVIS, J. W., PHILLIPS, P. E., and YEU, K. T. N. (1978) Platelet Aggregation, Adult Onset Diabetes Mellitus and Coronary Artery Disease. *JAMA*, 239, 732–734.

DRAKE, T. A., HANNANI, K., FEI, H., LAVI, S. and BERLINER, J. A. (1991) Minimally Oxidized Low-density Lipoprotein Induces Tissue Factor Expression in Cultured Human Endothelial Cells. *Am. J. Pathol.*, 138, 601–607.

DUSSAULT, P. (1995) *Reactions of Hydroperoxides and Peroxides in Active Oxygen in Chemistry* (C. S. Foote, J. S. Valentine, A. Greenberg, and J. F. Liebman, Eds.) Chapman and Hall, pp. 141–203.

ESTERBAUER, H., GEBICKI, J., PUHL, H., and JURGENS, G. (1992) The Role of Lipid Peroxidation and Antioxidants in Oxidative Modification of LDL. *Free Rad. Biol. Med.*, 13, 341–390.

GROSSI, Z. M., FITZGERALD, L. A., UMBARGER, L. A., NELSON, K., DIGLIO, C. A., TAYLOR, J. D., and HONN, K. V. (1989) Bidirectional Control of Membrane Expression and/or Activation of the Tumor Cell IRGpIIb/IIa Receptor and Tumor Cell Adhesion by Lipoxygenase Products of Arachidonic Acid and Linoleic Acid. *Cancer Res.*, 49, 1029–1037.

HABERLAND, M. E., FONG, D., and CHENG, L. (1988) Malondialdehyde-altered Protein Occurs in Atheroma of Watanabe Heritable Hyperlipemic Rabbits. *Science*, 241, 215–218.

HAJJAR, D. P., and POMERANTZ, K. B. (1992) Signal Transduction in Atherosclerosis: Integration of Cytokines and the Eicosanoid Network. *FASEB J.*, 6, 2933–2941.

HALLIWELL, B., GUTTERIDGE, J. M., and CROSS, C. E. (1992) Free Radicals, Antioxidants, and Human Disease. Where Are We Now? *J. Lab. Clin. Med.*, 119, 598–619.

HAMBERG, M., SVENSSON, J., and SAMMUELSSON, B. (1975) Thromboxanes: A New Group of Biologically Active Compounds Derived from Prostaglandin Endoperoxides. *Proc. Natl. Acad. Sci. U.S.A.*, 72, 2994–2998.

HAWKINS, R. I. (1972) Smoking, Platelets, and Thrombosis. *Nature*, 236, 450–452.

HEINECKE, J. W., ROSEN, H., and CHAIT, A. (1984) Iron and Copper Promote Modification of Low Density Lipoprotein by Human Arterial Smooth Muscle Cells in Culture. *J. Clin. Invest.*, 74, 1890–1894.

HEINECKE, J. W., ROSEN, H., SUZUKI, L. A., and CHAIT, A. (1987) The Role of Sulfur-containing Amino Acids in Superoxide Production and Modification of Low Density Lipoprotein by Arterial Smooth Muscle Cells, *J. Biol. Chem.*, 262, 10098–10103.

HEINECKE, J. W., LI, W., FRANCIS, G. A., and GOLDSTEIN, J. A. (1993) Tyrosyl Radicals Generated by Myeloperoxidase Catalyze the Oxidative Cross-linking of Proteins. *J. Clin. Invest.*, 91, 2866–2872.

HUMPHREY, D. M., MCMANUS, L. M., HANAHAN, D. J., and PINCKARD, R. N. (1983) Morphological Basis of Increased Vascular Permeability Induced by Acetyl Glyceryl Ether Phosphorylcholine. *Lab. Invest.*, 50, 16–25.

HUMPHREY, D. M., MCMANUS, L. M., SATOUCHI, K., HANAHAN, D. J., and PINCKARD, R. N. (1982) Vasoactive Properties of Acetyl Glyceryl Ether Phosphorylcholine (AGEPC) and AGEPC Analogues. *Lab. Invest.*, 46, 422–427.

JENNINGS, P. E., MCLAREN, M., SCOTT, N. A., SANIABADE, A. R., and BELCH, J. J. F. (1991) The Relationship of Oxidative Stress to Thrombotic Tendency in Type 1 Diabetic Patients with Retinopathy. *Diabetic Med.*, 8, 860–865.

JOHNSON, M., RAMEY, E., and RAMWELL, P. W. (1975) Sex and Age Differences in Human Platelet Aggregation. *Nature*, 253, 355–357.

JORGENSEN, K. A., DYERBERG, J., OLESEN, A. S., and STOFFERSEN, E. (1980) Acetylsalicylic Acid, Bleeding Time, and Age. *Thromb. Res.*, 19, 799–805.

KINSELLA, J. E. (1987) Effect of Polyunsaturated Fatty Acids on Factors Related to Cardiovascular Disease. *Am. J. Cardiol.*, 60, 23G–32G.

KRAMP, W., PIERONI, G., PINCKARD, R. N., and HANAHAN, D. J. (1984) Observations of the Critical Micellar Concentration of 1-O-alkyl-2-acetyl-sn-glycero-3-phosphocholine and a Series of its Homologs and Analogs. *Chem. Phys. Lipids*, 35, 49–62.

KU, G., THOMAS, C. E., ADESON, A. L., and JACKSON, R. L. (1992) Induction of Interleukin-1 Beta Expression from Human Peripheral Blood Monocyte-derived Macrophages by 9-Hydroxyoctadecadienoic Acid. *J. Biol. Chem.*, 267, 14183–14188.

LANCHBURY, J. S., and PTZALIS, C. (1993) Cellular Immune Mechanisms in Rheumatoid Arthritis and Other Inflammatory Arthritides. *Curr. Opin. Immunol.*, 5, 918–924.

LANDS, W. E. M. (1985) Interactions of Lipid Hydroperoxides with Eicosanoid Biosynthesis. *J. Free Rad. Biol. Med.*, 1, 97–101.

LERNAM, A., HALLET, J. W., HEUBLEEIN, D. M., and BERNETT, J. C. (1991) The Role of Endothelin as a Marker of Diffuse Atherosclerosis in the Human. *N. Engl. J. Med.*, 325, 997–1001.

LUDWIG, J. C., MCMANUS, L. M., CLARK, P. O., HANAHAN, D. J., and PINCKARD, R. N. (1984) Modulation of Platelet-activating Factor (PAF) Synthesis and Release from Human Polymorphonuclear Leukocytes (PMN): Role of Extracellular $Ca^{2+}$. *Arch. Biochem. Biophys.*, 232, 102–110.

MARCUS, A. J. (1984) The Eicosanoids in Biology and Medicine. *J. Lipid Res.*, 25, 1511–1516.

MARX, J. L. (1987) Oxygen Free Radicals Linked to Many Diseases. *Science*, 235, 529–531.

MCMANUS, L. M., HANAHAN, D. J., and PINCKARD, R. N. (1981) Human Platelet Stimulation by Acetyl Glyceryl Ether Phosphorylcholine (AGEPC). *J. Clin. Invest.*, 67, 903–906.

MONCADA, S. (1982) Biological Importance of Prostacyclin. *Br. J. Pharmacol.*, 76, 3–31.

MONCADA, S., PALMER, R. M. J., and HIGGS, E. A. (1987) Prostacyclin and Endothelium-derived Relaxing Factor: Biological Interactions and Significance, in *Thrombosis and Hemostasis* (M. Verstraete, J. Vermylin, R. Lijnen, and J. Arnout, Eds.), Leuven University Press, Leuven, Belgium, pp. 505–523.

MONCADA, S., and VANE, J. R. (1979) Arachidonic Acid Metabolites and the Interactions Between Platelet and Blood Vessels. *N. Engl. J. Med.*, 300, 1142–1147.

NASEEM, K. M., GOODALL, A. H., and BRUCKDORFER, K. R. (1993) The Effects of Native and Oxidized Low Density Lipoprotein on Platelet Activation. *Biochem. Soc. Trans.*, 21, 140S.

O'FLAHERTY, J. T. (1985) Neutrophil Degranulation: Evidence Pertaining to its Mediation by the Combined Effects of Leukotriene $B_4$, Platelet Activating Factor, and 5 HETE. *J. Cell. Physiol.*, 122, 229–239.

O'FLAHERTY, J. T., MILLER, C. H., LEWIS, J. C., WYKLE, R. L., BASS, D. A., MCCALL, E. E., WAITE, M., and DEDHATELET, L. R. (1981) Neutrophil Responses to Platelet-activating Factor. *Inflammation*, 5, 193–201.

PALMER, R. M. J., FERRIGE, A. G., and MONCADA, S. (1987) Nitric Oxide Release Accounts for the Biological Activity of Endothelium-derived Relaxing Factor. *Nature*, 327, 524–526.

PARTHASARATHY, S. (1987) Oxidation of Low Density Lipoprotein by Thiol Compounds Leads to its Recognition by the Acetyl LDL Receptor. *Biochim. Biophys. Acta*, 917, 337–340.

PETTERSEN, K. S., BOBERG, K. M., STABERSVIK, A., and PRYDZ, H. (1991) Toxicity of Oxygenated Cholesterol Derivatives Toward Cultured Human Umbilical Vein Endothelial Cells. *Arterioscler. Thromb.*, 11, 423–428.

PINCKARD, R. N., MCMANUS, L. M., and HANAHAN, D. J. (1982) Chemistry and Biology of Acetyl Glyceryl Ether Phosphorylcholine (Platelet Activating Factor), in *Advances in Inflammation Research* (G. Weissmann, Ed.), Raven Press, New York, 4, pp. 147–180.

PRITCHARD, K. A., TOTA, R. R., STEMERMAN, M. B., and WONG, P. Y. (1990) 14,15 EET Promotes Endothelial Cell Dependent Adhesion of Human Monocytic Tumor U937 Cells. *Biochem. Biophys. Res. Commun.*, 167, 137–142.

ROSENBERG, R. D. (1987) The Heparin–Antithrombin System: A Natural Anticoagulant Mechanism, in *Hemostasis and Thrombosis: Basic Principles and Clinical Practice* (R. W. Colman, J. Hirsh, V. J. Marder, and E. W. Salzman, Eds.), J. B. Lippincott, Philadelphia, pp. 1373–1392.

ROSENFELD, M. E., PALINSKI, W., YLÄ-HERTTUALA, S., BUTLER, S., and WITZTUM, J. L. (1990) Distribution of Oxidation Specific Lipid-protein Adducts and Apolipoprotein B in Atherosclerotic Lesions of Varying Severity from WHHL Rabbits. *Arteriosclerosis*, 10, 336–349.

ROSS, R. (1986) The Pathogenesis of Atherosclerosis—An Update. *N. Engl. J. Med.*, 314, 488–500.

ROSSI, A. G., and O'FLAHERTY, J. T. (1991) Bioactions of 5-Hydroxyeicosatetraenoate and its Interaction with Platelet-activating Factor. *Lipids*, 26, 1184–1188.

SABA, S. R., and MASON, R. G. (1975) Some Effects of Nicotine on Platelets. *Thromb. Res.*, 7, 819–824.

SERHAN, C. N. (1991) Lipoxins: Eicosanoids Carrying Intra- and Intercelluar Messages. *J. Bioenerg. Biomemb.*, 23, 105–122.

SERHAN, C. N., and SAMUELSSON, B. (1988) Lipoxins; A New Series of Eicosanoids (Biosynthesis, Stereochemistry, and Biological Activity). *Adv. Exp. Med. Biol.*, 229, 1–14.

SEVANIAN, A., BERLINER, J. A., and PETERSON, H. (1991) Uptake, Metabolism and Cytotoxicity of Isomeric Cholesterol-5,6-epoxides in Rabbit Aortic Endothelial Cells. *J. Lipid Res.*, 32, 147–155.

SEVANIAN, A., and HOCHSTEIN, P. (1985) Mechanisms and Consequences of Lipid Peroxidation in Biological Systems. *Annu. Rev. Nutr.*, 5, 365–390.

SIE, P., MONTAGUT, J., BLANC, M., BONEU, B., CARANOBE, C., CAZARD, J. C., and BIERME, R. (1981) Evaluations of Some Platelet Parameters in a Group of Elderly People. *Thromb. Haemostas.*, 45, 197–199.

SMILEY, P. L., STREMLER, K. E., PRESCOTT, S. M., ZIMMERMAN, G. A., and MCINTYRE, T. M. (1991) Oxidatively Fragmented Phosphatidylcholines Activate Human Neutrophils Through the Receptor for Platelet Activating Factor. *J. Biol. Chem.*, 266, 11104–11110.

SMITH, M. J. H., FORD-HUTCHINSON, A. W., and BRAY, M. A. (1980) Leukotriene B: A Potential Mediator of Inflammation. *J. Pharm. Pharmacol.*, 32, 517–518.

SPERBER, K. (1993) Asthma: An Inflammatory Disease. *Mount Sinai J. Med.*, 60, 218–226.

STEINBERG, D., PARTHASARATHY, S., CAREW, T. E., KHOO, J. C., and WITZTUM, J. L. (1989) Beyond Cholesterol. Modifications of Low Density Lipoprotein That Increase its Atherogenicity. *N. Engl. J. Med.*, 320, 915–924.

STEINBRECHER, U. P. (1988) Role of Superoxide in Endothelial Cell Modification of LDL. *Biochim. Biophys. Acta*, 959, 20–30.

STEINBRECHER, U. P., PARTHASARATHY, S., LEAKE, D. S., WITZTUM, J. L., and

STEINBERG, D. (1984) Modification of Low Density Lipoprotein by Endothelial Cells Involves Lipid Peroxidation and Degradation of Low Density Lipoprotein Phospholipids. *Proc. Natl. Acad. Sci. U.S.A.*, 81, 3883–3887.

SURYA, I. I., GORTER, G. MOMMERSTEEG, M., and AKKERMAN, J. W. N. (1992) Enhancement of Platelet Functions by Low Density Lipoproteins. *Biochim. Biophys. Acta*, 1165, 19–26.

SZCZEKLIK, A., and GRGLEWSKI, R. J. (1991) Low Density Lipoproteins Are Carriers for Lipid Peroxides and Inhibit Prostacyclin Biosynthesis in Arteries. *Artery*, 7, 488–495.

TERKELTAUB, R. A. (1993) Gout and Mechanisms of Crystal-induced Inflammation. *Curr. Opin. Rheumatism*, 36, 1274–1285.

THIEMERMAN, C., MITCHELL, J. A., and FERNS, G. A. A. (1993) Eicosanoids and Atherosclerosis. *Curr. Opin. Lipidol.*, 4, 401–406.

VARICEL, E., CROSET, M., SEDIVY, P., COURPRON, P., DECHAVANNE, M., and LAGARDE, M. (1988) Platelets and Aging. I—Aggregation, Arachidonate Metabolism, and Antioxidant Status. *Thrombos. Res.*, 49, 331–342.

VERBEUREN, T. J., JORDAENS, F. H., ZONNEKEYN, L. L., VAN HOVE, C. E., COENE, M. C., and HERMAN, A. G. (1986) Effect of Hypercholesterolemia on Vascular Reactivity in the Rabbit. *Circ. Res.*, 58, 552–564.

WALKER, F. J. (1984) The Control of Prothrombin Activation, in *The Thrombi* (R. Machovich, Ed.), CRC Press, Boca Raton, FL, pp. 57–88.

WARSO, M. A., and LANDS, W. E. M. (1983) Lipid Peroxidation in Relation to Prostacyclin and Thromboxane Physiology and Pathophysiology. *Br. Med. Bull.*, 39, 277–280.

WITZTUM, J. L., and STEINBERG, D. (1991) Role of Oxidized Low Density Lipoprotein in Atherogenesis. *J. Clin. Invest.*, 88, 1785–1792.

YLÄ-HERTTUALA, S., PALINSKI, W., ROSENFELD, M. E., PARTHASARATY, S., CAREW, T. E., BUTLER, S., WITZTUM, J. L., and STEINBERG, D. (1989) Evidence for the Presence of Oxidatively Modified Low Density Lipoprotein in Atherosclerotic Lesions of Rabbit and Man. *J. Clin. Invest.*, 84, 1086–1095.

ZAHAVI, J., JONES, N. A. G., LEYTON, J., DUBIEL, M., and KAKKAR, V. V. (1980) Enhanced *in vivo* Platelet "Release Reaction" in Old Healthy Individuals. *Thromb. Res.*, 17, 329–336.

ZHANG, H., YANG, V., and STEINBRECHER, U. (1993) Structural Requirements for the Binding of Modified Proteins to the Scavenger Receptor. *J. Biol. Chem.*, 268, 5535–5542.

# Index

## A

Aconitase 321
Acquired immunodeficiency syndrome (AIDS) 317
Actinomycin 223, 368
Actinomycin D 362
Activated organic peracids 30
Active-site synthesis 95
Acyl CoA dehydrogenases 42
Adamantane-6, 6-cyclophane heme-1,5-DCI 100
Adamantane heme cyclophane, 120, 123
Adenine (A) 340
Adenosine monophosphate (AMP) 38
S-Adenosylmethionine 301
Aliphatic hydrocarbons 26
Alkenes 145
  epoxidation 144–147
  vertical ionization potentials 145
Alkyl hydroperoxides 28
Allene oxide 403
Allylglycine 260
Allylic hydroperoxides 284, 296
Alzheimer's disease 206, 433
Amide basket-handle hemes 106
Amione oxidases 217
2-Aminophenols 222
2-Aminophenoxazinone 222
Amines 48, 151
D-Amino acid oxidase 43, 315
δ-(L-α-aminoadipoyl)-L-cysteine-D-valine (ACV) 259
Ammonia monooxygenase 210–211
Amyotrophic lateral sclerosis (ALS or Lou Gehrig's disease) 189, 318
Anaerobic microorganisms 1, 313
Anthracene cyclophanes 111
Anthranilate hydroxyl 63
Antioxidant defense systems 313
Apoprotein B 439
Apoprotein E 439
Aquo complexes 13
Arachidonate metabolism 437

Arachidonic acid(s) 437, 440
2-Arachidonoly (2-Linoleoyl phosphatidylcholine) 438
Arginine 64
Aromatase 26
Aromatic hydrocarbons 26
Aromatic hydroxylases 55, 61, 64
Artherial endothelium 435
Arthritis 276
*Arthrobacter* 412
Artificial proteases 379
Ascorbate (Vitamin C) 317, 361, 371, 380, 382
Ascorbate oxidase 213
Asthma 276, 435
Atherosclerosis 276, 328, 433, 435, 437, 440
Atherosclerotic lesion 434
Atherosclerotic plaque 442
ATP (Adenosine triphosphate) 2
Autooxidation 279

## B

Bacterial luciferases 51, 53
Baeyer–Villiger reactions 50, 55
Basic/Nucleophilic peroxide 200
Benzene 401, 411–413, 423
*trans*-Benzene-1,2:4,5-dioxide 408
Benzene oxide 414, 416, 419–420, 423
Benzooxepins 425–426
Benzoylacetoneethylenediimine 6
1,4-Benzyne (1,4-Benzendiyl) 409–410
Biciyclo [5.1.0] octa-2,4-diene 408
Binding dynamics of chelated hemes 97
Biotin 384
Bleomycin (BLM) 233–237, 239, 342–343, 346, 350, 355, 357, 369
Bleomycin-cobalt (Co$^{II}$ BLM) 355
Bleomycin-copper (BLM Cu$^{I}$), 358–359
Bleomycin-iron (BLM Fe) 346–347, 349–351, 355, 357, 363, 372–373
Bovine Serum Albumin (BSA) 379–380
1,3-Butanediol 402

## C

Calcitonin 208
Calicheamicin 411
Calmodulin 383
d-Camphor 25
Cancer 433
Capped hemes 112
ε-Caprolactone 55
Carbanions 282
Carbonyl oxide 58
β-Carotene 141, 439
Catalase(s) 87, 93, 126, 132, 136, 318
Catechol 412
Catecholase activity 195
Catechol dioxygenase enzymes 30
Catechol dioxygenases 253–254, 286
Catechol 1,2-dioxygenase 254
Catechol 2,3-dioxygenase 256
Ceruloplasmin 439
C—H activation 198
Chelated 18-cyclophane 114
Chelated heme system 96, 102
  reactions with carbon monoxide 102
  reactions with oxygen 102
Chelated heme-*t*-butyl isocyanide complex 118

*m*-Chloroperbenzoic acid  47
Cholest-5-en-3β,7α-diol  441
Cholest-5-en-3β,26-diol  441
Cholest-5-en-3β-ol-7-one  441
Cholesta-3,5-diene-7-one  441
Cholestan-5α,6α-epoxy-3β-ol  441
Cholestan-5β,6β-epoxy-3β-ol  441
Cholestan-3β-5α-6β-triol  441
Cholesterol  439, 441, 444
Chromate ($CrO_4^{2-}$)  12
Chromomycin $A_2$  363
Chemically induced electron exchange luminescence (CIEEL)  53
Clavaminate synthase  252
$Co(DIP)_3^{3+}$  375
Collagen  217, 434–435
Continuous wave–electron nuclear double resonance (CW–ENDOR)  217
Coodination complex  12
  copper  5, 188
  hydroxide  12
  oxide  12
$Co(phen)_2^{2+}$  374
Copper amine oxidases  218
Copper homeostasis  189
Copper monooxygenase model system  197
Copper monooxygenases  210
Copper oxidases  191, 213
  amine oxidase  191
  ascorbate oxidase  191
  'blue' multicopper oxidases  191
  cerulopasmin  191, 213
  cytochrome *c* oxidase  191
  galactose oxidase  191
  laccase  191, 213–214
  phenoazinone synthase  191
Copper oxygenases  191, 195
  ammonia monooxygenase (AMO)  191

  dopamine β-monooxygenase (DβM)  191
  methane monooxygenase (MMO)  191
  peptidylglycine α-amidating monooxygenase (PAM)  191
  phenylalanine hydroxylase (PAN)  191
  tyrosinase (Tyr)  191
Copper–zinc SOD (Cu ZnSOD)  318–319
Corey mechanism  301
Cresolase activity  195, 197, 203
18-Crown-6  366
$Cr(phen)_2^{3+}$  374
{$Cu^I[HB(3,5-iPr_2pz)_3]$}  200–201
{$Cu [HB(3,5-iPr_2pz)_3]$}$_2(O_2)$  193–194, 196, 200–202
$Cu(phen)_2^+$  374
$Cu(phen)_2^{2+}$  374
$Cu^{II/I}(phen)_2$  369–374
Cupric-peroxo complexes  27
Cupric-water complex  19
Cyclohexanone  55
Cyclohexanone monooxygenase  49–50, 54, 56
Cyclooxygenase  437, 439–440
Cyclophane(s)  95
  rate constant for binding carbon monoxide  113
  rate constant for binding oxygen  113
Cyclophane heme  99
5,5,5-Cyclophane heme  114
7,7,7,7-Cyclophane heme  114
Cyclophane heme compounds  111
β-Cyclopropylanine  260
Cyclopropylcarbinyl radicals  292, 299
Cysteine  24, 340
Cytochrome $b_5$  88
Cytochrome *c*  88, 93
Cytochrome *c* oxidase  2, 16, 19–20, 191, 219–220

Cytochrome c oxidase (four-electron reduction by cytochrome c oxidase)   20
Cytochrome c peroxidase   87, 90, 93, 128, 132–133
  mechanism of proton-transfer catalysis   132–133
Cytochrome oxidase   88, 90
Cytochrome P-450   18–26, 68, 88, 93, 143, 147, 151–152, 154, 232–233, 235–236, 239, 243–244, 248, 299–300, 315, 325, 437
Cytochrome P-450 monooxygenase   190, 215
Cytochromes   88

## D

Daunomycin   362
Deacetoxycephalosporin C synthase   252
14,15-Dehydroarachidonic acid   293
Denitrifying bacteria   189
DNA hydrolases   337
DNA ligases   337
Deoxyhemerythrin   238
Deoxyhemoglobin   16, 86
Desferrioxamine   326
Diabetes   433
Dialkyldioxiranes   137
Dihydroxyacid dehydratase   321
3,4-Dihydroxyphenylalanine (L-DOPA)   64
Diiron enzymes   237
Diiron-oxo proteins   237
1,1-dimethoxyethylene   406
trans-3,4-Dimethyl-1,2-dioxetane   408, 418
Dimethyldioxirane   415–416
2,7-Dimethyloxepin   415
5,5-Dimethyl-1-pyrroline-N-oxide (DMPO)   339

2,8-Dioxabicyclo[5.1.0]octa-3,5-diene   408
4,8-Dioxabicyclo[5.1.0]octa-2,5-diene   422
7,8-Dioxabicyclo[2.2.2]octa-2,5-diene   408
7,8-Dioxabicyclo[4.2.0]octa-2,4-diene   408
Dioxetane(s)   280–281, 417
Dioxetane   52–53
Dioxygen-activating enzymes   18
Dioxygen-activating metalloenzymes   22
Dioxygen activation   21, 23, 126
Dioxygenases(s)   2, 18, 211, 277, 286, 413
Dioxygenase quercetinase   212
Dioxygen binding kinetics   117–125
  environmental effects on   121–125
Dioxygen bonding   8, 9
Dioxygen carriers   86
Dioxygen complex   5
Dioxygen ligand   5
Dioxygen reduction   9–10
Dioxygen reduction by complexes   9
Dioxygen reduction by metal ions   9
Dioxygen storage proteins   15
Dioxygen toxicity   30
Dioxygen transport   15
Direct reactions of peroxidic intermediates with substrates   24
Diradicals   280
Distal side effects   117
  hydrogen bonding   117
  local polar environment   117
  pocket steric effects   117
Distal (reactive) side steric effects   110
  dynamics   117–120
  electronic effects on   101–103
  internal polar effects on   103

reversibility 98–100
solvent effects on 103
steric effects on 103–117
Distamycin 365, 368
Distamycin A 362
Disulfides 48
DNA 340, 342–343, 348–349
  affinity, cleavage probes 383
  B DNA 341, 364, 370–372, 377
  B-form helices 375
  circular supercoiled DNA (form I) 340
  closed circular, superhelical DNA (form I) 341
  double-stranded B DNA 344
  linear duplex (form III) 341
  nicked circular (form II) 341
  right-handed helical B-form 340
  single stranded DNA 344
  supercoiled DNA 359, 362
  triple helices 366
  Z DNA 371, 377
DNA cleavage 340–378
DOPA 245, 250
Dopamine 205
Dopamine β-monooxygenase (DβM) 205
Dopamine quinone 218–219
Down's syndrome 319

E

Eicosanoids 436, 439, 442
Elastin 217
Electron paramagnetic resonance (EPR) 41, 59, 212, 214, 236, 241–242, 255, 285, 289, 339, 356
Electron spin echo envelope modulation (ESEEM) 217
Electrophilic reactivity of ferric-hydroperoxide complexes 30

Endoperoxides 417–418
Enzymes-catalyzed oxidation of cyclopropylamines 152
Epinephrine 250
Epoxyeicosatrienoic acids (EEIs) 437
14,15-Epoxyeicosatrienoic acid 437
Exo-epoxynorbornane 137
2,3-Epoxyoxepin 414–417, 419, 422–424
7-Ethoxycoumarin 152
Ethylmethyldioxirane 137
Extended X-ray absorption fine structure (EXAFS) spectroscopic study 206, 255, 261, 285, 288
External monooxygenases 38–39, 46, 52
Extradiol-cleavage 258
Extradiol-cleaving catechol dioxygenases 257
Extradiol-cleaving enzymes 254

F

FAD (Flavin adenine dinucleotide) 37–38, 46, 48, 55, 64
Fatty acid hydroperoxides 276, 292, 298
Fatty acid radicals 294–295, 298
[Fe$_2$(BIPhMe)$_2$(O$_2$CH)$_4$] 239
[Fe$_2$(BPMP)(μ-O$_2$CC$_2$H$_5$)$_2$]$^+$ 242
Fe(5,5,5)cyclophane 111
Fe(6,6,6)cyclophane 111
Fe(7,7,7)cyclophane 111
Fe(7,7,7,7)cyclophane 111
Fe(8,8,8,8)cyclophane 111
[Fe(EDTA)(OH$_2$)]$^-$ 360
[Fe{HB(3,5-$i$-Pr$_2$-pz)$_3$}(OBz)(CH$_3$CN)] 247
[Fe$_2$(HPTP)OBz]$^{2+}$ 247
[Fe$_2$(6-Me-HPTP)OBz(H$_2$O)]$^{2+}$ 247

[Fe$_2$(N-Et-HPTB)(OBz)]$^{2+}$   247
Fenton reactions   322, 338
[Fe$_2$(OH)(OAc)$_2$(Me$_3$TACN)$_2$]$^+$   238
[Fe(EDTA)(OH$_2$)]$^-$   560
FeO—O dipole   105
  R-state   105
  T-state   105
Fe(phen)$_2^{2+}$   374
Fe(phen)$_2^{3+}$   374
Fe$^{III}$P(O$_2^-$)   22
Fe$^{IV}$P(O$_2^-$)   19
Fe$^{III}$P(OH)   19
Ferric complexes   27
Ferric-heme hydroperoxide intermediate   25
Ferric-hydroperoxide intermediates   25
Ferric lipoxygenase   286, 299
Ferric-peroxo species   22
Ferritin   327, 439
Ferrous cytochrome $c$   87
Ferrous lipoxygenase   288, 301
Fe$_2$[(salmp)$_2$]$^{2-}$   240
[Fe$_2$(TPA)$_2$(OAc)$_2$]   246–247
Fibroblasts   316
Flavin C (4a)-hydroperoxide   47
Flavin hydroperoxide(s)   40, 44, 46, 56–58, 62
Flavin hydroxylases   39
Flavin monooxygenase enzymes   21
Flavin monooxygenases   3
Flavin peroxide   42
Flavin reduction   40
Flavin semiquinone   39
Flavins   3, 37–40
  oxidized   40
  reduced   40
Flavocytochrome b$_2$   45
Flavoenzymes   38–39
  oxygen-O$_2^-$ utilizing   38
Flavoprotein hydroxylases   55
Flavoprotein oxidases   39, 43
Flavoproteins   38
Fluoride (F$^-$)   12

FMN (Flavin mononucleotide)   37–38, 46, 51, 64
FMO (FAD-containing monooxygenase)   48, 54
Free-radical intermediates   150

## G

D-Galactose   215
Galactose oxidase   215–217
Gastrin   208
Glucose oxidase   44
Glutathione peroxidases   318
Glyceraldehyde-3-phosphate dehydrogenase   322, 324
Glycolate oxidase   45, 315
Glyoxylic acid   208
Gout   435
Group-increment(s)   401–402, 411, 421
Guanine (G)   340, 344, 376

## H

Heats of reaction   4
  oxygenation of organic compounds   4
Heme   189
Heme-containing enzymes   21
Heme-containing proteins   15
Heme-containing system   26
Heme (FeTpivPP)-1,2-dimethylimidazole   103
Heme-dioxygen complex   22
Heme protein active sites   85
Heme proteins   84
Hemerythrin   17–18, 237, 240–241
Heme Fe$^{IV}$-oxo intermediate   25
Heme-oxo complex ([Fe$^{IV}$(P·)(O$^{2-}$)]$^+$)   23
Heme oxygenase   26
Hemin   136–137
Hemin catalysts   150

Hemin-catalyzed epoxidation 149
Hemin-catalyzed oxidation 151
Hemocyanin 17–18, 190–191, 196, 204
Hemoglobin 15–18, 86, 92, 94–95, 98, 120, 123, 125, 190, 327
 R-state 94, 98, 99, 105, 107, 110, 120, 124
 T-state 94, 99, 105, 107, 120, 124
Hemostasis 436
Z-1,5-Hexadiyne-3-ene 410–411
*trans*-2-Hexanal 276
*cis*-3-Hexanol 276
High density lipoproteins (HDL) 439
High-spin $d^5$ ferric hydroperoxide complexes 28
High-valent metal complexes 128
High-valent metal-oxo 19
High-valent metal-oxo complexes 27
High-valent metal-oxo intermediates 26
Horseradish peroxidase 87, 110, 125, 128, 131, 133–134, 143, 233
Horseradish peroxidase-hydroxamic acid complex 125
Horseradish peroxidase-oxygen complex 124
Human immunodeficiency virus I (HIV-I) 317
Hydrogen peroxide ($H_2O_2$) 5, 20, 40, 43, 46, 48, 55, 58, 61, 65, 69, 86–87, 126, 128, 130, 133–134, 136, 213, 217, 233, 244, 279, 314, 317–318, 321, 322–326, 338, 347–350, 359–360, 362, 371, 374, 378, 380–382
 toxicity 320–323

Hydroperoxide(s) (HOO$^-$) 10, 29, 133–135, 143
 free radical reactions 135–136
Hydroperoxo complxes 27, 29
Hydroperoxo-Cu$^{II}$ intermediate 202
Hydroperoxo intermediate 26
Hydroperoxo species 22
Hydroperoxyeicosatetraenoic acid (HPETE) 437
5-Hydroperoxyeicosatetraenoic acid 437
Hydroperoxyiron intermediate 126
Hydroperoxyl (·OOH) 336
13-Hydroperoxy-9,11-octadecadienoic acid 277
2(5)-Hydroperoxy phosphatidylcholine 438
C(4a)-Hydroperoxypterin 68
Hydroquinone(s) 3, 141, 418
Hydroxo complexes 13
2-Hydroxy-adenine 318, 373
*p*-Hydroxybenzoate hydroxylase 55
Hydroxycyclohexadienyl radical 59
6-Hydroxy dopa (topa) 217
11-Hydroxy-12-13-epoxy-9-octadecenoic acid 277
Hydroxyflavin 53, 60, 62
8-Hydroxyguanine 318, 350, 373, 376
Hydroxyl (OH·) 20, 22, 314–315, 318, 321–327, 336, 338–339, 349–350, 358, 361, 363, 367–369, 372, 378, 381, 416, 425
Hydroxylamines 48
Hydroxylation of unactivated alkanes 24
6-Hydroxynicotinate 57
4-Hydroxynonenal 438, 440–441
*p*-Hydroxyphenyl acetate 3-hydroxylase 62
*p*-Hydroxyphenylpyruvate dioxygenase 251–253, 300

C(4a)-Hydroxypterin   67

## I

Interferons   327
Interleukins   327
Internal entropies   407
Internal monooxygenases   38
Intradiol-cleaving catechol dioxygenases   256
Intradiol-cleaving enzymes   254
Iodosylbenzene(s)   27, 129–132, 134, 143, 146, 150, 236
Iron(II) porphyrin-oxygen complex   96
Iron (Fe)-complexes   5
Iron (Fe)-heme complexes   15
Iron (Fe$^{IV}$)-oxo complex   11, 237, 361
Iron-oxo intermediate   68, 350
Iron-peroxo species   237
Iron(IV) porphyrin cation radical   128
Iron (Fe)-porphyrin complex   15
Iron porphyrin system   27
Iron superoxide dismutase (FeSOD)   261
Iron(III) tetramesitylporphyrin   129, 142
Iron(III) tetrapentafluorophenyl-porphyrin   142
Isoalloxazine   38–39
Isoalloxazine ring system   37
Isopenicillin N   259
Isopenicillin N synthase (IPNS)   253, 259, 286

## K

a-Ketoacid-dependent hydroxylases   253
Kinetic isotope effect(s)   280, 288, 296–297

Kitajima mechanism   203

## L

Laccase(s)   191, 213–214
Lactate oxidase   44–46
Leukotriene B$_4$ (LTB$_4$)   437
Leukotrienes   327
Ligand-to-metal σ-bonding   8
Ligand-to-metal π-bonding   13, 27
*Limulus II* (horseshoe crab) hemocyanin   192
Linoleic acid   277, 287–289, 295–298, 440
Lipid hydroperoxides   321, 442
Lipid oxidation   433
Lipid peroxidation   327, 336
Lipoproteins   439
Lipoxin   437
Lipoxygenase(s)   30, 276–279, 284, 286–290, 292–297, 299–301, 437–440
Lipoxygenase inhibitors   276
Low-spin d$^5$ ferric hydroperoxide complexes   28
Low-spin d$^6$ metal(s)   27–28
   Co$^{III}$   27
   Ir$^{III}$   27
   Rh$^{III}$   27
Luciferases   51–52, 54
Lumazine   54, 57
Lumiflavin   47
Lymphocyte   315
Lysine monooxygenase   46

## M

Macroincrementation   403
Macrophages   435, 440, 442
Magnesium oxide (MgO)   12
Malondialdehyde   438, 440
Mammalian mitochondrial

cyctochrome *c* oxidase 220
Mechanisms of dioxygen
   activation 20
Metal-containing oxygenase
   enzymes 18
Metal-dioxygen biochemistry 15
Metal-dioxygen chemistry 4
Metal-dioxygen complexes 5–6
Metal-EDTA complexes 359–369
Metallobleomycins 342–343,
   347, 355, 359
Metalloproteins 5
Metal-oxo complexes 11–13, 338
Metal-oxo compounds 12
Metal-peroxo complexes 29
Metal-to-ligand π-backbonding
   8
Methane 18
Methane monooxygenase
   (MMO) 210, 237, 241–243,
   246, 300
Methanotrophic bacteria 210
2-Methyl-3-hydroxyl-5-
   carboxypyridine (MHPC)
   oxygenase 63
Methyl (Trifluromethyl)-
   dioxirane 415
Microperoxidase 131
Microperoxidase-8 134
Mn(phen)$_2^{2+}$ 374
Model porphyrin complexes 84
Mono-1-methylimidazole-*meso*-
   *tetra* (α,α,α,α-O-
   pivalamidophenyl) porphyrin
   iron(II) 96
Mononuclear monodentate metal-
   oxo complexes 15
Monooxygenase(s) 2, 18, 203,
   300, 413
Mössbauer spectroscopy 236,
   240, 244–245, 248, 255, 285,
   288
MPE-Fe (Methidiumpropyl-
   EDTA-Fe$^{II}$) 361–363, 365
Muconaldehyde 407, 413, 416,
   424

E,E 407, 413–414
Z,E 407
Z,Z 407, 413, 417–419
Muconic acid 407, 413
E,E 407, 417
Z,E 407
Z,Z 407, 412, 417
Myeloperoxidase 440
Myoglobin 16–18, 86, 90, 95, 98,
   110, 121, 123, 125, 190, 327

# N

NAD (Nicotinamide adenosine
   dinucleotide) 315
NADH 46, 48, 55, 62, 241,
   315, 323, 358, 371
NADH dehydrogenase 321
NADPH 46, 48, 55, 59, 65, 323
NADPH oxidase 439
Naphthalene 424
Naphthalene oxides 425–426
Netropsin 362, 373
NIH shift 151, 423
Ni(phen)$_2^{2+}$ 374
Nitric oxide (NO) 64, 85, 288,
   320–322
Nitric oxide synthase 64
Nitrite 189
Nitrones 48
Nitrous oxide 189
Non-basic/electrophilic peroxide
   200
'Nonblue' oxidases 215
Nonheme iron centers 232
Nonheme iron enzymes 189–
   190, 286
Nonheme iron monooxygenase
   enzymes 28
Nonheme metalloenzymes 26
Nonheme systems 26
Non-next-nearest neighbor
   effects 402
Noradrenaline 250

Norbornene 137, 141, 146
Norbornyl radicals 146
Norcaradiene 420
Norepinephrine 205
N-oxides 48
Nuclease activation 326
Nucleic acid affinity cleavage 365–366
Nucleophilic dioxygen species 283

O

Octadecadienoic acid 290
Z,Z-Octa-1,3,5,7-tetraene 408, 424
O—O stretching frequencies and separations 7
Organic peracids 29
σ-Organoiron 282
Ortho effect 107
7-Oxabicyclo[4.1.0]heptane 421
Oxene [Fe$^V$O] 90, 129, 131, 133–134, 136, 138–139, 141–142, 153–155, 236
Oxepin 414, 416, 419–420, 423, 426
Oxidases 38
Oxidases and internal monooxygenases 43
Oxidative stress, 323–328
Oxide (O$^{2-}$) 4, 9, 13, 15
Oxidized flavin 40
Oximes 48
Oxo complexes 13
Oxo-bridged complexes 11
Oxo-bridged iron species 11
E,E-6-Oxo-2,4-hexadienoic acid 414
Oxoiron [Fe$^{IV}$] 129
Oxoiron(V) porphyrin 232
Oxoiron(IV) porphyrin cation radical complex 233

Oxo iron species 129, 138, 322
Oxo ligands 10
Oxygenase enzymes 18, 21, 30
Oxygenases 16
Oxygen atom (O$^0$) 13
Oxygen rebound mechanism 24
Oxygen (O$_2$) storage 15
Oxygen (O$_2$) transport 15
Oxyhemerythrin 6–7, 238
Oxyhemocyanin 6–7, 17–18
Oxyhemoglobin 7, 22
Oxymyoglobin 22, 91
Oxy-picket fence heme 100
Ozone 137, 314

P

Particulate-methane monooxygenase (p-MMO) 211
E-2,4-Pentadienal 406
Pentafluoroiodosylbenzene (PFBI) 135, 140
Peptidylglycine α-amidating monooxygenase (PAM) 208
Peracids 27, 129
Perepoxides 280–281
Perhydroxyl radical (HO$_2$) 321
Permanganate (MnO$_4^-$) 12
Peroxidase(s) 87, 126, 277, 294, 359
Peroxide (O$_2^{2-}$) 4, 338
Peroxide shunt 23, 233
Peroxo complexes 27–28
  Co$^{III}$ 27
  Ir$^{III}$ 27
  Mo$^{VI}$ 27
  Rh$^{III}$ 27
  Ti$^{IV}$ 27
μ-peroxo complex 11, 17
μ-peroxo diheme complex 17
V$^V$ 27
Peroxodicopper (II) oxy-form 192

Peroxo ligands  10, 19
Peroxo-like complexes  5–6
Peroxyl radicals  321
Peroxynitrite  320–321
Phagocytes  315, 317, 320, 324
PHBH  60–61
Phenethylamine hydroxylation mechanism  207
Phenol  412, 416
Phenol hydroxylase  62
Phenoxazinone synthase  222
Phenylalanine  67
Phenylalanine hydroxylase (PAH)  64–69, 210, 250, 424–425
6-Phosphogluconate  321
Phospholipase $A_2$  438
Phospholipids  439
Phthalate dioxygenase  248–249
Picket fence heme  95, 98
Picket-fence porphyrin  96
Platelet activating factor (1-O-alkyl-2-acetyl-3-phosphocholine, PAF)  438
$P$-nitroperbenzoic acid  30
Pocket hemes  112, 115
Polyunsaturated fatty acid (PUFA)  436
Potassium monoperoxysulfate ($KHSO_5$)  350
Potassium superoxide  7
Procollagen  252
Prolyl hydroxylase  249, 251
Prostacyclin ($PGI_2$)  438
Prostacyclin synthase  437–438
Prostaglandin(s)  327
Prostaglandin $H_2$  437–438
Prostaglandin H synthase  300
Prostaglandin $I_2$  437
Protein cleavage  378–384
Protein kinases  324
Protocatechuate 3,4-dioxygenase (PCD)  254–256
Protoheme  84
Protohemin  84–85, 130
Protoporphyrin IX  16, 19
*Pseudomonas putida*  62, 412

Pterin-dependent hydroxylases  249
Pterin-dependent iron hydroxylases  250
Pterin hydroperoxide  66–68
Pterin hydroxylases  39, 64
Pulse radiolysis  44–46
Purple acid phosphatase  237
Purple lipoxygenase  295
Putidamonoxin  248–249
2H-Pyran  404–407
  2-carboxaldehyde  404, 407, 409
  2-carboxylic acid  404, 407, 409
  6-hydroxy  404, 407
  6-hydroxy-2-carboxaldehyde  407
  6-hydroxy-2-carboxylic acid  407
  6-methoxy  404, 407
4H-Pyran  405
Pyridine-5,5-cyclophane  115
3,5-Pyridine-5,5-heme cyclophane  112
Pyruvate  45

## Q

Quercetinase  191, 211
Quercetin-2,3-dioxygenase  211
Quercetin (3′,4′,5,7-tetrahydroxyflavonol)  211

## R

'Rebound' mechanism  154
Reduced flavins  39
Reduction of nitrogen-oxide species  189
Resonance Raman spectroscopy  238, 247–248, 255–256
Retrolental fibrophasia  314
Reversible dioxygen binding  98, 190

Rheumatoid arthritis  433, 435
Ribityl phosphate chain  38
Ribonucleotide reductases  300
Rieske-type $Fe_2S_2$ cluster  248
Ring correction factor  402
RNA  365
 A-form helices  376
 IVS RNA  365
 5 S ribosomal RNA  368
 16 S ribosomal RNA  365
 tRNA  365
 tRNA$^{phe}$  377
RNA polymerase  373
RNR (ribonucleotide reductase)  237, 243
RNR R1  239
RNR R2  239–242, 244–246
ROS (reactive oxygen species)  314, 316, 324–325, 327
Rubrerythrin  237
$(Ru(DIP)_3)^{2+}$  375
$(Ru(phen)_3)^{2+}$  375–377
$Ru^{II}(phen)_3^{2+}$  375
$(Ru(TMP)_3)^{2+}$  375–377

## S

Selenoproteins  318
Singlet dioxygen ($^1\Delta g\ O_2$)  7, 279, 281, 296, 299, 314, 318, 339, 377, 418
 enthalpy of formation  418
Singlet dioxygen-ene reaction  279, 298
Single-oxygen atom donors  22, 27
Sodium metaperiodate ($NaIO_4$)  350
Soybean lipoxygenase  284, 293
Spin traps  287, 294, 339
Standard reduction potential for one- and two-electron reduction of dioxygen species in water  3

Staphylococcal nuclease-Fe(EDTA) system  382
Steric strain effects on kinetic and equilibrium data  108
Stopped-flow methods  44–45, 49, 56, 148
Strain energy, 402
Streptavidin, 384
Substrate activation, 30
Substrate activation by oxygenase enzymes  30
Suicide-labeled (N-alkylated) hemins  149
Suicide labeling  147
Superoxide ($O_2^-$)  2–5, 10, 41, 284, 314, 316–318, 320–325, 327, 336, 338, 360, 362, 371, 382
 toxicity  320–323
Superoxide dismutase (SOD)  41, 65, 189, 191, 314, 318–319, 323, 337, 362, 371
Superoxo-complexes  5–6
Superoxo ligands  10
Symmetry number  407
Synthetic copper-dioxygen complexes  193

## T

TARF (Tetraacetyl riboflavin)  41
Tetra-2,6-dichlorophenyl porphyrin iron(III) chloride (TDCPPFeCl)  130, 137
Tetrahydrobiopterin  39, 64, 68–69
Tetrahydropterin cofactor  249
5,10,15,20-Tetrakis (2,4,6-triphenylphenyl porphinato) (FeTTPPP)-1,2-dimethyl-imidazole complexes  103
5,10,15,20-Tetrakis (2,4,6-triphenylphenyl) porphinato iron(II) complex  103

Tetramesityl heme   110
Tetramesityl iron(III) porphyrin chloride   135
Tetraphenyl borate   246
Tetraphenyl hemes   107, 110
Tetraphenyl porphyrin(s)   139
Thermochemical estimations   401
13-Thiaarachidonic acid   293
Thiolate rebound mechanism   261
Thiols   48–49
Thrombosis   433, 435, 437
Thromboxane (TxA$_2$)   438
Thromboxane A$_2$   437
Thromboxane synthase   437–438
Thymine (T)   340
[{(TMPA)Cu}$_2$(O$_2$)]$^{2+}$   193–194
α-Tocopherol   314, 317
Topa   218–219
Topaquinone   218–219
Transgenic mice   318
Trichlomethylperoxyl radical   325
Trichloromethyl radical   325
Trifluoroperazine (TFP)   383
Triglycerides   439
Tropylidene   420
Tryptophan monooxygenase   46
T-state hemoglobin model   99
Tyrosinase   18, 190, 195–196, 202–205
Tyrosine hydroxylase   64, 67, 69, 249–250

## U

Ubiquinol   439

## V

Vascular endothelial cells   316
Vasopressin   208
Very low density lipoproteins (VLDL)   439
Vitamin B$_{12}$   300–301
Vitamin C   439
Vitamin E   439
Vitamin K-dependent proteins   252

## Z

Zn(phen)$_2^{2+}$   374
Zwitterions   280–281